AF393815

Electronic Materials
From Silicon to Organics

Electronic Materials
From Silicon to Organics

Edited by
L. S. Miller
Coventry Polytechnic
Coventry, England

and
J. B. Mullin
Electronic Materials Consultancy, Malvern
Formerly at Royal Signals and Radar Establishment
Malvern, England

Springer Science+Business Media, LLC

Library of Congress Cataloging-in-Publication Data

Electronic materials : from silicon to organics / edited by L.S.
  Miller and J.B. Mullin.
       p.   cm.
    Includes bibliographical references and index.
    ISBN 978-1-4613-6703-1   ISBN 978-1-4615-3818-9 (eBook)
    DOI 10.1007/978-1-4615-3818-9
    1. Electronics--Materials.   I. Miller, L. S.   II. Mullin, J. B.
  TK7871.E437 1991
  621.381--dc20   91-11190
                                                           CIP

ISBN 978-1-4613-6703-1

© 1991 Springer Science+Business Media New York
Originally published by Plenum Press, New York in 1991

Contributors

Simon Allen • ICI Wilton Materials Research Centre, Wilton, Middlesbrough, Cleveland TS6 8JE, U.K.

David A. Anderson • Royal Signals and Radar Establishment, St. Andrews Road, Malvern, Worcestershire WR14 3PS, U.K.

A. Barr • Department of Applied Physical Sciences, Coventry Polytechnic, Priory Street, Coventry CV1 5FB, U.K.

J. W. Brightwell • Department of Applied Physical Sciences, Coventry Polytechnic, Priority Street, Coventry CV1 5FB, U.K.

D. A. Cardwell • Plessey Research Caswell Limited, Caswell, Towcester, Northamptonshire NN12 8EQ, U.K.

Michael G. Carter • Rank Xerox Limited, Mitcheldean, Gloucestershire GL17 0DD, U.K.

David Coates • BDH Limited, Broom Road, Poole, Dorset BH12 4NN, U.K.

Sally E. Day • Thorn EMI Central Research Laboratories, Dawley Road, Hayes, Middlesex UB3 1HH, U.K. *Present address*: Royal Signals and Radar Establishment, St. Andrews Road, Malvern, Worcestershire WR14 3PS, U.K.

J. R. Dodgson • Thorn EMI Central Research Laboratories, Dawley Road, Hayes, Middlesex UB3 1HH, U.K.

Malcolm H. Dunn • Department of Physics and Astronomy, University of St. Andrews, North Haugh, St. Andrews, Fife KY16 9SS, U.K.

D. J. Foster • Plessey Research Caswell Limited, Caswell, Towcester, Northamptonshire NN12 8EQ, U.K.

M. J. Goodwin • Plessey Research Caswell, Allen Clark Research Centre, Caswell, Towcester, Northamptonshire, NN12 8EQ, U.K.

Harry G. Heller • School of Chemistry and Applied Chemistry, University of Wales College of Cardiff, Cardiff CF1 3TP, Wales, U.K.

Robert M. Hill • Department of Physics, King's College London, The Strand, London WC2R 2LS, U.K.

D. E. Hookes • Electrical, Electronic, and Systems Engineering Department, Coventry Polytechnic, Priory Street, Coventry CV1 5FB, U.K.

D. T. J. Hurle • Royal Signals and Radar Establishment, St. Andrews Road, Malvern, Worcestershire WR14 3PS, U.K.

E. D. Jones • Department of Applied Physical Sciences, Coventry Polytechnic, Priory Street, Coventry CV1 5FB, U.K.

P. G. LeComber • Department of Applied Physics and Electronic and Manufacturing Engineering, University of Dundee, Dundee DD1 4HN, Scotland.

L. S. Miller • Department of Applied Physical Sciences, Coventry Polytechnic, Priory Street, Coventry CV1 5FB, U.K.

P. T. Moseley • Electrical Ceramics and Gas Sensors, AEA Industrial Technology, Harwell Laboratory, Oxfordshire OX11 0RT, U.K.

J. B. Mullin • Royal Signals and Radar Establishment, St. Andrews Road, Worcestershire WR14 3PS, U.K. *Present address*: EMC The Hoo, Brockhill Road, West Malvern, Worcestershire WR14 4DL, U.K.

R. W. Munn • Department of Chemistry and Centre for Electronic Materials, UMIST, Manchester M60 1QD, U.K.

A. W. Nelson • Epitaxial Products International Limited, Cypress Drive, St. Mellons, Cardiff CF3 0EG, U.K.

Brian Ray • Faculty of Applied Science, Coventry Polytechnic, Priory Street, Coventry CV1 5FB, U.K.

E. P. Raynes • Royal Signals and Radar Establishment, St. Andrews Road, Malvern, Worcestershire WR14 3PS, U.K.

A. M. Scott • Royal Signals and Radar Establishment, St. Andrews Road, Malvern, Worcestershire WR14 3PS, U.K.

J. A. Turner • Plessey Research Caswell Limited, Allen Clark Research Centre, Caswell, Towcester, Northamptonshire NN12 8EQ, U.K.

R. H. Wallis • Plessey Research Caswell Limited, Caswell, Towcester, Northamptonshire NN12 8EQ, U.K.

D. J. Walton • Department of Applied Physical Sciences, Coventry Polytechnic, Priory Street, Coventry CV1 5FB, U.K.

R. W. Whatmore • Plessey Research Caswell Limited, Towcester, Northamptonshire NN12 8EQ, U.K.

J. O. Williams • Solid State Chemistry Group, Chemistry Department, UMIST, Manchester M60 1QD, U.K.

A. C. Wright • Solid State Chemistry Group, Chemistry Department, UMIST, Manchester M60 1QD, U.K.

A. Yates • Lucas Automotive Limited, Advanced Eneineering Centre, Dog Kennel Lane, Shirley, Solihull, West Midlands B90 4JJ, U.K.

Preface

Electronic materials are a dominant factor in many areas of modern technology. The need to understand them is paramount; this book addresses that need. The main aim of this volume is to provide a broad unified view of electronic materials, including key aspects of their science and technology and also, in many cases, their commercial implications. It was considered important that much of the contents of such an overview should be intelligible by a broad audience of graduates and industrial scientists, and relevant to advanced undergraduate studies. It should also be up to date and even looking forward to the future. Although more extensive, and written specifically as a text, the resulting book has much in common with a short course of the same name given at Coventry Polytechnic.

The interpretation of the term "electronic materials" used in this volume is a very broad one, in line with the initial aim. The principal restriction is that, with one or two minor exceptions relating to aspects of device processing, for example, the materials dealt with are all active materials. Materials such as simple insulators or simple conductors, playing only a passive role, are not singled out for consideration. Active materials might be defined as those involved in the processing of signals in a way that depends crucially on some specific property of those materials, and the immediate question then concerns the types of signals that might be considered. This has been interpreted very generally, in particular to include optical as well as electronic functions, but also to include chemical sensing (though other sensor types would also have been of interest if space had permitted).

Having defined the range of functions that the materials under consideration might have, it becomes necessary to consider whether restrictions should be placed on the types of materials to be considered for those functions. From the outset, the view was taken that organic materials, already dominant for displays, may well bring about the greatest revolution in electronic materials since silicon, and attention is specifically given to organics in many of the chapters. It therefore becomes of interest to discuss both simple and compound semiconductors and both inorganic and organic materials.

In terms of the range of relevant science and technology, again a broad interpretation has been used. The chemistry, physics, and materials science relating to fabrication and processing steps, material properties, and device properties are all considered of relevance. Commercial, historical, and other aspects were also considered of interest.

There is then the question of the ordering of the material. Here the decision has been taken to begin with some important aspects of the basic science relevant to a wide range of electronic and optical materials and devices (though this does not include organics), then to look at some key devices in the light of this science. This provides pointers to the material types of greatest interest, and is followed by chapters discussing principally the science and technology of semiconductors, electrophotographic receptors, and nonlinear optical materials and liquid crystals. These are all areas considered to be of clear present or early relevance. Additional chapters discuss high-temperature superconductors, Langmuir–Blodgett films, electrically conducting polymers, and photochromics as examples of areas expected to play an important role in the medium term. Finally, three chapters on chemical (principally gas) sensing materials (and applications) reflect their present relevance and great future importance. But there is a good additional reason for putting sensors at the end: for many of them a proper treatment requires an understanding of much of what has gone before; they are a perfect example, true to a significant extent for many of the areas discussed in this text, of the need for scientists and technologists with wide-ranging knowledge and expertise to work together if the many new and exciting possibilities are to be properly realized.

Finally, of course, it has to be said that in a book of this scope there are bound to be omissions. Any attempt to be encyclopedic would have increased the book's length (and price) manyfold. This brings us to another organizational aspect: to give authors moderate freedom concerning the level of technical or scientific difficulty, whether they discussed commercial relevance, and so on. In this way, we have achieved a text which spotlights all these aspects in certain ways, and we hope gets across the overall scope and excitement, for the scientist, for the technologist, for the educationalist, and for the business man, of this demanding area of human endeavor.

L. S. Miller
J. B. Mullin

Contents

Chapter 6
Key Electrical Devices
R. H. Wallis

Chapter 7
Key Optoelectronic Devices
A. W. Nelson

Chapter 8
Thermodynamics and Defect Chemistry of Compound Semiconductors
D. T. J. Hurle

Chapter 9
Single Crystal Growth I: Melt Growth
J. B. Mullin

Chapter 10
Single Crystal Growth II: Epitaxial Growth
J. B. Mullin

Chapter 11
Amorphous Silicon-Electronics into the Twenty-First Century
P. G. LeComber

Chapter 12
Control of Semiconductor Conductivity by Doping
E. D. Jones

Chapter 13
Silicon Processing: CMOS Technology
D. J. Foster

Chapter 14
Technologies for High-Speed Compound Semiconductor ICs
J. A. Turner

Chapter 15
Phosphors and Luminescence
Brian Ray

Chapter 16

Microstructural and Compositional Characterization of Thin-Film Semiconductor Materials by Transmission Electron Microscopy (TEM)
A. C. Wright and J. O. Williams

Chapter 17

Dielectric Properties and Materials
Robert M. Hill

Chapter 18
Xerographic Photoreceptors
Michael G. Carter

Chapter 19
Piezoelectric and Pyroelectric Materials and Their Applications
R. W. Whatmore

Chapter 20
Principles of Nonlinear Optical Response
R. W. Munn

Chapter 21
Electro-optic Materials and Applications
Simon Allen

Chapter 22
Nonlinear Waveguides
M. J. Goodwin

Chapter 26
Electro-optic Effects in Liquid Crystals
E. P. Raynes

Chapter 27
Liquid Crystal Applications
Sally E. Day

Chapter 28
High-Temperature Superconducting Materials
D. A. Caldwell

Chapter 29
Langmuir–Blodgett Films
D. E. Hookes

Chapter 30
Electrically Conducting Polymers
D. J. Walton

Chapter 31
Photochromics for the Future
Harry G. Heller

Chapter 32
Materials for Sensing Flammable and Toxic Gases
P. T. Moseley

Chapter 33
Exploiting Semiconductor Oxides for Automative Exhaust Gas Oxygen Sensors
A. Yates

Chapter 34
Field-Effect Chemical Sensors
J. R. Dodgson

Structures of and Bonding in Electronic Materials

J. W. Brightwell

1. INTRODUCTION

The group IV semiconductor materials Si and Ge are both of the cubic, diamond, structural type in which the coordination number is four (tetrahedral) and the bond type is covalent since all atoms are the same and hence have identical electronegativities. In the III–V and II–VI systems the atoms alternate in the structure so that each atom is coordinated tetrahedrally by four of the opposite type; this leads to the zinc blende structure type, which is normally observed provided that the bond type is predominantly covalent.

When departure from covalent bonding becomes significant or when growth takes place under nonequilibrium conditions other structural types emerge, principally the hexagonal 4:4 coordinated wurtzite lattice. The interrelationship between zinc blende and wurtzite structures is often complex. The rates of interconversion by purely thermal means can be very slow (i.e., there is a high activation energy). Even if the enthalpy of transition is low, there may be a considerable effect on structure type from low levels of impurities and grinding can effect a degree of transition.

This chapter sets out to explore the interrelationships between structure and bond type in these 4:4 coordinate systems, to relate cell parameters to bond lengths, and to mention briefly other simple structural types that are important as electronic materials.

2. THE STRUCTURE OF THE GROUP IV ELEMENTS AND OF III–V AND II–VI SEMICONDUCTORS

These have structures that are related to cubic or hexagonal close packing. Figure 1 shows a close-packed plane and it is apparent that there are two sets of cusps between the spheres that form the plane, one set pointing upwards in the plane of the paper, shown lightly shaded in Fig. 1, and the other downwards, shown

---

J. W. Brightwell • Department of Applied Physical Sciences, Coventry Polytechnic, Priory Street, Coventry CV1 5FB, U.K.

heavily shaded in Fig. 1. It is also apparent that each set of cusps is the same in number as the set of spheres and the spacing between the cusps of each set is equal to the spacing of the spheres. The spacing between the two sets of cusps, however, is not a simple multiple of the radii of the spheres. It follows, therefore, that if an identical plane of spheres is placed above this reference plane at densest packing it will cover either all cusps of one set or all cusps of the second set but never some of each.

If three-dimensional sets of planes of spheres are built up then the commonest arrangements are

1. Where planes above and below the reference plane cover the same set of cusps, leading to close-packed hexagonal (cph); or
2. Planes above and below the reference plane cover different sets of cusps, leading to cubic close packing (ccp), i.e., face-centered cubic (fcc).

The respective unit cells are shown in Figs. 2 and 3; note that the trapezium shown in Fig. 1 is the top face of the unit cell in Fig. 2. In the cph cell the close-packed planes are the 00.1 and 00.2 sets and in the fcc cell the 111 sets.

The c dimension in the cph cell is equal to two times the height of a regular tetrahedron of side a, which is equal to $2R$, where R is the atomic radius. It is easy to show that c/a for such an ideal arrangement is $2 \times \sqrt{2}/\sqrt{3} = 1.633$.

The a dimension in the fcc cell is equal to $4R/\sqrt{2}$.

For elements that exhibit both structures the two values of a should be related by $a_{fcc} : a_{cph} = \sqrt{2} : 1$.

In the 4:4 coordinate structure types, zinc blende and wurtzite, both forms of ZnS, the three-dimensional structures are generated in a similar fashion but the sulfur atoms in the planes are expanded a little and have superimposed on them similar planes of zinc atoms (Fig. 4).

Figure 5 then shows the unit cell for wurtzite—note that one atom of Zn is above (or alternatively below) each atom of S (or the atom positions could be reversed).

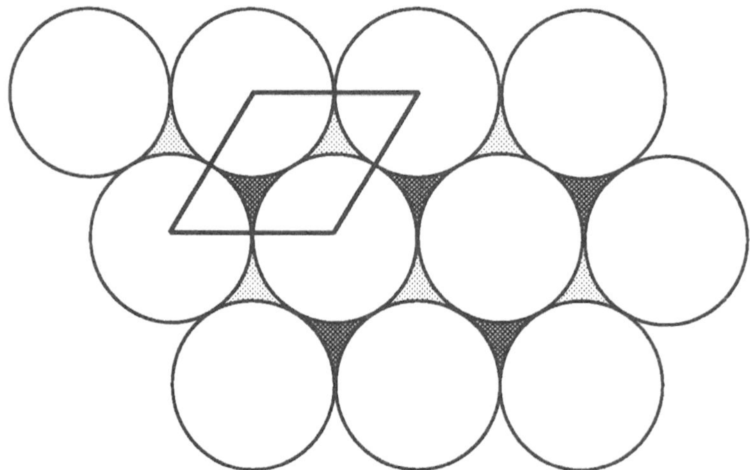

FIGURE 1.   Close-packing in a plane. The trapezium corresponds to the top face in Fig. 2.

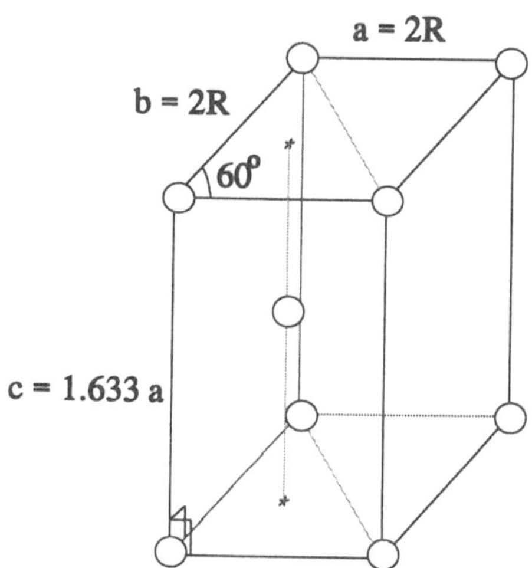

FIGURE 2. The hexagonal close-packed (hcp) unit cell.

Figure 6 shows the unit cell of zinc blende, in which the S atoms form a fcc array and within that unit cell there are eight tetrahedral sites. Half of them (four) are filled in an alternating fashion. (In the CaF_2, fluorite, structure all eight are filled.) Again the cell could equally well be drawn with the atom positions reversed.

The interatomic distances (○---◉) are calculated easily from cell parameters: ○---◉ $= a_{fcc}\sqrt{3}/4$ and, slightly less obviously, $= a_{hex} \sqrt{3}/(2\sqrt{2})$. (Note: a_{hex} = ○---○ in Fig. 6b.)

In the diamond structure C replaces both Zn and S in the zinc blende structure or, for the hexagonal form, in the wurtzite structure.

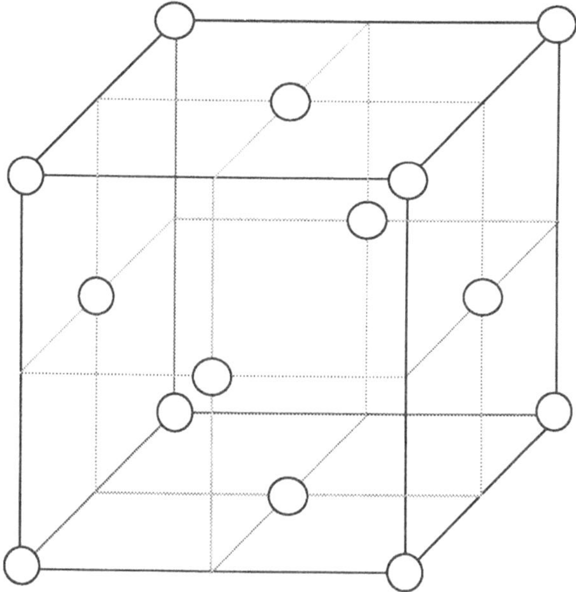

FIGURE 3.   The face-centered cubic (fcc) unit cell.

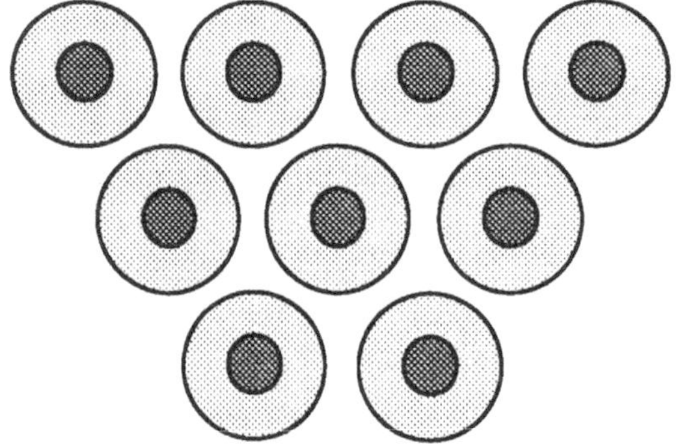

FIGURE 4.   The superposition of planes of zinc atoms on those of sulfur in the wurtzite lattice.

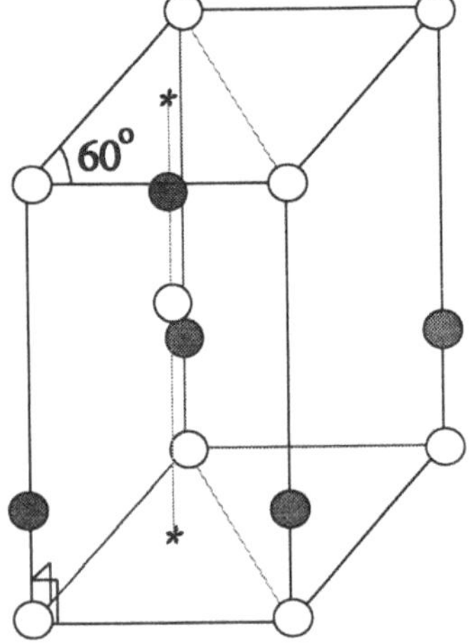

FIGURE 5.   The wurtzite unit cell. Either the unshaded atoms are zinc and the shaded sulfur, or the unshaded atoms sulfur and the shaded zinc.

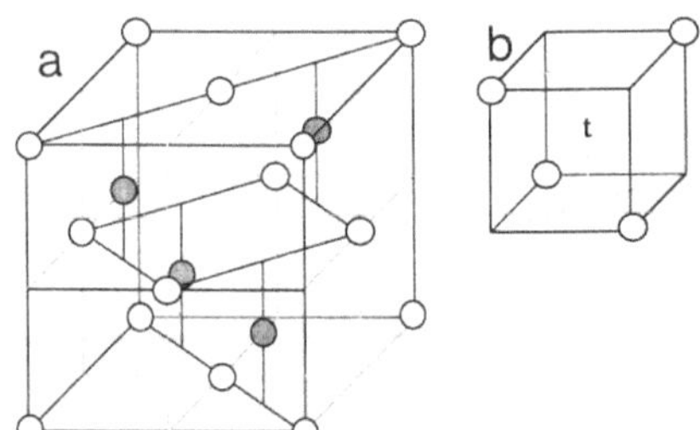

FIGURE 6.   (a) The unit cell of zinc blende. The sulfur atoms form an fcc array, and four of the possible eight tetrahedral sites are filled by sulfur atoms (shaded) (or vice versa). (b) One of the tetrahedral sites (t).

3. BONDING IN AND RELATIONSHIP BETWEEN ZINC BLENDE AND WURTZITE-TYPE COMPOUNDS

As a "rule of thumb" the zinc blende structure is favored for the more covalent systems whereas the wurtzite structure is observed for the more ionic systems, e.g., ZnO. Further increase in ionic character leads to the NaCl and the CsCl type

TABLE 1. Lattice Parameters of Semiconductor Materials with
Zinc Blende/Wurtzite Structures

Substance	$a_0(cub)/pm$	$a_0(hex)/pm$	$c_0(hex)/pm$
Diamond	356.67		
Si	543.0		
Ge	656.76		
GaAs	565.2		
ZnSe	566.7	399.6	653.0
AlP	545.1		
InSb	647.82	(and many other forms)	
Sn	648.9		
CdTe	648.1	458.0	750.0
AgI	649.5	459.22	751.0
ZnS	540.6	382.0	626.0
CdS	581.8	413.6	671.3
InP	586.9		

structures.[1] Usually, however, both zinc blende and wurtzite forms of a compound can be synthesized or a range of intermediate polytypes can be observed.

The group IV elements C, Si, Ge, and α-Sn all have the diamond structure, but synthetic diamond, which is grown relatively rapidly, is normally hexagonal.

An interesting illustration of the relationship between the two structure types is found in AgI, the "stable" β form of which is of wurtzite type; however, inclusion of 5% CuI into the lattice leads to the cubic, γ, form.[2] Rapid precipitation of AgI from aqueous solution gives a mixed product. Refluxing AgI in organic media at 80–120°C leads to pure β-AgI. When AgI is compressed it transforms successively to a tetragonal then to a cubic form, which, on relaxation of pressure, leaves γ AgI.[3]

In zinc sulfide itself the low-temperature zinc blende form transforms to the high-temperature wurtzite form at 1020°C. This process, however, is slow and takes several hours to complete at just above 1020°C. The reverse transition at say 950°C is extremely slow. For zinc sulfide prepared by reaction of $ZnCl_2$ with H_2S at 800°C, the product is, unexpectedly, predominantly wurtzite,[4] indicating that formation of this form is favored kinetically and that the rate of transition to the thermodynamically stable zinc blende form is slow. A table of lattice parameters is given in Table 1 for some substances of interest. When these compounds are related to the positions of the elements in the periodic table (Table 2) then the isoelectronic sets

TABLE 2. Section of Periodic Table

Be			B	C	N	O	F
Mg			Al	Si	P	S	Cl
Ca	Cu	Zn	Ga	Ge	As	Se	Br
Sr	Ag	Cd	In	Sn	Sb	Te	I
Ba	Au	Hg	Tl	Pb	Bi	Po	At

Si, AlP (543 and 545 pm);
Ge, GaAs, ZnSe (565.76m, 565.2, 566.7 pm);
α Sn, InSb, CdTe, AgI (648.9, 647.8, 648.1, 649.5 pm)

are seen to have very similar lattice parameters, consistent with a covalent bonding model.

Moreover, it can be shown that sums of ionic radii (from standard tables) do not relate closely to the observed lattice parameters for these 4:4 coordinate systems.

Isolated values for ionic radii have little meaning; ionic radii are derived from interionic distances (for predominantly ionically bonded compounds) which are measured and divided into r_{m+} and r_{x-} values in some way. These ions do not behave as hard incompressible spheres: they are readily polarized and hence their apparent size increases with coordination number, e.g., Zn^{2+}: $r(CN4) = 74$ pm, $r(CN6) = 88$ pm.

It is important to use only values from the same set of data. That referred to[5] is the most recent authoritative set which has cation values increased by 14 pm and anion radii decreased by 14 pm relative to earlier "traditional" values. Ionic radii are not given for I^-, S^{2-}, or Se^{2-} in coordination number 4. However, O^{2-} differs little in ionic radius in coordination numbers 4 and 6 (124 and 126 pm, respectively), so based on the assumption that this will always be so for anions a_0 values may be "calculated". For example:

For AgI, $rAg^+ + rI^- = 114 + 206 = 320$ pm; and hence $a_0 = 739$ pm (cf. observed value of 649.5 pm;

For ZnS (zinc blende), $rZn^{2+} + rS^{2-} = 74 + 170 = 244$ pm and hence $a_0 = 563.5$ pm (cf. observed value of 540.6 pm).

It is thus apparent that a covalent model is reasonable for these compounds whereas an ionic one is not.

For compounds where the rock salt lattice is exhibited there is a reasonable accord with an ionic model, e.g., MgS for which the sum of ionic radii is $86 + 170 = 256$ pm, which gives $a_0 = 512$ pm compared with the observed value of 520 pm.

4. OTHER STRUCTURE TYPES

There are many other structures of importance, which include perovskite, $CaTiO_3$, and a range where octahedral or octahedral + tetrahedral sites are occupied in fcc or hexagonal arrays of oxygen atoms.

The $CaTiO_3$ structure is shown in Fig. 7. The coordination numbers are Ti-6 (octahedral), Ca-12 (icosahedral), and O-2 (linear). A significant feature of this structure is that the Ca atom is at a rather large separation distance from the oxygen atoms. The O and Ca atoms together form an fcc array (assuming Ca and O to be indistinguishable); however, the Ca^{2+} ion is smaller than the O^{2-} ion and hence cannot be "in contact" with the 12 neighboring O^{2-} ions. This leads to ready polarization of Ca^{2+} ions in the lattice when under the influence of an applied potential. The superconductors of type $YBa_2Cu_3O_x$ are related to this structure type.

The unit cell of a fcc structure (Fig. 3) contains four atoms per unit cell; within it are eight tetrahedral interstitial sites and four octahedral interstitial sites, which

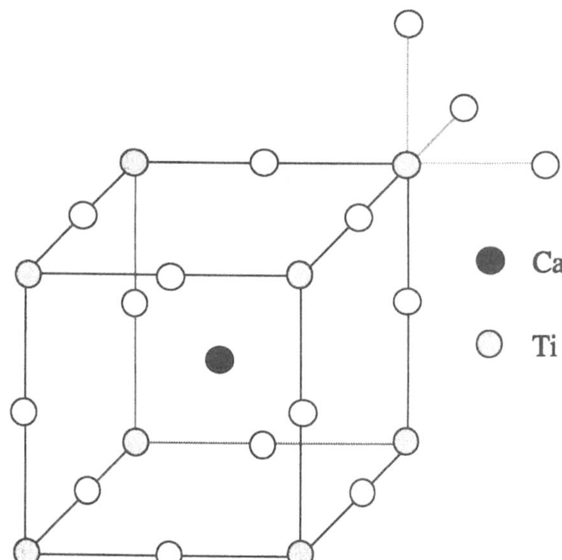

FIGURE 7.   The perovskite structure.
Oxygen atoms are unshaded.

are at the midpoints of the edges and the center of the cube. These sites are not all available since atoms placed in adjacent tetrahedral and octahedral sites would overlap and occupy some of the same space.

If all octahedral sites in a fcc array are filled, the NaCl structure results. The spinel, $MgAl_2O_4$, the O atoms are in such an fcc array and Mg occupies one eighth of the tetrahedral sites and Al half of the octahedral sites.

The hexagonal unit cell (Fig. 2) contains four tetrahedral and two octahedral sites, the positions of which are less obvious. If all octahedral sites are filled in this hexagonal arrangement the NiAs structure results. If two thirds are filled then the α Al_2O_3, corundum, structure is generated. $LiNbO_3$ has a structure based on α Al_2O_3 where two Al(III) atoms are replaced by Li(I) and Nb(V) in an alternating fashion.

For more detailed description or discussion the reader is referred to standard texts.[6,7]

REFERENCES

1. E. Mooser and W. B. Pearson, *Acta Crystallogr.* **12**, 1015 (1959).
2. J. W. Brightwell, C. N. Buckley, and B. Ray, *Phys. Stat. Sol.* (a) **69**, K11 (1982).
3. W. A. Bassett and T. Takahashi, *Am. Mineral.* **50**, 1576 (1965).
4. B. Ray, S. White, A. Murphy, and J. W. Brightwell, *J. Cryst. Growth* **72**, 179 (1985).
5. R. D. Shannon, *Acta Crystallogr.* **A32**, 751 (1976).
6. A. F. Wells, *Structural Inorganic Chemistry*, Oxford U.P., London (various editions, 1945–1984).
7. A. R. West, *Solid State Chemistry and Its Application*, Wiley, New York (1985).

2

Electron Energy Bands

L. S. Miller

1. INTRODUCTION

The previous chapter considered bonds in crystalline materials, particularly semiconductors. A feature of great significance in relation to the applications of semiconductors is the fact that the electrons involved in the bonding (or in other allowed states) can generally be considered to occupy delocalized states. That is, the electrons cannot be associated with one particular atomic orbital; the electron states extend throughout the whole crystal. This is of crucial importance in considering the electrical or optical properties of the crystal.

Two broad strategies for considering this problem are discussed below: the nearly free electron model and the tight binding model. The problem is a very complex one, and no attempt has been made to be rigorous or comprehensive.

The development assumes that the reader is familiar with the basic quantum mechanical idea that an electron may be represented by a wave function (ψ for the time-independent case), which has associated with it a wavelength, λ (and momentum $p = h/\lambda$), an angular frequency, ω (and energy $\hbar\omega$), and so on.

2. MODELS

2.1. The Nearly Free Electron Model

For simplicity, a one-dimensional treatment will be given, with x as the distance coordinate.

First consider a truly free electron: The electron is in a constant electric potential (zero field), so that there are no forces to confine the electron. The electron energy $E(=\hbar\omega$, where ω is the angular frequency) and momentum $\mathbf{p}(=\hbar\mathbf{k}$, where $\mathbf{k}$ is the wave vector) are related by $E = p^2/2m = \hbar^2 k^2/2m$: E is parabolic in k (Fig. 1). The time-independent wave functions (amplitude ψ) are of the form

$$\psi = A \exp(ikx) \tag{1}$$

L. S. Miller ● Department of Applied Physical Sciences, Coventry Polytechnic, Priory Street, Coventry CV1 5FB, U.K.

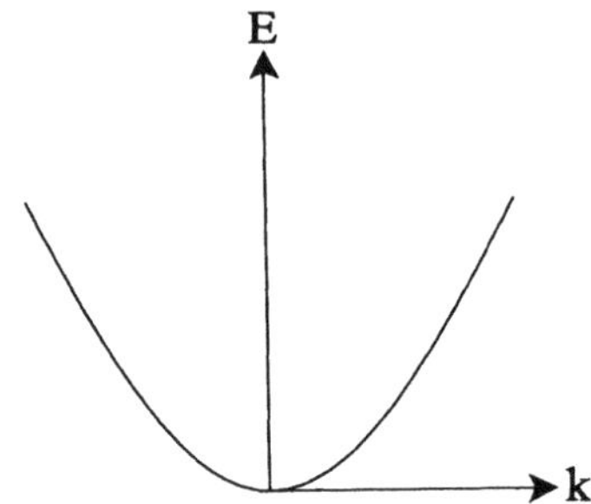

FIGURE 1.   The energy vs. wave vector (E–k) curve for a free electron.

where A is a constant and $i = \sqrt{-1}$. The probability density $\psi^*\psi$ is constant everywhere: over an indefinite period of time the electron is equally likely to be found anywhere. (It must be confessed that if the region of constant potential were infinite the wave function would be everywhere zero; it is reasonable, however, to consider merely a comparatively large region of constant potential, when a description in terms of the free-electron wave function is applicable to a good approximation, and the wave function is finite.)

The model above cannot describe a *moving* electron. To do this, a wave packet is needed. Figure 2 shows the real part of the wave function of a typical wave packet [which is complex, in totality, being constructed from the wave functions of Eq. (1)]. The velocity of the wave packet is the group velocity $d\omega/d\kappa$.

Now introduce a periodic potential (period a); this introduces discontinuities into the E–k curve (Fig. 3). The regions of k are called Brillouin zones; the regions of E are the well known energy bands.

2.2. The Tight-Binding Model

This model approximates the electron wave function in a solid by the sum of suitable atomic wave functions; the development here is a simpled one-dimensional approach as developed by Feynman.[1]

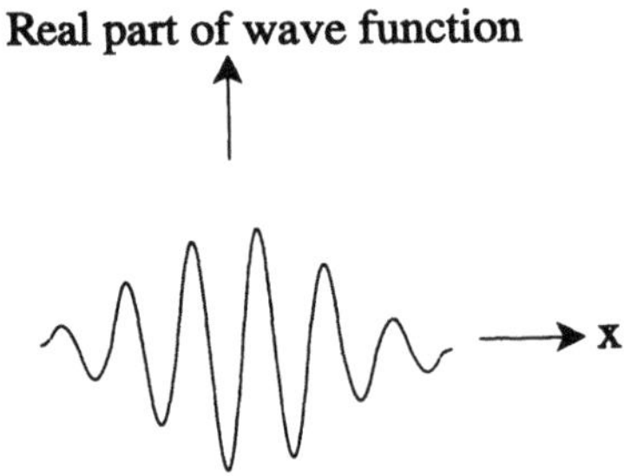

FIGURE 2.   An electron wave packet.

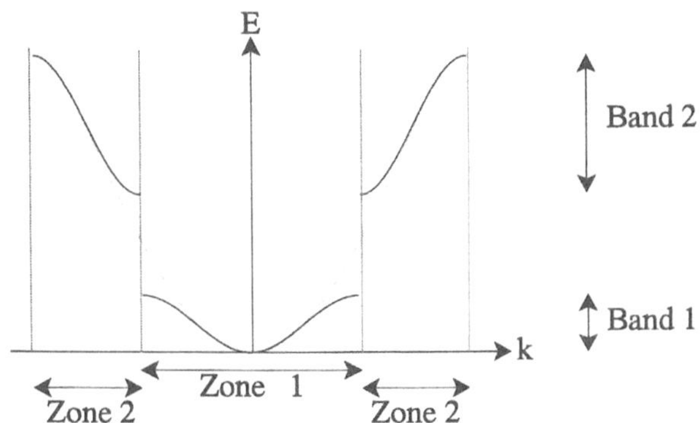

FIGURE 3. The form of the E–k curve for the nearly free electron model.

Figure 4 shows a row of atomic orbitals (shown as s-orbitals, which is the simplest case). For the stationary-state or time-independent solutions, the probability of finding an electron in a particular atomic orbital (P_n) is constant and the same for each orbital. Quantum mechanically, this can be described in terms of a "coefficient," C_n (which may be likened to the wave function ψ, but applies to the complete atomic orbital rather than to one point in space):

$$C_n = A_n \exp(-i\omega t) \tag{2}$$

The resultant probability for the electron to be (somewhere) in the orbital is the real square of this:

$$P_n = C_n C_n^* = |A_n|^2 \tag{3}$$

Because the orbitals overlap, there is a probability (and hence a coefficient) for direct transitions to the nearest neighbors in the line as shown. Assume that the (angular) frequency associated with (and determining the energy of) an isolated orbital is ω_0; that associated with the transitions between orbitals is ω_1. Thus the total rate of change of the coefficient C_n may be written

$$dC_n/dt = i\omega_0 C_n + i\omega_1 C_{n+1} + i\omega_1 C_{n-1} \tag{4}$$

An exponential solution is clearly a possibility. The solution is assumed to be

$$A_n = B \exp(ikx_n) \tag{5}$$

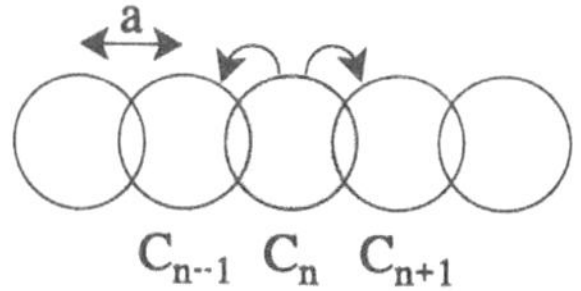

FIGURE 4. A linear array of atomic (s) states with a single electron in one of those (shaded); this forms a basis for discussing the tight-binding model (in one dimension).

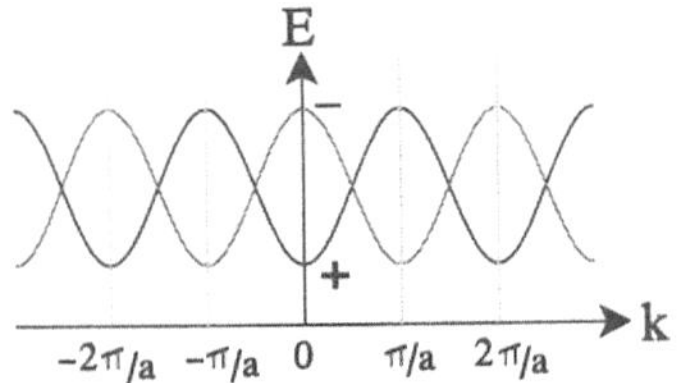

FIGURE 5.   *E–k* curves for the tight-binding model (in one dimension) with the sign of the interaction frequency positive (+) or negative (−).

This satisfies (4) provided

$$\omega = \omega_0 - 2\omega_1 \cos(ka) \tag{6}$$

Since ω_1 may be positive or negative in principle, this gives the two possible results, shown in Fig. 5.

2.3. The Relationship Between the Results of the Two Models

Figure 6 shows two E–k curves (remembering $E = \hbar\omega$) that might result if two different atomic base states (with different ω_0) are considered. Here it has been assumed that the ω_1 associated with the lower band is negative and that associated with the upper band is positive, a point that will be discussed further shortly. The first point to note is that *in the first zone* the upper band is a similar shape to the lower band in Fig. 3.

It may be shown that states with k separated by $2\pi/a$ are equivalent. This means that the upper band in Fig. 3 may be represented in the first or "reduced" zone by translating the E–k curve by this amount; the result would be similar to the lower band in Fig. 6. Thus the two models above give a similar set of E–k curves when represented in the first zone.

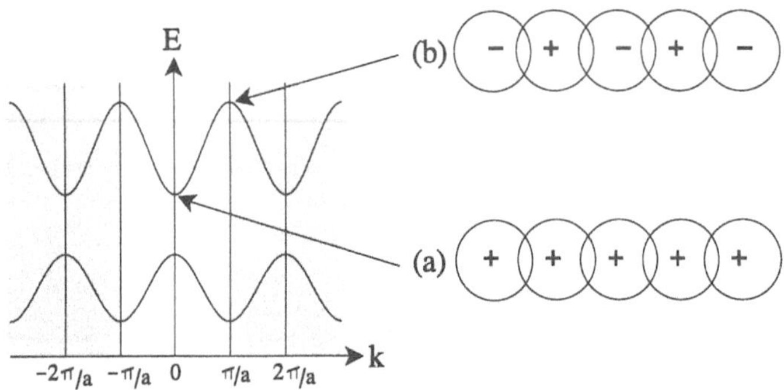

FIGURE 6.   On the left: *E–k* curves for two bands from the tight-binding model (in one dimension). On the right: *s*-orbitals giving (a) bonding and (b) antibonding.

2.4. The Relation Between Maximum Energy and k

In the above development, the sign of ω_1 was not discussed. A simple way of considering this is to relate to conventional ideas of bonding. Figure 6 shows s-orbitals (a) when the electron wavelength is infinity, so that the phase is everywhere the same, and (b) when the wavelength is $2a(k = \pi/a)$, so that the sign of the orbitals alternates. The case (a) gives all bonding orbitals, and (b) all antibonding. Thus the upper curve in Fig. 6 would be appropriate for an s-orbital. By a similar argument, p-orbitals are expected to have the form of the lower curve in Fig. 6.

2.5. Three-Dimensional Effects

The wave vector $k(=2\pi/\lambda$, where λ is the wavelength) has the dimensions of $(\text{length})^{-1}$ and the "scaling factor" 2π. In three dimensions, the k-value may be described by the appropriate point in k-space, which is the "space" defined by the axes k_x, k_y, k_z. The properties of the electron with wave vector $\mathbf{k}$ are related to the crystal *structure* via the *reciprocal lattice*. The basis vectors of the reciprocal lattice are constructed from the basis vectors of the direct lattice by the transformations

$$\mathbf{a}^* \equiv 2\pi(\mathbf{b} \times \mathbf{c})/(\mathbf{a} \cdot \mathbf{b} \times \mathbf{c}), \qquad \mathbf{b}^* \equiv 2\pi(\mathbf{c} \times \mathbf{a})/(\mathbf{a} \cdot \mathbf{b} \times \mathbf{c}),$$

$$\mathbf{c}^* \equiv 2\pi(\mathbf{a} \times \mathbf{b})/(\mathbf{a} \cdot \mathbf{b} \times \mathbf{c}) \tag{7}$$

The points of the reciprocal lattice are then generated by addition of integral multiples of the reciprocal basis vectors. If the real lattice is simple cubic, the reciprocal lattice is also simple cubic, with a "reciprocal lattice constant" of $2\pi/a$. On the other hand, an fcc (face-centered cubic) direct lattice leads to a bcc (body-centered cubic) reciprocal lattice (and vice versa).

The first Brillouin zone in three dimensions is constructed as follows. Choose a reference reciprocal lattice point. Construct lines joining this to all adjacent reciprocal lattice points. Construct planes perpendicularly bisecting these lines. The smallest volume enclosed by these planes is the first Brillouin zone; that for the diamond and zinc blende lattices is shown in Fig. 7. Some of the symmetry points are noted.

2.6. Real Materials

Approximate E–k curves for some real materials are shown in Figs. 8 and 9. They are plotted in two directions in k-space and the corresponding zone boundaries may be related to the symmetry points shown in Fig. 7. (The Miller direction indices are also indicated for those familiar with that idea.) It is immediately noticeable that our model does not predict the form very well: there is additional structure. Various approaches can be used to refine the simple one-dimensional models to predict such effects (for example, allowing for second nearest neighbor interactions), but the only satisfactory approach is to use a more elaborate and three-dimensional

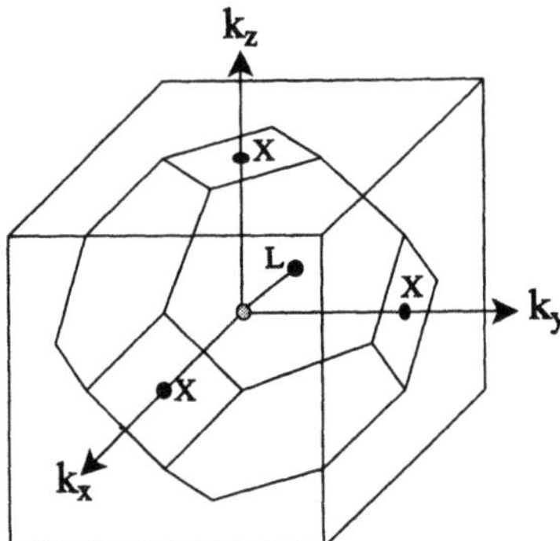

FIGURE 7.   The first Brillouin zone for the diamond and zinc blende lattices.

model. The distinction between Ge and Si shows that the actual *potentials are important*: the result is *not* purely geometrical.

3. EFFECTIVE MASS

The concept of effective mass is very important; the effective mass of (e.g.) an electron is deduced by assuming that Newton's laws are valid; that is, force = mass × acceleration. As for a free electron, it is the group velocity that must be used. On the basis also that force = rate of change of momentum, it is easily shown that the effective mass (m^*) is given (in one dimension) by

$$\frac{1}{m^*} = \frac{1}{\hbar^2} \frac{\mathrm{d}^2 E}{\mathrm{d}k^2} \tag{8}$$

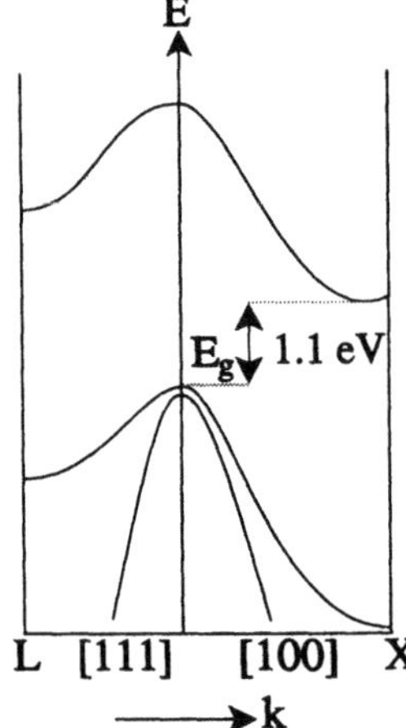

FIGURE 8.   Approximate *E–k* curves for the valence and first conduction band of silicon (indirect gap).

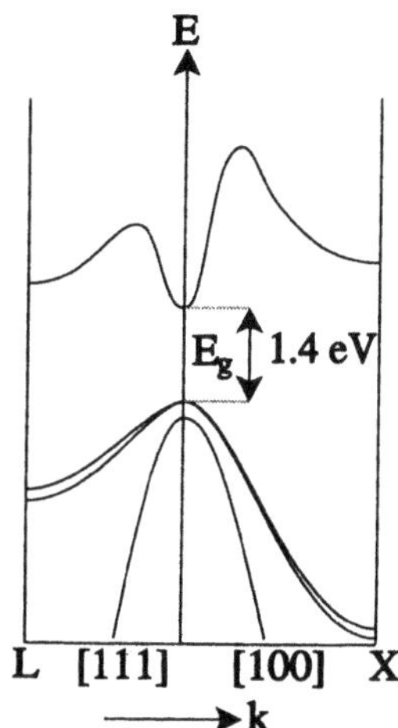

FIGURE 9. Approximate *E–k* curves for the valence and first conduction band of gallium arsenide (direct gap).

In some regions of the band the effective mass may be negative, but *near a minimum of energy* (such as the bottom of the band) the effective mass is always positive. Furthermore, since the band is approximately parabolic ($E \propto k^2$) in this region, the effective mass [Eq. (8)] is approximately constant but generally different from the free-electron mass. However, the effective mass will vary with the direction in k-space, and this variation will be a function of $\mathbf{k}$. In three dimensions $(1/m^*)$ is thus a tensor (relating the acceleration to the applied force).

4. POSITIVE HOLES

If a single electron is missing from a line of otherwise full orbitals (Fig. 10) then the mathematics for the missing electron (that is, the hole) may be developed just as for the electron in Fig. 4. One-hole solutions are found; the relation between hole energy and hole momentum is similar in principle to that for one electron, but the minimum hole energy corresponds to the maximum electron energy. In terms of the electron band, the hole "floats to the top" like a bubble in water. The net

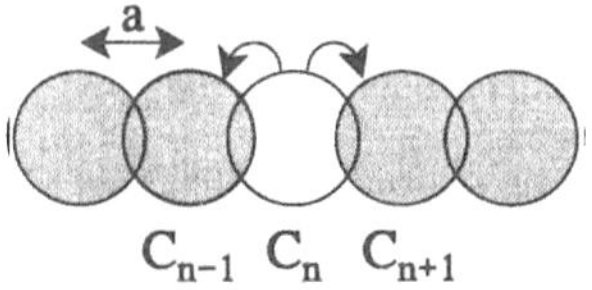

FIGURE 10. A linear array of atomic (s) states with a single electron missing (unshaded). This forms a basis for discussing positive holes.

charge of the hole is positive; its effective mass (near the top of the band) is also positive. In the presence of an electric field, such a positive hole accelerates in the direction of the field.

5. METHODS OF COMPUTING BAND STRUCTURE

The computation of band structures in real materials is very much a specialist problem. Basic strategies are to: use the "first principles" Hartree–Fock approach; alternatively, employ an "ad hoc" potential, solve Schrödinger's equation, and fit critical points of the curve to experimental data (see, for example, Ref. 2).

6. CONDUCTANCE, THE OCTET RULE, AND BANDS

Semiconductors (and certain insulators) are normal valence compounds: their composition and structures correspond to saturation of the chemical bonds within each unit cell. There are in simpler systems a total of eight valence electrons per reduced formula unit (the octet rule), a relation that can be generalized (see, for example, Ref. 3) to cover all semiconductors. In band terms, of course, such properties are expressed by there being (at 0 K) only filled bands and empty bands, neither of which contribute to conduction. The limited conductivity occurring in semiconductors then results from thermal excitation; this is the subject of the next chapter.

7. POSTSCRIPT: THE KRONIG–PENNY MODEL

One of the earliest models for the band structure of solids was that of Kronig and Penney, which assumed a square-well potential as a (rather poor) approximation to a real solid. This is amenable to analytical solution. Gross though the approximation is, it brings out the major features of the E–k curves, the results being very similar in form to those of the nearly free electron model (Fig. 3). In recent years, "low-dimensional structures" have been made in which the Kronig–Penney model is realized, and is is for this reason that it is mentioned here.

BIBLIOGRAPHY

Some of the key points in relation to bands are summarized in:
S. M. Sze, *The Physics of Semiconductor Devices*, Wiley, New York (1981).
A brief but useful discussion of various models is included in:
L. Solymar and D. Walsh, *Electrical Properties of Materials*, Oxford U.P., London (1979).

REFERENCES

1. R. P. Feynman, *Lectures in Physics*, Vol. III, Addison-Wesley, Reading, Massachusetts (1965).
2. G. Grosso, in *Crystalline Semiconducting Materials and Devices* (P. N. Butcher *et al.*, eds.), Plenum Press, New York (1985).
3. E. Mooser, in *Crystalline Semiconducting Materials and Devices* (P. N. Butcher *et al.*, eds.), Plenum Press, New York (1985).

3

Electrical Properties of Semiconductors

A. Barr

1. INTRODUCTION

This chapter provides a basis for a general understanding of some of the basic electrical properties of semiconductors such as conductivity and mobility. It is not intended to be a comprehensive survey.

2. INTRINSIC SEMICONDUCTORS

The electrical properties of semiconductors are determined mainly by those of electrons and holes in the conduction and valence bands, respectively. If conduction holes and electrons are created almost entirely by thermal excitation of electrons across the energy gap (i.e., by thermal excitation from the valence to the conduction band), then the semiconductor is termed intrinsic, and in this case the number of electrons equals the number of holes.

To calculate the electron density (i.e., the number of electrons per unit volume) and the hole density for an intrinsic semiconductor it is necessary to calculate the density of states functions in the conduction and valence bands [$g_c(E)$ and $g_v(E)$, respectively] and multiply by the appropriate probability of occupancy. The probability of occupancy for electrons is the Fermi function, $F(E)$, and $F(E)$ is given by

$$F(E) = \frac{1}{\exp[(E - E_F)/kT] + 1} \tag{1}$$

where E is the energy of the state considered, E_F is the Fermi energy and is the value of E when $F(E) = \frac{1}{2}$, k is Boltzmann's constant, and T is the absolute temperature.

A plot of $F(E)$ vs. E at temperature T has the form shown in Fig. 1. The rate of change of $F(E)$ in the region of E_F is dependent on the temperature T: the lower the value of T the sharper the change of $F(E)$ from one to zero.

A. Barr • Department of Applied Physical Sciences, Coventry Polytechnic, Priory Street, Coventry CV1 5FB, U.K.

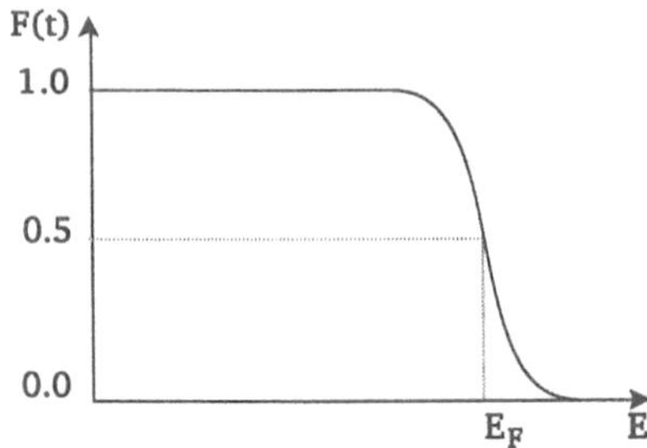

FIGURE 1. The Fermi function $F(E)$ plotted against the energy E, for an absolute temperature T, where T is greater than zero. The value of the Fermi energy, E_F, is that energy when $F(E) = 0.5$.

Further analysis of the Fermi function shows that values of E greater than E_F by approximately $4kT$ give a Maxwellian form (i.e., the $+1$ term in the denominator can be ignored). This approximation (often termed the Maxwell–Boltzmann approximation) simplifies much of the mathematics.

To find the total number of electrons, n, in the conduction band, the product $F(E)g_c(E)\,\mathrm{d}E$ is integrated over all available energies in the band. For electrons with energies near the bottom of the conduction band (for $E > E_c$) we have

$$g_c(E)\,\mathrm{d}E = Cm_e^*(E - E_c)^{1/2}\,\mathrm{d}E \qquad (2)$$

where m_e^* is the effective mass of electrons, E_c is the energy of the bottom of the conduction band, and C is a constant. The result obtained for n is

$$n = 2(2\pi m_e^* kT/h^2)^{3/2}\exp[-(E_c - E_F)/kT] \qquad (3)$$

In the case of holes in the valence band, the probability of occupancy for holes is $[1 - F(E)]$ with $g_v(E)\,\mathrm{d}E$ being given (for $E < E_v$) by

$$g_v(E)\,\mathrm{d}E = Cm_h^*(E_v - E)^{1/2}\,\mathrm{d}E \qquad (4)$$

The result obtained for p, the total number of holes in the valence band, is

$$p = 2(2\pi m_h^* kT/h^2)^{3/2}\exp[(E_v - E_F)/kT] \qquad (5)$$

It is seen that the density of states functions depend on m_e^* and m_h^*.

For an intrinsic semiconductor, n equals p; it is thus possible, by equating the respective expressions for n and p, to obtain an expression for E_{F_i}, the Fermi energy for an intrinsic semiconductor; E_{F_i} is found to be given by

$$E_{F_i} = \frac{E_c + E_v}{2} + \frac{3kT}{4}\ln\left(\frac{m_h^*}{m_e^*}\right)$$

It is seen that E_{F_i} depends on T, m_e^*, and m_h^*, and since m_h^* is usually greater than m_e^*, E_{F_i} increases with temperature. For the case where $m_e^* = m_h^*$, E_{F_i} lies midway between E_c and E_v.

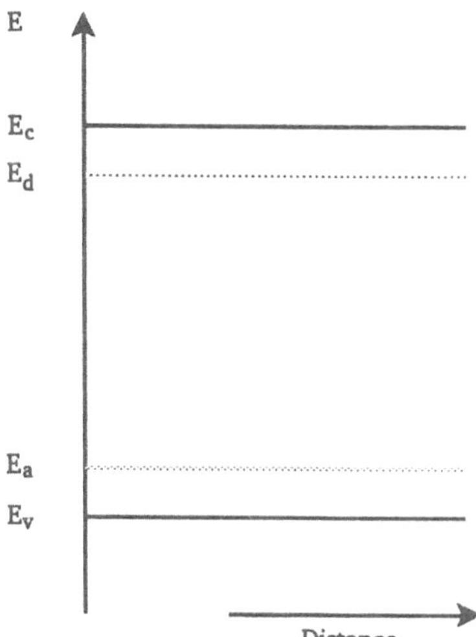

FIGURE 2. The energy diagram for an extrinsic semi-conductor showing the donor and acceptor energy levels E_d and E_a, respectively.

3. EXTRINSIC SEMICONDUCTORS

The introduction of very small amounts of V-valent or III-valent material into silicon (say) produces n-type and p-type silicon, respectively, with associated donor sites and acceptor sites. The donor sites are situated just below the bottom of the conduction band while the acceptor sites are just above the top of the valence band. The energy band diagram for such a semiconductor with both donor and acceptor levels is shown in Fig. 2, with the corresponding densities of states in Fig. 3.

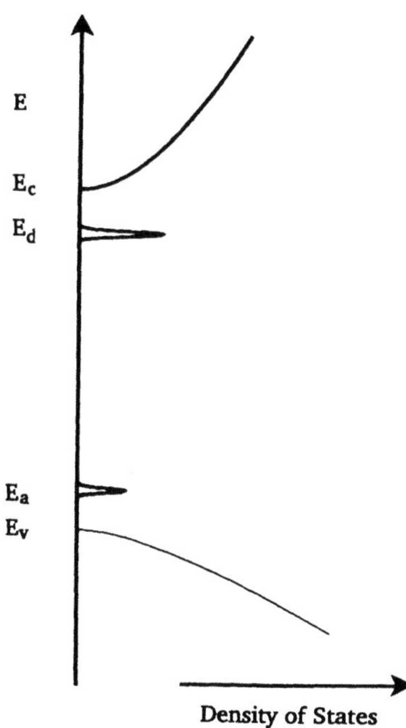

FIGURE 3. The density of states curves for electrons and holes in the conduction and valence bands, respectively, together with the associated curves for the donor and acceptor atoms.

The probabilities of occupancy of the donor and acceptor sites are governed by the modified Fermi function

$$F^*(E) = \{1 + \beta \exp[(E - E_F)/kT]\}^{-1}$$

where β is less than unity. Thus $F^*(E)$ is less than $\frac{1}{2}$ when $E = E_F$. Although donor and acceptor atoms have been added to the system and many of these atoms are ionized, overall there will be charge neutrality. That is, we have

$$p + (N_d - n_d) = n + (N_a - n_a)$$

Here p and n are the hole and electron densities in the valence and conduction bands, respectively, N_d and N_a are the densities of impurity donor and acceptor atoms, respectively, and $(N_d - n_d)$ and $(N_a - n_a)$ give the densities of ionized donors and acceptors, respectively. It is then possible to calculate E_F exactly but often graphical methods or use of approximations suffice.

Also, the charge carrier populations can be obtained from the product of the distribution functions and appropriate Fermi functions. It is found that the charge carriers are all within a few kT of their respective band edges, kT being around 0.025 eV at room temperature.

4. SCATTERING AND MOBILITY OF CHARGE CARRIERS

Once the charge carrier density in the conduction and valence bands has been established, it is then possible, in principle, to calculate the conductivity of the material under consideration since the conductivity, σ, is related to charge carrier density. Assuming an n-type material, for example, with carrier (electron) density n, we have

$$\sigma = e^2 \tau_n / m_e^*$$

Here m_e^* is the effective mass and τ is the mean free time or, more properly, the relaxation time.

The relaxation time τ is the time constant for the decay of current after the electric field is removed, and is a measure of the resistance to electron motion resulting from the scattering of the electrons. For a perfectly periodic lattice, τ is infinite; if the periodicity of the lattice of the sample is altered in any way, then the value of τ is reduced. The two most important effects are (i) the thermal motion of the atoms and (ii) the presence of impurities. Thus there is a relaxation time τ_{lattice} associated with lattice vibrations and τ_{ionic} associated with ionic impurities in the lattice. The increased lattice vibration with increasing temperature, T, means that τ_{lattice} is proportional to $T^{-3/2}$ while the effect of ionic impurities gives τ_{ionic} proportional to $T^{+3/2}$ (both these being approximations).

With both types of scattering present, the resultant relaxation time, τ, is given approximately by

$$\frac{1}{\tau} = \frac{1}{\tau_{\text{ionic}}} + \frac{1}{\tau_{\text{lattice}}}$$

TABLE 1. Mobilities and Bandgaps for Some Commonly Used Materials

Material	Electron mobility at 300 K $(\mu/\mathrm{m}^2\,\mathrm{V}^{-1}\,\mathrm{s}^{-1})$	Hole mobility at 300 K $(\mu/\mathrm{m}^2\,\mathrm{V}^{-1}\,\mathrm{s}^{-1})$	Bandgap at 300 K (E_g/eV)
Ge	0.39	0.19	0.66
Si	0.15	0.045	1.12
GaAs	0.85	0.040	1.40
CdTe	0.11	0.010	1.56

The application of an electric field $\mathscr{E}$ to the sample in conjunction with the scattering processes produces a drift velocity, v_d, and the constant of proportionality between these two is the mobility, μ, of the carriers. So we have

$$\mu = \frac{v_d}{\mathscr{E}} = \frac{e\tau}{m^*}$$

Thus there are different mobilities associated with different scattering processes. The resultant overall mobility, μ, is given approximately by

$$\frac{1}{\mu} = \frac{1}{\mu_{\text{lattice}}} + \frac{1}{\mu_{\text{ionic}}}$$

The conductivity, σ, of the sample is given by

$$\sigma = ne^2\tau/m^* = ne\mu$$

The above equation applies to the situation when one type of carrier (electrons) is present. With both holes and electrons present, σ is given by

$$\sigma = ne\mu_e + pe\mu_h$$

where μ_e and μ_h are the mobilities of electrons and holes, respectively. An analysis of the drift velocity of a "bunch" of one type of carrier (say electrons) in a "sea" of the other type (holes) is quite complex, the resultant "ambipolar" mobility being a function of the individual mobilities.

A number of techniques are available for measuring concentrations and mobilities of charge carriers in samples; some are briefly indicated in Chapter 16 by Wright and Williams. Table 1 lists values for some of the parameters for a number of different materials to show the range of values expected. Values for other materials can be found in various texts (see the Bibliography, for example).

5. HIGH-FIELD EFFECTS

If the electric field applied exceeds a certain value (say, around 100 V/mm) then the electrons and holes in wide-band semiconductors such as silicon and gallium arsenide are heated by the electric field above the temperature of the lattice (which is normally around room temperature). This leads to "hot carrier" effects, and results from the fact that the carriers cannot lose energy as fast as they gain it from the field; as they "heat up" their frequency of scattering increases (basically because they are moving faster): τ reduces and the rate of energy loss becomes equal to the rate of energy gain. Clearly, a reduced τ corresponds to a reduced mobility. In the case of silicon, this leads, at fields of around $10^4\,\mathrm{V\,mm^{-1}}$ to a saturation of the drift velocity at around $10^8\,\mathrm{mm\,s^{-1}}$ for electrons, and for gallium arsenide this figure is around $6 \times 10^7\,\mathrm{mm\,s^{-1}}$. Thus the saturation velocity for electrons in gallium arsenide is less than that in silicon, despite the much higher low-field mobility (Table 1). At intermediate fields, however (about $300\,\mathrm{V\,mm^{-1}}$), the velocity in gallium arsenide peaks at about $2 \times 10^8\,\mathrm{mm\,s^{-1}}$ and is about four times that of silicon at the same field. The actual drop in drift velocity in gallium arsenide results from the presence of the indirect band gap (see Chapter 2 by Miller) at only about 0.3 eV above the direct gap; electrons excited into this 'sub-band' at high fields have a reduced mobility.

These phenomena must be carefully considered in relation to the relative "speed" advantages of the two materials. The picture is further complicated by the fact that for very small devices the voltages involved can be insufficient to provide the energy involved in heating the carriers, although the field is very high (note the need for 0.3 eV in exciting to the gallium arsenide sub-bands). In this case, the carriers are not reaching equilibrium within the length of the device, and this can occur at low fields also; clearly care must be exercised in predicting the relative "speed" advantages of different materials.

BIBLIOGRAPHY

The following general text books are some of many that extend the ideas introduced in this section:

H. M. Rosenburg, *The Solid State*, Oxford U.P., London (1983).
S. M. Sze, *Physics of Semiconductor Devices*, Wiley, New York (1981).
S. M. Sze, *Semiconductor Devices, Physics and Technology*, Wiley, New York (1985).
H. E. Hall, *Solid State Physics*, Wiley, New York (1974).
L. Solymar and D. Walsh, *Lectures on the Electrical Properties of Materials*, Oxford U.P., London (1984).

4

Optical Properties

L. S. Miller

1. INTRODUCTION

The classical approach to understanding optical phenomena such as absorption is based on Maxwell's equations; this is still important, and some basic ideas are introduced here. It is perhaps even more important to consider some of the quantum processes involved in absorption and emission of electromagnetic radiation, and in particular to distinguish between those semiconductors that have applications in optical emission and those that do not.

2. THE CLASSICAL APPROACH

The optical properties of a medium may be described in terms of its complex refractive index. For isotropic linear media (the latter meaning that the optical properties are independent of the light intensities used) this is

$$n' = n - ik \tag{1}$$

where n and k are real positive constants, and i is $\sqrt{-1}$. The velocity of propagation is c/n, where c is the velocity in vacuo; the wavelength in the medium is λ_0/n, where λ_0 is the wavelength in vacuo. It is n that determines the angle of refraction.

It is easily shown that the intensity of the wave varies as $\exp(-\alpha x)$, where $\alpha = 4\pi k/\lambda$ and α is the absorption coefficient. Thus k determines the absorption.

Equations for parameters such as reflection, often presented assuming a real refractive index n ($k = 0$) may be universally applied by substituting n' for n with suitable interpretation. For example, the normal incidence reflection (R) at an interface is

$$R = \left| \frac{n' - 1}{n' + 1} \right|^2 = \frac{(n - 1)^2 + k^2}{(n + 1)^2 + k^2} \tag{2}$$

L. S. Miller • Department of Applied Physical Sciences, Coventry Polytechnic, Priory Street, Coventry CV1 5FB, U.K.

Note that if k is large, and thus the absorption is large, then R is large: a very absorbing medium such as a metal absorbs strongly and so k is very large and R approaches unity.

2.1. Relation to Conductivity

The Maxwell formalism relates k to the electrical conductivity (σ):

$$k = 2\pi\sigma/n\omega = \sigma/n\nu \tag{3}$$

This is the effective conductivity at the frequency of the e.m. wave, and does not require an absorbing medium to have a dc conductivity.

2.2. Optical Constants and Relative Permittivity

The relative permittivity (ε_r) may also be complex, the imaginary part giving the dielectric loss. It may be shown that

$$\varepsilon(= \varepsilon' - i\varepsilon'') = (n')^2 \tag{4}$$

For wavelengths around 0.8 μm, where silicon photodiodes are often used, k is small ($<10^{-2}$), but the high value of ε' (around 12) gives an n of around 3.5. [Note that even at these optical (near ir) frequencies, n approaches its low-frequency limit.] This value of n leads to a high normal-incidence reflectivity (R) of around 30%. This is typical of semiconductors and leads to problems in the efficiency of sources and detectors. The resultant small incident angle for the onset of total internal reflection is a further problem for emitters, resulting in only a small fraction of the emitted light being able to escape (in the absence of special coatings or geometries). In the visible region (for silicon) the high absorption leads to an even higher reflectivity from Eq. (2). (The extreme case is a metal: the absorption coefficient is very strong, but in practice little light is normally absorbed because most of it is reflected, and so does not penetrate the material and become absorbed.)

2.3. Resonance

Figure 1 shows the typical variation of n and k through an absorption peak. The value of n is always highest at the low-frequency limit. As the frequency is reduced from a very high value (x-ray region) the value of n (or ε) increases from its limit of unity through each absorption region and is highest at low frequencies [excepting the low-frequency resonant peak(s)]. Some processes (e.g., the orientation of permanent dipoles give a relaxation rather than a resonance.

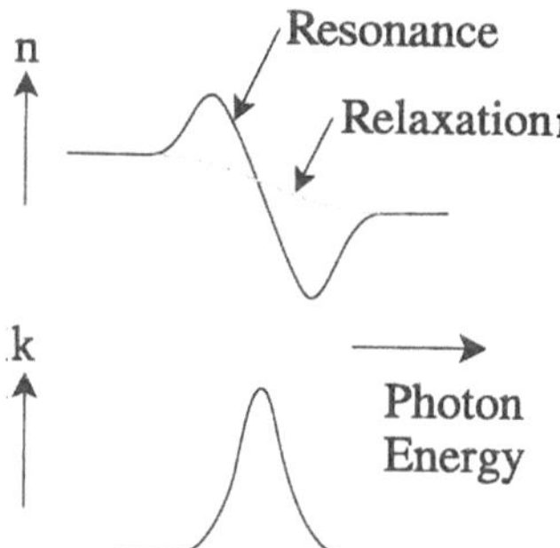

FIGURE 1.   The general form of the variation of *n* and *k* through an absorption peak (resonance or relaxation).

3. ABSORPTION MECHANISMS

3.1. Fundamental Absorption

This occurs when an electron is excited from the valence to the conduction band of a semiconductor. The momentum of the photon is negligible compared to the width of the Brillouin zone, so the process (for photon energies just greater than the band gap) is as shown in Fig. 2. For indirect gap materials such as silicon it involves the absorption or emission of a phonon for momentum conservation; this gives a smaller and less steeply rising absorption coefficient than for a direct gap. The corresponding absorption curves are shown in Fig. 3. The logarithmic plot shows clearly how the absorption coefficients rise rapidly for energies just greater than the band gaps, the steeper rise being for the direct gap (gallium arsenide).

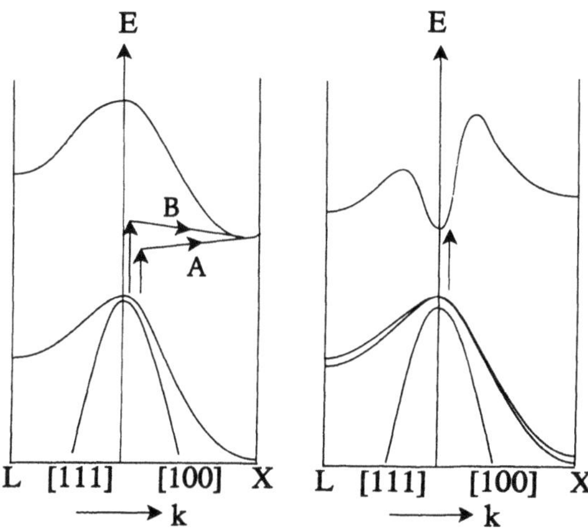

FIGURE 2.   Band-edge (fundamental) absorption for Si (left) and GaAs (right). For Si, which is indirect, the absorption of a photon must be accompanied by the absorption or emission of a phonon. (The phonon energy is exaggerated for clarity.)

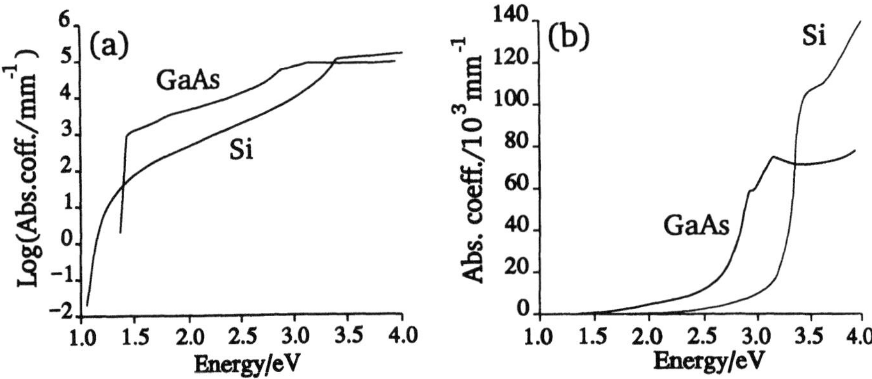

FIGURE 3. The absorption coefficient of Si and of GaAs, plotted on a logarithmic scale (a) and a linear scale (b).

However, the onset of direct transitions in silicon at around 3.3 eV leads ultimately to comparable absorption coefficients for both materials at around 4 eV. The significance of the rise at 3.3 eV is shown most clearly on the linear plot, where it dominates the absorption curve (the band-edge effects being barely visible).

3.2. Other Mechanisms

Figure 4 shows the above and some other mechanisms on a simple band diagram. These are as follows:

1. Fundamental absorption as above.
2. Exciton absorption: The electron and hole remain bound to each other, giving a reduced energy compared to that required to produce a free (band) electron and hole.
3. Free-carrier absorption: Free (band) electrons or holes are given energy within their respective band. This may be modeled classically as the absorption of a conducting medium; the highest frequency at which it occurs is typically in the infrared and increases with increasing conductivity. The absorption and reflection of metals in the visible region has similar origins, the higher conductivities leading to effects at higher frequencies.

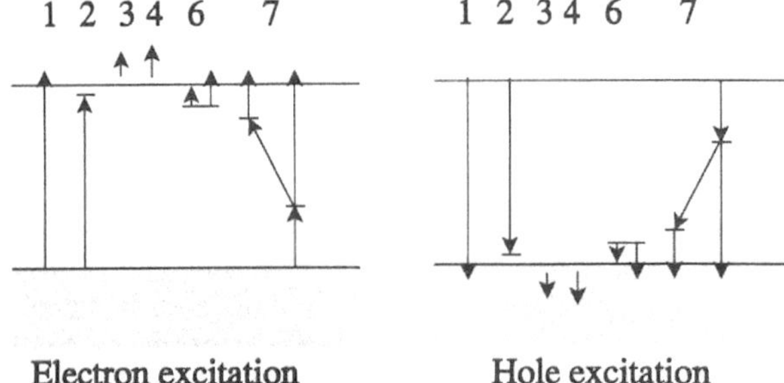

FIGURE 4. Some optical absorption mechanisms. Some processes are most naturally described in terms of electron excitation (left), others as hole excitation (right).

4. Interband absorption: Transitions between two conduction bands or two valence bands.

5. Optical phonon absorption: Certain lattice vibrations (optical phonons) are, in partially ionic materials, excited by the appropriate energy infrared photons. (This cannot be shown in Fig. 4.)

6. Long-wavelength absorption by impurity centers: Electrons or holes trapped, in particular, in donor or acceptor sites are excited or released into the conduction or valence band, respectively. Only at low temperatures (much lower than room temperature) will the carriers be trapped in these sites.

7. Impurity absorption: Various transitions to and from impurity sites give effects at various energies depending on the presence and occupancy of the sites.

For 1–6 the wavelength increases (with overlapping ranges) from (typically) around 1 μm to 25 μm, but the longer-wavelength effects are not normally observed except at reduced temperatures.

4. PHOTOCONDUCTIVITY

Any mechanism whereby electrons (or holes) are excited to the conduction (or valence) bands from sites where they were previously not mobile will yield a photocurrent. In the most common device (the silicon photodiode) the fundamental absorption is utilized. To achieve efficient detection of long-wavelength ir requires (for example) a narrow-band-gap material and may require operation below room temperature to reduce the resultant intrinsic conductivity, which may be too high at room temperature.

5. EMISSION

5.1. Spontaneous Emission

A simple way of considering emission is to relate the absorption to the emission by the principle of detailed balancing: The densities of electrons, holes, photons, etc. must each be constant in thermal equilibrium. This leads to a (conduction band to valence band) recombination coefficient given approximately by

$$R(\nu) \propto np\alpha\nu^2 \tag{5}$$

where n and p are the electron and hole densities, and α the absorption coefficient.

It is seen that efficient emission is associated with efficient absorption. Direct-gap materials are ideal. If appropriate impurities are present, efficient recombination may occur through these (Fig. 5). The "third body" relaxes the constraints imposed by momentum conservation, so that efficient recombination may occur in indirect-gap materials, particularly when there is a direct gap at only slightly greater energy. The classic case is the (isoelectronic) nitrogen trap in $GaAs_{1-x}P_x$ which enables green emission in a range of x for which the gap is indirect. An advantage of such a

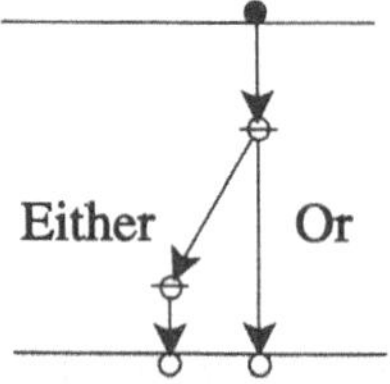

FIGURE 5.   Some electron–hole recombination routes involving impurity or defect levels.

situation is that although the emission via the impurity centers is usefully strong, the related absorption coefficient is much less than that for fundamental (conduction to valence band) emission; this gives reduced reabsorption.

5.2. Stimulated Emission

Figure 6 illustrates the concept of stimulated emission, which is the process involved in a laser (*light amplification by stimulated emission of radiation*, discussed further in Chapter 7 by Nelson). A necessary condition is for the probability of occupation of some energy states to be higher than the probability of occupation of other states at lower energy (population inversion); this is the opposite of the situation occurring in thermal equilibrium, and requires appropriate "pumping" of the upper states.

The consequence of this situation is that as a suitable photon passes through the medium, an electron may be stimulated to move down from the upper to the lower state and in doing so emit another photon of the *same energy and phase.* The process is related in an obvious way to absorption, where the electron is stimulated to move up in energy and absorb the photon energy; the difference is in the electron population.

For the process to be efficient, direct-gap materials, or those showing efficient impurity recombination, are needed. In typical semiconductor diode lasers, the region of recombination contains high densities of both donors and acceptors, forming impurity bands close to the band edges which effectively become part of

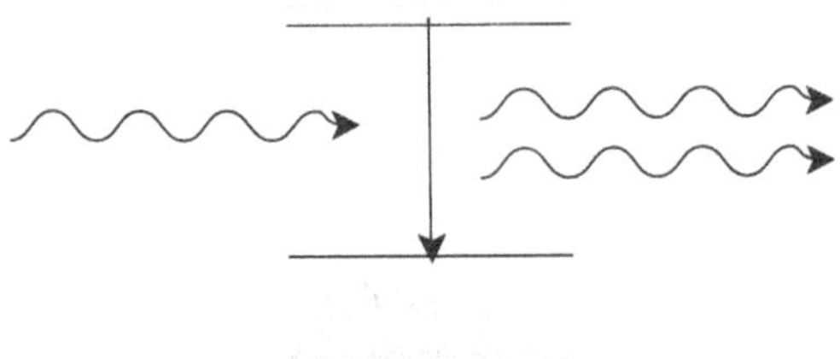

FIGURE 6.   Stimulated emission. An incoming photon stimulates the transition of an electron from a filled conduction-band state to an empty valence-band state (effectively recombination of an electron and a hole). The emitted photon has the same energy and phase as the incoming photon.

the band. As far as the simple representation of Fig. 6 is concerned, however, this would simply lead to a slightly reduced effective band gap compared to pure material. This in turn corresponds to a slightly smaller emission frequency (longer wavelength).

Further theoretical and practical aspects of achieving lasing in diodes are discussed in Chapter 7 by Nelson.

5.3. Nonradiative Recombination

In many cases, nonradiative recombination of electrons and holes (often through impurity centers) limits the efficiency of emission by competing with the radiative processes; poor emitters such as silicon thus give no useful emission: all the recombination is nonradiative.

6. ANISOTROPIC MATERIALS

In many materials, it is possible to identify one or more directions that are distinct and lead to a variation of optical properties with direction. In ZnS, for example, there is a cubic form, giving isotropic properties, and a hexagonal (wurtzite) form, giving a (small) variation of n with direction. For some materials such as $LiNbO_3$ this orientation dependence is marked. Anisotropic (and indeed nonlinear) materials are discussed later in this book.

7. POLARIZED LIGHT

In other chapters, discussion of anisotropic and nonlinear materials will utilize the concept of polarization of light, and it is also important in measurements of isotropic materials; it is therefore introduced briefly below.

The electric vector of an e.m. wave oscillates transversely and the plane of oscillation defines the plane of polarization. From a normal source the plane of polarization varies randomly: the light is unpolarized.

Plane polarized light has a unique plane of polarization. When light is incident on the surface of a sample, light polarized in the plane of incidence is known as the "p" ray; that polarized perpendicularly to the plane of incidence is known as the "s" ray.

By combining two (normally orthogonal) plane-polarized beams at different phases, circularly or elliptically polarized light is produced; the former is a special case of the latter when the components have equal amplitudes. Normally, such light is obtained by appropriate manipulation of light from a single source, to ensure a constant phase relationship between the components.

8. THIN-FILM SYSTEMS

The reflection, transmission, and absorption of single or multiple thin films may be calculated, and the optical properties of thin films may be determined by a variety of techniques. The most frequently used is ellipsometry, which measures in effect the changes in relative amplitudes and phase of the p and s components of the beam (normally from a laser). These parameters can be related to their theoretical values to determine optical parameters.

The calculation of the properties assumes an addition of the amplitudes of waves reflected and transmitted at all interfaces; by using matrix methods this can be conveniently performed for any number of layers. The properties of such structures may be determined as much by the film thicknesses as by the optical constants of the components. In particular, a single nonabsorbing film on a substrate such as silicon has a color characteristic of its thickness; for many such films n is close to 1.5 and an approximate thickness is obtained simply by observing the color. One feature of thin surface films of technical importance is that they can be used to reduce reflection and improve the efficiency of detectors and emitters.

BIBLIOGRAPHY

A discussion of measurement methods, related theories, and detailed optical data for a range of materials in bulk and thin-film form is given in:

Handbook of Optical Constants of Solids (E. D. Palik, ed.), Academic, New York (1985), from which the data for Fig. 3 were derived.

Basic optical processes are discussed in an accessible way by:

J. I. Pankove, *Optical Processes in Semiconductors*, Prentice-Hall, Englewood Cliffs, New Jersey (1971).

The interaction of polarized light with isotropic and anisotropic thin-film systems is discussed in:

R. M. Assam and N. M. Bashara, *Ellipsometry and Polarized Light*, North-Holland, Amsterdam (1977).

5

Interfaces and Low-Dimensional Structures

David A. Anderson

1. INTRODUCTION

The concept of an energy band structure as described in previous chapters is derived on the basis of an infinite periodic crystal lattice. In practice this is a very good approximation since the dimensions of the materials we work with are generally large compared with the lattice spacing and the defects to the crystal periodicity few enough to be treated as perturbations. One area where these assumptions are not valid, however, is at the boundaries of the material, at free surfaces (i.e., the semiconductor–air interface), contacts with metals, interfaces with insulators, and the junction between one semiconductor and another. In devices, the influence of these boundaries can rarely be ignored. In many cases it is the change in the band structure induced by the boundary that produces the device effect. In this chapter we will discuss in detail just one subject of this very wide field, that of semiconductor–semiconductor heterojunctions. The technology of epitaxial growth of semiconductors has developed to the point that not only can layers of a material be grown one atom layer at a time, but the composition of the crystal can be changed abruptly within one monolayer giving hyperabrupt interfaces. Furthermore, because this interface is grown continuously without interrupting the growth process it can be made almost ideal, provided the lattice constants of the two materials forming the heterojunction are not too dissimilar.

2. BAND STRUCTURE AT A HETEROJUNCTION INTERFACE

Three energy levels can be used to characterize the important aspects of the band structure of a semiconductor in relation to heterojunction interfaces: the lower edge of the conduction band E_c, the upper limit of the valence band E_v, and the Fermi energy E_f. On an absolute scale the values of E_v for the technologically

David A. Anderson • Royal Signals and Radar Establishment, St. Andrews Road, Malvern, Worcestershire WR14 3PS, U.K.

33

important semiconductors all lie within 0.5 eV of one another and the band gaps $(E_c - E_v)$ range from about 0.5 eV to 2.0 eV. The Fermi energy is controlled by the doping level in the semiconductor. So, in general, in two isolated semiconductor materials the conduction and valance bands will lie in different places in energy and the Fermi energies will be different. When these two semiconductors are brought together they form a single system in thermodynamic equilibrium and so must be described by one single Fermi energy. For this to happen, charge must flow from one material to the other to create a dipole whose potential cancels out the difference in the electrochemical potentials of the two isolated semiconductors generated by the Fermi energy difference. The net effect is that the conduction and valence bands are distorted in the region of the interface by the dipole space charge. This so-called "band-bending" effect is illustrated in Fig. 1 for the particular case of the gap of one semiconductor nested within the gap of the other. Many other permutations exist, but this is the one that turns out to be of most practical significance. The figure shows two situations: (a) in which the wide gap material is n-doped and the narrow gap material p-doped and (b) the reverse. The difference in the band gaps produces an abrupt step in the bands at the interface ΔE_c and ΔE_v independent of Fermi level position. In addition, in case (a) one can think of negative charge having to be transferred from semiconductor 1 to semiconductor 2 to bring the Fermi levels into coincidence. This results in a depletion region (of reduced negative charge) on the left of the interface and an accumulation of negative charge on the right. A characteristic spike is produced in the conduction band. In case (b) the charge transfer is such as to produce a similar spike in the valence band.

The band edge offsets ΔE_c and ΔE_v represent potential energy barriers for electron or hole motion perpendicular to the interface in one direction and as

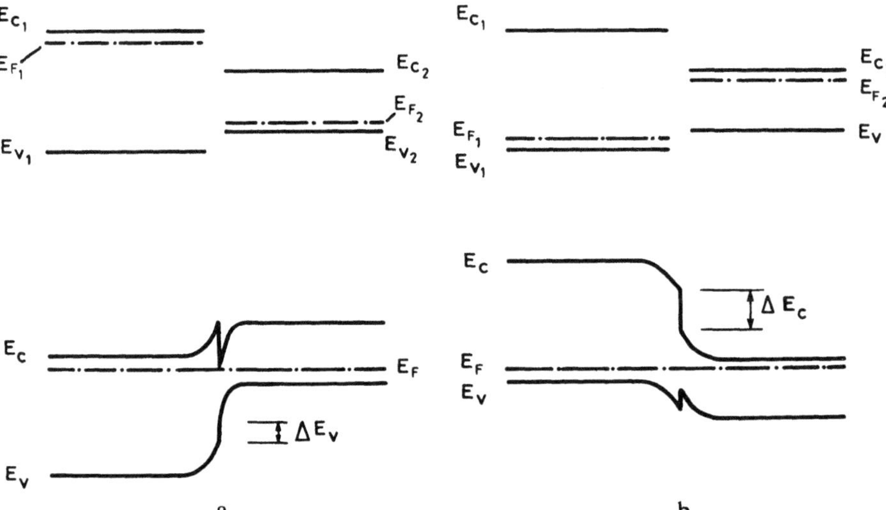

FIGURE 1. Band bending in the vicinity of a heterojunction interface between a wide gap semiconductor (1) and a narrow gap semiconductor (2). In (a) semiconductor (1) is n-doped and semiconductor (2) is p-doped. In (b) the doping is reversed. This is only one of many permutations depending on the relative gap positions and energies of the two semiconductors.

injectors of energetic or hot electron or holes in the other. The use of the band edge offset to confine carriers to the narrower gap material is exploited in photodetectors, in semiconductor lasers, and in the heterojunction bipolar transistor.

If in Fig. 1a the wide gap material is doped heavily n-type and the narrow gap material is lightly doped n- or p-type, then the band bending causes the lower end of the conduction band spike to dip below the Fermi level (see Fig. 2b). When this happens the conduction band fills with carriers immediately adjacent to the interface and the semiconductor is said to be degenerate in this region. Calculations of the

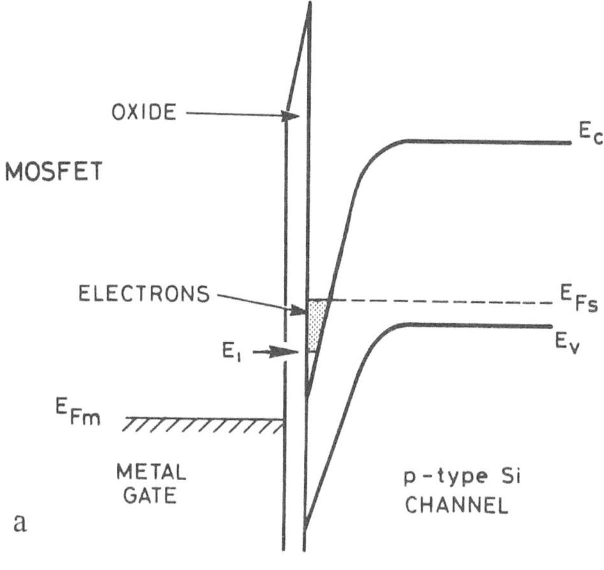

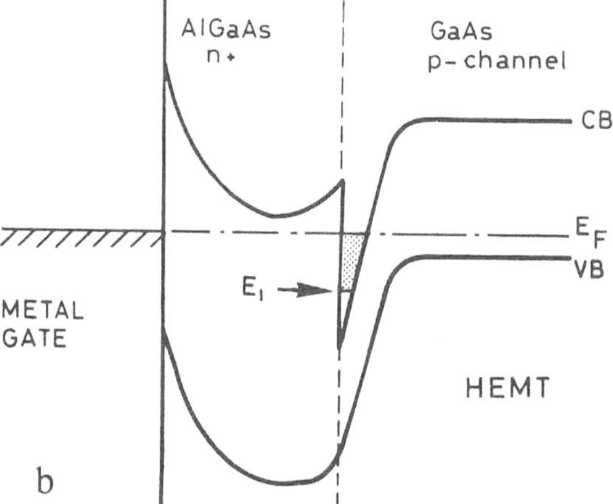

FIGURE 2.   (a) Schematic diagram of the band bending in the channel of a Si MOSFET device under positive gate bias. (b) Schematic diagram of the band bending through a HEMT field effect transistor using GaAs/AlGaAs heterojunctions. The diagram emphasizes the broad similarity with the MOSFET.

precise form of the potential show that the electrons are confined in a roughly triangular well within less than 10 nm of the interface. The carrier density can easily be increased to about 10^{12} per cm^2, equivalent to 10^{18} cm^{-3}. Note, however, that while the electrons are in the narrow gap material, the donors that produced them are in the wide gap material. This remote doping phenomenon means that the electrons can move parallel to the interface under an applied electric field with very little scattering from the ionized donor atoms. The resultant electron mobilities are larger than in conventional semiconductor layers, particularly at low temperature, and the effect has been exploited in a device known as a high electron mobility transistor (HEMT) which has produced the highest-frequency microwave amplification to date.

3. LOW-DIMENSIONAL EFFECTS

In the HEMT potential well the electrons are confined to within about 30 atomic layers of the interface but free to move parallel to it. Motion is highly anisotropic and for all practical purposes it can be described as two-dimensional rather than three-dimensional. The conventional assumption of an infinitely repeating periodic lattice is no longer applicable for the direction normal to the layers. One immediate consequence of this reduced or low dimensionality is that the energy levels available to the electrons in the conduction band, which form a virtual continuum in the three-dimensional case, are restricted by quantization effects arising from confinement in the direction normal to the interface; this is more pronounced for increasing confinement. In Fig. 2, this leads to the minimum electron energy E, as shown, and a two-dimensional electron gas (or 2DEG) occupying a range of energies above it.

HEMT structures have been used to study these two-dimensional effects in great detail, but it is worth noting in passing that the earliest experiments were performed on Si metal-oxide-semiconductor devices (MOSFETS). A schematic diagram of the band structure for this device is shown in Fig. 2a and shows some striking similarities with the HEMT. Like the HEMT, the MOSFET also has a degenerate two-dimensional electron well but in this case it is induced by applying a positive potential to a metal contact on the other side of a dielectric oxide layer. The two-dimensional effects are not so prominent in the MOSFET device, partly because the electron mobility is lower in Si and partly because the semiconductor-oxide interface is not a single-crystal heterojunction as it is in the HEMT.

One of the most remarkable effects observed in both the MOSFET and the HEMT is the appearance at low temperatures of very large amplitude oscillations in the resistance as a function of magnetic field. In a Hall effect geometry the Hall voltage exhibits broad plateaus, flat to 1 part in 10^7. This observation has become known as the quantum Hall effect and earned its discoverers the Nobel Prize in Physics in 1985. In the plateaus the Hall resistance defined as the Hall voltage/sample current is given by $h/e^2 i$, where i is an integer; i.e., it can be expressed simply in terms of fundamental constants. This fact, coupled with the very high accuracy, has prompted standards laboratories around the world to adopt the quantum Hall effect as the basis of a new resistance standard.

Although arguably the simplest structure to make, the single heterojunction interface is not the most suitable structure for the study of the full range of two-dimensional phenomena. This is because the shape of the potential well in which the carriers sit is controlled by the carrier density. The more carriers there are, the higher the electric field at the interface, the more tightly confined the carriers are, and the greater the splitting between the available energy levels. Also, the band bending at the heterojunction interface is such as to produce a well in either the conduction band or the valence band, depending on the doping, but not in both at the same time.

If two heterojunction interfaces are brought close enough back to back it is possible to have carriers confined in a rectangular well rather than the triangular one above. This situation is shown schematically in Fig. 3, where, for simplicity, the band bending has been neglected. For the particular case of the gap of the narrow gap material lying entirely within the gap of the other semiconductor it can be seen that both electrons and holes sit in rectangular potential wells within the same material. This structure is known as a single quantum well and behaves optically and electrically as a two-dimensional semiconductor.

The carrier confinement within the quantum well gives rise to discrete energy levels in both conduction and valence bands for carrier motion normal to the layer. In the plane of the two-dimensional sheet the energy bands are approximately parabolic as in the three-dimensional case. The energies of the discrete levels can be derived fairly simply in the case of an infinite quantum well. The allowed states are those for which the width of the well is an integral number of half wavelengths of the electron (λ), i.e.,

$$L = \frac{n\lambda}{2} = \frac{n\pi}{k}$$

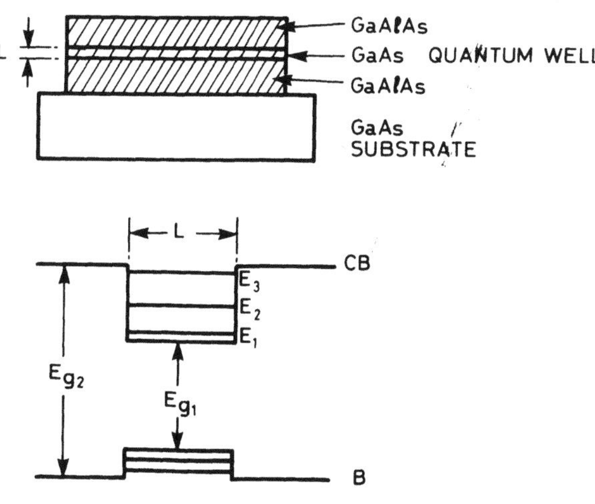

FIGURE 3. Schematic diagram of growth of GaAs quantum well between AlGaAs cladding layers (top). Below that is a sketch of the band structure in the vicinity of the quantum well. The quantized energy levels are shown in conduction band and valence band.

The energy and wavelength of the electron are related through the momentum p, and the effective mass m^*:

$$E = \frac{1}{2}\frac{p^2}{m^*}, \qquad \text{where } p = \hbar k$$

$$E = \frac{h^2 n^2 \pi^2}{2m^* L^2}, \qquad n = 1, 2, 3 \ldots$$

In practice we cannot achieve infinitely deep wells but this equation for the energy is still a good approximation, particularly in revealing the dependence on the well width L and the effective mass m^*. The energy splitting varies inversely as L^2 and becomes significant enough to be observed experimentally for well widths less than about 15 nm. Note also that materials with lower m^* will exhibit larger splitting of the energy levels. At the top of the valence band of a three-dimensional semiconductor two bands with different effective masses merge. In two dimensions these two bands split and two sets of discrete levels are seen, known as the heavy and light hole levels, respectively.

Effective masses of the electrons in the conduction band of the materials of interest are always light. For example, in GaAs $m^* = 0.07 m_e$, where m_e is the rest mass of an electron in free space. To get a feeling for the size of the up-shift in the bottom of the conduction band in going to two dimensions, ΔE for a 10-nm well in GaAs is 45 meV. For a 5-nm well, $\Delta E = 190$ meV.

The two dimensionality of the quantum well also has a major effect on the energy dependence of the density of states in the conduction band and valence band as shown in Fig. 4. In three dimensions the density of states N_s increases from zero at the band edge with an $E^{1/2}$ dependence. In two dimensions the density of

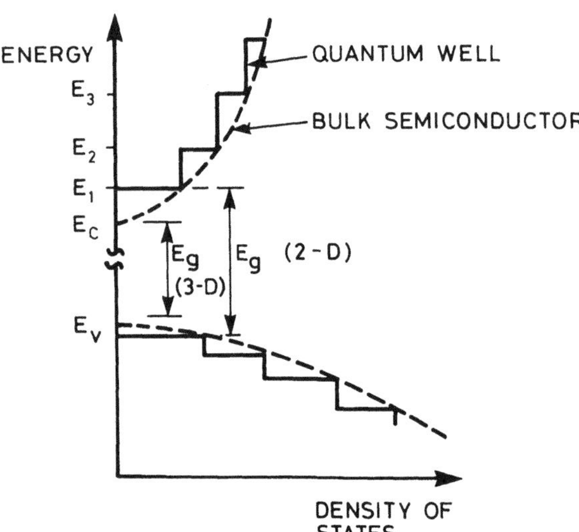

FIGURE 4. Density of states available to the electrons in a two-dimensional semiconductor (solid curve) compared with the smooth variation found in a three-dimensional material (broken curve). The diagram is intended to compare the functional dependence on energy only and is purely schematic.

states is independent of energy so it rises abruptly to some finite value at the band
edge and then remains constant until the next discrete energy level is reached, when
it rises abruptly again. Each additional energy level causes further steps leading to
a staircase form for the density of states.

Since virtually all devices involve either carrier transport within the states at
the band edge or optical transitions between E_v and E_c this change in the density
of states near the edges of the bands has important implications.

4. NEW EFFECTS IN LOW-DIMENSIONAL STRUCTURES

Much interest has been generated by low-dimensional structures. At a fundamental level quantum wells have produced a long list of new effects, many of which
are not yet fully understood. From a practical point of view the technology to grow
quantum wells is essentially incremental on existing semiconductor growth and the
materials in which quantum confinement effects can be observed are well established
in different semiconductor device fields. The following is a brief list of some of the
new effects capable of device exploitation in low-dimensional structures:

- Discrete quantized energy levels for motion normal to the plane of the layer.
- Two-dimensional motion in the plane of the layer.
- Density of states with a staircase energy dependence.
- Energy gap between conduction and valence bands which depends on the
  well width.
- Remote doping across the heterojunction to produce high mobilities free
  from ionized impurity scattering.
- Tunneling interaction between neighboring wells.
- Very large degeneracy of the electrons in the conduction band giving
  quasimetallic behavior.

Above all, the properties of the material are controllable by changing the geometry
of the well. This makes available to the device designer an unprecedented degree
of tailorability.

5. MATERIALS GROWTH

Figure 5 shows a transmission electron micrograph of a cross section through
a multilayer sample of alternating GaAs and AlGaAs layers. In the sequence of
layers of decreasing thickness the widest bright layer is 10 nm wide and each
successive layer is half the width of the one before. The thinnest layer is less than
a monolayer thick, but such is the precision with which the material can be grown
that it is clearly discernible as a uniform, continuous, and linear sheet. The technique
for producing the micrograph uses transmission through a thickness of some 100 nm
in the cross section so the line is the integrated intensity from a submonolayer
distribution of atoms in a single atomic plane. Any disorder would simply smear
it out.

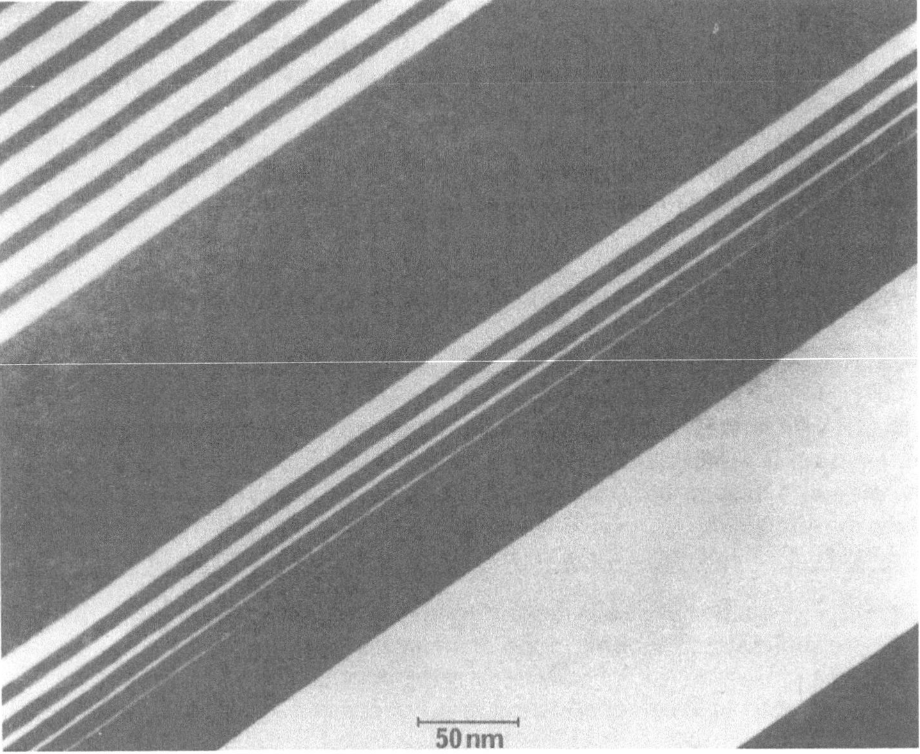

FIGURE 5. Transmission electron micrograph of a cross section through a multilayer sample of alternating GaAs and AlGaAs layers. The sample was grown by molecular beam epitaxy (MBE) and the thinnest layer is less than a monolayer thick.

The structure shown in Fig. 5 was grown using a technique known as molecular beam epitaxy. Sources of the elements Ga, Al, and As are heated in small furnaces bolted onto a large flange of an ultra-high-vacuum chamber. When hot, material from each furnace is evaporated into the chamber, but since the pressure in the chamber is kept low enough that the evaporated material suffers no collisions with gas atoms, this material travels in a straight line as an atomic or molecular beam. A heated substrate, consisting in this case of a wafer of single-crystal GaAs, is placed where the different molecular beams converge and ultra-high-purity layers of GaAs or GaAlAs are grown on its surface. Each furnace has a shutter to block the beam, so switching from GaAs to AlGaAs requires only the opening and closing of the Al shutter.

In this case the addition of Al to GaAs hardly changes the lattice constant at all (lattice parameters are GaAs: 5.653; AlAs: 5.661). The materials are said to be lattice matched. Extremely high-quality multilayers can be grown as almost perfect single crystals. For many years it was thought that this degree of high-quality crystal epitaxy could be achieved only with lattice matched materials, and this imposed quite a severe restriction on the combination of materials that might be suitable. Combinations of lattice matched materials that have been studied include

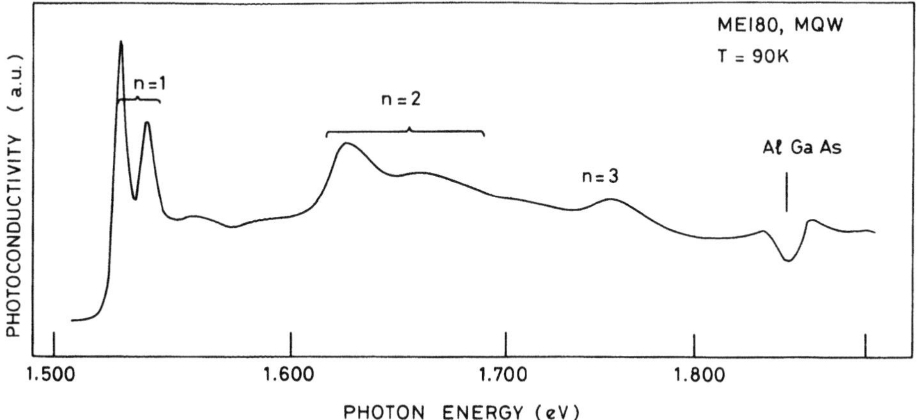

FIGURE 6.   Photoconductivity spectrum from a multi-quantum-well sample of GaAs/AlGaAs showing clearly optical transitions between the various quantized energy levels in the wells. This is just one of the many optical and electrical techniques that have been used to characterize these very thin layers.

InP/InGaAs(P)/InAlAs and CdTe/HgTe/InSb. Recently, however, it has been shown that some mismatch in lattice constants ccan be tolerated provided the layers are thin enough. The lattice of the whole structure strains elastically. The greater the lattice mismatch between the constituents the thinner the layers must be to be able to accommodate the elastic strain. This finding has opened up the range of materials that might be considered, to include GaAs/InAs, GaAs/GaP, Si/SiGe, and even GaAs/Si. Strain further perturbs the energy levels and so can be used as an additional free parameter in tailoring the material properties for a particular task.[1]

Returning to the GaAs/AlGaAs quantum wells grown by MBE, a great many techniques have been applied in the study and characterization of their properties. There is no room in this brief account to discuss this huge body of data but details can be found in a number of excellent reviews.[2-4] By way of confirmation of the existence of the discrete energy levels in quantum wells, Fig. 6 shows a photoconductivity spectrum obtained from a GaAs/GaAlAs quantum well sample. At the band edge the doublet labeled $n = 1$ corresponds to optical absorption between the first electron level and the light and heavy hole levels, respectively. A second doublet is seen corresponding to the $n = 2$ transitions and an unresolved feature is observed in the region of the $n = 3$ transition. The AlGaAs band gap, and hence the Al composition, can also be determined.

6. OTHER LOW-DIMENSIONAL STRUCTURES

The technology to grow quantum wells is also capable of growing a wide range of other complex multilayer structures. The inverse of a quantum well is a very thin rectangular barrier, so thin that electrons can be made to tunnel right through it. Both quantum wells and barriers can be made in any combination to enhance effects that might be weak in the single structure alone. Multi-quantum-well samples have been grown, for example, with many hundreds of wells in them to create what is essentially a bulk solid with two-dimensional properties.

If the separation of the wells in a multi-quantum-well (MQW) structure is reduced sufficiently, the energy levels in one well start to interact with the identical levels in neighboring wells in the same way as the discrete energy levels of an atom interact in the solid to produce energy bands. When this happens the MQW is called a superlattice. It is an artificial man-made crystal with properties that can be designed in during growth.

Figure 7 shows a schematic energy band diagram of a superlattice. The discrete levels of the individual wells broaden out into narrow energy bands. Also shown is the effect of the new periodicity on the $E-k$ diagram. New gaps are opened up at multiples of π/d.

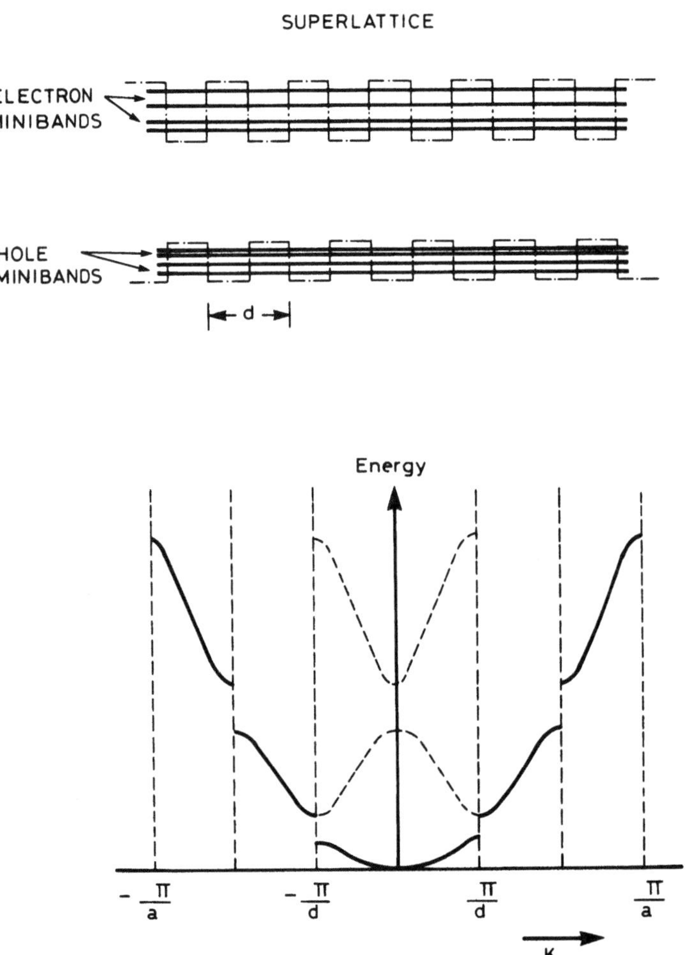

FIGURE 7. The upper diagram shows the formation of superlattice minibands for electrons and holes through the interaction of the discrete levels of the quantum wells when the barrier thickness is reduced. The lower diagram shows the corresponding electron dispersion curves of E vs. k. a is the original lattice period, d is the imposed superperiod of the superlattice.

7. APPLICATIONS

Figure 8 shows another schematic picture of a single quantum well identifying some important transitions that can be exploited in devices. The most studied materials listed at the top of the figure are those used most widely in virtually all microwave and optoelectronic devices at the moment, and therefore this technology is compatible with present day device fabrication technology. The figure shows three key transitions or excitations that can be exploited. The fundamental gap transition occurs in the visible and near infrared. In a quantum well this gap is tunable and has found application in shifting the operating wavelength of a quantum well laser

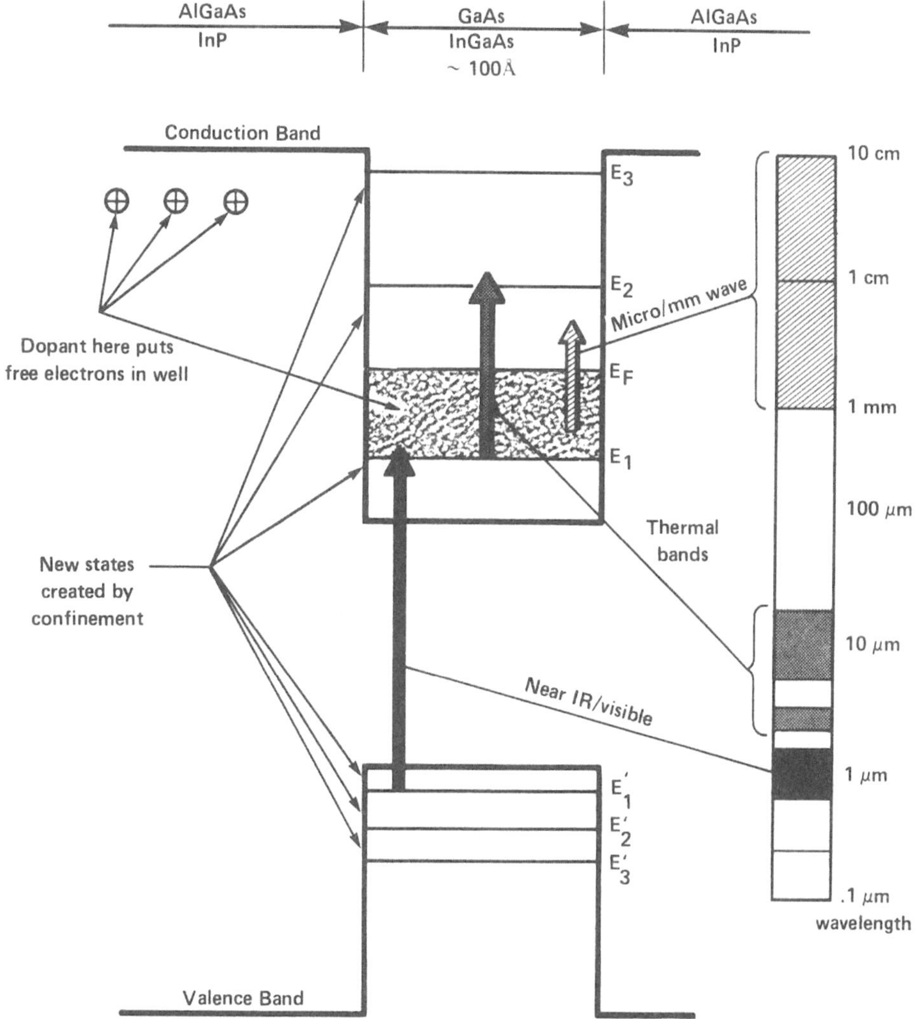

FIGURE 8. Device exploitation of quantum wells. The figure shows how various spectral regions can be accessed by using different excitations. All transitions are determined by grown structure and tunable by electric fields.

from the gap of GaAs at approximately 0.9 μm in the infrared toward 0.7 μm or below in the red part of the visible spectrum. Optical storage systems based on compact disk technology require solid state lasers with as short a wavelength as possible to maximize the information density achievable.

The second transition is unique to two-dimensional systems. Electrons in the lower level in the well can be excited to an upper level—a so-called intersubband transition. The energy difference between the two levels can be tuned to correspond to wavelengths in the mid-infrared from 5 to 20 μm. The transition can be exploited in modulators and detectors for this part of the spectrum.

Electrons present in the well through remote doping of the barrier material behave in many ways as a two-dimensional metal. In particular they exhibit collective oscillations known as plasmons which have characteristic energies in the microwave and millimeter wave regions of the spectrum.

Thus quantum wells can span the whole usable range of the electromagnetic spectrum in just one single materials system. In the past it has been necessary to choose a different material for each region. Figure 9 shows another structure that has been exciting a lot of attention. It consists of two tunneling barriers separated by a quantum well. At zero bias no current flows because there is no available state in the quantum well at the same energy as the electrons in the GaAs contact layers. When a positive bias is applied the lowest level of the well is brought into coincidence with the GaAs conduction band and a current can flow. At higher bias the resonance is removed and the current decreases, giving a negative differential resistance region in the characteristic. Several groups are investigating the potential of this device for very high-frequency oscillation.

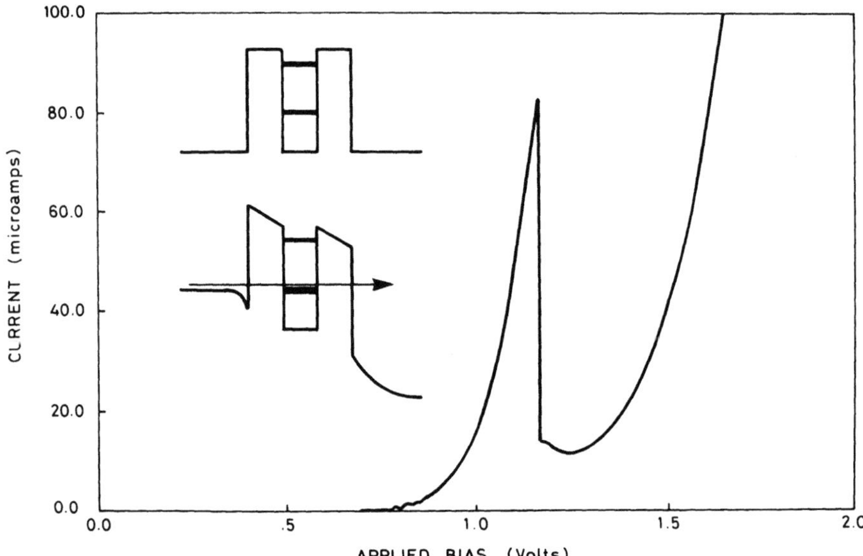

FIGURE 9. Current–voltage characteristic (at 77 K) of a GaAs/AlAs resonant tunneling device. The band diagram at zero bias and at resonance is shown in the inset. The strong negative differential resistance and the intrinsically high speed of the tunneling process are exciting considerable device interest in this effect.

8. CONCLUSIONS

Because of its basic compatibility with existing devices, LDS has found its way already into a number of device areas where performance enhancement results. Examples include the HEMT,[5] the quantum well laser,[6] low noise optical detectors, and Gunn diodes for microwave oscillation. There is also, however, an increasing number of completely new ideas to exploit the new properties of these systems. It is probably no exaggeration to say that LDS will bring about a major revolution in semiconductor devices in the years to come.

REFERENCES

1. E. P. O'Reilly, Valance band engineering in strained layer structures, *Semicond. Sci. Technol.* **4**, 121 (1987).
2. R. J. Nicholas and M. J. Kelly. The physics of quantum well structures, *Rep. Prog. Phys.* **48**, 1699 (1985).
3. *Physics and Applications of Quantum Wells and Superlattices* (E. E. Mendez and K. von Klitzing, eds.), Plenum Press, New York (1987).
4. K. von Klitzing, The quantized Hall effect, *Rev. Mod. Phys.* **58**, 519 (1986).
5. H. Sakaki, Physical limits of heterostructure field effect transistors and possibilities of novel quantum field effect devices, *IEEE J. Quantum Electron.* **QE-22**, 1845 (1986).
6. Y. Arakawa and A. Yariv, Quantum well lasers, gain, spectra, dynamics, *IEEE J. Quantum Electron.* **QE-22**, 1887 (1986).

6

Key Electrical Devices

R. H. Wallis

1. INTRODUCTION

This chapter describes the basic operating principles of the key semiconductor devices that lie at the heart of modern electronics.

Historically, the subject of semiconductor devices can be traced back to the work of Braun in 1874, long before the concept of a semiconductor was identified. While studying the current flow at a metallic point contact on semiconducting sulfide crystals, he observed that the current passed more easily in one direction than the other. This first observation of rectification eventually led to the "cat's whisker" of early radios. Far more significant, however, was the demonstration of the first transistor by Bardeen, Brattain, and Shockley in 1947, showing that amplification of an electrical signal, previously requiring cumbersome vacuum-based devices, could be achieved in a semiconducting solid. Subsequently many different semiconductor structures capable of transistor action have been proposed and demonstrated, the most significant being the bipolar junction transistor (BJT), which was first developed in the 1950s, and the metal-oxide-silicon field-effect transistor (MOSFET), developed in the 1960s. In parallel with the development of the basic devices have come sustained advances in the growth of semiconducting materials, in process technology, and in integration techniques, making possible the complex devices and integrated circuits of today.

This chapter will be concerned only with the basic principles of the most important devices: materials growth and processing technologies are covered elsewhere in this book. The most important devices are unquestionably transistors. Accordingly, the emphasis will be on the fundamental operation of the different types of transistor, with some consideration of basic types of diodes since, as well as being important devices in their own right, their operation is crucial to an understanding of transistors. More specialized devices such as thyristors or various forms of microwave diodes will not be discussed, while optoelectronic devices are discussed in the next chapter. Nevertheless, even with these restrictions, only an outline description of the operation of diodes and transistors can be given, and for a more detailed description the reader is referred to one of the many excellent

R. H. Wallis • Plessey Research Caswell Limited, Caswell, Towcester, Northamptonshire, NN12 8EQ, U.K.

standard texts on the subject, some of which are listed at the end of the chapter.[1-4] In keeping with the overall theme of the book, the emphasis will be on the demands that these devices make on materials. In particular, this chapter will try to explain what materials properties are required for device operation, why silicon enjoys such a dominant position in the field of semiconductor devices, and what role other materials play.

The organization of this chapter is as follows. In the next three sections the principles of the p-n junction diode and metal–semiconductor diode (Section 2), the junction transistor (Section 3), the MOSFET, and the junction field-effect transistor (Section 4) are described, with little reference to the material from which they are fabricated. The next section (Section 5) discusses their material requirements and shows that silicon has properties that make it nearly ideal for these devices. The next section (Section 6) discusses the role of compound semiconductors such as gallium arsenide for specific applications at very high frequencies. Finally, Section 7 describes some special cases, and some conclusions are drawn in Section 8.

2. BASIC SEMICONDUCTOR DIODES

2.1. The p–n Junction Diode

The p-n junction constitutes perhaps the simplest semiconductor device. Its most basic property is that of rectification, passing current easily in one direction only. Detailed descriptions of the operation of a p-n junction are given in standard texts[1-4]; here only an outline is given.

A p-n junction is created by introducing p-type and n-type dopants into adjacent regions of a single semiconducting crystal. An idealized representation of such a structure is shown in Fig. 1a. Such a junction may be formed in a number of ways: for example, by changing the dopant type during growth, by implanting donor and acceptor ions to different depths, or by diffusion of one type of dopant into a crystal already doped with the other. The different methods give rise to different profiles

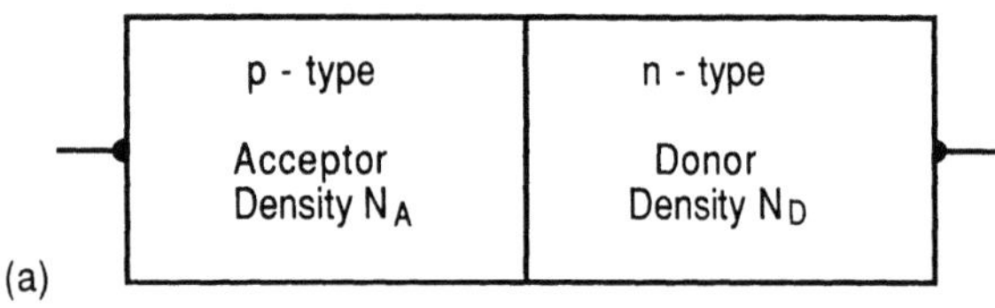

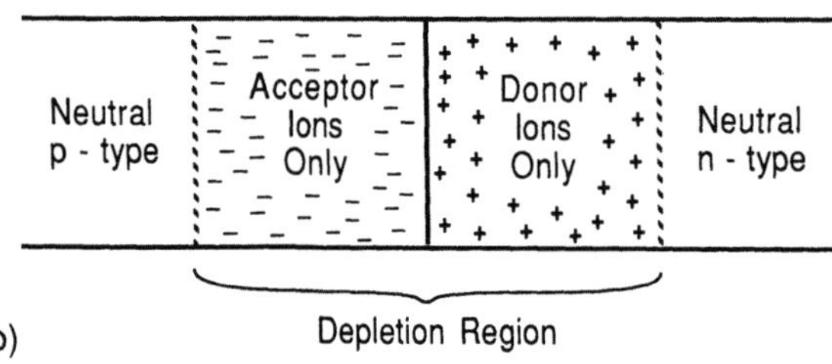

FIGURE 1. (a) Schematic representation of an abrupt p–n diode. (b) Representation of the space charge or depletion region near the metallurgical junction. + and − represent the ionized donor and acceptor ions, respectively.

for the donor and acceptor atoms around the p-n junction. For brevity we consider here only the idealized "abrupt structure," as illustrated in Fig. 1a, with a uniform donor density N_D on one side of the junction and a uniform acceptor density N_A on the other side.

The concentration gradient across the junction causes electrons from the n-type side to diffuse across to the p-type side, leaving behind a positive space charge region close to the junction due to the fixed donor ions. Similarly, holes diffuse across from the p-type side to the n-type side, leaving behind a negative space charge due to the fixed acceptor ions. The two equal and opposite space charge regions constitute a dipole layer, which sets up a built-in potential difference V_{bi} across the junction. In equilibrium (that is, with no external bias applied to the junction) V_{bi} produces drift currents of electrons and holes that are exactly equal and opposite to the diffusion currents, so that no net current flows across the junction. (This is equivalent to saying that the value of V_{bi} is that required to bring the Fermi levels on opposite sides of the junction into coincidence. A similar situation for heterojunctions has already been discussed in the previous chapter.) The value of V_{bi} may be shown to be given by

$$V_{bi} = (kT/q) \ln(N_A N_D / n_i^2)$$

where n_i is the intrinsic carrier density, k Boltzmann's constant, T the absolute temperature, and q the magnitude of electronic change.

The space charge region near the metallurgical junction is devoid of free carriers, due to the strong electric field, and is known as the depletion region. The situation is illustrated schematically in Fig. 1b, with the band diagram for zero applied bias shown in Fig. 2a.

When an external bias voltage is applied across the junction, the balance between the drift and diffusion currents for each type of carrier is destroyed, and a net current flows. The drift currents are, to a good approximation, unaffected by the applied bias, but the diffusion currents are strongly modified. For a forward bias (that is, with the p-type side of the junction positive with respect to the n-type side), the applied potential tends to oppose the built-in potential, reducing the total barriers "seen" by the diffusing carriers. The band diagram for this situation is shown in Fig. 2b. Electrons from the n-type side will diffuse more easily into the p-type side, where they are then minority carriers, and will recombine with the majority carriers (holes) as they diffuse away from edge of the depletion region. Similarly, holes from the p-type side will diffuse more easily into the n-type side, where they will recombine with electrons. These diffusion currents together give rise to a large net current flow: note that although the electrons and holes diffuse in opposite directions, since they carry opposite electrical charge then their contributions to the total electric current are additive.

Conversely, a reverse applied bias adds to the built-in potential, as illustrated in the band diagram of Fig. 2c, suppressing the diffusion of electrons and holes across the depletion region. For reverse biases greater than about $3kT/q$, the diffusion currents are negligible and only a leakage current due to the drift currents flow.

The form of the current-density-voltage (J-V) characteristic for an ideal p-n junction is shown schematically in Fig. 3. It may be shown that this characteristic

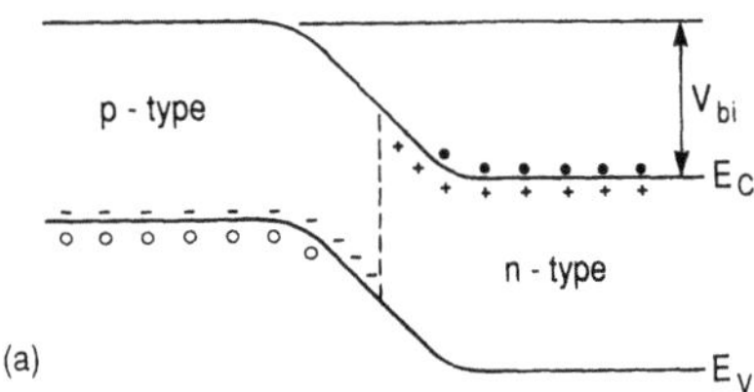

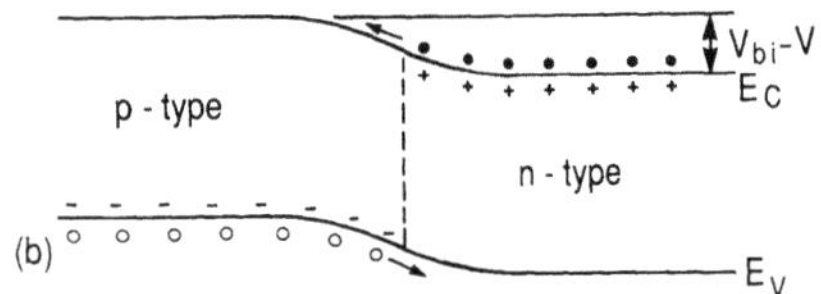

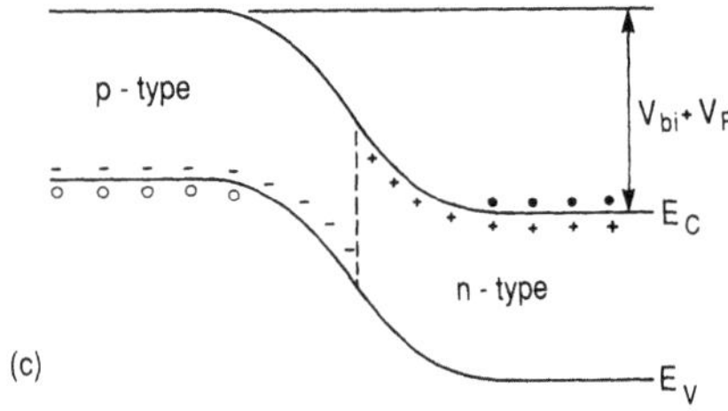

FIGURE 2. Band diagrams for a p–n junction under (a) zero bias, (b) forward bias, and (c) reverse bias. E_C and E_V denote the conduction band and valence band edges, respectively, V_{bi} denotes the built-in voltage, $+$ and $-$ denote donor and acceptor ions, and ● and ○ denote free electrons and holes.

is described by the diode equation:

$$J = J_0[\exp(qV/kT) - 1] \tag{1}$$

This expression shows that the current rises exponentially by large forward biases and saturates to a constant value of $-J_0$ at large reverse biases. J_0 is given (approximately) by

$$J_0 = qn_i^2[(D_n/L_nN_A) + (D_p/L_pN_D)] \tag{2}$$

where D_n, D_p, L_n, and L_p are the diffusion coefficients and diffusion lengths for electrons and holes, respectively.

The derivation of Eq. (1) assumes that the recombination and thermal generation of carriers takes place only in the neutral regions outside the depletion region. This

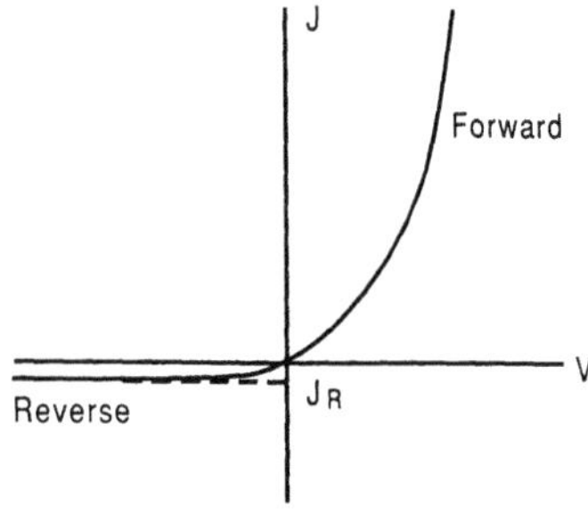

FIGURE 3. Form of the current density (J)–bias (V) characteristic for an ideal p–n diode. J_R is the saturation current density in reverse bias.

assumption is generally only valid for semiconductors with relatively small energy gaps (e.g., germanium) or at temperatures above room temperature. At lower tempeatures, or for wider gap semiconductors such as the technologically more important materials silicon and gallium arsenide, J_0 is very small, because n_i is small. Under these circumstances, recombination and generation processes within the depletion region cannot be ignored. These processes generally involve the participation of energy levels close to midgap, which may be due to impurities or crystalline defects, and can alternatively emit or capture electrons and holes. In forward bias, recombination within the depletion region gives a contribution to the diode current that varies approximately as $\exp(qV/2kT)$, rather than as $\exp(qV/kT)$ as for Eq. (1). In reverse bias, the generation of electron–hole pairs within the depletion region gives rise to an additional component of leakage current. This component does not saturate but increases with reverse bias, since the depletion region width increases with reverse bias.

The properties of a p–n junction diode are exploited in a wide range of applications. The most important is as a device capable of rectification, since the I–V characteristics expressed by Eq. (1) correspond to a very low resistance in forward bias and a very high resistance in reverse bias. But in addition, p–n junctions may be used in a number of other applications. The recombination of injected carriers can involve the emission of a photon, especially in direct-gap materials, a property that is exploited in the light-emitting diode and the diode laser. Conversely, the absorption of light in or near the depletion region creates electron–hole pairs which give rise to a current in a reverse-biased diode: this is the basis of the photodiode. Light-emitting diodes, diode lasers, and photodiodes are discussed fully in the following chapter. Under high reverse bias, a p–n junction may break down, either due to avalanching or to band-to-band tunneling. This breakdown is exploited in a number of device concepts (Zener diodes, avalanche photodiodes, microwave devices such as IMPATTs), which will not be discussed in this brief survey: details may be found in Sze.[2] The depletion region under reverse bias acts as an insulator of bias-dependent width, thus enabling a reverse-biased p–n junction to be used as a voltage-variable capacitor (varactor diode). But perhaps the most important aspect of the p–n junction diode is its operation as a fundamental part of the bipolar transistor, described in Section 3 below.

2.2. The Metal–Semiconductor or Schottky Diode

While the rectifying properties of a p–n junction arise from a change of doping within the bulk of the semiconductor, rectification can also be obtained at the contact between a uniformly doped semiconductor and a metal at its surface. Braun's earliest observation of rectification referred to earlier was on a structure of this type, made by contacting the semiconductor with a metallic point. Nowadays such diodes would be formed by deposition of the metal, for example by evaporation, onto the cleaned semiconductor surface. The structure of such a diode is depicted schematically in Fig. 4a.

As for a p–n junction, the rectifying properties of a metal–semiconductor diode arise from the presence of an internal potential barrier, and again this barrier is due to a fixed space charge created by the movement of free carriers to achieve

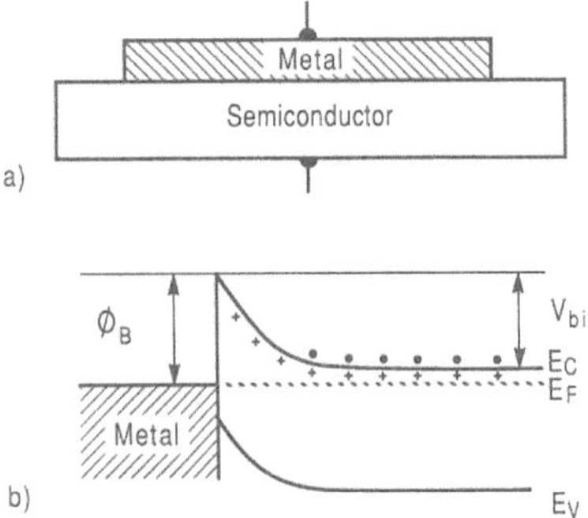

FIGURE 4. (a) Schematic representation of a metal–semiconductor diode (Schottky diode). (b) Band diagram for a Schottky diode at zero bias. ϕ_B is the Schottky barrier height and E_F represents the Fermi level. The other symbols have the same meaning as in Fig. 2.

equilibrium throughout the system. In general the Fermi levels of the metal and the semiconductor in isolation will not be equal, and so charge must flow to bring them into coincidence when the contact is established. To be specific we consider an n-type semiconductor, and further assume that its Fermi level is above that of the metal. When contact is established, electrons will flow from the semiconductor into the metal, accumulating at the metal surface and leaving an equal and opposite charge due to the positive donor ions near the semiconductor surface. This dipole layer therefore creates an energy barrier in the semiconductor so that the Fermi level is constant throughout. The band diagram for this situation is depicted in Fig. 4b. This energy barrier is frequently referred to as a Schottky barrier and the metal–semiconductor diode as a Schottky diode.

From this simple description it might be supposed that the height of the Schottky barrier ϕ_b (see Fig. 4b) is, for a given semiconductor, directly dependent on the work function of the metal (that is, the energy required to lift an electron from the metal Fermi level to the vacuum level). In fact for most semiconductors (including the technologically important materials silicon and gallium arsenide) this dependence is weak. The reason for the discrepancy is the role played by the energy states associated with the semiconductor surface, which were ignored in the above description but which tend to "pin" the Fermi level and give a barrier height almost independent of the metal used. A lucid explanation of surface states in Schottky barriers may be found in Rhoderick and Williams.[5]

As for the p–n junction, the space charge region in the semiconductor is referred to as the depletion region. When the diode is forward biased (metal positive with respect to the semiconductor, for a n-type semiconductor), the depletion region is narrowed, and conversely under reverse bias it is widened. For a uniformly doped n-type semiconductor the width W of the depletion region as a function of the applied bias V is given by

$$W = [(2\varepsilon_s/qN_D)(V_{bi} - V)]^{1/2} \tag{3}$$

where ε_s is the permittivity of the semiconductor and V_{bi}, the built-in potential, is equal to the barrier height ϕ_b less the difference between the semiconductor conduction band and Fermi level (see Fig. 4b). It may be shown that a reverse-biased Schottky barrier behaves like a parallel-plate capacitance with a dielectric of width W. Since W is bias-dependent, the device can act as a voltage-variable capacitance.

In an ideal Schottky diode, the dominant current flow mechanism is by thermal excitation of majority carriers (electrons for an n-type semiconductor) over the barrier. (In this respect the Schottky barrier differs from the $p-n$ junction, for which the dominant current transport is due to minority carriers.) Under zero bias, when no net current flows, there are equal and opposite currents I_{ms} from the metal to the semiconductor and I_{sm} from the semiconductor to the metal. Under forward bias the potential difference that electrons in the semiconductor must surmount is reduced, so the current I_{sm} from the semiconductor to the metal increases, while the barrier "seen" by electrons in the metal remains ϕ_b, so I_{ms} is unchanged. A net current therefore flows. Conversely in reverse bias the potential barrier to electrons leaving the semiconductor is increased, so I_{sm} is reduced, while again I_{ms} is unchanged. The current density–voltage characteristic has the same form as for a $p-n$ diode, Eq. (1):

$$J = J_0[\exp(qV/kT) - 1] \tag{4}$$

but the expression for J_0 is different

$$J_0 = A^* T^2 \exp(-q\phi_b/kT) \tag{5}$$

where A^* is a constant known as the effective Richardson constant.

Deviations from Eq. (4) may arise from a number of additional current transport mechanisms. In particular, for very thin barriers [corresponding, through Eq. (3), to very heavy doping], or for very low temperatures, the dominant current path may be tunneling through the barrier rather than thermal excitation over it.

The fact that the Schottky barrier is a majority carrier device makes it useful for applications requiring a very fast response. If a $p-n$ junction is switched from forward to reverse bias, current continues to flow until the injected minority carriers have recombined. Since this limitation does not apply to the Schottky barrier, it is used for very fast switching and microwave applications. It is also an important element of the MESFET, described in Sections 4 and 6 below.

More detailed descriptions of Schottky barriers are given in the books by Rhoderick and Williams[5] and Sze.[2]

2.3. Ohmic Contacts

Although an ohmic contact cannot be regarded as a device, it is obvious that all devices require contacts in order to be connected to metallic wires to the outside world or interconnects to other devices. But, as described above, a simple metal-semiconductor contact is generally rectifying! The technology of ohmic contacts therefore seeks to destroy these rectifying properties and to produce a metal-semiconductor junction that has a linear current–voltage relationship with as low a resistance as possible. In practice this is generally achieved by using a metallization scheme that gives a low barrier height on a highly doped surface layer of the semiconductor. The low barrier height allows thermionic currents to flow easily and the high doping makes the barrier thin and so the tunneling component of current is significant. Frequently such contacts are formed by metal deposition followed by alloying to diffuse in dopants from the deposited metal and to form intermetallic phases. For the technologically important semiconductors such as silicon and gallium arsenide, recipes for the fabrication of low-resistance ohmic contacts are well established. In contrast, it may be difficult or impossible to fabricate good ohmic contacts on certain wide gap semiconductors which are difficult to dope and for which low barrier heights cannot be achieved.

3. BIPOLAR JUNCTION TRANSISTORS

It was the development of the bipolar junction transistor in the 1950s that began the era of modern solid state electronics. Continuous development of the technology since that time has resulted in bipolar transistors based on silicon being the key devices in the highest-speed digital integrated circuits, as well as being widely used in analogue integrated circuits and as discrete devices, especially for power and high-frequency applications. Here we shall be concerned only with the basic operating principles of the device itself, which were first established in the 1950s. More detailed descriptions of the operation of bipolar transistors may be found in the books by Grove,[1] Sze,[2,3] Streetman,[4] and Ashburn.[6]

The structure of a bipolar junction transistor comprises a thin layer with one layer of doping sandwiched between two layers of the opposite type, forming two p–n junctions back to back. Either n–p–n or p–n–p structures may be used: here for brevity we consider only the n–p–n structure. (The operation of the p–n–p structure is identical, if holes are substituted for electrons and vice versa.)

The basic device structure is illustrated very schematically in Fig. 5. The central p-type region is known as the base and the n-type regions on either side of it as the emitter and collector. The structure is not, however, symmetrical, since the emitter is much more heavily doped than the collector. The base is generally only lightly doped. (Note that, in a real device, there will also be heavily doped regions adjacent to the base and collector, to minimize the series resistance to their ohmic contacts. However, these do not affect the fundamentals of device operation and need not concern us here.)

When the bipolar transistor is operated in its "active" mode, external voltages are applied so that the emitter–base junction is forward biased while the base–collector junction is reverse biased. The forward bias across the emitter–base junction will cause a current I_E to flow into the emitter contact, composed of a current I_{En} of electrons injected from the emitter into the base and a current I_{Ep} of holes injected from the base into the emitter. (Note that these add because the direction of electrical current flow is the same as the flow of holes, but opposite to the flow of electrons.) In a well-designed transistor, the light doping of the base and heavy doping of the

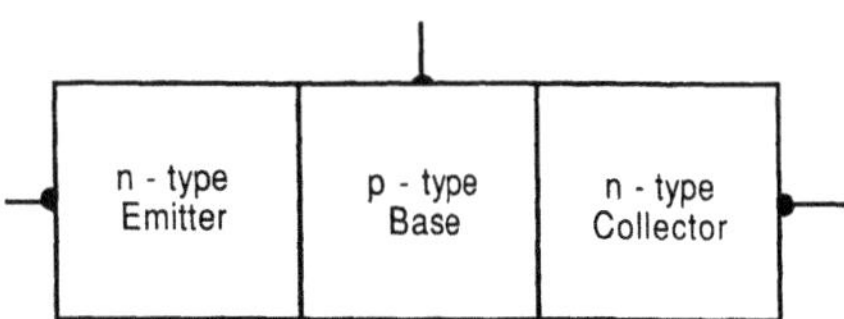

FIGURE 5. Schematic representation of an n–p–n bipolar junction transistor structure.

emitter will ensure that I_{En} is much greater than I_{Ep}. Since the base is narrow, most of the injected electrons will diffuse across it without recombining and reach the base–collector junction, where they will be swept into the collector by the strong reverse bias. The collector current I_C will therefore be almost equal to the emitter current I_E, the ratio I_C/I_E generally being denoted by α, known as the common base current gain. α can be regarded as a product of two factors, the emitter efficiency γ, which denotes the ratio of the emitter current due to injected electrons to the total emitter current, and the base transport factor α_T, the fraction of injected electrons that cross the base without recombining and so reach the collector. (In this simplified description it has been assumed that I_C is only due to injected electrons. The contribution to I_C of the reverse leakage current from collector to base has been ignored.) In a well-designed transistor both γ and α_T approach unity and so their product α is also close to unity. The small difference between I_C and I_E is made up by the base current I_B. A number of physical processes contribute to I_B, including the injection of holes from base to emitter (I_{Ep}), the hole current due to recombination with those injected electrons that do not diffuse right across the base, and the hole current for recombination in the emitter–base depletion region. Collectively these processes may give a base current that is less than 1% of the emitter current in a good transistor.

Expressing the above description mathematically, we have

$$I_C = \gamma\alpha_T I_E = \alpha I_E = I_E - I_B$$

from which we have $I_C = [\alpha/(1 - \alpha)]I_B$.

The bipolar transistor can therefore exhibit large current gains, because a small change in base current ΔI_B gives a large change in collector current ΔI_C, since

$$\Delta I_C/\Delta I_B = \alpha/(1 - \alpha) = \beta$$

where β is called the common-emitter current gain. Since α may be 0.99 or greater, B may exceed 100. "Common-emitter" refers to the configuration used most often in transistor circuits, where the input is applied to the base and the output taken at the collector, with the emitter terminal common to input and output.

The collector current I_C as a function of emitter–collector voltage V_{CE} is shown schematically for several values of base current I_B in Fig. 6. From the description

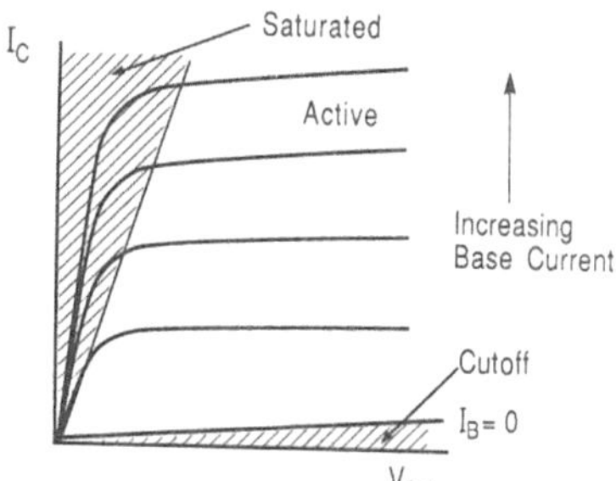

FIGURE 6.   Form of the collector current (I_C) vs. collector–emitter voltage (V_{CE}) characteristics for a bipolar transistor.

above it might be expected that the collector current in the "active" region depends only on I_B and is independent of V_{CE}. Figure 6 shows that this is not exactly the case: I_C rises with increasing V_{CE}. This is due to the depletion region of the base–collector junction widening with increasing applied voltage, so reducing the width of the neutral region of the base. The probability of an injected electron crossing this region without recombining therefore increases, so α_T becomes even closer to unity and, from the above expressions, β rises, giving the observed increase in I_C.

In addition to operating in the active mode, two other modes of operation are important. These are also shown in Fig. 6. When both the emitter–base and base–collector junctions are forward biased, the transistor is said to be in saturation mode: collector current flows with only a small collector–emitter voltage, and so the transistor is "on." Conversely in the cutoff mode, when both junctions are reverse biased, there is almost no collector current and so the transistor is "off." The bipolar transistor can thus function as a switch. Digital circuits involve switching the transistor between these two states.

4. FIELD EFFECT TRANSISTORS

4.1. MOSFETs

Field effect transistors (FETs) constitute the second of the two major classes of transistor. FETs may be further subdivided into three principal types: junction FETs (JFETs), metal–semiconductor FETs (MESFETs), and metal–insulator–semiconductor FETs (MISFETs). The MOSFET is a specific type of MISFET in which the insulator is an oxide layer, in the case of the silicon MOSFET being thermally grown SiO_2. The silicon MOSFET is the most important of all types of transistor, being the active device in most large-scale digital integrated circuits such as microprocessors and memories. Such circuits may contain upwards of a million MOSFETs.

All the different types of FET operate according to broadly the same physical principles. A field effect transistor may be considered as a parallel plate capacitor, in which one "plate," known as the channel, is a lightly conducting semiconducting layer, and the other "plate," known as the gate, is a metal or a very heavily doped semiconductor. In any capacitor, when a voltage is applied to the plates, one plate becomes positively charged and the other negatively charged. In the FET, the "plate" represented by the semiconducting channel has only a low density of mobile charge, and so the change in this charge density induced by applying a bias between the channel and the gate can be relatively large. The conductivity of the material making up the channel layer is therefore changed. If a current is made to flow along the channel, this current will be controlled by the voltage applied between the channel and the gate.

The different types of FET differ principally in the nature of the "dielectric" used to form this capacitor. In the MOSFET it is a layer of insulating oxide and in the MISFET some other insulating layer, while in the JFET and MESFET it is the depletion layer of a p–n junction and a metal–semiconductor Schottky barrier, respectively. JFETs and MESFETs have to be operated with the gate junction reverse-biased or only slightly into forward bias, to prevent the flow of current between the gate and the channel: this restriction does not apply to MOSFETs and MISFETs.

For all types of FET, if the current in the channel is transported by electrons, then the device is said to be n-channel, and conversely in a p-channel device the current is carried by holes. If the channel is conducting with no bias voltage applied to the gate, and gate bias must be applied to close off the channel, then the device is described as being "normally on" or a depletion mode device. If the channel is not conducting at zero gate bias and requires an applied gate voltage to create a conducting channel, then the device is described as "normally off" or an enhancement-mode device.

The simplified structure of an n-channel enhancement-mode silicon MOSFET is shown schematically in cross section in Fig. 7. The device consists of a p-type substrate into which two n^+ ohmic contacts, called the source and drain, have been alloyed. The source and drain ohmic contacts represent the ends of the channel. A layer of silicon dioxide covers the region between these contacts and slightly overlaps them. Above this oxide is the metallic gate electrode. [For technological reasons, in modern Si MOSFETs the gate is generally no longer made of metal but instead of heavily doped polycrystalline silicon (polysilicon). This does not affect the basic operation of the device, and the acronym MOSFET is still used.] The length of the gate in the direction from source to drain is generally small (of the order of 1 μm in modern devices), but to enable sufficient current to pass its width (i.e., perpendicular to the plane of the figure) is several times greater than its length.

In operation, both the drain and the gate electrodes will be biased positively with respect to the source. The voltages on the drain and gate, with the source as reference, are denoted by V_{DS} and V_{GS}, respectively. To describe the operation, we first consider the case when $V_{GS} = 0$ and V_{DS} is very small. In this situation no current can flow from the source to the drain, because the n^+ ohmic contacts and the p-type substrate form two back-to-back p–n junctions. If V_{GS} is now raised to make the gate metal positive, holes at the semiconductor–oxide interface will be repelled, depleting the semiconductor surface region and bending the energy bands downwards. Further increasing V_{GS} will eventually lead to the bands being bent so much that a layer of electrons is induced at the semiconductor surface. The energy band diagram for this situation has already been given in the previous chapter (Fig. 2a). This layer of electrons at the surface of the p-type substrate is known as an inversion layer, and the value of V_{GS} required to form it is called the threshold voltage V_T. The n-type inversion layer now joins the n^+ source and drain ohmic contacts allowing a current to flow along the channel. Further increase in V_{GS} will increaase the density of electrons in the inversion layer. So long as V_{DS} is kept

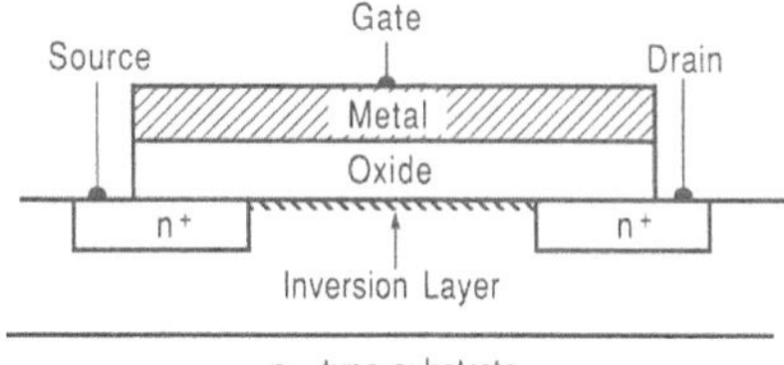

FIGURE 7. Schematic diagram of the structure of a MOSFET.

small, the current I_D flowing out of the drain will be proportional to V_{DS}, so the channel behaves like a resistance whose value is controlled by V_{GS}. This region of operation is called the linear region.

The linear region corresponds to the difference in potential between the gate and the channel being almost constant along the channel, so that the electron density in the channel is constant. If V_{DS} is now raised, this condition will no longer be true: there will then be a smaller potential difference between the channel and gate at the drain end, and so the density of electrons will decrease along the length of the channel. The drain current I_D will therefore rise sublinearly as V_{DS} is increased, as shown in Fig. 8. As V_{DS} is increased further, eventually the electron density at the drain end of the channel will fall to zero. At this point the channel is said to be pinched-off. The value of V_{DS} required to reach pinch-off is called the saturation voltage. Any further increase in V_{DS} causes the point at which pinch-off occurs to move back along the channel towards the source, but the drain current remains constant because it is limited by the rate of arrival of electrons at the point of pinch-off. This region of the device characteristics is referred to as the saturation region and the corresponding value of I_D as the saturation current. Clearly the saturation current will be greater the larger the value of V_{GS}, giving a set of characteristic I_D-V_{DS} curves of the form illustrated in Fig. 8.

The operation of a p-channel enhancement-mode device is exactly the same as that of the n-channel device described qualitatively above, with holes replacing electrons and the signs of all voltages changed throughout. In the corresponding depletion mode devices, the conducting channel between source and drain is present even for $V_{GS} = 0$. This may be achieved by local doping of the semiconductor beneath the oxide layer, generally by ion implantation.

The importance of MOSFETs for silicon integrated circuits stems from several factors. Foremost among these is the basic simplicity and planar nature of the device. This has allowed device dimensions to be shrunk (by over two orders of magnitude since the earliest devices) and high densities to be achieved, permitting very complex integrated circuits. Of particular importance is the use of both n-channel and p-channel devices in combination in so-called complementary MOS (CMOS) circuits. In CMOS circuits the channels of the two types of device are connected in series and the same logic signal applied to both gates, which are connected together. The threshold voltages of the two devices are fixed during fabrication so that one device is biased "off" when the other is "on," and vice versa, so that in either logic

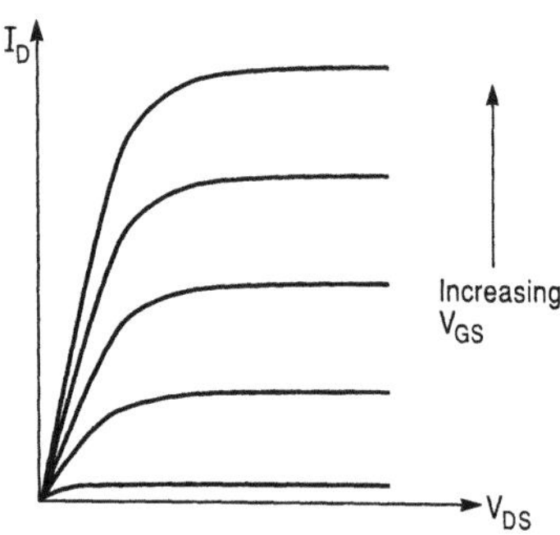

FIGURE 8. Form of the drain current (I_D) vs. drain–source voltage (V_{DS}) characteristics for a MOSFET.

state their series combination draws no current: current is only drawn during switching from one state to the other. CMOS circuits therefore have very low power dissipation. The technology of CMOS circuit fabrication is described in Chapter 13 by Foster.

4.2. JFETs and MESFETs

The n-channel JFET consists of a thin layer of n-type material forming the channel on a p-type or high-resistivity substrate, with two n^+ ohmic contact stripes to the n-type layer forming the source and drain and a narrow p^+ stripe between them forming the gate. Because the n-type channel already extends from the source to the drain, the gate does not need to overlap the source and drain contacts as it does in a MOSFET. The asymmetric doping of the gate–channel p^+-n junction means that the depletion region is much wider on the channel side than on the gate side of the junction. When this junction is reversed biased, by making the gate negative with respect to the source, the depletion region expands further into the channel and pinches it off. As for the MOSFET, pinch-off will occur first at the drain end of the channel when the drain is biased positively with respect to the source, leading to saturation of the drain current I_D. The I_D-V_{DS} characteristics for the JFET are therefore of the same form as those for the MOSFET represented in Fig. 8. The operation of the p-channel device is once again the same as that of the n-channel device, with p-type replacing n-type replacing n-type material and vice versa.

Despite their superficial similarity, the JFET is much less widely used than the MOSFET. This is because the latter possesses important advantages over the former in respect of both circuit design and processing technology.

The MESFET resembles the JFET except that instead of the current in the channel being controlled by a p-n junction, a metal–semiconductor Schottky barrier is used. This restricts the range of semiconductors that can be used for MESFETs to those to which useful Schottky barrier heights ($\geqslant 0.6$ eV) can be formed. MESFETs are most commonly implemented in gallium arsenide (GaAs), for reasons that will be explained in Section 6 below.

5. MATERIALS FOR ELECTRONIC DEVICES: THE SIGNIFICANCE OF SILICON

In describing the operating principles of the basic types of diode and transistor in the last three sections, little has been said about the materials from which they are fabricated. This section discusses these requirements, and in particular explains why, of the many hundreds of materials that may be classified as semiconductors, transistors are fabricated almost exclusively from only one material—silicon.

The dominant position of silicon does not arise from a single property, but rather from a combination of electrical, mechanical, structural, chemical, and thermal properties that make it the preferred choice for most transistors. In addition, silicon also possesses advantageous technological properties that have permitted the rapid progression from discrete devices to today's complex integrated circuits. Nevertheless, the advantages of silicon over other materials were identified before integration was considered. What, then, are these advantages?

The first requirement of a semiconductor is that it has an energy gap of an appropriate value. Too small a value leads to too high a density of minority carriers and so to large leakage currents in reverse-biased p–n junctions. This is particularly critical in the reverse-biased base–collector junction of bipolar transistors. One of the reasons for the early demise of germanium (the semiconductor used for the first bipolar transistors) in favor of silicon was the smaller energy gap (0.66 eV at room temperature, compared with 1.12 eV for silicon). On the other hand, for too large an energy gap the injection of minority carriers by a forward biased p–n junction becomes difficult, owing to the competing process of electron–hole recombination via traps in the depletion region becoming dominant. The 1.12 eV energy gap of Si therefore represents a good compromise.

High-speed device operation and low parasitic resistances require high values of carrier mobility. The electron and hole mobilities of pure silicon at room temperature are 1450 and 450 $cm^2\,V^{-1}\,s^{-1}$, respectively, although these values are reduced by up to a factor of about 10 in heavily doped material. While certain other semiconductors have values exceeding these figures, the mobilities in silicon are generally adequate for all but the highest-speed devices.

An obvious requirement for device applications is that the material can be controllably doped, both p-type and n-type, over a very wide range. This further requires that the material can be obtained with a very low background of electrically active impurities. Silicon has many advantages over other semiconductors in this respect. Atoms from groups III and V of the periodic table form simple, substitutional acceptors and donors, respectively, their binding energies being sufficiently small to be fully ionized at room temperature. The elemental nature of silicon means that these dopants cannot be amphoteric. (In comparison, for compound semiconductors such as GaAs, group IV impurities such as silicon may be a donor if substitutionally incorporated onto a Ga lattice site or an acceptor if incorporated onto an As site.) The usual donors (P, As) and acceptor (B) have solubilities well in excess of 10^{20} atoms/cm^{-3}, allowing very high carrier densities and low-resistivity material to be achieved. Dopants may be introduced either during bulk and epitaxial growth or by diffusion and ion implantation. Bulk and epitaxial growth are discussed elsewhere in this volume. Diffusion and ion implantation are essential techniques for introducing localized doped regions for device structures. Both require high-temperature treatments to redistribute the impurities, and here silicon has the advantage of thermal stability and low vapor pressures up to high temperatures. Again this contrasts with many compound semiconductors: for example, the thermal annealing of GaAs after ion implantation requires special precautions to prevent arsenic evaporation from its surface. Additionally, the diffusion coefficients of common dopants in silicon are low at normal device operating temperatures, another essential requirement for stable and reliable device operation.

A quantity related to overall material quality is the lifetime of minority carriers, since this is determined by recombination occurring through deep traps due to chemical impurities or crystalline defects. Long lifetimes are required for bipolar transistors in order to achieve high base transport factors. Values in excess of 100 μs may be achieved in Si, corresponding to electron diffusion lengths of over 0.5 μm. and so making the base transport factor α_T in silicon bipolar transistors very close to unity. For most other semiconductors (germanium being a notable exception), lifetimes are orders of magnitude shorter. For applications that require silicon devices with short lifetimes, such as fast-switching p-n diodes, deep levels which act as recombination centers may be deliberately introduced. Doping with Au is frequently used for this purpose.

The properties of thermally grown SiO_2 and the Si–SiO_2 interface are of crucial importance for the operation of silicon MOSFETs. The description of MOSFET operation given in Section 4 assumed that all the charge induced by the gate voltage was mobile charge located near the semiconductor–oxide interface. However, changes in the gate voltage will also change the occupancy of surface states at the semiconductor–oxide interface and traps in the oxide. These changes represent a modulation of charge that is essentially immobile and so cannot contribute to the channel conductivity. Most semiconductor–oxide or semiconductor–insulator systems have very high surface state densities, making the fabrication of practical MOSFETs and MISFETs impossible. The Si–SiO_2 system is exceptional in that the thermal growth of SiO_2 can result in surface state densities as low as about 10^{10} cm^{-2}, which compares with an electron sheet density in the inversion layer which can be about three orders of magnitude higher. The high value of electron sheet density results from another advantageous property of SiO_2, its high dielectric breakdown strength (10^7 V/cm), which allows gate voltages of a few volts to be applied across very thin (typically 200 Å) oxides.

A further advantageous property possessed by SiO_2 is that the diffusion coefficients of common dopants are very low, much lower than in Si itself. SiO_2 can therefore act as a diffusion barrier, and locally doped areas of silicon can be produced by oxidizing its surface, etching windows in the oxide, and diffusing in p-type or n-type dopants from gaseous sources. Similarly locally doped areas may be created by ion implantation through windows in SiO_2. These processes are the basis of planar technology, which more than any other factor has contributed to the expansion of the semiconductor industry.

In addition to these electrical and processing advantages, silicon also has excellent mechanical and thermal properties. Its thermal conductivity of 1.5 W K^{-1} cm^{-1} is the highest of all the commercial semiconductors. This is an important consideration for power devices and high-density integrated circuits. Mechanically, it is sufficiently robust to be pulled as large-diameter single crystals and cut, lapped, and polished into thin wafers of commonly 100–200 mm diameter for processing.

In view of all these factors, it is not surprising that silicon has achieved its pre-eminent position for transistor fabrication. Broadly, silicon MOSFETs are the dominant devices for most digital circuits, especially for complex circuits, while silicon bipolar transistors are used for most analogue devices and circuits, power devices, and digital circuits requiring very high-speed operation.

In passing it is worth noting that p-n diodes are fabricated from a much wider range of materials than transistors. This is because transistors are generally used only for the functions of amplification and switching, whereas diodes fulfill a much wider range of functions. These include, in addition to rectification, switching, and voltage limiting, functions such as optical generation (LEDs, lasers), optical detectors (photodetectors, solar cells), and microwave generation (e.g., IMPATT diodes). Accordingly, the list of materials used for $p-n$ diodes includes Si, Ge, GaAs, GaP, InP, InGaAs, InGaAsP, and GaAsP. The technology for most of these materials is, however, rudimentary compared with Si technology.

6. GALLIUM-ARSENIDE-BASED TRANSISTORS

6.1. The GaAs MESFET

Although the advantages of silicon make it the preferred material for mainstream transistor fabrication, there are some more specialized device applications for which silicon is less well suited. Foremost among these is for transistors to operate at microwave or millimeter wave frequencies.

The complex subject of the frequency limitations of transistors will not be discussed here: It is sufficient for present purposes to recognize that the intrinsic speed of a transistor is determined by the transit time for a carrier to pass through its active region. Higher-frequency operation in silicon devices has therefore been achieved by shrinking the critical dimensions. Technology developments have led to Si bipolar transistors with base widths less than 0.1 μm and MOSFETs with gate lengths less than 1 μm. But transit times may also be reduced by moving to materials in which the carriers travel faster than in silicon. Most attention has been focussed on the III–V compound semiconductor GaAs, which has an electron mobility when pure of about 8600 cm^2 V^{-1} s^{-1}, about six times greater than silicon.

A second reason for the use of GaAs in high-frequency transistors is the availability of so-called semi-insulating substrates. In any real device the upper frequency of operation is determined not only by carrier transit time limitations but also by parasitic capacitances between the device electrodes. Minimizing these parasitic capacitances requires the active device to be on a very high-resistance substrate. Gallium arsenide has an energy gap of 1.42 eV at room temperature, and would have a resistivity as high as 10^8 Ω cm if it could be obtained in pure enough form to be intrinsic. This would, however, require the concentrations of shallow impurities to be reduced to about 10^6 cm^{-3}, or about ten orders of magnitude lower than currently achievable in bulk crystals! Fortunately it is possible to introduce deep levels during the growth of bulk GaAs which effectively compensate the residual shallow donors or acceptors and pin the Fermi level close to midgap, thereby giving material with the resistivity of intrinsic material. For such material the description "semi-insulating" is more appropriate than "semi-conducting."

Transistors fabricated from GaAs are almost exclusively MESFETs. Unlike Si, it is not possible to grow an oxide layer on GaAs with a low surface state density, thereby ruling out GaAs MOSFETs, while it is difficult to fabricate JFETs with very short gates. Fortunately the Schottky barrier height of most metals on n-type GaAs is about 0.7–0.8 eV, high enough for MESFET gates with low leakage currents. The structure of a GaAs MESFET is shown schematically in Fig. 9. The active n-channel layer is created either by epitaxial growth on the semi-insulating substrate or by ion implantation directly into the substrate. (GaAs MESFETs are always n-channel, because holes have a low mobility.) Doped material outside the device area is

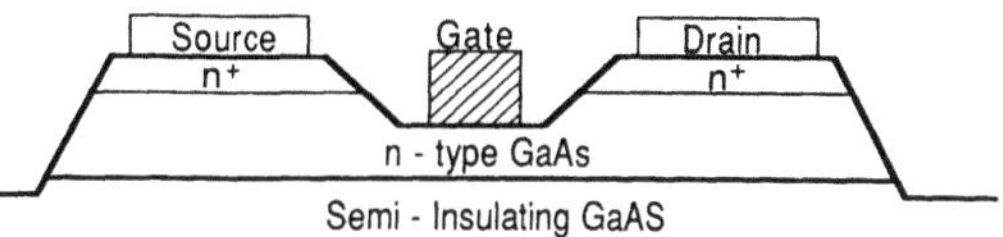

FIGURE 9. Schematic diagram of the structure of a GaAs MESFET.

removed by etching back to the substrate. Typically the channel is doped in the range 10^{17}–10^{18} cm^{-3}. As shown in Fig. 9, the regions under the source and drain ohmic contacts are generally more heavily doped, to reduce their resistances, and the gate is deposited into an etched recess. For high-frequency operation the gate is always very short, as small as 0.25 μm in commercial devices, while sub-0.1-μm devices have been reported in the laboratory.

For microwave applications, GaAs MESFETs are generally depletion-mode devices, with the drain current controlled by pinching the channel between the gate depletion region and the semi-insulating substrate. Although the operation of the GaAs MESFET is broadly similar to that of a Si JFET, an important difference is that the channel in a MESFET does not completely pinch off, current saturation being due to the electrons reaching their saturation velocity as they pass under the gate. The transit time of electrons in a MESFET is therefore determined by their saturation velocity rather than their mobility, and the notion that the higher speed of GaAs devices over Si devices results from the higher mobility of GaAs, although often quoted, is an oversimplification. This point is fully discussed by Ladbrooke.[7]

As well as discrete devices, integrated circuits operating at microwave frequencies are fabricated on GaAs, using MESFETs for active devices and the semi-insulating substrate as a dielectric in microstrip transmission lines and for supporting integrated passive components. MESFET-based digital circuits, generally using enhancement-mode devices, are also finding increasing acceptance for high-speed applications. The technology of GaAs-based integrated circuits is reviewed in Chapter 14 by Turner.

6.2. Heterojunction-Based Devices

The possibility of using the properties of heterojunctions to enhance the performance of electronic devices has received considerable attention during the last decade. Generally, but not exclusively, the favored material system is AlGaAs/GaAs, this being the most mature system for epitaxial growth. The device that has so far received the most attention is the high electron mobility transistor or HEMT (also known under other acronyms). The basic idea of the HEMT has already been described in the previous chapter. In brief this is to retain the high electron mobility of pure GaAs while simultaneously obtaining the high electron density required for a device. This is achieved by using the heterojunction to separate the electrons from their parent donors, the donors being introduced only into the wider gap material (AlGaAs) while the electrons accumulate in the narrow gap material (GaAs) close to the interface (see Fig. 2b of the previous chapter). Physically the construction of the HEMT is similar to that of the MESFET shown in Fig. 9, except that the n-type channel layer of the MESFET is replaced by the selectively doped AlGaAs/GaAs heterojunction layer. The gate metal is placed on the doped AlGaAs layer, and the

AlGaAs/GaAs heterointerface is typically 200–400 Å beneath the gate. The HEMT has demonstrated the highest-frequency performance of any transistor. It also has better noise performance than the conventional MESFET, a result that has led to its rapid acceptance for use in domestic systems for satellite television reception.

The use of heterojunctions to enhance bipolar transistor performance has also attracted considerable attention. As explained in Section 3, a conventional bipolar transistor requires a heavily doped emitter and lightly doped base to achieve good emitter efficiency. However, for high-speed operation a low base resistance is desirable, and this is incompatible with low base doping. In the heterojunction bipolar transistor (HBT) the emitter is made from a wider band gap material. The step in the valence band blocks the injection of holes from the base to the emitter and so allows high emitter efficiencies to be achieved even with a heavily doped base. HBTs are showing great promise for very high-speed digital circuits and for microwave power devices. Most work to date has again used the AlGaAs/GaAs materials system, although the Si/SiGe system is also being explored with the aim of making HBTs compatible with advanced silicon technology. It is interesting to note that the concept of the HBT was first patented as long ago as 1951, but it is only in recent years that materials growth techniques have made the device a practical proposition.

7. OTHER MATERIALS

This penultimate section briefly mentions two situations requiring transistors to be fabricated in materials systems imposed by their applications. The first of these is for switching on and off the picture elements in flat panel liquid crystal displays. For large-area displays this requires a transistor to be associated with each picture element, and so necessitates the fabrication of transistors in semiconductors which can be deposited in the form of thin films over the whole area of the display. The application of FETs fabricated in amorphous silicon for this application is described in Chapter 11 by LeComber.

The second situation relates to transistors for optoelectronic integration. As described in the following chapter, lasers and photodetectors for operation at the preferred optical fiber communication wavelengths of 1.3 and 1.55 μm are fabricated in the materials systems $Ga_xIn_{1-x}As_yP_{1-y}$. The compositions of this quaternary alloy, determined by the values of x and y, are chosen to give an energy gap corresponding to the wavelength of interest while maintaining the lattice parameter of the alloy equal to that of InP, so that the material may be grown epitaxially on InP substrates. In current optical fiber systems the lasers and photodetectors are discrete components. There are cost, performance, and system complexity advantages to be gained by integrating these optoelectronic devices with transistors, thus producing optoelectronic integrated circuits (OEICs). For example, a laser could be combined with its driver transistors or a photodetector with a preamplification and pulse regeneration stage. Such OEICs will therefore require high-performance transistors to be fabricated on InP substrates. The simplest option would be InP MESFETs, but this is ruled out by the low Schottky barrier height (typically ~0.2 eV) of common metals on InP. Many options are under investigation, based on either

InP, $In_{0.53}Ga_{0.47}As$, $In_{0.52}Al_{0.48}As$, (both of which are lattice-matched to InP), or some combination. Devices being explored include InP and $In_{0.53}Ga_{0.47}As$ MOSFETs, MISFETs, and JFETs, $In_{0.53}Ga_{0.47}As/InP$ heterojunction JFETs and HBTs, and $In_{0.52}Al_{0.48}As/In_{0.53}Ga_{0.47}As$ HBTs and HEMTs. No clear front-runner has yet emerged, although very promising results have been reported for several systems. InGaAs FETs have recently been reviewed by Heime.[8]

8. CONCLUSIONS AND FUTURE PROSPECTS

This chapter has given a brief review of the principles of basic electronic devices, concentrating on transistors and their materials requirements. The importance of silicon for bipolar and MOSFET devices and circuits has been emphasized. At present there is no indication that the position of silicon as the dominant material for electronic devices will ever be challenged, nor is there any indication that the development of silicon technology is slowing down. Gallium arsenide has a specialized and expanding role to play in high-speed and high-frequency circuits, the performance of which will increasingly benefit from the introduction of "second generation" devices involving heterojunctions. Devices in prospect are refinements of the basic bipolar transistor or FET, rather than devices based on radically new concepts. The extension of existing device concepts to new materials systems, for applications such as OEICs, is likely to be a major theme for device research and development in the coming years.

REFERENCES

1. A. S. Grove, *Physics and Technology of Semiconductor Devices*, Wiley, New York (1967).
2. S. M. Sze, *Physics of Semiconductor Devices*, 2nd Ed., Wiley, New York (1981).
3. S. M. Sze, *Semiconductor Devices: Physics and Technology*, Wiley, New York (1985).
4. B. G. Streetman, *Solid State Electronic Devices*, Prentice-Hall, Englewood Cliffs, New Jersey (1980).
5. E. H. Rhoderick and R. H. Williams, *Metal–Semiconductor Contacts*, 2nd Ed, Clarendon Press, Oxford (1988).
6. P. Ashburn, *Design and Realization of Bipolar Transistors*, Wiley, New York (1988).
7. P. H. Ladbrooke, *MMIC Design: GaAsFETs and HEMTs*, Artech House, Boston and London (1989), pp. 69–83.
8. K. Heime, *InGaAs Field-Effect Transistors*, Research Studies Press, Somerset, England (1988).

Key Optoelectronic Devices

A. W. Nelson

1. INTRODUCTION

Light in its various forms is one of the most powerful tools of nature. Yet despite its fundamental influence on life, we are only now just learning how to utilize more fully the extensive range of properties it offers. Optical fiber communications systems that begin to utilize the enormous bandwidth (10^{13}–10^{15} Hz) of light are now being installed worldwide, the energy of light is being harnessed by fabricating solar cells, and the power of light is concentrated in lasers for use in cutting, welding, medicine, and a host of other applications. In addition light is used in a multitude of forms for display, indicator, and control purposes.

One of the major reasons for the delay in more fully utilizing the properties of light has been the basic lack of suitable materials technologies. Although in principle there are an almost infinite number of materials that can influence light's properties, there are only a few that can be used together with electronic circuits to generate, detect, and manipulate light signals, a fundamental requirement in order to fully access light's phenomenal capacity. The family of devices that are able to perform the aforementioned functions are known generally as optoelectronic devices and with a few notable exceptions are generally manufactured from semiconductor materials.

The number of potential materials that can be used to produce optical devices is vast and would be outside the scope of a single chapter. This chapter therefore focuses on those devices and materials that are capable of integrating optical and electronic functions to produce such devices as light-emitting diodes, solid state lasers, solar cells, infrared detectors, light modulators, and waveguides.

By far the most important and useful group of semiconductors used in the manufacture of optoelectronic components are those of the III–V family, so called because the atoms making up the compounds are derived from the group III and group V columns of the periodic table. Other semiconductors that are also important are silicon, germanium, and compounds belonging to the II–VI family of materials. Table 1 shows the most important semiconductor materials currently used in optoelectronic device technology.

A. W. Nelson ● Epitaxial Products International Limited, Cypress Drive, St. Mellons, Cardiff CF3 0EG, U.K.

TABLE 1. Important Semiconductor Materials for Optoelectronics

Material	Type	Substrate	Devices	Wavelength range (μm)
Si	IV	Si	Detectors, solar cells	0.5–1
SiC	IV	SiC	Blue LEDs	0.4
Ge	IV	Ge	Detectors	1–1.8
GaAs	III–V	GaAs	LEDs, lasers, detectors, solar cells, imagers, intensifiers	0.85
AlGaAs	III–V	GaAs	LEDs, Lasers, solar cells, imagers	0.67–0.98
GaInP	III–V	GaAs	Visible lasers, LEDs	0.5–0.7
GaAlInP	III–V	GaAs	Visible lasers, LEDs	0.5–0.7
GaP	III–V	GaP	Visible LEDs	0.5–0.7
GaAsP	III–V	GaP	Visible LEDs	0.5–0.7
InP	III–V	InP	Solar cells	0.9
InGaAs	III–V	InP	Detectors	1–1.67
InGaAsP	III–V	InP	Lasers, LEDs	1–1.6
InAlAs	III–V	InP	Lasers, detectors	1–2.5
InAlGaAs	III–V	InP	Lasers, detectors	1–2.5
GaSb/GaAlSb	III–V	GaSb	Detectors, lasers	2–3.5
CdHgTe	II–VI	CdTe	Long detectors	3–5 and 8–12
ZnSe	II–VI	ZnSe	Short LED	0.4–0.6
ZnS	II–VI	ZnS	Short LED	0.4–0.6

In many cases, especially for the III–V compounds, the powerful combination of light generation, detection, and manipulation properties, together with the ability to produce highly sophisticated electronic devices, gives the III–V compounds an almost unassailable position as the ideal group of optoelectronic materials.

The use of semiconductors in preference to other materials for use in optoelectronics is due to the numerous advantages they impart to device operation; devices can be made very small, leading to a high degree of compactness and compatibility with other electronic and optical functions; they are robust, leading to high reliability; they are highly efficient as light-generating sources with internal efficiencies approaching 100% in some cases; they are capable of large power dissipation, of very high-frequency performance, and they can access a huge range of wavelengths. In addition, because of their specialized properties, performances can be tuned over wavelength, frequency, and efficiency, by the application of appropriate electric fields.

These numerous advantages have led to the rapid development of optoelectronics as a major industry spanning everything from simple display LEDs and complex compact disk players, to entirely new industries—a prime example of this being optical fiber communications. Currently the world optoelectronic market is worth some £7 billion per annum and the growth rate is approximately 30% compound per annum for many areas. A list of commercial applications of optoelectronic materials is shown in Table 2.

Underpinning the technological advances is also a great deal of research and development into new devices and systems that will further utilize the potential of

TABLE 2. Commercial Applications of Optoelectronic Devices

Materials	Devices	Applications
GaAs/AlGaAs	Detectors, infrared LEDs, and lasers	Remote control TV etc., video disk players, range-finding, solar energy conversion, optical fiber communication systems (local networks), image intensifiers
InP/InP	Solar cells	Space solar cells
InP/InGaAsP	Infrared LEDs, lasers (1.0–1.6 μm)	Optical fiber communications (long-haul and local loop)
InP/InGaAs	1–1.67 μm detectors	Optical fiber communications, instrumentation
InGaAlAs/InGaAs	1.67–2.4 μm detectors	Military applications, medicine, sensors
GaAs/GaInP/GaInAlP	0.5–0.7 μm LEDs and lasers	Displays, control, compact disk players, laser printers/scanners, optical disk memories, laser medical equipment
Si	Detectors, solar cells	Solar energy conversion, e.g., watches, calculators, heating, cooling, 0.4–1.0-μm detectors
Ge	Detectors	1.0–1.6 μm
SiC	Blue LEDs	Displays, optical disk memories, etc.
GaSb/GaAlSb/InSb	Long-wavelength detectors/emitters	2–3.5 μm
CdTe/HgCdTe	Infrared detectors at 3–5 and 8–12 μm	Infrared imaging, night vision sights, missile seekers, and many other military applications
ZnSe/ZnS	Visible LEDs	Commercial applications (R&D stage only)

light; in particular, a major field of research at present is optoelectronic integration, where many optical and electronic functions can be combined on a single chip.

The remainder of this chapter deals with the various key optoelectronic devices and the importance of materials technology to their operation and advancement. Emphasis throughout is placed on the materials requirements for the various devices described.

2. MATERIALS TECHNOLOGIES

Prior to dealing with the key optoelectronic devices in more detail it is worthwhile reviewing the materials considerations and present technology available to prepare the complex heterostructures that are now required by many of today's advanced devices.

2.1. Important Optoelectronic Materials

Increasingly the materials used originate from the group III and group V columns of the periodic table, atoms being mixed in such proportions that the average atomic spacing (or lattice parameter) a of the active material is equal to that of a binary (two element) substrate. Figure 1 details the relationship between the various ternary (three-element) and quaternary (four-element) compounds and the binary substrates making up the III–V family of materials. The figure displays lattice constant a as a function of energy gap with direct band gap materials illustrated by solid lines and indirect band gap materials by dotted lines. This diagram is extremely important since a vital factor in the efficient, long-term operation of many devices is that the overall device structure is free from any strain effects, which occur when materials of different lattice parameters are deposited upon one another. It is therefore important that any materials structure used to make optoelectronic devices consist of materials combinations which lie on a vertical line in Fig. 1.

The difference between direct band gap materials and indirect gap materials results in an enormous difference in optical device efficiency; this is discussed in Chapter 4 by Miller.

One key parameter that illustrates this difference in optical efficiency is the minority carrier lifetime. For indirect semiconductors such as Si and Ge minority carrier lifetimes are exceedingly long (~1 ms), compared with direct band gap materials such as GaAs, which have minority carrier lifetimes of <1 ns.

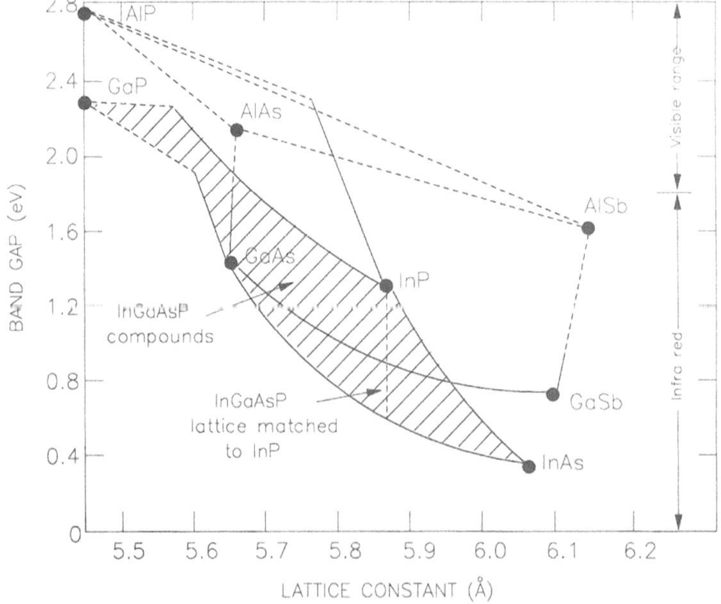

FIGURE 1. Lattice constant as function of energy gap for III–V optoelectronic materials.

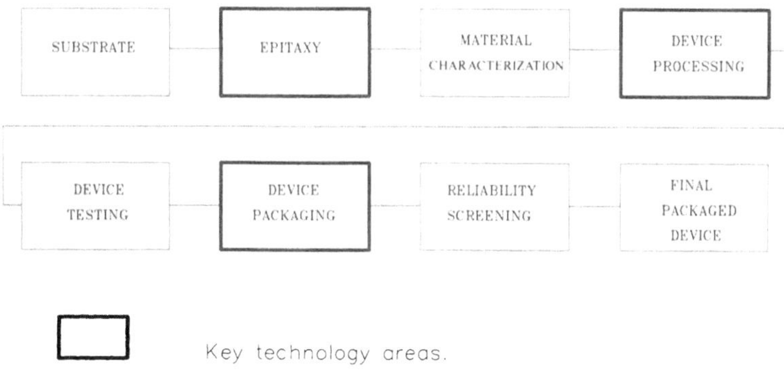

FIGURE 2. Stages of optoelectronic device manufacture.

Efficient recombination can be achieved in some indirect gap materials by the introduction of impurities. This occurs when isoelectronic nitrogen impurities are introduced in GaP. It is not possible in Si or Ge owing to the large difference in energy between the direct and indirect band gaps. Thus almost all other emitters are direct band gap III–V materials.

2.2. Epitaxy

In general the devices described in this chapter are produced by depositing very thin layers of various III–V and II–VI compounds onto binary substrates. Figure 2 illustrates the various stages involved in the preparation of optoelectronic devices. One of the key technologies required in this overall process is that of epitaxy, which is the process of depositing thin layers of single-crystal compounds onto the binary substrate. This is dealt with more fully in a separate chapter, but it is worth emphasizing that the control, reproducibility, and large-area uniformity coupled with the ability to make compositional change with atomic abruptness has been the key technological breakthrough in the general development of optoelectronic components. A great deal of basic device technology development was carried out using a technique known as liquid phase epitaxy (LPE), but the more critical and sophisticated devices of today use technologies such as metal organic vapor phase epitaxy (MOVPE) or molecular beam epitaxy (MBE). For further information on these techniques, please refer to Chapter 10 by Mullin.

3. LIGHT-EMITTING DEVICES

3.1. Basic Principles

Most semiconductor light-emitting devices are made by forming a p–n junction within the semiconducting material, usually by growing a sequence of p- and n-doped epitaxial layers. By applying a forward bias voltage to the diode so formed, electrons and holes are forced to recombine. In suitable materials, such as those discussed earlier, this results in radiative recombination, leading to useful light output from the device. This sequence is illustrated in Fig. 3 for a double heterostructure p–n junction where the active layer is made from a slightly different composition of

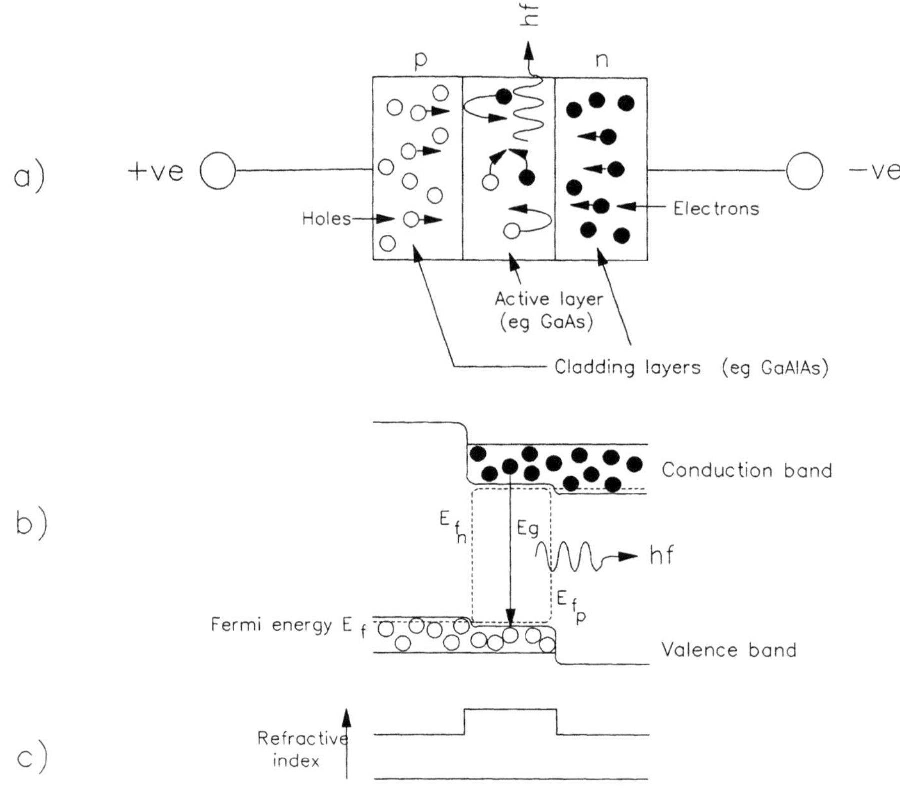

FIGURE 3.   Schematic operation light-emitting devices. (a) *p–n* junction. (b) Energy band diagram. (c) Refractive index profile.

material (e.g., GaAs/AlGaAs) to aid in both light and injected carrier confinement; a simple *p-n* junction device operates in exactly the same way except that injected carriers are less well confined. The wavelength of the light generated is related to the forbidden energy gap simply by

$$\text{Energy } E = hf \quad \text{or} \quad E(\text{eV}) = \frac{1.24}{\lambda(\mu\text{m})} \tag{1}$$

where f is frequency, h is Plank's constant, and λ is wavelength.

There are basically two forms of light emission, as discussed in Chapter 4 by Miller. The first is termed spontaneous emission and occurs randomly when electrons and holes recombine at the band edge energy. Light is emitted at a wavelength corresponding to the band energy, but is random in phase. This type of emission is typical of the process that occurs in light-emitting diodes (LEDs).

The second type of light emission is known as stimulated emission and is the process that is fundamental to the operation of the laser (*light amplification by stimulated emission of radiation*). Stimulated emission occurs when a photon of light traveling through the semiconductor interacts with the electron and hole population to cause the radiative recombination of another electron–hole pair. The emitted photon in this case is of exactly the same phase, amplitude, and energy as the stimulating photon, resulting in effective amplification of the original photon. This process only occurs under specific conditions which are described later in the section on lasers (see Section 3.3).

There is a third process, which is independent of the two emission processes, but which can have an important influence in the operation of lasers. This process is known as absorption and is the fundamental process utilized in photodetectors. A photon of light of energy larger than the semiconductor band gap is absorbed within the semiconductor with the creation of an electron–hole pair. Light below the band gap energy (i.e., of longer wavelength) is able to pass through the semiconductor completely unmodified. This process is discussed in Chapter 4 by Miller.

3.2. Light-Emitting Diodes (LEDs)

Light-emitting diodes are now found in a vast range of applications ranging from simple indicator lamps and huge display screens to optical fiber communication links. LEDs are made from a wide variety of semiconductor materials depending on the wavelength requirement. Although the most efficient emitters are direct band gap III-V materials, Fig. 1 illustrated that as the band gap of the semiconductor increases (i.e., the wavelength of emitted light penetrates further into the visible spectrum) almost all of the III–V compounds become indirect gap materials. Consequently, the most commonly used materials for visible LEDs are gallium phosphide and gallium arsenide phosphide. A variety of wavelengths can be obtained from these materials by doping with various impurities, such as oxygen and nitrogen. This creates energy levels within the forbidden energy gap, which act as recombination centers and cause these indirect band gap materials to behave like direct gap materials. Table 2 lists all the common materials that are used in the manufacture of light-emitting diodes, together with a list of the uses in the various wavelength ranges.

The structure of an LED can take many forms, ranging from a very simple p–n junction to the very complex double heterostructure design. In practice, however, there are four types of structure that dominate, as illustrated in Fig. 4; the planar and dome LED structures are usually encapsulated into inexpensive plastic packages for visible device applications; the edge-emitting and surface-emitting diodes are used mainly in optical fiber communication systems.

The planar and dome designs are used in most display devices and applications such as intruder alarms, remote controls, etc. where the interest is in extracting the maximum amount of light from the device. In this case, light is emitted in all directions, and using a lens arrangement light is focused in the general direction in which display or control is required.

In the edge-emitting case the structure is similar to that of a laser, where light is confined in the active region of the device. Light is emitted at either or both ends of the device, and because of the narrow dimension of the active layer this design is ideal for launching light into the narrow core region of an optical fiber.

Operation of an LED is effected simply by applying a suitable voltage (usually 5 V) to the device, resulting in a forward current of between 5 and 100 mA. This

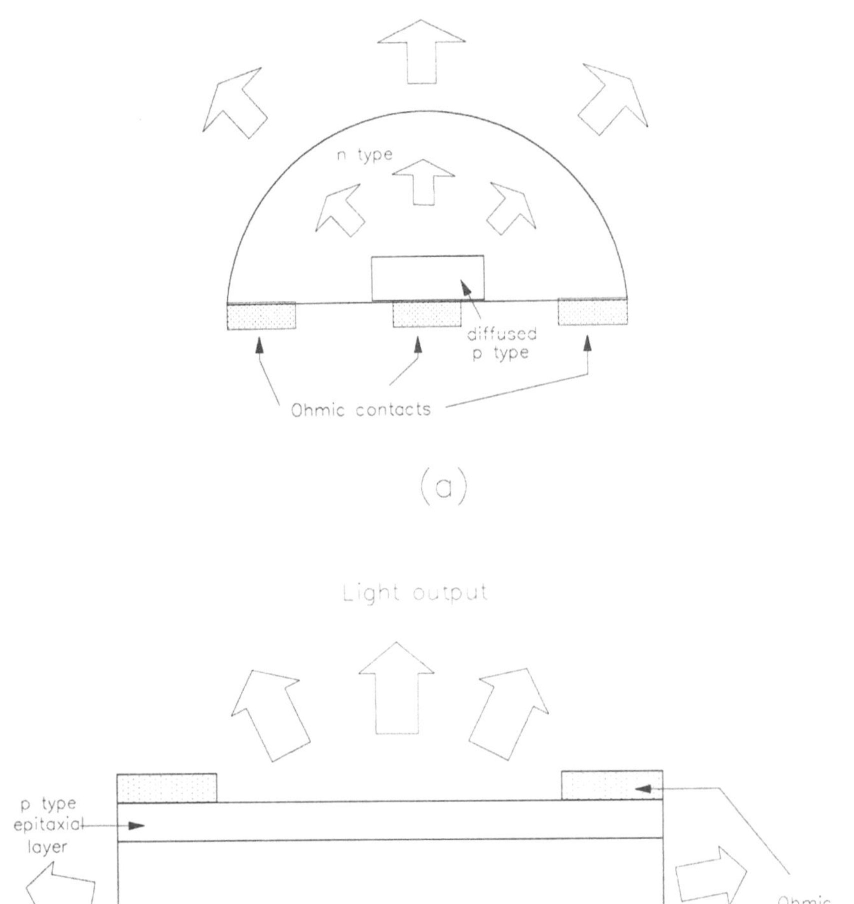

FIGURE 4.   Important LED structures. (a) Dome LED. (b) Planar LED. (c) Burrus LED. (d) Edge-emitting LED.

current determines the light output from the device, which is normally linearly related to the drive current. A typical light output vs. current plot is shown in Fig. 5. The advantages of these devices are their ease of modulation, long lifetime, low cost, and high yield. However, they do have limitations, especially when high levels of light output are required, and in these cases a more appropriate device is usually the semiconductor laser diode.

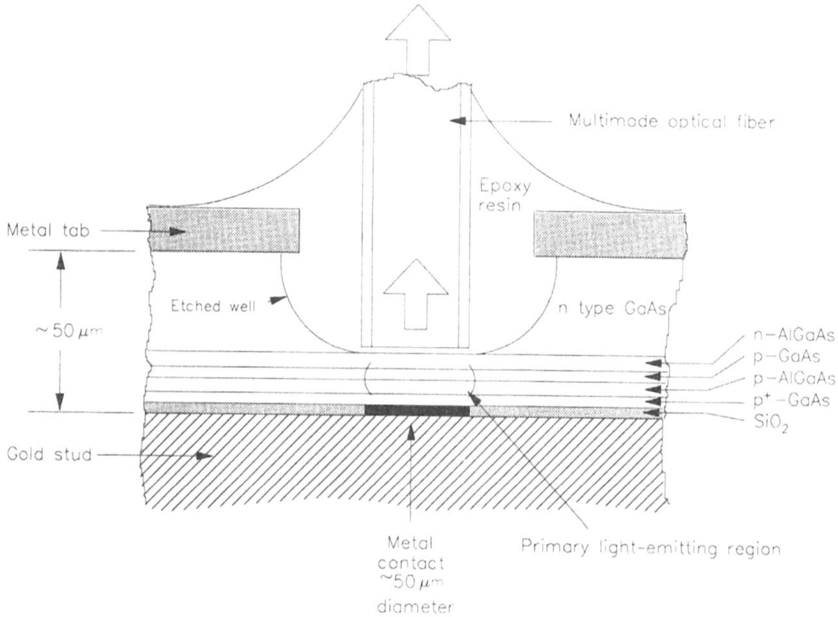

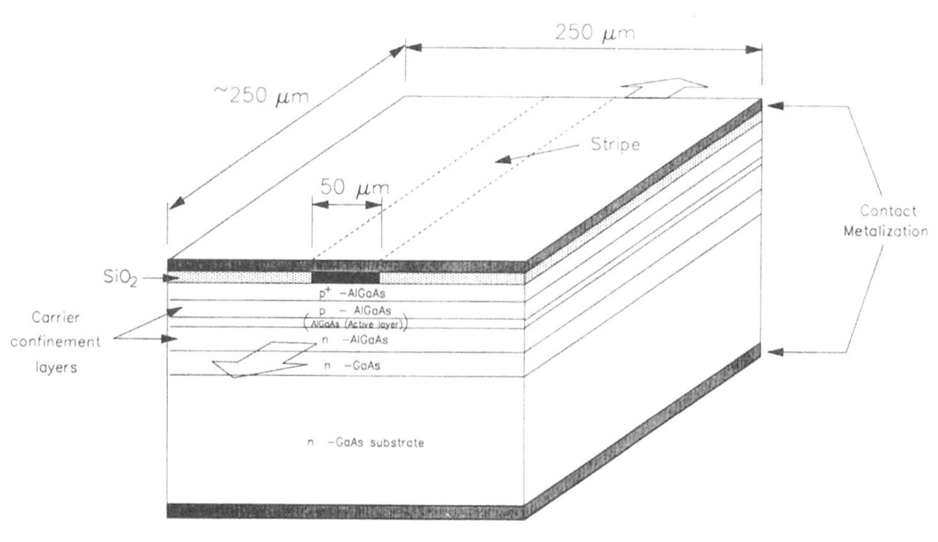

FIGURE 4. *Continued.*

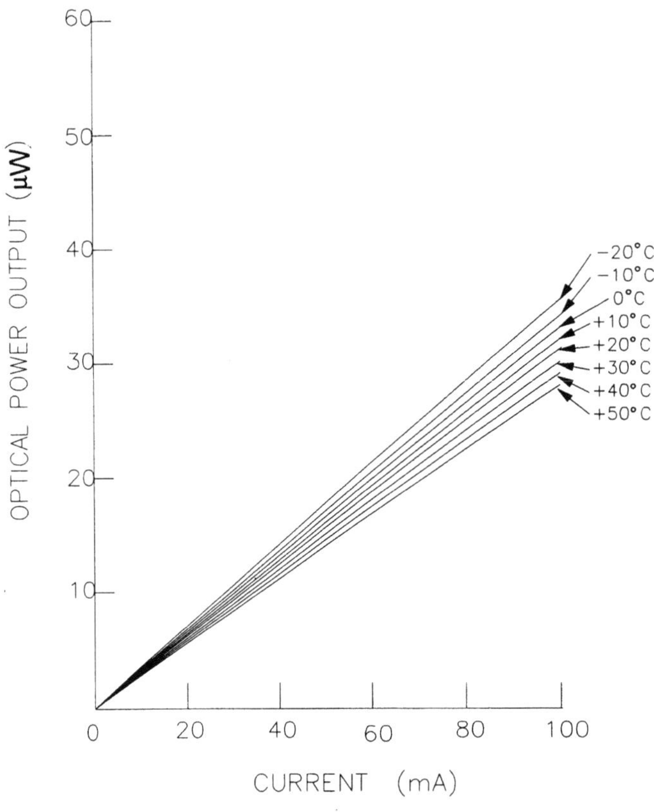

FIGURE 5.   Light output—current characteristic for a typical LED.

3.3. Semiconductor Lasers

The semiconductor laser was first developed in the 1960s from GaAs and early devices were highly inefficient with operation only possible at very low temperatures. However, with improvements to carrier and optical confinement by the use of GaAlAs cladding layers and GaAs active layers (the double heterostructure design) devices were able to function continuously at 300 K by the early 1970s. Since that time the applications of semiconductor lasers have increased dramatically and they can be found in many commercial, military, and consumer products. The importance of semiconductor lasers cannot be overstated, and their development, especially in the wavelength range of 1.3–1.67 μm has led directly to the rapid implementation and the burgeoning market in optical fiber communication systems. They are also used, however, in a vast range of other applications such as compact disk players, data storage devices, sensing systems, medical instruments, amplifying devices, range-finding equipment, and many other areas too numerous to mention.

Although very similar in structure and principle to the light-emitting diode, there is a very significant difference in light output characteristics between the laser and LED, as illustrated in Fig. 6. As is shown, the laser has a much narrower spectral range (<3 nm compared with 75–100 nm for the LED) and a much more intense light output (at least 100 times more intense than the LED output).

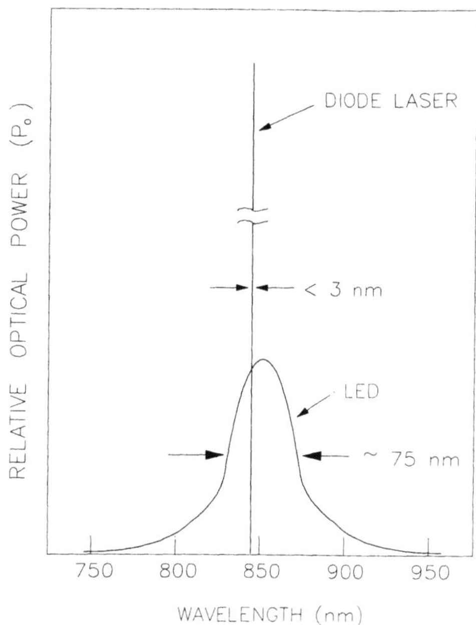

FIGURE 6.   Comparison of laser and LED spectral outputs.

The main reason for these improved characteristics is the formation of a lasing cavity that acts as an extremely high-Q resonator. This cavity is usually created by the formation of mirrors at each end of the laser device (a typical semiconductor laser is between 400 and 800 μm long). These mirrors are made by cleaving precisely along a crystallographic plane [usually the (110) planes] resulting in an atomically abrupt refractive index change at the semiconductor–air interface. Typically the mirror so formed has a reflectivity of approximately 0.33. The semiconductor between the two mirrors then forms the laser cavity. Optical gain in the cavity is achieved by the injection of electron–hole pairs (i.e., by the injected current) resulting in both spontaneous and stimulated emission. The basic rate equations for the processes occurring within the laser cavity are given below by the Einstein relations:

$$B_{12} = \frac{g_2 B_{21}}{g_1} \tag{2}$$

$$\frac{A_{21}}{B_{21}} = \frac{8hf^3}{c^3} \tag{3}$$

where B_{12}, B_{21}, and A_{21} are the Einstein coefficients of absorption, stimulated emission, and spontaneous emission, respectively, from an energy level 2 to a lower energy level 1. g_2 and g_1 are the degeneracies of the two levels. For $g_1 = g_2$ the ratio

$$\frac{\text{Stimulated emission rate}}{\text{Spontaneous emission rate}} = \frac{1}{\exp\{hf/kT\} - 1} \tag{4}$$

This shows that for thermal equilibrium, spontaneous emission completely dominates stimulated emission. For thermal equilibrium we also know that the upward transitions rate R_{12} must equal the downward transition rate R_{21}; therefore we have

$$\text{Absorption} = \text{spontaneous emission} + \text{stimulated emission}$$

$$N_1 p_f B_{12} = N_2 A_{21} + N_2 p_f B_{21} \tag{5}$$

where p_f is the spectral density of the radiation energy, N_1 and N_2 are the populations of the lower and upper energy levels. Equation (5) illustrates that for stimulated emission to dominate, the radiation density must be increased, and the population of the upper level must be increased substantially above that of the lower level. This is known as population inversion and is a fundamental precondition for lasing to occur.

A high rate of stimulated emission results in optical gain g and for lasing to occur a further threshold condition must be met, which is that the round trip gain of a photon is greater than unity. Under these circumstances light amplification occurs and by a regenerative process the light traveling within the laser cavity is narrowed in line width and enhanced in intensity.

The current level at which lasing occurs is known as the threshold current (I_{th}) and is given by the following formula:

$$\frac{I_{th}}{A} = J_{th} = \frac{g_{th}}{\beta} = \frac{1}{\beta}\left[\alpha + \frac{1}{2L}\ln\left(\frac{1}{r_1 r_2}\right)\right] \tag{6}$$

where I_{th}, J_{th}, g_{th}, are the threshold current, threshold current density, and threshold gain coefficient, respectively, and β is the gain constant, which is dependent on materials properties. L is the device length, r_1 and r_2 are the reflectivities of the end facets, and A is the cross-sectional area of the active layer.

A typical light output–current characteristic for a single mode laser is shown in Fig. 7, illustrating the threshold current I_{th} and device external quantum efficiency.

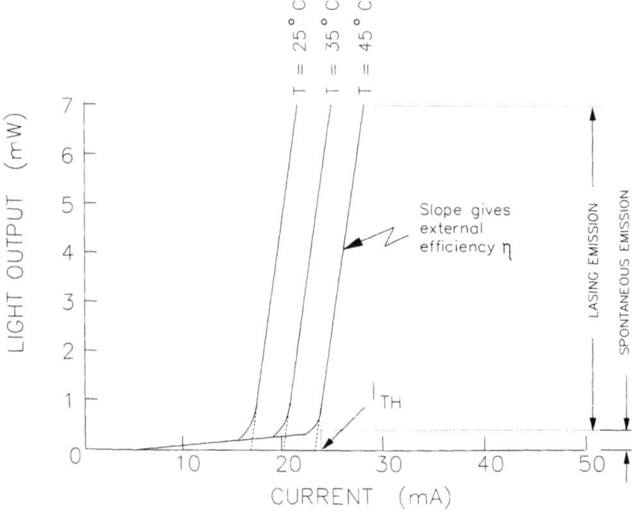

FIGURE 7.   Light output–current characteristic for a typical semiconductor laser.

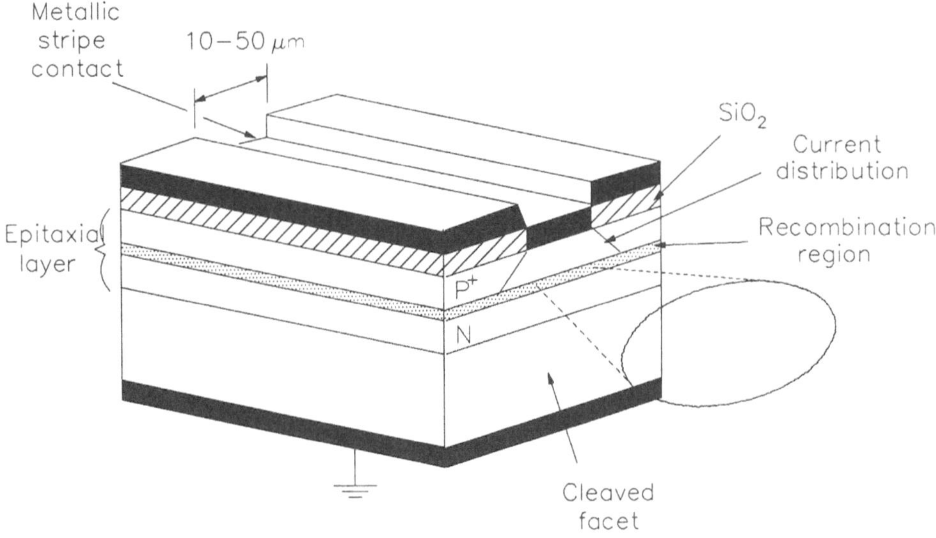

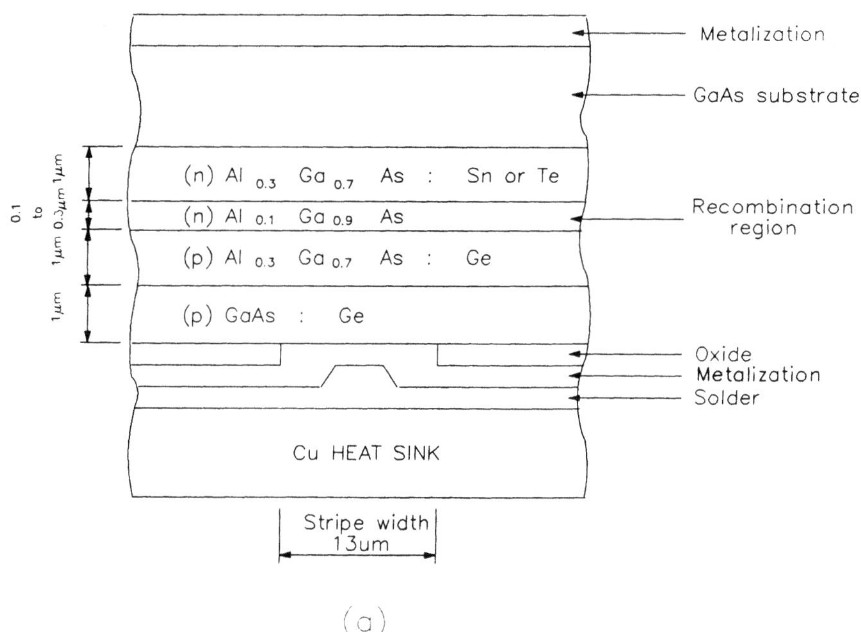

FIGURE 8. **Types of semiconductor lasers: (a) Simple oxide stripe DH; (b) classical buried heterostructure (BH) laser; (c) double channel planar buried heterostructure (DCPBH); (d) distributed feedback (DFB) laser.**

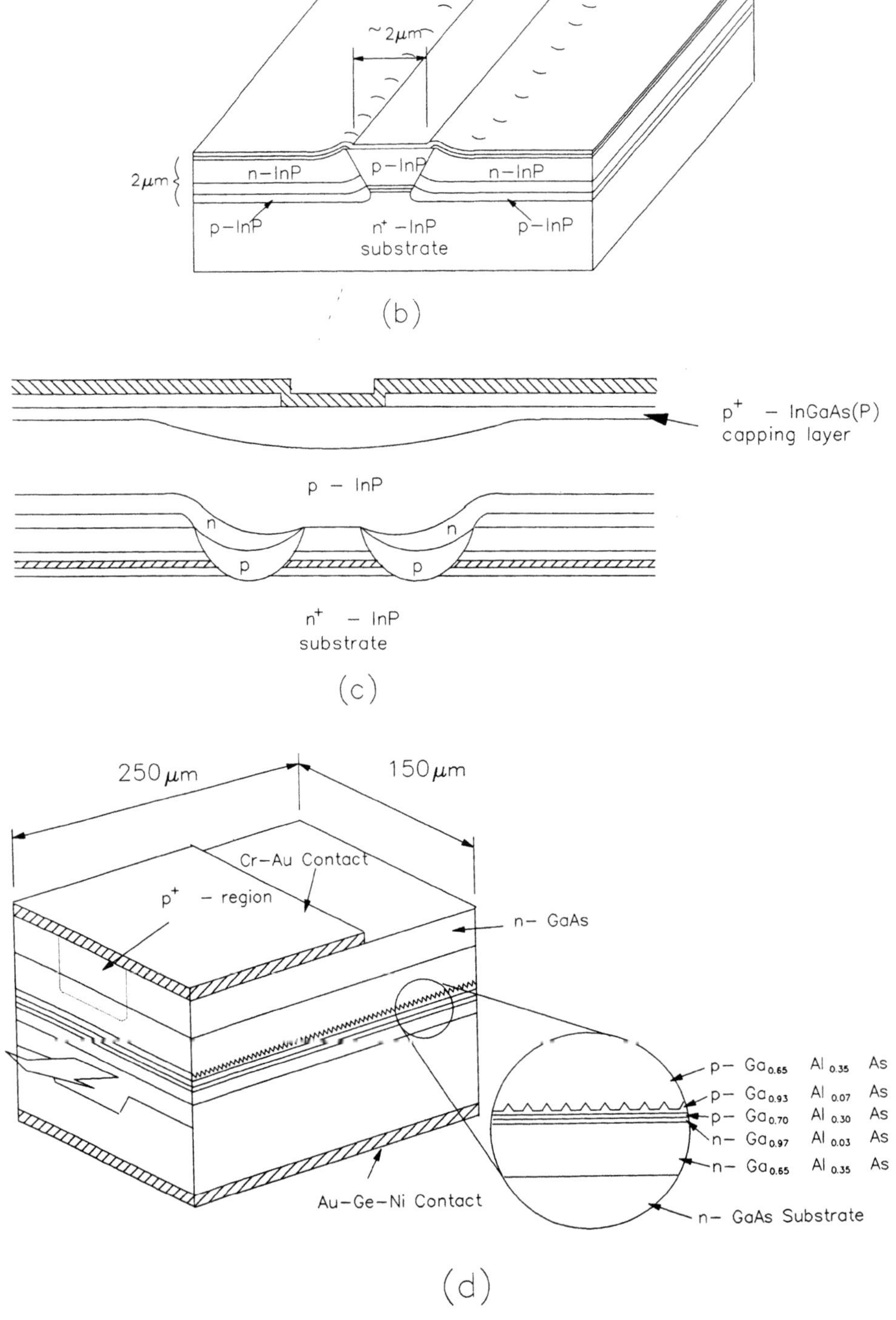

FIGURE 8. *Continued.*

The most important parameters to be optimized in this output characteristic depend to some extent on the application, but generally a reduction in the threshold current, an increase in the total light output, and an increase in the external quantum efficiency all lead to improved lasing devices.

There are a multitude of different designs that can be used to effect and improve these various parameters, and some examples are given in Fig. 8. Present day designs are almost exclusively the double heterostructure type with many variations used to constrain the device to operate in a single lateral mode. More sophisticated designs such as the GRINSCH (*graded index single confined heterostructure*) are used in the GaAs/AlGaAs system to achieve extremely low threshold current densities of 120 A/cm^2. These devices use a quantum well to aid carrier confinement and recombination and the light is confined to a wider optical area using the graded index design, to reduce optical damage at the laser facets, enabling high-power output from such lasers.

Single longitudinal mode operation can also be achieved in order to obtain extremely narrow linewidth emission by using diffractions gratings placed adjacent to the active layer of the laser. These gratings can replace the end mirrors of the laser by providing distributed optical feedback into the laser cavity, the device being known as the distributed feedback laser (DFB). There are many other variants to modify laser performance but by far the most common are the buried heterostructure (BH) and strip geometry lasers illustrated in Fig. 8.

4. OPTICAL DETECTORS

The simplest definition of an optical detector is a device that changes its properties by the absorption of light. Consequently there are a great variety of different types of optical detector ranging from thermal and pneumatic detectors to pyroelectric detectors. To describe each of these types of detector is well beyond the scope of this chapter, but by far the most important family of devices are semiconductor photodiodes. This type of optical detector is made simply by forming a p–n junction within the semiconducting material, applying a suitable reverse bias voltage to the device to create an electric field profile which separates the photogenerated electrons and holes. Absorption of light within the semiconductor causes the creation of electron–hole pairs, which results in a build up of electrical charge, which can be used appropriately by the external electronic circuit connected to the device. Designs of such photodetectors are relatively simple, as illustrated in Fig. 9a, but various parameters can be optimized depending on the most important characteristics required of the final device. Speed of response, for instance, is determined by the device capacitance, which in turn is governed by the thickness of the depletion region within the device and of the surface area. Therefore, for high-speed photodetectors one should design very small area devices having very low doped active regions, and low parasitic capacitance. If noise, on the other hand, is of paramount importance, then the device should be designed such that any surface leakage is minimized, by having a large band gap material on the surface. For example, a typical design used to overcome large surface leakage currents in low band gap InGaAs detectors is shown in Fig. 9b. The large band gap InP capping layer is

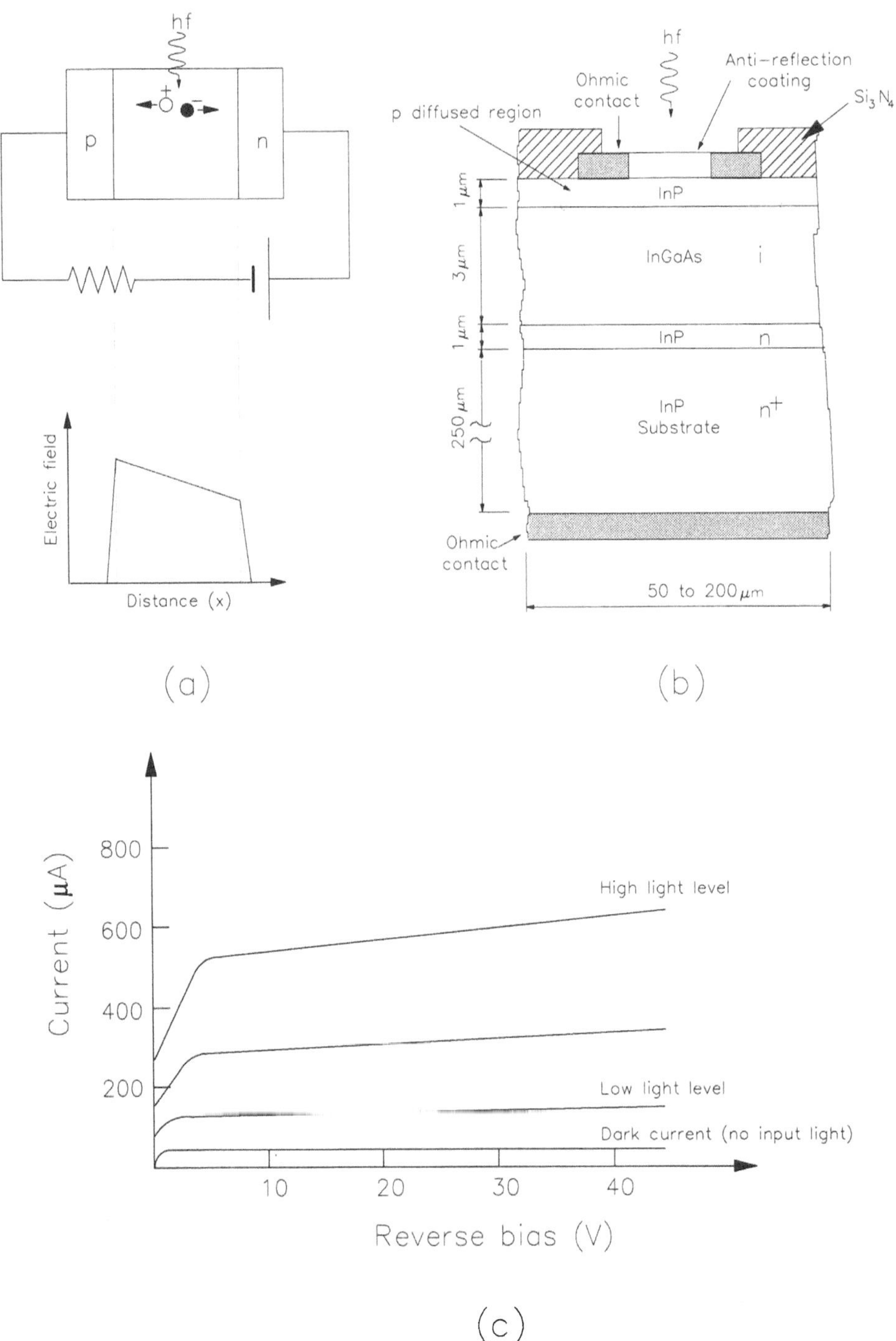

FIGURE 9. Schematic of (a) a simple *p–i–n* photodiode; (b) planar diffused InGaAs photodiode; (c) output characteristic of PIN photodiode.

extremely effective in reducing surface leakage current, allowing devices of much lower noise figures than equivalent germanium detectors to be made.

The cutoff wavelength of a detector and the photoresponse above the band gap energy is mainly a function of the material used. Figure 10 illustrates the various materials that are commonly used in semiconductor photodiodes together with their spectral photoresponse and cutoff wavelengths.

The trade-off between price and performance is an extremely critical one in determining the general applications of photodetectors. For instance, silicon is used in most photodetector applications where relatively low power can be tolerated but where price is of paramount importance, e.g., solar-powered consumer devices, large-area detectors for a multitude of applications, terrestrial solar panels, etc.

In applications where performance outweighs the need for very low cost, for example, in optical communication systems, then more sophisticated detector materials and structures are used. A good example of this is the replacement of germanium photodiodes by InGaAs-based devices for long-wavelength optical communications where the low dark current of InGaAs detectors (<1 nA) completely outweighs the lower-cost benefit of germanium photodiodes which have high dark currents (>100 nA). In other applications choice of materials can be somewhat limited, depending on wavelength range and other parameters of importance. In space satellite power supply applications, for instance, InP-based solar cells are becoming increasingly important because of their high radiation resistance, enabling greatly enhanced lifetimes.

In most detector structures, the absorption of a single photon of light leads to the creation of a single electron–hole pair. This means that the signal generated is normally amplified by electronic means in the external circuit. This can also lead, however, to the amplification of unwanted noise in the system. More sophisticated

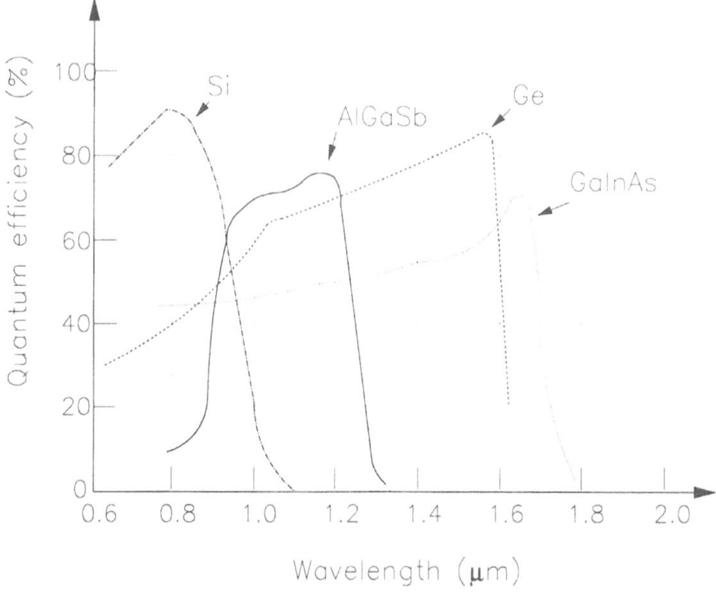

FIGURE 10.   Spectral responses of various detector materials.

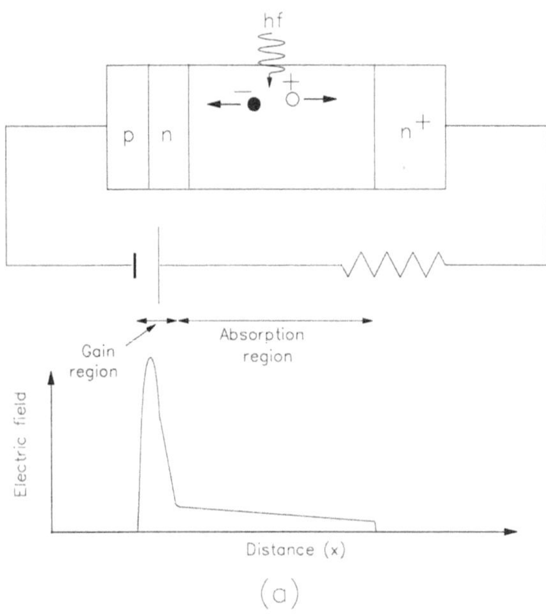

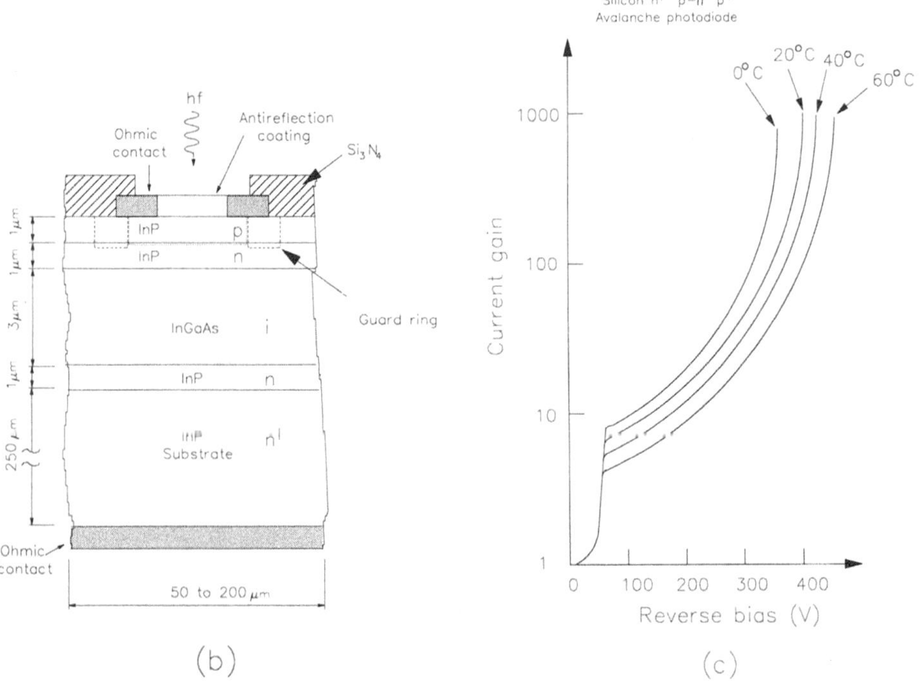

FIGURE 11.   Schematic of (a) avalanche photo diode (APD); (b) practical device; (c) output characteristic of APD.

detector structures have therefore been designed, which incorporate gain as well as the detection process, enabling greater sensitivity and signal-to-noise ratio to be obtained under certain circumstances. A good example of this type of device is the avalanche photodiode (APD) illustrated in Fig. 11. In this structure part of the device is biased such that the breakdown electric field of the semiconductor is exceeded. Electron–hole pairs created by the absorption of a single light photon are then subjected to avalanche multiplication leading to many electron–hole pairs being generated for the absorption of a single photon, i.e., internal multiplication within this device. The general drawback of avalanche photodiodes, however, is their much more sophisticated structure, resulting in significantly lower production yields and therefore higher cost and also their limited speed of response in certain frequency ranges and high voltage requirements.

5. WAVEGUIDE COMPONENTS

Semiconductor waveguides for the manipulation of optical signals are becoming increasingly important. An optical waveguide is simply a materials structure that constrains light to travel in a certain path. This is normally achieved by designing the structure such that there is a refractive index change between the material and its surroundings, thus confining the light by total internal reflection. The best example of this is probably silica fibers, used in optical fiber communications, where the light is constrained completely to the very narrow core of the optical fiber. The entire fiber is made up from glass but the refractive index of the core region is changed by the addition of impurity species. This then has a refractive index slightly higher than the cladding glass, with the result that any light launched into the optical fiber is confined to the core region.

Exactly the same operating principle applies to semiconductor devices, and in fact the active region of a laser is a good example of a semiconducting waveguide. In this case nature has been exceedingly kind in that the semiconducting materials generally used as laser active layers have (compared to adjacent layers) a higher refractive index, enabling confinement of the light to the active region while at the same time having a lower bandgap energy, enabling the confinement of electrical carriers to the active region.

There are, however, many other types of waveguide (see Chapter 22 by Goodwin) and a typical semiconductor guide is made simply by fabricating a rib-type structure shown in cross section in Fig. 12. Light passing through such a waveguide structure can also be strongly modified by the application of an electric field applied perpendicularly to the direction of the waveguide structure. Application of an electric field affects the refractive index of the semiconductor by the well-known electro-optic (Pockels) effect. This can be used in active waveguide devices in a variety of ways. Phase modulators, for instance, are obtained simply by applying an electric field of varying size to the waveguide, resulting in a change of refractive index with time and thus phase modulation of the light signal. Light signals can also be modulated in intensity and split into a variety of forms by the use of waveguide couplers and the aforementioned modulators. Some examples of these structures are shown in Fig. 12. Intensity modulation is achieved by splitting a light signal and applying

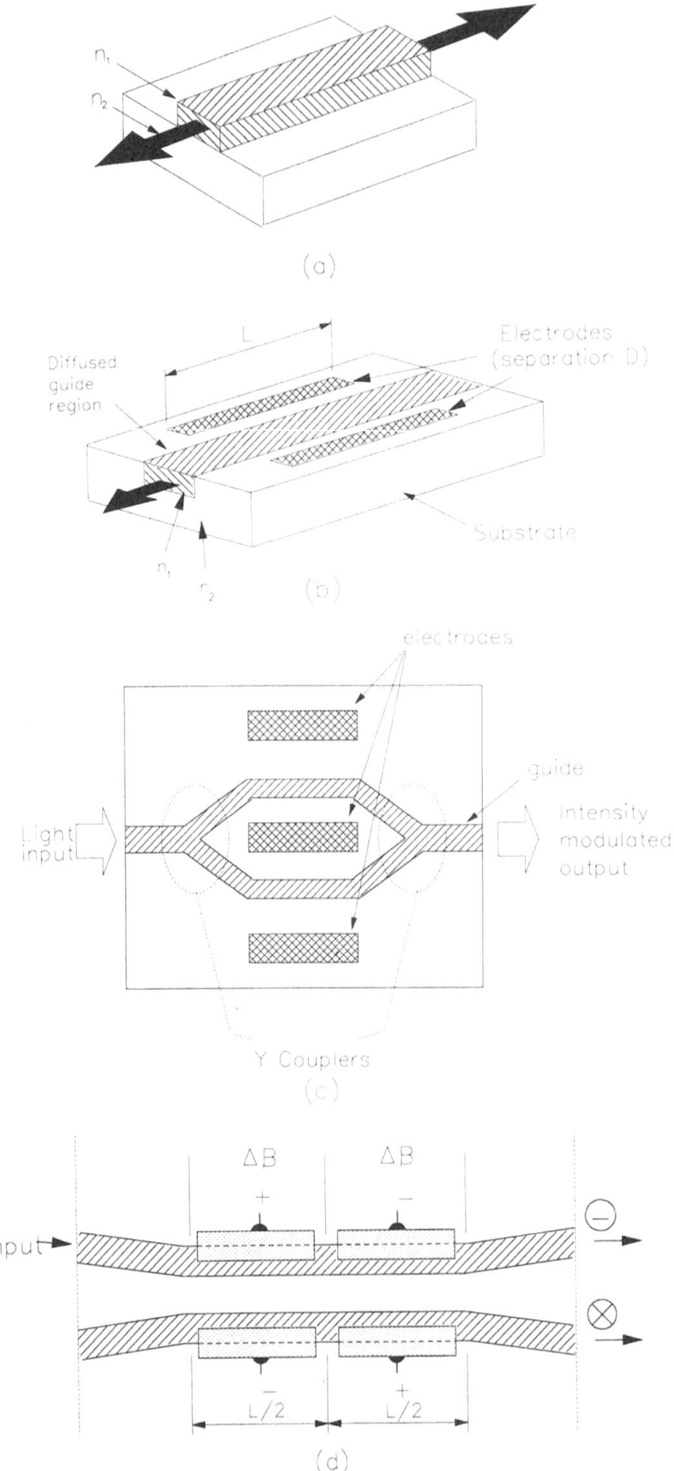

FIGURE 12. Typical waveguide designs: (a) Simple rib waveguide; (b) waveguide phase shifter; (c) Mach–Zehnder interferometer (intensity modulation); (d) stepped APD directional coupler switch.

modulation of the phase to one arm (as shown in Fig. 12c). When the two signals (one with phase modulation the other untouched) are combined via a Y coupler, constructive or destructive recombination takes place, which is modulated by the application of the electric field.

Critical parameters in the design of these waveguide components are:

a. The achievement of minimal loss within the material, which is governed by free carrier absorption, light scattering, and loss of light into the substrate.
b. The magnitude of the Pockels effect, enabling the devices to operate at as low a voltage as possible.
c. Ease of fabrication.

Many of these waveguide components, in addition to forming useful discrete passive and active devices, are also of great importance for optoelectronic integrated circuits.

6. OPTOELECTRONIC INTEGRATED CIRCUITS

Many systems today that utilize optical devices also require electronic signal-processing devices to manipulate the information contained in the light signals. Presently this is done by hybrid technology by combining discrete devices that perform different functions on a ceramic substrate (e.g., a discrete detector device with a GaAs discrete integrated circuit). By interconnection of these discrete components by thin bond wires the overall circuit function is effected. The most significant drawbacks of this methodology are firstly that the system performance can be compromised by the introduction of parasitic capacitance, inductance, and resistance, and secondly that, because many components are interconnected by thin bond wires, the overall circuit is less reliable than a fully integrated circuit. In order to access the huge power of light, one ultimate goal of many workers is to eventually achieve a significant level of signal processing by optical means alone, using light to influence the optical properties of materials and thereby use light itself to manipulate other light beams. Recently there have been clear demonstrations of such devices with the aim of using these so-called optical integrated circuits (OICs) in all optical computers, thereby taking advantage of the fundamental speed of light.

These types of systems will, however, also require electro-optical conversion devices, and a highly significant area of research and development at present is the integration of electronic and optoelectronic devices on a single semiconductor substrate. This type of circuit is known as an optoelectronic integrated circuit (OEIC) (Fig. 13), and over the last two years significant progress has been made in this area with demonstrations of OEICs that have comparable performances to their hybrid equivalents. At present, these types of circuits have major applications in the area of optical fiber communications where vast amounts of information transmitted using light need to be processed by electronic circuitry. However, there are a host of other applications that will benefit from the development of such circuits.

The materials requirements for OEICs are many and varied and depend to a large extent on the final application. For optical fiber communications, where InP-based devices such as InGaAsP (1.3 and 1.55 μm) lasers and InGaAs detectors

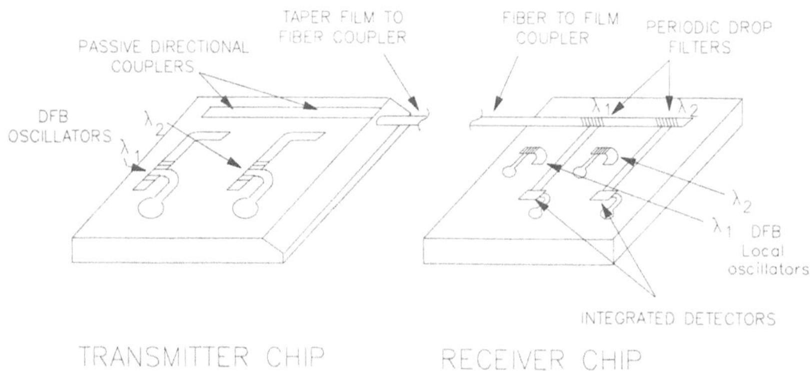

FIGURE 13. Optoelectronic integrated circuitry.

are the fundamental optoelectronic devices, work is focusing on the integration of these optical devices with high-speed electronic processing circuitry. Ultimately, this will be achieved by the development of very high-speed ICs made from materials lattice matched to InP such as GaInAs (which has an extremely high mobility) and AlInAs, which has highly compatible electronic characteristics for the development of extremely high-speed ICs. Over the last year or so there have been impressive demonstrations of this technology, even though the processing technology for these circuits is very immature. Consequently, work on OEICs over the last two to three years has been focused on the integration of InP-based optoelectronic devices with the more mature GaAs based ICs. Although ultimately not the best combination, impressive progress has been made using such integration. Another area of significant interest is that of GaAs on silicon. The materials technologies for both of the aforementioned systems are difficult because of the large difference in lattice parameter between the various materials. The application for GaAs-based devices such as those of lasers and LEDs used in compact disk players or data storage devices with the extremely mature electronic processing capabilities of silicon integrated circuits.

Clearly materials and associated technology is of fundamental importance in realizing the above combinations of materials and circuit functions, and areas that are receiving a great deal of attention in the development of OEICs are selective area epitaxy (SAE), where epitaxial material is deposited only in certain areas of a single-crystal substrate—the rest of the substrate usually being masked with a refractory material. Other technologies important to the full realization of OEICs are ion beam etching for the formation of laser cavities and waveguide components, the deposition of insulating materials, and metalization methodology. Increasingly, because many of these processes are required to be done in extremely clean conditions and limiting the exposure of critical interfaces to air and oxygen is thus important, fundamental research is being conducted into the integration of many of these deposition technologies into one multipurpose deposition system. Such a system would allow the manufacture of OEICs without ever exposing any of the interfaces to the outside environment. OEICs remain a considerable materials challenge, but extremely encouraging progress is presently being achieved.

7. CONCLUSIONS

Optoelectronic devices are playing an increasingly important role in commerce, industry, and medicine. In these fields, the development of new semiconductor materials and structures has played and will continue to play a dominant role. Increasingly, as the technology improves, systems designers are able to choose the characteristics of devices they would ideally like for a system and these are engineered by a variety of means. In addition, new effects and device characteristics are continually being discovered as the critical dimensions of different materials compositions that can be deposited together become ever smaller. This is allowing completely new devices and circuits to be conceived, allowing more efficient and novel circuit functions to be achieved.

One of the most important materials systems emerging for optoelectronics systems is that of InP and related compounds that are lattice matched to it, including InGaAsP for lasers and LEDs, InGaAs for detectors and electronic devices, and AlInAs for electronic circuits. These materials are critical in the rapidly emerging technologies of optical fiber communications and information technology. GaAs and AlGaAs will continue to be important materials since the technology is now well established such that lasers for commercial purposes (for example, compact disk players) are available at very low cost. Obviously as cost is decreased, the range of application is increased, especially in consumer areas. One materials system receiving a great deal of attention at present is that of GaInAlP lattice matched to GaAs for visible laser diode and LED applications. Here the areas for exploitation are vast, from data storage devices to a host of display technologies.

Combinations of the above materials and their integration into new device structures incorporating quantum wells, strained layer super lattices, and other quantum effect devices are playing increasingly important roles in the attainment of novel optical and electronic functions. This ongoing development of "band gap engineering" is continuing to make fundamental changes to the designs of the new systems.

The field of optoelectronics is an extremely exciting one, and as materials and processing technologies further mature, we will find optoelectronic devices and systems playing an ever increasingly important role in our everyday lives.

8

Thermodynamics and Defect Chemistry of Compound Semiconductors

D. T. J. Hurle

1. INTRODUCTION

An understanding of the conditions of thermodynamic equilibrium between a crystal and the nutrient phase from which it is to be grown is a valuable asset for the crystal grower for a number of reasons. Firstly, it enables one to determine the range of experimental conditions under which the required crystalline phase is the thermodynamically most stable phase. Secondly, where the purpose is to grow an alloy crystal (such as a ternary III–V compound), it enables one to predict the composition of the nutrient phase required to yield a given composition of crystal. Thirdly, it provides a basis for rationalizing and understanding dopant segregation behavior, such as the dependence of the dopant segregation coefficient on crystal stoichiometry and on doping level. Fourthly, it provides a framework for understanding native point defect incorporation and nonstoichiometry in compound crystals.

This list is far from comprehensive; it is simply a list of the four topics to be described in this chapter, illustrated with specific reference to the growth of gallium arsenide and to alloys containing it. The purpose is to give the reader a feel for the utility of the thermodynamics of multicomponent phase equilibria in this field rather than to make him an expert in it.

2. ELEMENTS OF MULTICOMPONENT PHASE EQUILIBRIA

For the necessary grounding in the subject the reader is referred to the very adequate literature of the subject. Books by Swalin, Guggenheim, Prigogine and Defay, and Kroger are particularly recommended.[1–4]

We confine ourselves here to a statement of the basic results needed in the rest of the chapter. This whole subject is firmly founded on the outstanding work at the

---

D. T. J. Hurle ● Royal Signals and Radar Establishment, St. Andrews Road, Malvern, Worcestershire WR14 3PS, U.K.

end of the last century by a single man—J. Willard Gibbs. His collected works on thermodynamics are available as a book.[5]

2.1. Gibbs's Phase Rule

Gibbs showed that the number of phases (n) that can coexist under given conditions of temperature (T), total pressure (P), and molar concentrations of the components $(x_i : i = 1, 2, \ldots, c)$ of a system containing c components is given by

$$n = c + 2 - f \tag{1}$$

where f is the number of degrees of freedom, i.e., the number of independent variables (such as T, P, and x_i) that can be altered while preserving the n phases in equilibrium.

Thus, for the crystal-melt equilibrium of a unary system (i.e., an element such as silicon: $c = 1$), there is one degree of freedom. For example, we can change the system pressure and still maintain crystal-melt equilibrium, the equilibrium temperature (that is, the melting point of the crystal) changing slightly with pressure, as given by the well-known Claperyon equation. (See, for example, Guggenheim,[2] p. 119.)

In a two-component system, such as Ga–As, the number of components is $c = 2$ and therefore the crystal-melt equilibrium is characterized by two degrees of freedom. The molar fractions of gallium and arsenic in the melt x_{Ga} and x_{As} are related by

$$x_{Ga} + x_{As} = 1 \tag{2}$$

so that there is only one independent concentration.

If we consider the three variables—pressure (P), absolute temperature (T), and molar fraction of gallium (x_{Ga})—then we may, within limits, vary any two of these and this will define a value for the third when thermodynamic equilibrium exists between solid and liquid phases. For example, if we fix the total system pressure and the temperature there is then a specific value of x_{Ga} for which the solid and liquid coexist in equilibrium (see Fig. 1). Actually, in the case that we have chosen, there are *two* values of x_{Ga} corresponding, respectively, to equilibrium with the gallium-rich liquidus curve and the arsenic-rich liquidus. This makes the point that the equilibrium conditions are not necessarily single valued. If we lower T down to T_e, the temperature of the gallium-rich eutectic, then we now have a liquid phase coexisting with *two* solid phases (gallium arsenide and metallic gallium). The number of phases (n) has therefore increased to three and the number of degrees of freedom, by applying the phase rule [Eq. (1)], is reduced to one. Thus the eutectic temperature and eutectic composition are fixed for any chosen total pressure and, in fact, vary only slowly as the total pressure is changed.

2.2. Pressure–Temperature Equilibrium

In Section 2.1 above we considered solid–liquid phase equilibria where a hydrostatic pressure P was imposed on the system in the absence of a gaseous

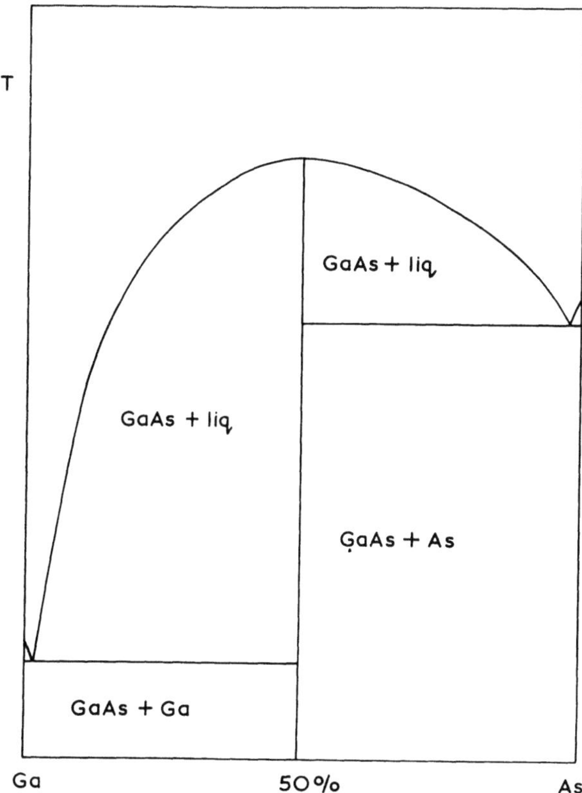

FIGURE 1. Schematic representation of the gallium–arsenic solid–liquid phase diagram.

phase. Full equilibrium involves equilibrium with the gaseous phase also. The question posed is "what are the partial vapor pressures of gallium and arsenic that are in equilibrium with gallium-arsenic liquid along the liquidus curve?." The answer to this is slightly complicated by the fact that the arsenic pressure is composed of the sum of the partial pressures of several differing arsenic species; principally As_2 and As_4. In Fig. 2 we show the dependence of the gallium and As_2 partial pressures along the liquidus. To calculate the As_4 pressure we apply the law of mass action to the equilibrium:

$$2As_2(g) = As_4(g)$$

viz.,

$$K_{As} = p_{As_4}p_{As_2}^{-2} \tag{3}$$

where p_i $(i = As_2, As_4)$ are the partial pressures of the constituent gases.

The mass action constant K_{As} has been determined thermochemically (data reviewed by Hurle[6]):

$$\ln K_{As} = 27.3 \times 10^3/T - 19.8 \tag{4}$$

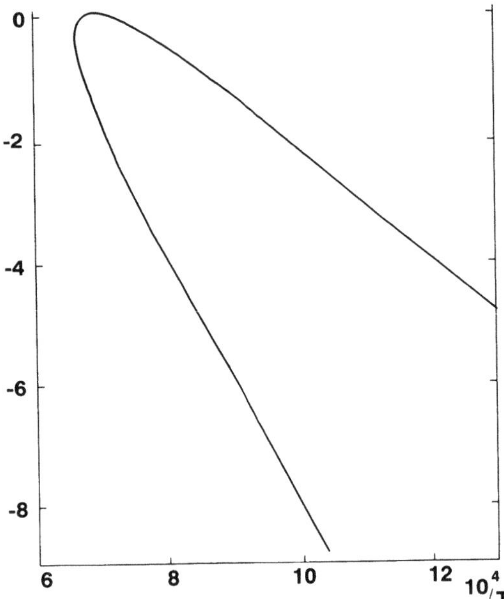

FIGURE 2. Partial pressure of the arsenic dimer in equilibrium with the gallium–arsenic liquidus. The upper portion of the curve corresponds to the arsenic-rich liquidus.

If no inert gas is present (e.g., if the materials were sealed in an ampoule and then heated), then the total system pressure P is

$$P = p_{Ga} + p_{As_2} + p_{As_4} \tag{5}$$

2.3. Solid–Liquid Equilibria in Multicomponent Systems

In multicomponent systems, full thermodynamic equilibrium implies equality of the chemical potentials of each of the components in each of the coexisting phases, viz.,

$$\mu_i^\alpha = \mu_i^\beta = \mu_i^\gamma \qquad \text{etc.} \tag{6}$$

where $i = 1, 2, \ldots, c$ are the components and where $\alpha, \beta, \gamma \ldots$ represent the coexisting phases. If we know the dependence of these chemical potentials upon the temperature, pressure, and composition parameters then we can compute the equilibria by solving the set of Eq. (6).

The chemical potential of a component is its partial molar-free energy and the above equilibrium relations follow from minimization of the total free energy of the system. For the condensed phases we can relate the chemical potential of each component to its composition by the relation

$$\mu_i = \mu_i^0 + RT \ln \gamma_i x_i \tag{7}$$

where R is the gas constant and $\mu_i^0\,(T, P)$ is the partial molar-free energy of the ith component in its reference state at temperature T and pressure P. γ_i is an activity coefficient expressing the deviation of the solution behavior from ideal random mixing ($\gamma_i = 1$). Thus a knowledge of the chemical potentials requires a knowledge of the reference state values (μ_i^0) and of the activity coefficients (γ_i).

Where the condensed phase is a compound of narrow phase extent, such as a III–V compound, then it is less appropriate to think of the chemical potential of, say, gallium in gallium arsenide than of the partial molar-free energy of the compound itself. We therefore introduce the concept of the chemical potential of the compound (e.g., μ_{GaAs}^S, where the superscript denotes the solid state). Thus we have:

$$\mu_{GaAs}^S = \mu_{Ga}^L + \mu_{As}^L \tag{8}$$

which defines the solid–liquid phase equilibria (the superscript L denotes the liquid state).

If we go further and consider a mixed III–V compound (such as, for example, the ternary alloy GaAlAs) then we will need to take account of the non-ideality of the solid solution (of GaAs in AlAs). We therefore define an activity coefficient (for example, γ_{GaAs}^S) for the compound as follows:

$$\mu_{GaAs}^S = \mu_{GaAs}^{SO} + RT \ln \gamma_{GaAs}x \tag{9}$$

where x represents the mole fraction of gallium arsenide in the solid solution, i.e., $Ga_xAl_{1-x}As$.

2.4. Representation of the Activity Coefficients

In principle, the activity coefficients could be measured throughout the parameter space using standard thermochemical methods. In practice, the daunting magnitude of this task leads one to use a much more limited set of experimental data to specify the parameters of an appropriate model solution. Models of varying sophistication are to be found in the literature (see, for example, Prigogine and Defay[3] and Guggenheim[2]). Here we describe the use of one of the simplest and one that is widely used to describe III–V mixed compounds—the regular solution model.

The free energy of mixing of a solution can be split into its enthalpic and entropic terms:

$$\Delta G^M = \Delta H^M - T\Delta S^M \tag{10}$$

In the regular solution model the entropy of mixing is supposed to be just the ideal configurational entropy of mixing:

$$\Delta S^M = R[x \ln x + (1 - x)\ln(1 - x)] \tag{11}$$

The enthalpy of mixing is taken to be the appropriate summing of the nearest neighbor bond energies, giving the result

$$\Delta H^M = x(1 - x)\Omega \tag{12}$$

where

$$\Omega = ZN^0[H_{AC} - \tfrac{1}{2}(H_{AA} + H_{CC})] \tag{13}$$

where H_{AC}, H_{AA}, and H_{CC} are the nearest neighbor bond energies for the AC mixture. Z is the number of nearest neighbors and N^0 is Avogadro's number. If the bond energies are all equal then $\Omega = 0$, which is an ideal solution. Ω is determined in practice by fitting to experimentally measured liquidus and solidus curves.

The activity coefficient is related to the enthalpy of mixing by

$$RT \ln \gamma_i = (1 - x_i)^2 \Omega \tag{14}$$

If the data can be fitted to a constant value of Ω then the solution is said to be *strictly* regular. A better representation of III-V mixtures is obtained by taking Ω to be linearly dependent on temperature, i.e., of the form

$$\Omega = a - bT \tag{15}$$

The above enables us to fully describe the solid–liquid phase equilibria of a single binary III-V compound. To extend this to multicomponent alloys we need expressions for activity coefficients of component group III and group V elements in the ternary liquid and of the activity coefficients of the binary compounds in their solid solution. The latter are obtained by using the regular solution theory just described but applied to the two binary compound end members (e.g., GaAs and AlAs), values for the interaction parameter Ω again being obtained by fitting to the pseudobinary solid-liquid phase equilibria.

The activity coefficients of the elements in the ternary liquid are obtained from known values of those for the three binary solutions (AB, BC, CA in the binary solution A–B–C). Extension of the regular solution theory gives for the activity coefficient of a component in the ternary liquid

$$RT \ln \gamma_i = (1 - x_i)(\Omega_{ij} x_j + \Omega_{ik} x_k) - \Omega_{jk} x_j x_k \qquad (i, j, k = 1, \ldots, 3; \quad i \neq j \neq k) \tag{16}$$

The ternary solid–liquid phase equilibria are obtained by inserting Eq. (16) into Eq. (6) and solving numerically.

3. SOLID–LIQUID PHASE EQUILIBRIA IN TERNARY III–V COMPOUNDS

The theory outlined in Section 2 above has been widely applied to binary, ternary, and quaternary mixtures of the III-V systems. In this section we briefly illustrate its use in the construction of a ternary diagram such as $Ga_x Al_{1-x} As$.

An example of a pseudobinary system is shown in Fig. 3. The heavy line is a fit to the data of Steininger[7] using strictly regular solution theory. The thin liquidus line is the experimental one. The fit is obtained for a small negative value ($\Omega_l = -0.2$ kcal mol^{-1}) for the liquid interaction parameter and a relatively large positive

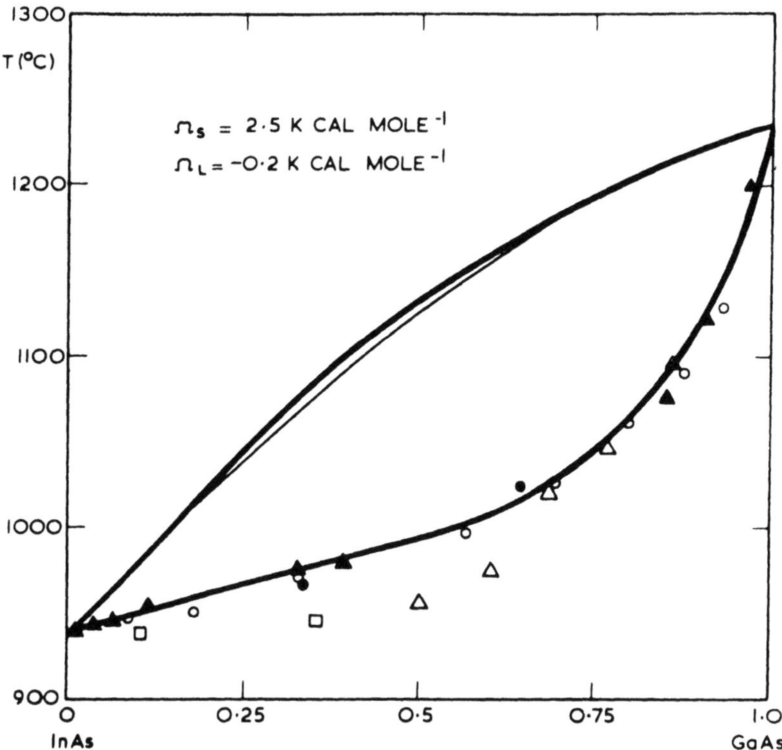

FIGURE 3.   The InAs–GaAs pseudobinary phase diagram. Thick lines show the theoretical fit to the experimental liquidus (thin line) and solidus (data points). Experimental data are taken from Ref. 7.

value (Ω_s = 2.5 kcal mol^{-1}) for the solid. This large positive interaction energy represents a repulsive interaction between the gallium and indium in the lattice and leads to the "fatness" of the lens shape of the phase diagram. For sufficiently large repulsive interactions the solidus becomes near horizontal and solid state immiscibility between the two compounds will set in.

A schematic ternary phase diagram for a III–III–V ternary is shown in Fig. 4. The could be, for example, the $Ga_xIn_{1-x}As$ system or the $Ga_xAl_{1-x}As$ system. From this diagram one can predict, for a chosen liquid phase epitaxial growth temperature, the composition of the melt needed to grow a solid solution of a given composition. To do this we slice the diagram into a number of horizontal sections (i.e., each section corresponding to a fixed temperature). The intersection of the liquidus surfaces with these isothermal sections can then be drawn on a single triangular graph, as shown in Fig. 5, for the system $Ga_xAl_{1-x}As$. (The As-rich section of the diagram is omitted.) The light lines are the liquidus intersections at the labeled isotherms and the heavy lines are lines of constant values of $(1 - x)$. Thus, to grow the composition $Ga_{0.05}Al_{0.95}As$ at a temperature of 1200°C requires a melt of approximate composition $Ga_{0.7}Al_{0.2}As_{0.1}$. The very large difference in the Ga to Al ratio in the liquid to that in the crystal is evident from this example and shows how necessary it is to have a model phase diagram as a starting point for liquid phase epitaxial growth of ternary semiconductors.

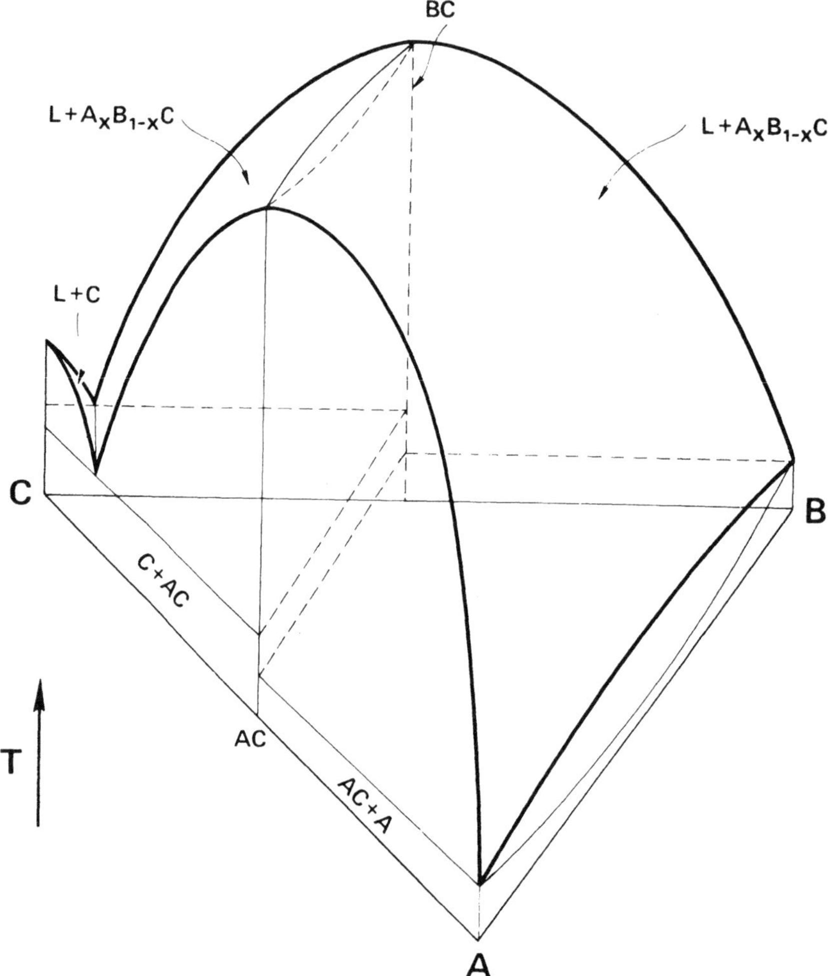

FIGURE 4.   Schematic representation of a ternary III–V phase diagram (taken from Stringfellow[15]).

4. SOLID–GAS PHASE EQUILIBRIA IN MULTICOMPONENT III–V COMPOUNDS

The same principles as outlined above apply to establishing the solid–gas equilibria relevant to consideration of chemical vapor epitaxy. The additional complication is that we have to consider chemical reactions which take place.

We will illustrate this with a simple case: the growth of gallium arsenide by the chloride transport process. In this process arsenic trichloride gas in high dilution in a hydrogen carrier is made to flow over a heated boat of molten gallium. The arsenic trichloride reacts with the gallium to form principally gallium monochloride and arsenic dimers, both existing as gaseous species. If this gaseous mixture is now passed over a gallium arsenide substrate held downstream at a slightly lower temperature, then deposition of gallium arsenide onto the substrate can occur.

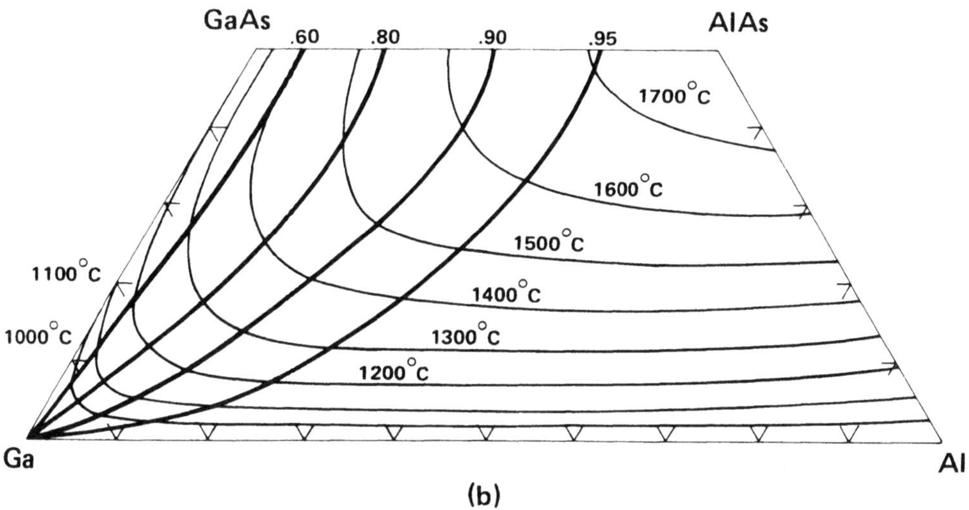

FIGURE 5. Section of the Ga–Al–As ternary diagram showing liquidus isotherms (thin lines) and lines of constant solidus composition (x) (heavy lines). From Ref. 15.

The pick-up reaction can be written

$$3Ga + AsCl_3 = 3GaCl + \tfrac{1}{2}As_2 \qquad (17)$$

[The true chemistry is more complicated than this, but Eq. (17) will serve to illustrate the principles.]

Applying the law of mass action to this reaction we have a mass action constant

$$K_{pu} = \exp(-\Delta G_{pu}/RT) = p_{As_2}^{1/2}p_{GaCl}^3/p_{AsCl}^3 \qquad (18)$$

where the p_i are the partial pressures of the gaseous species. The arsenic and the chlorine is conserved in this reaction so that we can write

$$3p_{AsCl_3}^0 = 3p_{AsCl_3} + p_{GaCl}$$

and

$$p_{AsCl_3}^0 = p_{AsCl_3} + 2p_{As_2} \qquad (19)$$

where the superscript 0 denotes the input concentration and the nonsuperscripted variables denote the values after reaction. The mass action constant K_{pu} can be readily obtained from the known tabulated values for the free energy of formation of the various species. [From which the free energy change for the reaction (ΔG_{pu}) can be calculated.]

The (simplified) deposition reaction can be written as

$$2GaCl + As_2 + H_2 = 2GaAs + 2HCl \tag{20}$$

A mass action relation can be written for this reaction also [equivalent to Eq. (18)] and also a conservation equation for the arsenic and chlorine species [equivalent to Eq. (19)].

Equations (18)–(19) can be solved simultaneously to yield the concentrations of GaCl, As_2, and $AsCl_3$ downstream of the pickup reaction. These concentrations provide the input data for the deposition reaction where the equivalent set of three equations can be solved to yield the concentrations of GaCl, As_2, and HCl downstream of the deposition reaction. The technique can readily be extended to the growth of mixed III–V compounds by including the activity (that is, the product of activity coefficient and concentration) of each binary component of the multicomponent crystal in the expression for the mass action constant.

An example is given in Fig. 6. This shows the composition of $In_{1-x}Ga_xAs$ obtained by chloride transport at a growth temperature of 1023 K as a function of

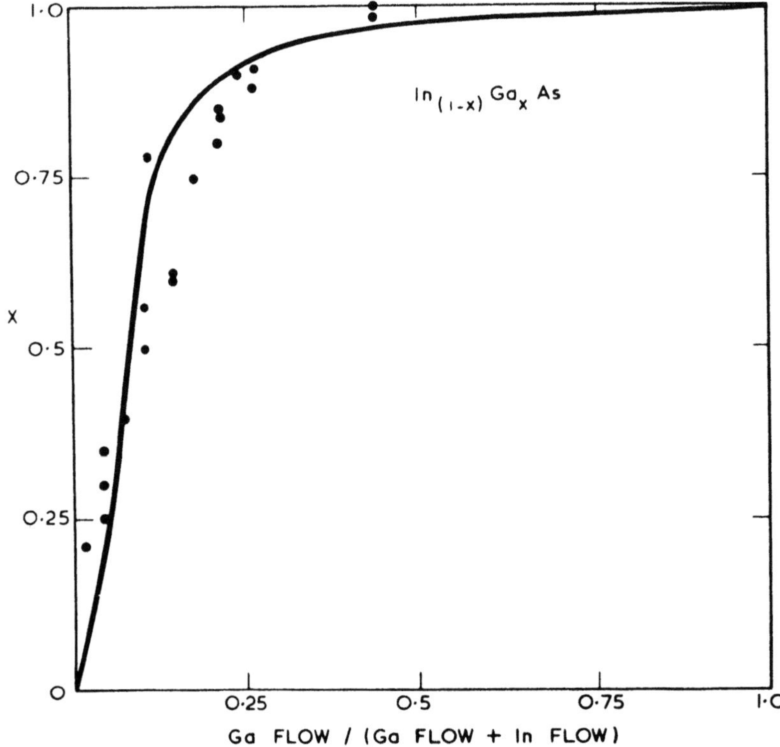

FIGURE 6. Composition of an $In_{1-x}Ga_xAs$ ternary alloy as a function of the ratio of the flow rates of gallium monochloride to gallium plus indium monochlorides. The solid line is the calculated curve, the data points are taken from Ref. 8. Data relate to a reactor working at atmospheric pressure with substrate temperature of 1023 K with a chlorine-to-hydrogen ratio of 10^{-2}.

the ratio of flow rates of gallium monochloride to gallium plus indium monochlorides for given input dilution. The points are the experimental data of Conrad *et al.*[8]

5. NATIVE POINT DEFECTS IN COMPOUND SEMICONDUCTORS

In Fig. 1 we have drawn the gallium arsenide solidus as having zero phase extent, i.e., as a vertical line. That would imply that a crystal grown at some temperature between the melting point and the arsenic eutectic temperature would have the same composition, irrespective of whether it was grown from a gallium-rich or arsenic-rich solution. In reality, the composition of the crystal will vary slightly, depending on the melt composition from which it is grown; that is to say, the solidus has a finite phase extent. Since this phase extent is very small we can consider the crystal as having small concentrations of native point defects. Suppose for the moment that the only point defects present are arsenic atoms on interstitial sites (As_i) and arsenic vacancies (V_{As}). The amount by which the crystal deviates from stoichiometric composition will be given by the difference in these two point defect concentrations, viz.,

$$\delta = [As_i] - [V_{As}] \tag{21}$$

The total number of these point defects will increase with increasing temperature since their formation by the Frenkel reaction

$$0 = As_i + V_{As} \tag{22}$$

is thermally activated. The law of mass action can be applied to Eq. (22), which shows that the product of the concentration of interstitials and vacancies is equal to the Frenkel mass action constant (K_{As}).

$$[As_i][V_{As}] = K_{As} \tag{23}$$

The equilibrium value of δ will be determined by the activity of arsenic in the ambient phase. If this is gaseous then we can write the following incorporation reaction:

$$\tfrac{1}{2}As_2(g) = As_i \tag{24}$$

The mass action constant for this reaction is

$$K_{As2i} = [As_i]p_{As_2}^{-1/2} \tag{25}$$

where p_{As_2} is the partial pressure of arsenic dimers. Noting from Fig. 2 that the arsenic dimer pressure can differ by several orders of magnitude between the gallium- and arsenic-rich liquidus surfaces, we see that the arsenic interstitial concentration will also vary by a large ratio. However, since the product of the interstitial and vacancy concentrations is a constant at a given temperature [Eq. (23)] then the vacancy concentration is reciprocally related to the interstitial concentration. In consequence the value of δ varies strongly, being >0 on the arsenic-rich liquidus and <0 on the gallium-rich liquidus.

This is shown schematically in Fig. 7, which is a much expanded portion of Fig. 1 in the region of the congruent melting point. The first point to notice is that the congruent point is not, in general, at the stoichiometric composition. In the case of gallium arsenide, the congruent point is likely to be to the gallium-rich site of stoichiometry but some uncertainty on this point remains at present. Suppose we grow a crystal from a gallium-rich melt at a temperature T_g yielding a crystal that is also gallium-rich, as shown in Fig. 7, then we see that as the crystal is cooled through the temperature T_s the solid solution of the excess arsenic vacancies become supersaturated. If the temperature T_s is sufficiently low for the native defects to be immobile within the lattice then this supersaturation is preserved and the crystal is, at room temperature, in a metastable state. If, however, the defects are sufficiently mobile and the cooling rate sufficiently slow then precipitation processes can occur so that, at room temperature, the crystal lattice itself is nearly stoichiometric but embedded within it are precipitates associated with the excess component.

Thus far we have assumed the presence of only interstitials and vacancies on the arsenic sublattice. In a real crystal, of course, such defects will also be present on the gallium sublattice. Moreover, a few gallium atoms will become misplaced onto arsenic sites and arsenic atoms onto gallium sites. Such antisite defects are attracting much interest since they appear to control the electrical properties of undoped semi-insulating gallium arsenide.

How can these (small) concentrations of native defects in the crystal be determined? Bublik and co-workers[9] have made careful density and lattice parameter measurements of as-grown crystals. The density measurements yield a value for the mass per unit volume, while the lattice parameter measurements yield the volume per unit cell. The quotient of this is therefore the mass per unit cell, which may be compared with the theoretical mass of an ideal crystal consisting of a perfect stoichiometric lattice. Some uncertainty arises in obtaining a value for the mass per unit cell of the ideal lattice since this requires, to an accuracy at the 10 ppm level, the atomic mass of gallium, which depends on its isotopic abundance ratio. A further uncertainty arises in our knowledge of the value of Avogadro's number. Bublik's

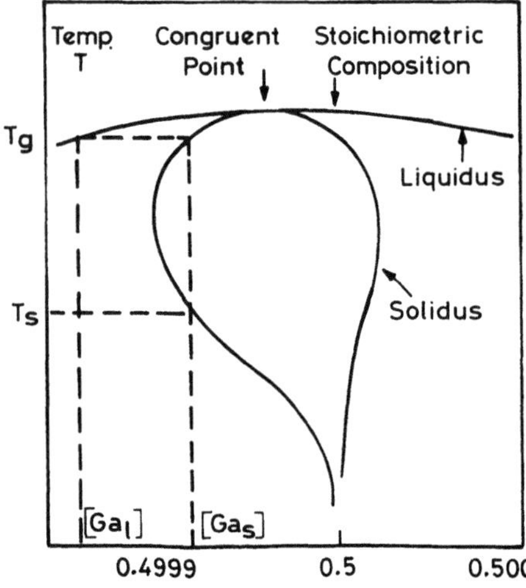

FIGURE 7. Schematic representation of a conceptually possible solidus curve for the Ga–As system. The congruent melting point is shown to be to the gallium-rich side of stoichiometry.

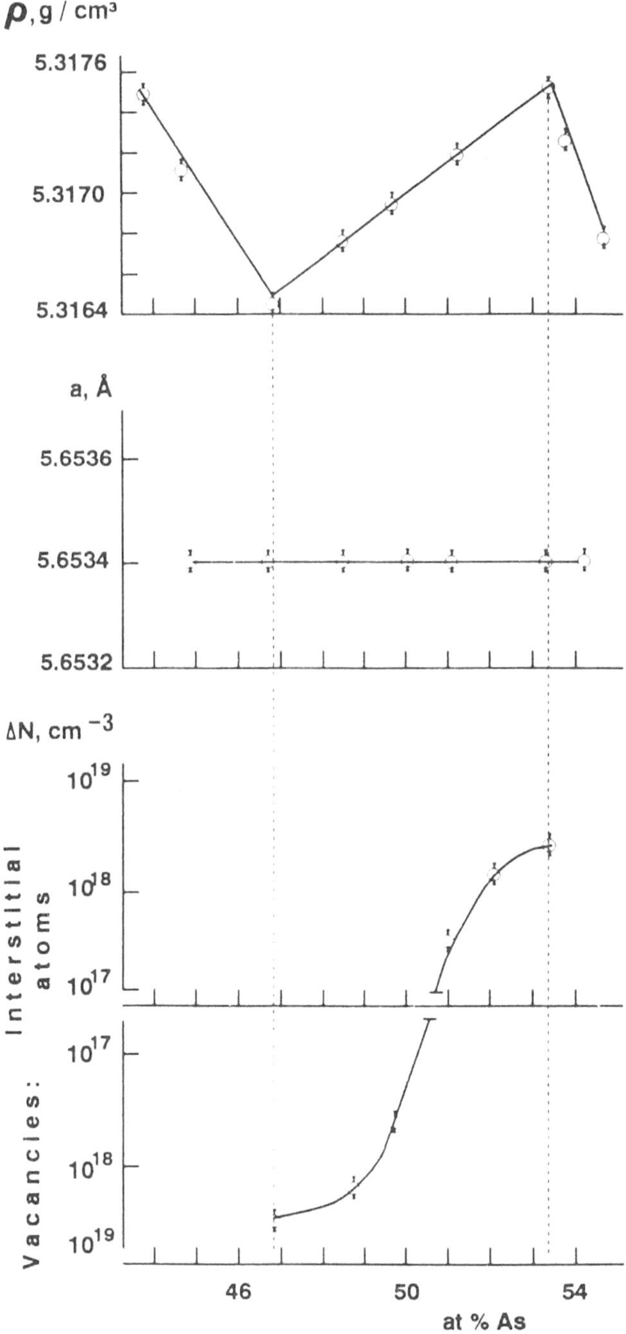

FIGURE 8.   Lattice parameter and density measurements on gallium arsenide crystals grown from melts of varying arsenic content (from Bublik *et al.*[9]).

results are shown in Fig. 8, from which it is seen that the mass per unit cell increases steadily as the arsenic concentration in the melt from which his crystals were grown is increased. This is most simply interpreted as being due to the arsenic Frenkel reaction, Eq. (22) above, where, in gallium-rich melts, arsenic vacancies dominate and there is a mass deficiency, while in arsenic-rich melts arsenic interstitials dominate and there is a mass excess. A crystal having the theoretical mass per unit cell is obtained by growth from a melt containing 50.5 atm. % arsenic. This implies that the congruent point is to the gallium-rich side of stoichiometry, as indicated on Fig. 7. An alternative explanation for the data is that antisite defects dominate, such that in gallium-rich material gallium atoms on arsenic sites give rise to a mass deficiency (gallium being of lower atomic mass than arsenic) whereas from arsenic-rich melts, arsenic antisites dominate giving a mass excess. However, for this explanation to be correct the concentration of these antisite defects would need to be unrealistically large (of the order of a few per cent). We infer, therefore, that the arsenic Frenkel reaction is the dominant one. This is not to say that there are no defects on the gallium sublattice or any antisite defects, or indeed that such defects are not of importance. It is only to say that, from Bublik's data, the arsenic Frenkel defects are present in concentrations much in excess of all others.

The deviation from stoichiometry ranges up to around $\pm 2 \times 10^{19}$ cm^{-3} (there are 2.2×10^{22} atm/cm^{-3} on each sublattice of the crystal). Bublik's data are replotted in Fig. 9. The full lines are the fitted to a model by Hurle.[10]

The above reactions have been written assuming that the point defects are uncharged. In reality most of these defects will also exist in more than one charge state, depending on the position of the Fermi level. In semi-insulating gallium arsenide the Fermi level is believed to be held at midgap by a deep donor, known as EL2, which is present in excess of all residual shallower acceptor levels. EL2 is known to be related to the arsenic antisite (As_{Ga}), though whether it is the isolated antisite or the antisite complexed with some other point defect or defects is unresolved. EL2 can exist in the neutral state and as a singly or doubly ionized donor. It appears to be present only in arsenic-rich material, up to a concentration of around

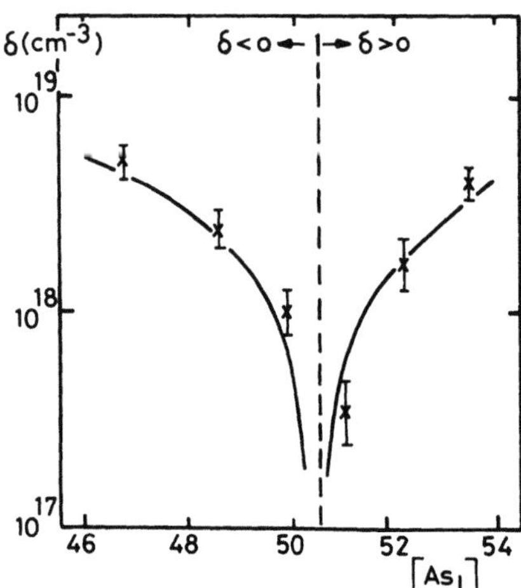

FIGURE 9. Deviation from stoichiometry of a gallium arsenide crystal as a function of arsenic content in the melt. The data points are from Bublik *et al.*[9] The solid curves are a fit to the data using a model solidus.[10]

2×10^{16} cm^{-3}—some two orders of magnitude less than the lattice nonstoichiometry. It appears to occur during cooling of the crystal from the melting point rather than by incorporation during the growth process. The formation reaction during cooling can be written

$$\text{As}_i + V_{\text{Ga}} = \text{As}_{\text{Ga}} \qquad (26)$$

Since the arsenic interstitial concentration is at least two orders of magnitude greater than the EL2 concentration, then it would appear that it is the gallium vacancy concentration which controls, through Eq. (26) above, the EL2 concentration. It has recently been found that annealing the material at a temperature around or in excess of 1100°C followed by rapid quenching of the crystal removes the EL2, presumably by dissociating the antisite defects. Reannealing at a lower temperature (of, say, 850°C) followed by a slow cooling reestablishes the EL2 concentration.

In the purest material there appear to be several acceptor levels due to native point defects rather than to chemical impurities. One of these appears to be the gallium antisite (Ga_{As}). This is currently an active subject of research and emphasizes how important is an understanding of the incorporation and the annealing kinetics of native point defects.

6. THE INCORPORATION OF SOLUTE (DOPANT) ATOMS

Group VI dopant atoms incorporated on to arsenic lattice sites and group IV dopant atoms incorporated onto gallium lattice sites form donors in the gallium arsenide crystal. Similarly, group II atoms incorporated on the gallium site and group IV atoms incorporated on the arsenic site form acceptor levels. The group IV atoms are frequently amphoteric—that is, they are incorporated on both sites and the net electrical result depends on the arsenic activity, i.e., on the native point defect equilibria and on the composition of the nutrient phase from which the crystal is grown.

As an illustration, we develop a thermodynamic model for the incorporation of a typical group VI donor—tellurium.

Tellurium incorporated on the arsenic site forms a donor state according to the reaction

$$\text{Te}_I + V_{\text{As}} = \text{Te}_{\text{As}}^+ + e^- \qquad (27)$$

having a mass action constant

$$K_{\text{Te}} = n[\text{Te}_{\text{As}}^+] / \gamma_{\text{Te}}[\text{Te}_I][V_{\text{As}}] \qquad (28)$$

γ_{Te} is the activity coefficient of tellurium in the liquid gallium-arsenic solution. Combining Eqs. (23) and (25) we see that the arsenic vacancy concentration is related to the arsenic pressure in equilibrium with the nutrient phase (p_{As_2}) by

$$[V_{\text{As}}] = \frac{K_{\text{As}}}{K_{\text{As}_{2i}}} \, p_{\text{As}_2}^{-1/2} \qquad (29)$$

Combining Eqs. (28) and (29) we see that the segregation coefficient for donors in the crystal, which is the ratio of the donor concentration to the melt concentration, is

$$k_D = \frac{[\text{Te}_{\text{As}}^+]}{[\text{Te}_I]} = \frac{K_{\text{As}}K_{\text{Te}}}{K_{\text{As}_2i}} \frac{\gamma_{\text{Te}}}{n_{gi}} (p_{\text{As}_2})^{-1/2} \tag{30}$$

For growth at a constant temperature, when p_{As_2} is constant, we see that the segregation coefficient should be a constant while n_{gi}, the electron concentration at the growth temperature, remains constant. If the crystal is intrinsic at the growth temperature then

$$n_{gi} = n_i = K_{cv}^{1/2} \tag{31}$$

where $K_{cv} = np$ is the mass action constant for electron–hole pair generation. However, once the crystal becomes extrinsic at the growth temperature so that the electroneutrality condition becomes

$$n = [\text{Te}_{\text{As}}^+] \tag{32}$$

then from Eq. (30) we see that the segregation coefficient should then vary as the (melt concentration).$^{-1/2}$ A plot of carrier concentration versus melt concentration should therefore be linear up to a donor concentration of the order of $K_{cv}^{1/2}$, above which it should fall off quadratically. Experimental data obtained by Goodwin *et al.*[11] are shown plotted in Fig. 10 for layers grown by liquid phase epitaxy at a temperature of 1123 K. The curve is indeed linear up to some fixed donor concentration, above which it falls away as predicted. However, the concentration at which

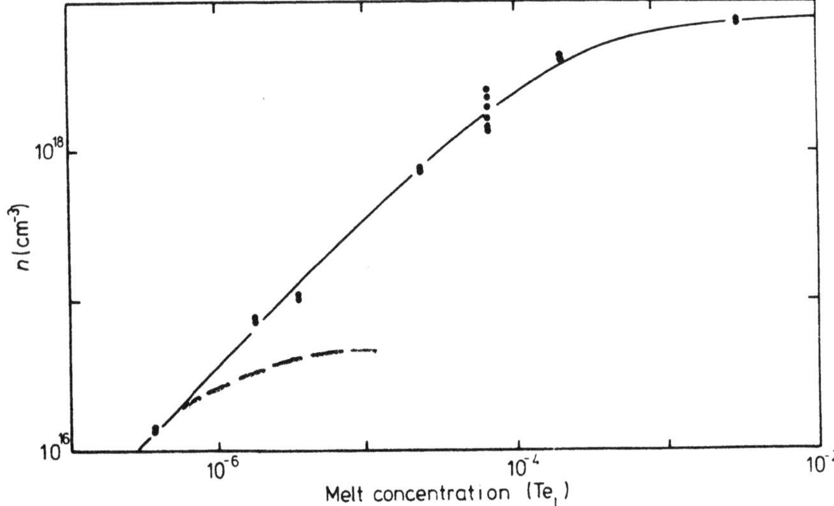

FIGURE 10. Room temperature carrier concentration plotted against melt tellurium concentration for liquid phase epitaxial growth at a temperature of 1123 K. Data from Goodwin *et al.*[11]

the shoulder occurs is some two orders of magnitude greater than the intrinsic electron concentration at the growth temperature. The shoulder should occur at the position shown by the dashed line if the above simple theory were valid. Clearly, then, thermodynamics has let us down or, alternatively, we are applying it to the wrong physical model! A similar, but smaller, discrepancy exists in the case of donor-doped material grown by vapor phase epitaxy. Figure 11 shows data taken for the growth of tin-doped layers by chloride epitaxy from Wolfe and Stillman.[12] Again, the dashed curve shows the position at which the shoulder would be expected to form.

Several explanations for this discrepancy have been proposed. Zschauer and Vogel[13] suggested that, at the growth interface, the Fermi level is pinned by a relatively high density of surface states which control the effective value of n_{gi} at a level significantly greater than the intrinsic carrier concentration, n_i. Thus the crystal *interface* is in equilibrium with the melt but equilibrium between the melt and the *bulk* crystal is not obtained because diffusion processes are too slow to readjust the incorporated donor concentration. Little is known about the behavior of such

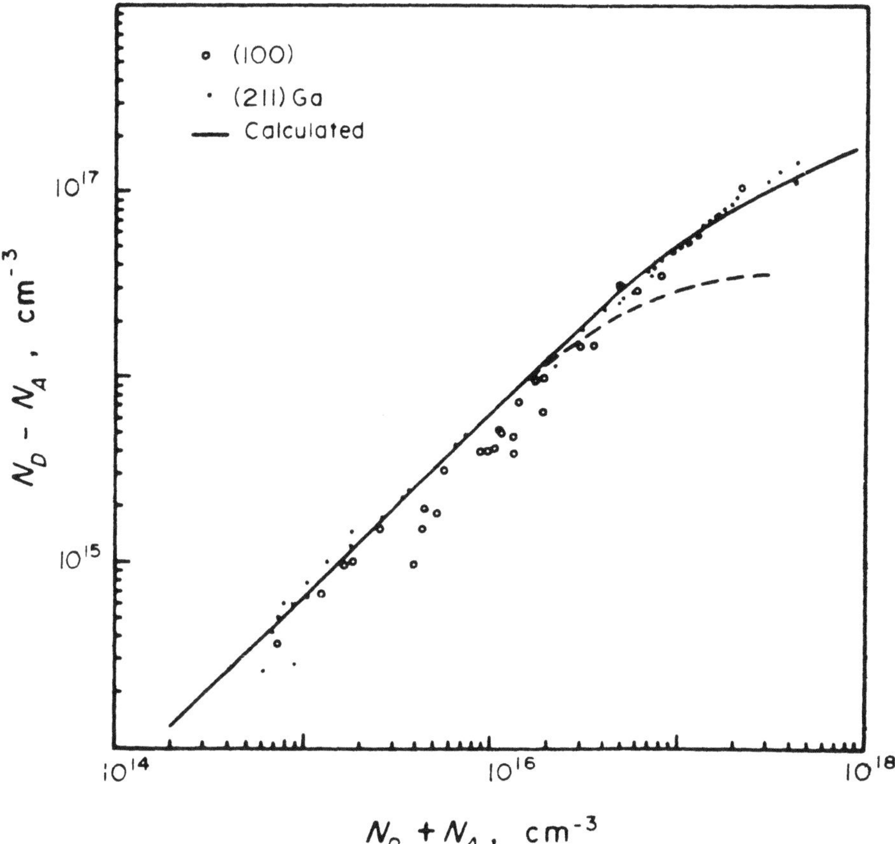

FIGURE 11. Room temperature carrier concentration against total ionized impurity concentration for tin-doped VPE layers. Data from Wolfe and Stillman.[12] The dashed curve shows the roll-off point expected if electroneutrality is governed by electron–hole pair generation.

Schottky barriers at high temperature. For this explanation to be correct then the surface state density has to be greater in the case of LPE growth than of VPE growth in order to account for the results in Figs. 10 and 11. This anomalous solubility curve remains an unsolved problem.

It is not the only unsolved problem for not only does the donor concentration show this superlinearity but it is also found that, for approximately every four donors incorporated into the crystal, there is incorporated an acceptor. This partial self-compensation is shown in Fig. 12, taken from Wolfe and Stillman, for the case of group VI donor-doped layers grown by chloride epitaxy. From this figure it can be seen that the acceptor to donor ratio is fixed at approximately 0.25 up to a donor concentration at which the shoulder occurs (see Fig. 11) and above this concentration the compensation rises steadily towards unity.

This self-compensation implies that the acceptor is composed of the single tellurium atom complexed with a native point defect. It has been suggested that this entity is the tellurium donor bound to a gallium vacancy:

$$Te_{As}^{+} + V_{Ga}^{2-} = Te_{As} V_{Ga}^{-} \tag{33}$$

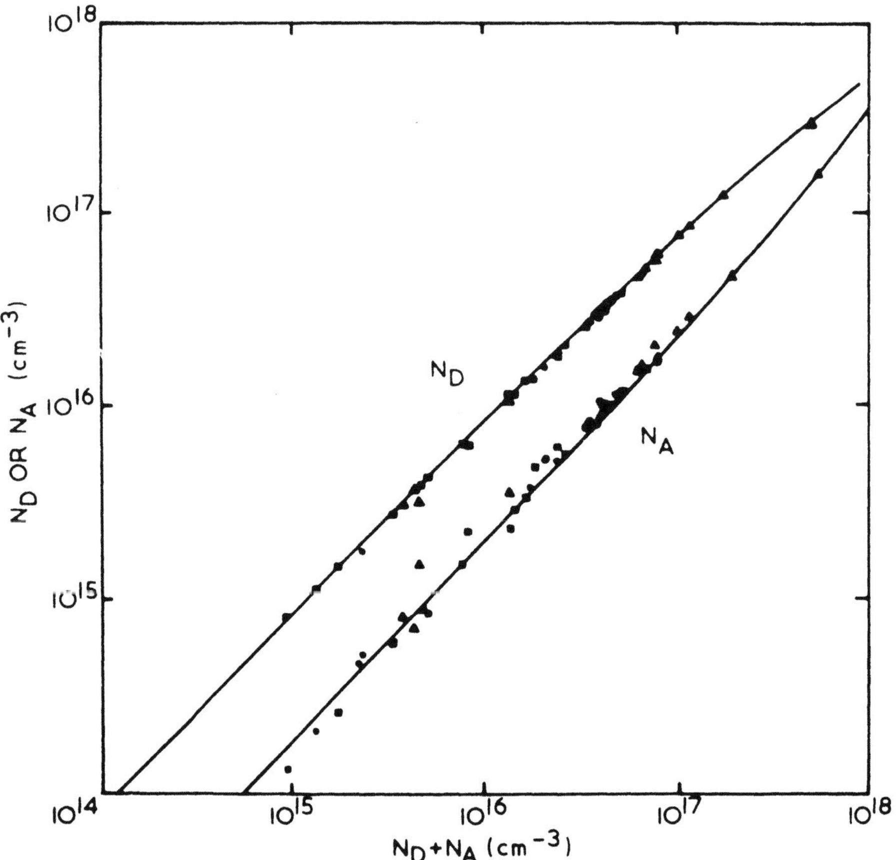

FIGURE 12. Electrical compensation data for vapor phase epitaxial layers doped with sulfur, selenium, or tellurium. Data from Wolfe and Stillman.[12]

having a mass action constant

$$K_{\text{TeV}} = [\text{Te}_{\text{As}} V_{\text{Ga}}^-]/[\text{Te}_{\text{As}}^+][V_{\text{Ga}}^{2-}] \tag{34}$$

The incorporation of ionized gallium vacancy acceptors can be represented by the equation

$$\tfrac{1}{2}\text{As}_2(V) = \text{GaAs} + V_{\text{Ga}}^{2-} + 2\text{h}^+ \tag{35}$$

having a mass action constant

$$K_{\text{inc}} = [V_{\text{Ga}}^{2-}]p^2(p_{\text{As}_2})^{-1/2} \tag{36}$$

where p is the hole concentration.

Combining Eqs. (28), (34), and (36) we obtain for the donor to acceptor ratio the following expression:

$$N_D/N_A = [\text{Te}_{\text{As}}^+]/[\text{Te}_{\text{As}} V_{\text{Ga}}^-] = K_{cv}^2/(K_{\text{TeV}} K_{\text{inc}} n_{gt} p_{\text{As}_2}^{1/2}) = K(T)/n_{gt} p_{\text{As}_2}^{1/2} \tag{37}$$

If n_{gt} is a constant then, at a fixed temperature, we would expect the compensation ratio to vary as the square root of the arsenic dimer pressure. Since this differs by some 3 orders of magnitude for growth by chloride VPE as against growth by LPE, the properties of donor-doped LPE and VPE layers should be very different. However, this is experimentally not the case and LPE material is found to have a similar compensation ratio. This dilemma also is not resolved. It may be that the Schottky barrier model is correct and the density of surface states is controlled by some mechanism that keeps, at a given temperature, the product $n_{gt} p_{\text{As}_2}^{1/2}$ constant or it may be that we are wrong in supposing that the acceptors are grown into the crystal. It might be that they are formed by some reaction during cooling of the crystal to room temperature. A third possibility is that the mobility theory used to obtain values of N_D/N_A from the conductivity and Hall coefficient data is wrong and the material is, in fact, not self-compensated. This possibility has been rendered unlikely by the results of Brozel et al.,[14] who have shown, using radio tracer techniques, for one donor at least (tin) that the material is genuinely autocompensated and that mobility theory is in accord with the radio tracer results.

So the above two paradoxes remain unresolved but it is hoped that the reader will have appreciated the utility of thermodynamics to explore the consequences of different physical effects which may be important in determining the electrical behavior.

7. SUMMARY AND CONCLUSIONS

We have shown the utility of thermodynamics in predicting the complex multicomponent phase equilibria which exist with ternary and quaternary III–V semiconducting compounds. The construction of model phase diagrams is of the greatest importance in developing epitaxial growth techniques for the fabrication of multicomponent epitaxial layers. This has been illustrated here in respect of growth by liquid phase epitaxy and for vapor phase epitaxial techniques. For a more extensive review of the application of thermodynamics to such systems, the reader is referred to the excellent reviews by Stringfellow.[15]

It has also been shown how thermodynamic models can be constructed to provide a rationale for understanding nonstoichiometry and native point defect incorporation during crystal growth. This can be extended to a consideration of the defect aggregation and complexing reactions which occur as the crystal is cooled to room temperature.

Finally, dopant incorporation and its relationship to the electrical properties of the semiconductor have been explored using the "simple" example of a group VI donor. This has highlighted the importance of dopant–native point defect interactions and has revealed a basic lack of understanding of the details of donor incorporation in gallium arsenide.

REFERENCES

1. R. A. Swalin, *Thermodynamics of Solids*, Wiley, New York (1962).
2. E. A. Guggenheim, *Thermodynamics*, North-Holland, Amsterdam (1967).
3. I. Prigogine and R. Defay, *Chemical Thermodynamics*, translated by D. H. Everett, Longmans, Green & Co., London (1954).
4. F. A. Kroger, *The Chemistry of Imperfect Crystals*, Vols. 1–3, North-Holland, Amsterdam, (1974).
5. J. Willard Gibbs, *The Scientific Papers*, Vol. 1: *Thermodynamics*, Dover, New York (1961).
6. D. T. J. Hurle, Revised calculation of point defect equilibria and nonstoichiometry in gallium arsenide, *J. Phys. Chem. Solids* **40**, 613–626 (1979).
7. J. Steininger, Thermodynamics and calculation of the liquidus–solidus gap in homogeneous, monotonic alloy systems, *J. Appl. Phys.* **41**, 2713–2724 (1970).
8. R. W. Conrad, P. L. Hoyt, and D. D. Martin, Preparation of epitaxial $Ga_xIn_{1-x}As$, *J. Electrochem. Soc.* **114**, 164–167 (1967).
9. A. N. Morozov and V. T. Bublik, Use of precision density and lattice parameter measurements for study of point defects of single crystals in semiconductors, *J. Crystal Growth* **75**, 497–503 (1986).
10. D. T. J. Hurle, in *Semi-insulating III–V Materials, Malmo, 1988* (G. Grossmann and L. Ledebo eds.), Adam Hilger, Bristol (1988), pp. 11–19.
11. A. R. Goodwin, C. D. Dobson, and J. Franks, in *Institute of Physics Conference Series No. 7*, published IOP, Bristol (1969), p. 36.
12. C. M. Wolfe and G. E. Stillman, *Solid State Commun.* **12**, 283 (1973).
13. K. H. Zschauer and A. Vogel, *Institute of Physics Conference Series No. 9*, IOP, Bristol (1971), p. 100.
14. M. R. Brozel, E. J. Foulkes, I. R. Grant, and D. T. J. Hurle, Tin segregation and donor compensation in melt grown gallium arsenide, *J. Crystal Growth* **80**, 323–333 (1987).
15. G. B. Stringfellow, in *Crystal Growth: A Tutorial Approach* (W. Bardsley, D. T. J. Hurle, and J. B. Mullin eds.), North-Holland, Amsterdam (1979), pp. 217–239; and *Vapour Phase Epitaxy of III–V Compounds: Fundamental Aspects. Proceedings 5th International Summer School on Crystal Growth, Davos, 1983* in *Crystal Growth of Electronic Materials* (E. Kaldis, ed.), North-Holland, Amsterdam (1985).

9

Single Crystal Growth I: Melt Growth

J. B. Mullin

1. INTRODUCTION: GENERAL PRINCIPLES

This chapter is concerned with growth of single crystals from melts of semiconductors. In the case of compound semiconductors, these melts are stoichiometric or nearly so. The production of semiconductors in single-crystal form is a core requirement of almost all semiconductor device technology. As a consequence, the whole field of crystal growth technology is strongly device driven and sensitive to commercial pressures related to production efficiency, quality, etc. Thus a fuller understanding of the evolution of a particular crystal growth technology requires an appreciation of the commercial background of the subject. Only cursory consideration can be given to these commercial aspects in this chapter. Here we are primarily concerned with the principles and concepts underlying the growth of single crystals from the melt.

Crystal growth involves the controlled deposition of atoms onto a single-crystal seed by making use of a phase change: liquid to solid, vapor to solid, or even solid (in a metastable form, for example) to solid. The phase changes can involve chemical reactions as is the case of metalorganic vapor phase epitaxy (MOVPE). The process of crystallization is schematically illustrated in Fig. 1. Here atoms in random motion, for example from a liquid in a higher energy state, are caused to nucleate on a single crystal seed where they lose energy, by latent heat of crystallization, and assume the crystallographic arrangement of the atoms of the seed. The driving force for growth is the difference of chemical potential between the atoms in the liquid phase μ_L from that in the crystalline phase μ_S, where μ_L is greater than μ_S. In melt growth the essence of the technique is to provide a suitable chemical potential difference between the liquid and the solid, and this is essentially achieved by adjusting the chemical potentials (temperature gradients) as illustrated in the diagram. Although this simple concept is universal for all melt growth techniques, the wide variety of material properties found in different crystals has resulted in the evolution of a variety of different growth technologies. These have evolved in

---

J. B. Mullin • Royal Signals and Radar Establishment, St. Andrews Road, Malvern, Worcestershire WR14 3PS, U.K. *Present address*: EMC, The Hoo, Brockhill Road, West Malvern, Worcestershire WR14 4DL, U.K.

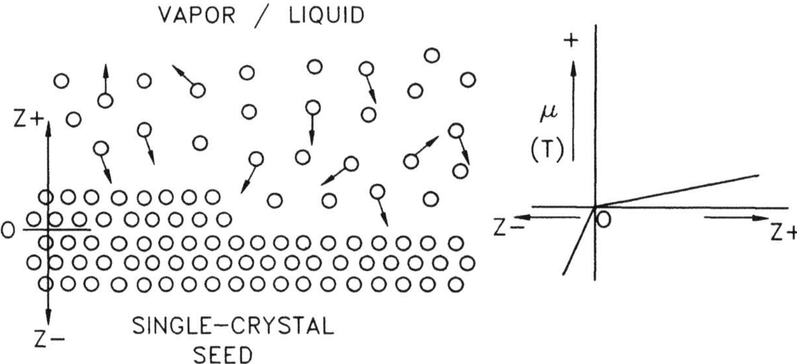

FIGURE 1.   Schematic illustrating the concept of crystal growth through the use of a single-crystal seed and the development of appropriate gradients of chemical potential μ through, in the case of melt growth, temperature.

order to cope with the difficulties and growth constraints imposed by the different properties of various classes of materials. These are considered in later sections.

2. ROLE OF MELT GROWTH

The main purpose of melt growth is to produce appropriately doped single crystals, having the minimum number of defects. An exceptionally high standard of chemical purity and physical perfection, so-called semiconductor quality, is required. Thus the electrically sensitive residual impurity atom concentration generally needs to be less than one part per billion atomic (ppba) for many semiconductor applications. In terms of physical perfection, single crystals are, with few exceptions, essential. Further the principal defect, the dislocation, generally needs either to be eliminated to give dislocation-free material or with the more difficult compound semiconductors reduced to an acceptable minimum.

The process of crystallization in melt growth generally results in significant purification in the first part of the crystal pulled. The maximum change in impurity content that can be achieved during crystallization is determined by the value of the distribution or segregation coefficient k. The equilibrium distribution coefficient is defined as $k_0 = c_S/c_L$, where c_S and c_L are the concentrations of solute in the solid and liquid at equilibrium. Thus the purification achieved by a single crystallization for impurities having small values of k_0, say 10^{-3}-10^{-5}, from a high-purity melt, impurity content <1 ppma, can result in semiconductor grade crystals provided the diffusion coefficients in the solid are not large, that is 10^{-5} cm^2 s^{-1} or less. However, for impurities with values of k_0 close to unity such as boron in silicon and zinc in indium antimonide, etc, specific chemical purifications are required to achieve semiconductor grade purity. Chemical purification with respect to specific elements is a major topic and will not be developed here. However, for many semiconductors such as germanium and for many starting elements, zone refining or the repeated passage of a molten zone through a bar of material can be a very effective way of producing semiconductor grade quality material. A combination of purification techniques, however, is generally required to achieve the ultimate standard of purity for most semiconductor materials.

3. CONSTRAINTS TO MELT GROWTH

A variety of melt growth techniques have been developed for growing single crystals. However, the suitability of any technique for growing a particular semiconductor is constrained by a variety of factors, which, for convenience, may be classified under the headings of (a) chemical reactivity; (b) vapor pressure; (c) mechanical; and (d) fundamental. An appreciation of the significance of these factors will help in understanding the reasons for the development of the different techniques discussed in later sections.

3.1. Chemical Reactivity

By way of example we can consider first the case of germanium. This archetypal semiconductor, though no longer of major commercial importance, has proved the pioneering scientific and technical vehicle for the evolution of so much of our knowledge on semiconductor materials. Germanium has a relatively low melting point, 937°C, and benign properties. It has no significant reaction with silica or carbon and can be grown from crucibles made from either of these materials. It is virtually involatile at the melting point and can be grown in vacuum or in an inert atmosphere including nitrogen, forming gas, etc. Naturally, as with all other semiconductors, it reacts with oxygen and water vapor, hence air and moisture must be completely excluded. Almost all melt growth techniques that have been developed can be used to grow single crystals for germanium, although the favored techniques are the horizontal Bridgman technique and the pulling technique.

In marked contrast, silicon melts at a relatively high temperature of 1417°C. It is an exceptionally reactive element and will react readily with carbon to form silicon carbide. It can be grown in silica crucibles, although they have very little mechanical rigidity at these temperatures and need to be supported with another crucible, usually graphite. Molten silicon reacts with SiO_2 causing the introduction of oxygen into the melt. The limitation and control of this reaction is a primary technical difficulty in the growth of crystals pulled from the melt. Silicon can either be pulled from melts in silica crucibles under controlled conditions or grown by the float zone technique, but it cannot be grown by the horizontal Bridgman technique using silica boats since the solid Si sticks tenaciously to the boat and differential contraction breaks up the crystal on cooling. Silicon can, however, be grown by the horizontal Bridgman method using a water-cooled high-conductivity metal boat, e.g., silver, provided the zone is levitated out of contact of the boat by using rf induction heating.

In summary, then, all aspects of the chemical reactivity of the melt with its crucible and its gaseous environment must be carefully analyzed and controlled.

3.2. Vapor Pressure

Control of vapor pressure is a major problem in the growth of many of the compound semiconductors. For example, the vapor pressure that results from the dissociability of the arsenides and phosphides of the III–V compounds is an important constraint that has limited the rate of development of these materials. In contrast, indium antimonide, which has a very low equilibrium vapor pressure (4×10^{-8} atm) at its melting point of 525°C was available in single-crystal form in semiconductor quality within ten years of its discovery. Indium antimonide can be grown by any of the methods that have been developed for germanium, including float zoning, although here the ingot diameter is limited by the density of the material. Gallium arsenide, on the other hand, exerts a vapor pressure of one atmosphere of arsenic at its melting point of 1238°C and cannot be grown by the standard vertical pulling technique since a melt will simply dissociate at this temperature leaving a gallium-rich melt. A special technique, liquid encapsulation, to be described in Section 4.4, has been developed to cope with this particular problem.

3.3. Mechanical

The mechanical constraints stem from the need to establish a steady state crystallization process which avoids fluctuations in growth rate. Any mechanical fluctuation, for example the vibration of the seed or melt, may induce thermal fluctuations at the solid–liquid interface. This can induce growth rate fluctuations, which may introduce defects into the growing crystal such as impurity fluctuations or striae or possibly initiate twin formation or dislocation formation. The mechanical constraints impose minimum vibration specifications on all moving parts associated with the crystal growth process which can be transmitted to the crystal melt interface. Additionally, in the rotated crystal there is a need to arrange for the crystal ideally to rotate concentrically with the thermal axis so as to avoid the solid–liquid interface experiencing a sinusoidal thermal oscillating temperature fluctuation which could induce growth rate fluctuations. The design of advanced crystal growth apparatus that takes into account the interaction of mechanical constraints on the static or moving thermal environment is clearly a very complex subject and one that attracts considerable research.

3.4. Fundamental

The growth of high-quality crystals is also subject to a number of critically important fundamental constraints. Here the crystalline quality can be affected either by phenomena related to the transport of solute or heat in the region of the solid–liquid interface, or by kinetic phenomena occurring on the growth interface. In the former situation a failure to adequately remove rejected dopant, for example, in the crystallization process, can result in phenomena like constitutional supercooling,[1] which leads to poor crystallinity of the grown crystal. Also, the kinetic effects like the facet effect[2] can lead to localized enhancement or reduction of dopant incorporation. Some of the more important fundamental constraints are considered after the following sections, which deal with the more important melt growth techniques.

4. TECHNIQUES OF MELT GROWTH

4.1. Vertical Pulling or Czochralski Growth

The vertical pulling technique is one of the first techniques that was developed in the semiconductor industry for growing single crystals of germanium. It is historically related to the technique used by Czochralski for pulling crystals of metals from melts. Czochralski, however, did not use rotation, and the modern vertical pulling technique owes much for its development to the pioneering work of Teal and Little,[3] who developed it out of all recognition from the original process. In terms of scale of production, the vertical pulling technique is clearly the dominant method used in the semiconductor industry for obtaining single crystals of substrate material. Other techniques, like the zone melting technique, are vital for obtaining high-quality material for specialized applications. Substrates can be used either directly in the fabrication of semiconductor device structures, such as those made by ion implantation, or they can be used indirectly as a means of establishing epitaxial structures that are used in the fabrication of devices.

The essentials of the crystallization process are illustrated in Fig. 2. The semiconductor starting material, germanium for example, is melted in a silica crucible E using radiofrequency induction heating (typically 0.5 MHz) via the coil M. The heat is generated in the carbon/graphite susceptor F, which is an ideally good fit to the silica crucible E. The process involves melting the germanium to just above its melting point and running the single-crystal seed A into the melt in order to establish a solid–liquid (S/L) interface just above the melt surface. Power is adjusted to the melt while rotating the seed in order to achieve thermal symmetry and develop a uniform boundary layer at the S/L interface. The thermal conditions at the S/L interface are illustrated in the insert of Fig. 2. In the steady-state pulling process one simply maintains the S/L interface at a substantially fixed position so as to match the velocity of crystallization in the direction of the pulling axis. The shape of the crystal is primarily determined by the shape of the meniscus at the S/L interface. In the case of germanium the crystal grows as a right cylindrical crystal when the contact angle between the liquid and the crystal, $\theta_L = 13°$ (defined as θ_L^0). The growth shape is thus determined by the balance of surface tension forces in the solid and liquid. However, if θ_L is $<\theta_L^0$, for example when the power to the melt is raised, the S/L interface rises and the melt under the seed "waists in." This causes a reduction in crystal diameter. In the reverse situation reducing the power to the melt will lower the position of the solid–liquid interface, the angle θ_L will be $>\theta_L^0$, and the crystal grows out to follow the shape of the meniscus.

Under steady state conditions the growth rate will be given by the heat flux condition at the solid–liquid interface,

$$\sigma_s G_s - \sigma_L G_L = v\rho_s H_f \tag{1}$$

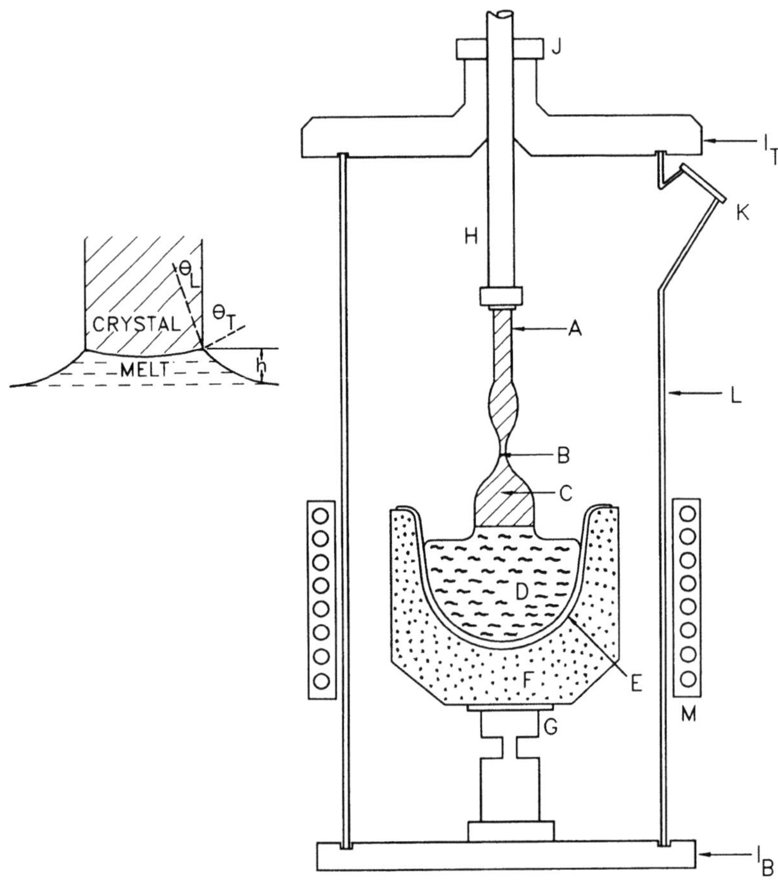

FIGURE 2.   Schematic of a crystal growth chamber showing a crystal C being pulled from a melt, D, using a single-crystal seed B. The crystal diameter is determined by the shape of the meniscus at the crystal/melt interface. See text for a discussion of the crystal melt inset figure.

where σ is the thermal conductivity, G the temperature gradient, v the growth velocity, ρ_s the density of the crystal, and H_f the latent heat of crystallization. G refers to the gradients normal to the solid–liquid interface.

Diameter control is an important parameter in the pulling process. It may be used initially to propagate a single-crystal seed from a polycrystalline seed simply by reducing the diameter below the typical grain size and then expanding the single-crystal grain in size by increasing the diameter. However, with the common semiconductors, single-crystal seeds are fairly readily available and the above process is usually a feature of the development of new semiconductors. The reduction in diameter following seed-on to the melt is, however, an important process called "necking in." This "necking in" is used to grow out dislocations and produce a dislocation-free seed for the subsequent growth of the dislocation-free crystal. The dislocation elimination is achieved primarily as a result of the vacancy supersaturation that occurs on crystallization and causes dislocations to climb and not propagate.

After "necking in," the crystal diameter is carefully increased to the required diameter by suitable adjustment of power to the melt. A feature of modern research and commercial crystal pulling equipment is the use of monitors that can measure the diameter of the crystal and provide a parameter to feed back into the power system in order to automatically control the diameter. A method used in the growth of silicon crystals is to locate the position of maximum radiation coming from the melt surface. A bright halo occurs where the liquid column of the crystal tapers into the melt. This is a phenomenon associated with the high latent heat of crystallization of silicon. The variation of this position can be used to monitor the crystal diameter. The technique, however, cannot be used for diameter control of III–IV compounds, not only because of their small heats of crystallization but also in the case of the arsenides and phosphides because they are normally grown using encapsulants. Here a universal method, the weighing method, has been developed. This involves weighing either the crystal plus pull rod or the melt plus crucible and its support. The rate of change in weight is monitored. This, together with the knowledge of the rate at which the crystal is pulled from the melt, can be used to drive a servo system operating the power supply. In this way one can control and program the crystal diameter.

The growth of single crystals of silicon from the melt is a major industrial activity. Crystals up to 150 mm in diameter and up to a meter in length are grown routinely. In the case of the III–V compounds, crystal diameters are typically 50 mm. The trend for all materials is for the growth of even larger diameter crystals.

4.2. Float Zone

The crystal pulling technique for Si, as noted earlier, introduces oxygen into the melt and crystal. This oxygen can be used to advantage for certain applications. For example, it can be made to precipitate just below the surface of the epitaxial slice or wafer where it acts as a gettering source for transition elements, leaving a high-purity oxygen impurity-free zone at the surface of the wafer, which is suitable for the fabrication of advanced integrated circuits.

However, for many device applications involving the use of large currents there is a need for large-area diodes, transistors, etc., and here an oxygen-free single crystal is required. This requirement has stimulated the development of a float zone technique. In this process, which is illustrated in Fig. 3, a molten zone is formed in a rod of silicon using induction heating. The rod is held by chucks on the ends of pull rods which can be rotated, usually in opposite directions, and raised or lowered at the same or differential speeds. The use of a single-crystal seed in one chuck can enable the propagation of a single crystal. Growth in vacuum can be used to remove oxygen. The very highest quality material can be grown in this way. The maximum stable height L_{max} of an unsupported zone that can be grown is given by $L_{max} = 2.8(\gamma/\rho g)^{1/2}$, where γ, ρ, and g are the surface tension, the density, and the gravity, respectively. Hence the larger the density the more difficult it is to maintain stable zones. The technique is highly advantageous for Si, which has a low density $(2.3\ \mathrm{g\ cm^{-3}})$. Very large quantities of crystals are grown in this way; even crystals up to 100 mm in diameter are routinely grown in production. But for large diameter crystals it is necessary to prevent the zone falling out. This is achieved by making

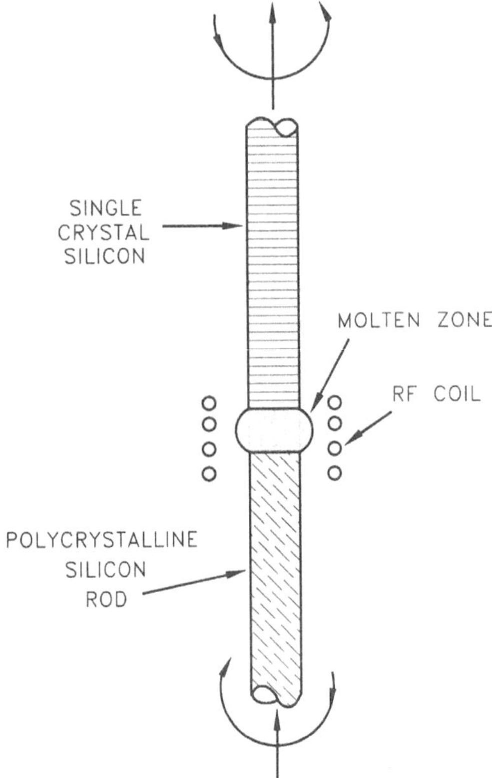

FIGURE 3. Schematic illustrating the principles of float zoning in which a molten zone is generated and caused to move vertically through counter-rotating rod of polycrystalline starting material.

use of the repulsive forces between the induced current in the melt and the rf coil current (at 24 MHz) using a coil narrower in diameter than the solid region. This is a highly skilled technology. The technique has not, however, been developed for the III–V compounds. Here of course, apart from the increased density (ρ for GaAs is 5.315 g cm^{-3}) the major difficulty is the problem of the high As vapor pressure at the melting point of GaAs.

4.3. Horizontal Bridgman

The horizontal Bridgman technique was one of the earliest methods to be applied to the growth of compound semiconductors and is still an important process. Horizontal Bridgman involves the progressive crystallization of a melt in a horizontal boat or tube. A schematic of an apparatus used for the preparation and growth of a dissociable compound such as GaAs is shown in Fig. 4. In order to prevent As loss from the molten ingot (the equilibrium vapor pressure of As over molten GaAs at 1238°C is ≈ 1.0 atm) the boat A used for growing the GaAs is sealed in a tube B. This whole tube must be kept at a temperature in excess of 615°C to prevent As$_4$ vapor condensing as solid on the walls of the tube. The preparation of GaAs from its elements can also be carried out in such a system. Initially, the growth boat A

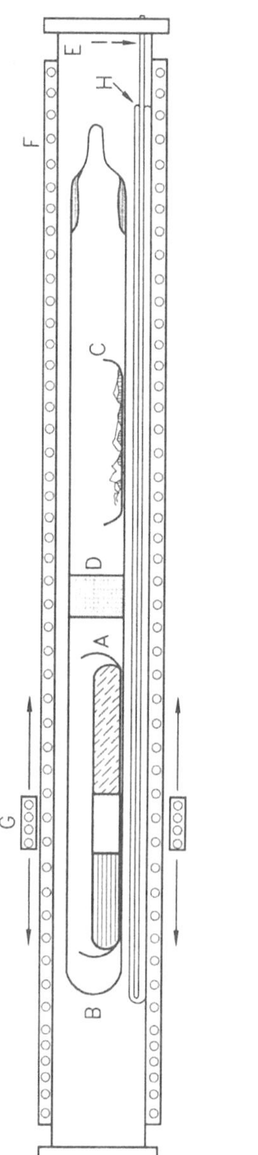

FIGURE 4. Horizontal Bridgman apparatus that can be used for the synthesis, zone refining, zone leveling, and crystal growth of dissociable compounds. The thermal distribution used in crystal growth is illustrated.

is partially filled with Ga metal and the reservoir boat C with crystalline As. Arsenic from the reservoir boat is then distilled through an anticonvection baffle D into the boat of liquid Ga held at a temperature a little in excess of the melting point of GaAs. On completion of the distillation process the melt is progressively frozen from one end to form a bar of GaAs. Such an ingot frequently contains two or three large single crystals. Completely single crystals can also be grown, but here a single-crystal seed is generally employed. A molten zone formed between the seed and a preformed bar of GaAs is moved through the polycrystalline bar so as to achieve a single crystal. The technique can be used to grow semi-insulating material as well as doped crystals. By moving a molten zone, first in one direction and then in the reverse direction, a process known as zone leveling, a uniformly doped crystal can be prepared.

The major virtue of the horizontal Bridgman technique, especially as applied to GaAs, is the relative ease with which very low dislocation content crystals (10^2 cm^{-2} or less), an essential prerequisite for the growth of semiconductor lasers, can be grown. The reason for this is associated with the very low temperature gradients which tend naturally to occur in the horizontal Bridgman technique. The two main disadvantages of the technique are, firstly, that the crystal grows in confined contact with the silica boat, which can result in stress and misnucleation causing twins and polycrystals. Secondly, the long period of contact of the melt with the silica boat can result in impurity incorporation. Contamination typically of the order of a few ppma of Si can be introduced from the silica boat into the grown crystal. The growth of compounds with melting points in excess of that of GaAs–GaP at 1465°C, for example—or compounds having high equilibrium vapor pressures at the melting point—InP at 27.5 atm and GaP at 32 atm—make the horizontal Bridgman either very difficult to apply in the case of InP or exceptionally difficult in the case of GaP. The solution to these growth problems has been found in the liquid encapsulation technique.

4.4. Liquid Encapsulation

The concept of liquid encapsulation is elegantly simple. It is illustrated in Fig. 5, which shows a crystal seed being pulled using a seed A from the melt D, say, of GaAs. The crucible arrangement of Fig. 4 would be used in a conventional Ge-type crystal puller using a growth chamber as illustrated in Fig. 2. A low-melting-point transparent liquid encapsulant floats on the surface of the melt. Provided the inert gas pressure p_i is greater than the dissociation pressure p_d exerted by the components of the melt, the encapsulant will act as a liquid seal and prevent vapor loss from the melt. The ideal encapsulant should possess additional properties. It should be immiscible with the melt and chemically unreactive towards it and the crucible, but most importantly the encapsulant should wet the crystal and the crucible. Further, its viscosity and the temperature dependence of this viscosity should be such as to allow it to be drawn up and coat the emergent crystal with a thin film and be retained as such throughout the growth process. This latter requirement is to prevent the decomposition of the hot crystal. In spite of a detailed study of very many glasslike systems, only B_2O_3 and related mixtures have been found to fulfill sufficiently well these ideal characteristics.

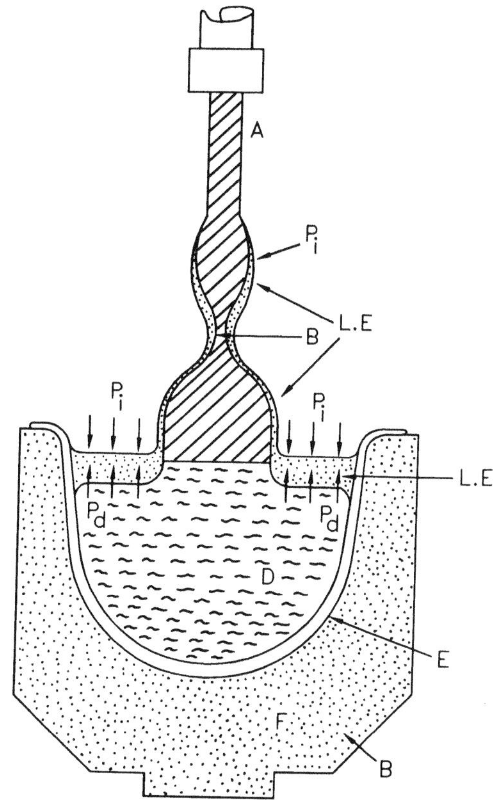

FIGURE 5. An illustration of the principle of liquid encapsulation. The encapsulant LE floats over the surface of the melt, the crucible, seed, and crystal and prevents component volatilization provided the pressure of inert gas p_i is greater than the dissociation pressure p_d of the melt.

The application of the concept of liquid encapsulation to the growth of III–V compounds was initially reported for the growth of InAs and GaAs by the author and his colleagues.[4] The use of B_2O_3 is well known metallurgically and has a long history of use in protecting metals from oxidation and vapor loss. In the case of the group IV/VI semiconductors, Metz *et al.*[5] used B_2O_3 in the crystal growth of volatile compounds like PbTe and PbSe. However, the significant advance in III–V compounds came with the application of liquid encapsulation to the concept of high-pressure pulling in steel pressure vessels. Liquid encapsulation high-pressure pulling was initially applied to the growth of InP and GaP[6] and represented a breakthrough in the growth of these materials as high-quality, uniformly doped single crystals. The liquid encapsulation technique is generally called liquid encapsulation Czochralski in the literature, and abbreviated as LEC.

4.4.1. Low-Pressure LEC

The low-pressure system is applied to compounds whose equilibrium vapor pressure is not greatly in excess of 1 atm. The growth apparatus is similar to that shown in Fig. 2. However, B_2O_3 would be added to the charge and crucible so that it could form a liquid encapsulant (LE) over the surface of the melt, crucible, seed and crystal as illustrated in Fig. 5. The process of growing a crystal is quite similar to that of growing Ge crystals. The seed is simply rotated and lowered through the B_2O_3 and by programming the appropriate thermal conditions, the crystal is pulled from the melt. Good visibility can be achieved and the crystal can be grown manually or under automatic diameter control.

4.4.2. High-Pressure LEC

In high-pressure LEC the concept is similar to that used in low-pressure LEC, except that the whole system is encased in a steel pressure vessel, as illustrated in the photograph shown in Fig. 6. The critical advance achieved in high-pressure LEC pulling is the result of the containment of the reactive materials at high temperature to the crucible. The steel pressure chamber is required merely to withstand the high pressure of inert gas. The seals and rotating parts are not subject to corrosive attack from hot elemental vapors. The growth process can be viewed using a TV monitor. Also, using a weighing system which is visible in the photograph in Fig. 6, it is possible to carry out the growth process automatically. The use of high-pressure technology represented a major breakthrough in the growth of crystals of GaP and InP, which were made available for the first time as large single crystals. In the early phase of the development of high-pressure LEC it was common practice to use semiconductor starting material compounded from the elements in a different apparatus from that used for crystal growth using the principles discussed in Section 4.3. However, an important breakthrough was made when the high-pressure apparatus itself was used for both processes, formation and growth. The most important application was in the preparation of semi-insulating (SI) GaAs. This is the starting material that is used in the fabrication of GaAs integrated circuits.

The major problem with the horizontal Bridgman technique in the production of SI GaAs is the silicon contamination from the boat used in its preparation, which can prevent the material becoming semi-insulating. It is possible to avoid silicon contamination by compounding semiconductor grade Ga and As in a boron nitride (BN) crucible under B_2O_3 using an insert gas pressure ≈ 100 atm. Pure unreacted As would have a pressure ≈ 80 atm at 1238°C and it is important to have a pressure sufficiently large to prevent arsenic evaporation in the compounding stage. The original work was carried out by Swiggard et al.,[7] who showed that stable high-resistivity ($10^8\ \Omega$ cm) GaAs resulted. This material can be grown as high-quality single crystals having a similar high resistivity using the liquid encapsulation technique provided the melt is slightly As-rich. The effects of defects on the behavior and properties of SI GaAs are complex. Indeed, there is sufficient science and technology associated with this material that it is now the sole subject of some international conferences.

The *in situ* compounding of InP and GaP is more difficult than for GaAs because of the very high pressures that can be generated. It is more feasible to distill P into the Ga in a pressure vessel than simply melt the elements together. The LEC technique is now the standard commercial technology for the growth of large-diameter single crystals of GaAs, InP, and GaP. The horizontal Bridgman technique is still used for the production of GaAs having a low dislocation content; such material is required for lasers.

FIGURE 6.   Photograph showing a high-pressure crystal puller. The steel growth chamber is opened at the base showing the rf induction coil and the crucible arrangement. At the base of the main growth chamber there is an additional horizontal chamber which houses the weighing cell. The TV camera and monitor enable remote viewing. The crystal puller is housed in a ventilated cage, which has a safety lining of chain mail.

5. FUNDAMENTALS

In the last four decades the semiconductor revolution has stimulated major research programs on the fundamental science, as well as the technology of melt growth. Our understanding of melt growth has advanced enormously. In the author's opinion, those areas of research that have been most fruitful in advancing the growth of better-quality crystals have been research related to the study of the transport of mass and heat adjacent to the solid liquid interface and in the interpretation of postcrystallization phenomena and point defects in general and dislocations in particular, which play such an important part in controlling the properties of the grown crystal.

Constraints of space forbid formal analysis of these topics, and the reader will find many excellent texts in the literature. The author's review[8] lists many of these. The international series of conferences on crystal growth[9] provide a splendid cross section of research approximately every three years not only about melt growth but about crystal growth in general. The international journal *Progress in Crystal Growth and Characterization*[10] is a valuable review journal on the subject.

Melt growth is still a dynamic area of research. One might argue that many of the more critical problems in the melt growth of Si have been solved in 30 years. This is true in relation to past device technology but, as has been stressed in this chapter, much of the stimulus for melt growth research comes from new device technology. The increase in size from 150–200 mm plus immediately presents additional problems of uniformity, wafer distortion in handling, etc. The newer technologies involving O_2 precipitation have generated new research on fluid flow and the role of magnetic fields in controlling mixing in the melt, especially the O_2-rich liquid Si which is adjacent to, and responsible for, partial solution of the silica crucible. Sufficiently large magnetic fields can be used to restrict fluid flow. This is a consequence of the fact that any motion of the liquid metal in a magnetic field causes induced currents that create their own magnetic field that opposes the applied magnetic field. Many of the more critical melt growth problems have been found and researched using semiconductors. The nonstoichiometry of the melt caused, for example, by group V volatilization, can give rise to a phenomenon known as constitutional supercooling.[1] The excess component is rejected at the solid–liquid interface (see Fig. 7) since it is virtually insoluble in the compound semiconductor; that is, k_{III} or k_V in the III–V compounds is $\approx 10^{-6}$. This excess solute lowers the freezing point of the solute and causes marked perturbations of the solid–liquid interface, as illustrated in Fig. 7. Also as a consequence of the break-up of the interface into cells, excess solute of group III or group V gives rise to marked nonuniformities in the grown crystal, including precipitates, dislocations, etc.

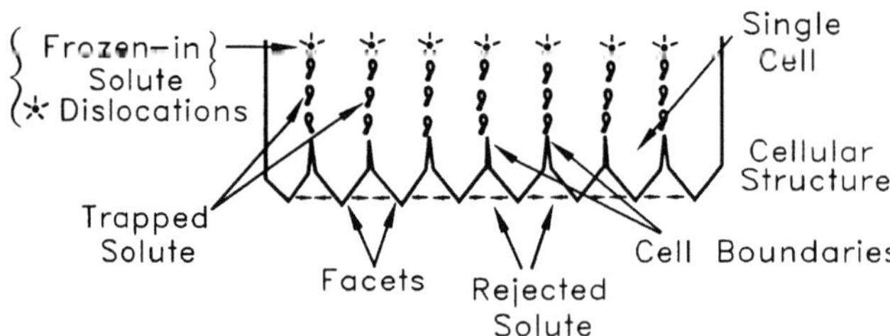

FIGURE 7.   Schematic illustrating the development of constitutional supercooling in the melt caused by excess component of a compound semiconductor or by heavy doping. The initially flat interface develops a faceted structure that results in individual growth cells. In the cell boundaries the rejected solute causes second phase precipitation, dislocations, grain boundaries, and very poor crystallinity generally.

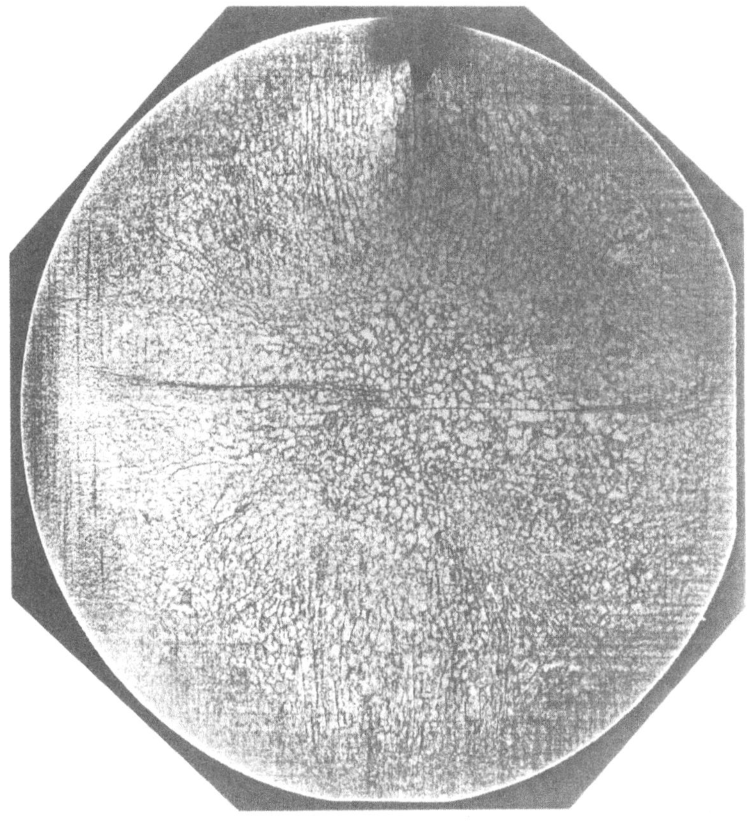

FIGURE 8. (a) Transmission x-ray topograph of a wafer of a 50-mm-diam (001) GaAs undoped semi-insulating LEC crystal showing the dislocation structure; (b) A B etched micrograph of the cellular structure near the center of a wafer similar to (a) above. Note the cellular dislocation structure. Courtesy D. Stirland, Plessey Caswell.

Constitutional supercooling readily occurs in the growth of dissociable compounds. It has been well researched and can be controlled but it appears not to be as widely appreciated as one might have thought.

Dominant problems of nonuniformity in crystals also arise during the postcrystallization phenomena. While this is true of elemental semiconductors, it is most important and significant in compound semiconductors. In these materials the concentrations of point defects can be very high. In a GaAs crystal at the melting point there are some 10^{18}–10^{19} point defects per cm^3, and probably more in CdTe at its melting point. These point defects interact with dopants, impurities, dislocations, etc. on cooling down the crystal after growth. The understanding of this behavior is a crucial requirement in, for example, controlling the nonuniform electrical properties of crystals. Figure 8a shows an example of a slice of a crystal cut from a GaAs crystal that had been grown by LEC. The x-ray topograph shows the so-called "cellular" dislocation structure. "Cellular" as used here does not imply cellular growth as in constitutional supercooling. Figure 8b shows the center of such a crystal etched to show up the cellular dislocation structure in the grown crystal. The dislocations probably originate from the thermal hoop stresses that arise as the crystal cools. The nonuniformities that are generated by the interaction between these dislocations and the point defects are a major source of nonuniformity and can directly affect, for example, the properties of FET devices used in integrated circuits.

The solution to these and many related problems is the key to new device markets. The problems are complex and their solution depends upon the evolution of new and more advanced preparation and crystal growth technologies.

REFERENCES

1. D. T. J. Hurle, Constitutional supercooling during crystal growth from stirred melts: I: Theoretical, *Solid State Electron.* 3, 37 (1961).
2. K. F. Hulme and J. B. Mullin, Facets and anomalous solute distributions in InSb crystals, *Phil. Mag.* 4, 1286 (1959).
3. G. K. Teal, Preparation of germanium single crystal by the pulling method: Introduction in *Transistor Technology* (H. E. Bridgers, J. H. Scaff, and J. N. Shive, eds.), Vol. I, Chap. 4. Van Nostrand, New York (1955).
4. J. B. Mullin, B. W. Straughan, and W. S. Brickell, Liquid encapsulation techniques: the use of an inert liquid in suppressing dissociation during the melt growth of InAs and GaAs crystals, *J. Phys. Chem.* 26, 782 (1965).
5. E. P. A. Metz, R. C. Miller, and R. Mazelsky, A technique for pulling single crystals of volatile materials, *J. Appl. Phys.* 33, 2016 (1962).
6. J. B. Mullin, R. J. Heritage, C. H. Holliday, and B. W. Straughan, Liquid Encapsulation crystal pulling at high pressures, *J. Crystal Growth* 314, 281 (1968).
7. E. M. Swiggard, S. H. Lee, and F. W. Van Batchelder, GaAs synthesised in pyrolytic boron nitride (PBN), in *Proc 6th Int Symp on Gallium Arsenide and Related Compounds, St. Louis, 1976, Inst. Phys. Conf. Series No. 33b,* 23, Institute of Physics, London (1977).
8. J. B. Mullin, Melt growth of III-V compounds by the Liquid Encapsulation and Horizontal Growth techniques, in *III–V Semiconductor Materials and Devices* (R. J. Malik, ed.), Chap. 1, pp. 1–72, Elsevier, New York (1989).
9. International Conferences on Crystal Growth, ICCG 1, *Crystal Growth,* Pergamon (1967); ICCG 2–9, *J. Crystal Growth,* Elsevier North-Holland (1968–1990).
10. *Progress in Crystal Growth and Characterization,* Pergamon, New York.

Single Crystal Growth II:
Epitaxial Growth

J. B. Mullin

1. INTRODUCTION: GENERAL PRINCIPLES

This chapter is concerned with the growth of single-crystal epitaxial layers of semiconductors. The term "epitaxy" was originally used to describe the regular growth of one material on another different material. With the introduction of vapor phase growth in the 1960s the use of the technology to grow one semiconductor on a substrate of the same material prompted the use of the term "homoepitaxial growth" in order to distinguish it from the original specialist term "epitaxy." Subsequently, however, the explosion of interest and widespread development of the technology has resulted in the crystal growth community abbreviating homoepitaxy to simply epitaxy. The growth of one material on a single-crystal surface of a different material is generally referred to in the crystal growth area as "growth on alternative" or, sometimes, "foreign substrates."

It is very important to realize that in all epitaxial processes there is a vital need for very high-quality substrates. It is essential that the epitaxial surface prior to crystal growth be defect-free and contamination-free. This need for high-quality substrates has accentuated the development of melt growth techniques.

The crystallization process and the driving force required to achieve epitaxy have been briefly referred to in the previous chapter on melt growth. It is interesting that in a period from the late 1960s a variety of different epitaxial techniques have evolved. They include liquid phase epitaxy (LPE), vapor phase epitaxy (VPE), metalorganic vapor phase epitaxy (MOVPE), molecular beam epitaxy (MBE), metalorganic molecular beam epitaxy (MOMBE), atomic layer epitaxy (ALE), and, more recently, Photo-MOVPE. This prompts the question: Why so many different epitaxial techniques and what is their role?

J. B. Mullin • Royal Signals and Radar Establishment, St. Andrews Road, Malvern Worcestershire WR14 3PS, U.K. *Present address*: EMC, The Hoo, Brockhill Road, West Malvern, Worcestershire WR14 4DL, U.K.

2. ROLE OF EPITAXY

Epitaxial growth is used in the provision of single high-quality semiconductor layers. These can be used as single layers for subsequent device fabrication, or in the provision of structures involving multiple layers. Epitaxial growth has emerged as a device fabrication process in its own right. This contrasts with bulk material, which is the starting point in device fabrication. The importance of epitaxial growth was rapidly appreciated with the initial work on the development of integrated circuits. These make use of planar technology for which epitaxy is ideally suited. Of course, the passivating properties of naturally grown silicon oxide was also a major feature that enabled the development of the integrated circuit.

Epitaxial growth has had, and continues to have, a major role in the development of virtually all GaAs, InP and related alloy devices. One of the first applications came with the development of liquid phase epitaxy (LPE), where it was found that growth from Ga-rich solution could produce very high-purity n-type GaAs which was suitable for producing high-performance devices such as high-frequency microwave devices, including transferred electron oscillators. Here the invention of the sliding boat technology (Section 4.1) was instrumental in the fabrication of suitable device structures.

More recently a whole new area of solid state physics based on the properties of quantum wells, superlattices, etc. (see Chapter 5 by Anderson) has emerged. The device exploitation and scientific development of this topic, that of so-called low-dimensional solids, is a direct consequence of the development of epitaxial structures using advanced techniques like MBE, MOVPE, etc. It is evident, then, that although the development of different melt growth techniques has primarily been in response to the need to grow materials with different physical and chemical properties, in the case of epitaxy the evolution of new techniques has been a consequence of the need to create new devices and produce device structures with ever more stringent requirements. The principles behind and the use of these epitaxial techniques will now be considered.

3. CONSTRAINTS TO EPITAXIAL GROWTH

In the case of melt growth, the constraints imposed on the suitability of a particular growth technique stem principally from the properties of the semiconductor. In the case of epitaxial growth, however, while the material's properties do form constraints to growth, the constraints that normally determine the suitability of a particular technique are more directly associated with the ability of the technique to cope with a particular device specification. Thus it is virtually impossible, for example, to grow good superlattices by LPE because one needs here to grow many alternate thin layers nanometers in thickness and atomically flat. The relatively high temperatures involved in most LPE techniques would prevent this by the interdiffusion that would occur in such circumstances. Further, it is very difficult in the LPE technique to carry out repeated layer growth, let alone maintain planar interfaces over large areas. It is appropriate then to consider here the general constraints associated with LPE and VPE.

3.1. Liquid Phase Epitaxy (LPE)

The chemical constraints to LPE are similar to those found in melt growth. Thus the exceptionally high chemical reactivity of silicon has prevented the development of a suitable LPE technique. However, the situation for most of the other semiconductors is that the lower temperatures of growth from solution compared with melt growth generally result in reduced chemical reactivity of the solutions and hence reduced contamination. Additionally the use of liquid metal solvents used to dissolve the semiconductors results in smaller distribution coefficients of impurities. As a consequence, LPE tends to favor the growth of high-purity epitaxial layers.

Vapor pressure is less of a problem in LPE, since it is very much lower than in melt growth, a consequence principally of the significant reduction in growth temperature. Thus all the III–V compounds can be grown without resort to encapsulation techniques. Mechanical and fundamental constraints, however, are similar to those in melt growth. LPE, however, attracts special problems associated with surface tension effects between the solid and solution. Thus inadequate wipe-off, the removal of the solution after epitaxial growth, is often a difficult problem, which can affect epitaxial layer quality and requires considerable development work in its avoidance.

3.2. Vapor Phase Epitaxy (VPE)

As in LPE, chemical constraints are also less severe in VPE techniques, the lower temperatures again being helpful. However, there are some situations where chemical reactivity can be a problem and needs to be controlled. The use of chloride transport in the growth of high-purity gallium arsenide is a notable example. Without proper control silicon can be introduced by the reaction of HCl with the walls of silica reactors. However, in the case of MOVPE and MBE, lower wall temperatures, additionally cooled in the case of MBE, are instrumental in removing this constraint in these technologies.

Pressure—that is, the operating pressure in a system, as opposed to the vapor pressure of a component over a semiconductor—is a fundamental parameter in VPE. Indeed, the use of very low background pressures, $\approx 10^{-11}$ torr, is the hallmark of MBE, a major crystal growth technology. However, pressure has important practical implications in VPE techniques. Atmospheric MOVPE is a favored technology for some firms for the production of many GaAs devices such as FETs, laser diodes, etc. However, other firms advocate and use reduced pressure MOVPE, typically working at a few tenths of an atmosphere. Such lower pressures are claimed to reduce premature reaction in MOVPE (see Section 4.4) and eliminate turbulent instability. The high flow rates provide a steady streamline flow, which is a factor in establishing dopant uniformity. However, well-designed MOVPE can avoid these undesirable effects. A disadvantage of higher pressures is that the consumption of precursors may be higher. Doping is also susceptible to pressure effects. Low pressure assists desorption processes. Thus it is not possible to dope Zn in the MBE of GaAs. A full understanding of pressure constraints in VPE would require a chapter on its own. However, some of the major implications of pressure conditions in systems will be dealt with as appropriate under each technique.

Mechanical constraints tend to be associated with equipment design. For example, there is a need to avoid the welding of clean contacting metal surfaces in the ultrahigh vacuum of MBE. In another example the design parameters in epitaxial reactors, especially the shape of the susceptor and its configuration and effect on flow, are very important. There is an overriding need to establish laminar flow in order to achieve doping uniformity. Whether one needs to rotate substrates in the vapor stream in order to achieve doping uniformity in MOVPE or in MBE is also an important consideration. In the latter case the use of confined area sources and the nature of molecular beams invariably means that large-area substrates need to be rotated. These latter aspects, of course, are intimately connected with the fundamentals of the hydrodynamics of gases. This science is of immense technical importance in VPE.

The development of epitaxy as a method of crystal growth of semiconductors has, even in such a short history, about one and a half decades, resulted in the evolution of a whole range of specialized techniques. This is a clear indication of the importance of the subject. Epitaxy is now an important commercial production activity and one can reasonably expect further rapid developments as the commercial demands increase. It will be of great interest to see which technologies will win out in the long run and whether the dominant technologies in research and production will be the same or different. So much, however, will depend on the evolution of device science and technology.

4. TECHNIQUES OF EPITAXIAL GROWTH

The following techniques have evolved from the beginning of the 1960s and the order in which they are considered is approximately the order in which they evolved.

4.1. Liquid Phase Epitaxy

The concept of LPE is simple and straightforward. Its practical implementation, however, is experimentally very demanding and requires considerable crystal growth knowledge, including detailed information on the thermodynamics of the system under investigation.[1] Essentially, a saturated solution of a semiconductor is brought into contact with a substrate at an appropriate temperature. On cooling the system epitaxial deposition occurs. The growth of GaAs from Ga-rich solutions provides a classic example of LPE. An essential requirement in LPE is a detailed knowledge of the phase diagram. The Ga:As system is reproduced as Fig. 1 of Chapter 8 by Hurle. The maximum liquidus temperature 1238°C occurs at a melt composition of 50:50 Ga:As under ≈ 1.0 atm of As. The addition of excess of either Ga or As reduces the liquidus temperature, or freezing point, of the solution. Lowering the temperature below the liquidus temperature at a fixed composition in the melt will cause GaAs to be precipitated or, in the presence of a substrate, result in epitaxial growth. As noted in Hurle's chapter, the excess solubility of Ga in GaAs is very small, at room temperature probably less than one part per million. Further, the excess point defects appear to have no deleterious effect on the electrical conductivity

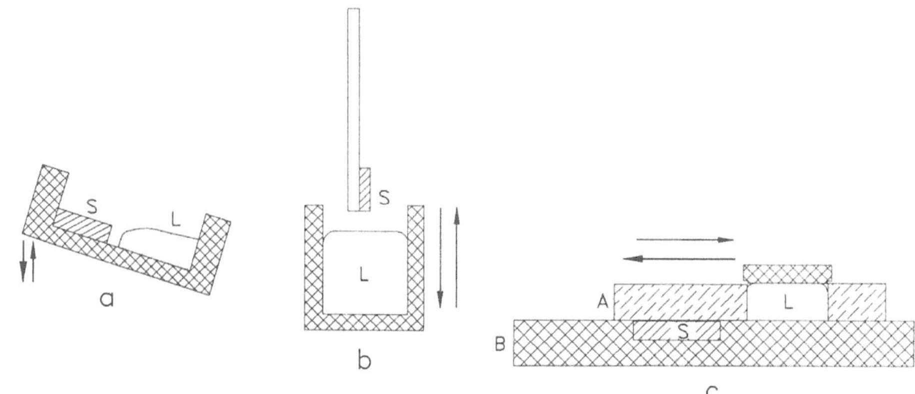

FIGURE 1. LPE: Schematic illustrations showing the three main ways by which the substrate and melt L are contacted and dewetted: (a) by tipping; (b) by dipping; and (c) by using a sliding boat A containing a well for the liquid solvent on a block B. B contains a recess in which the substrate S is located.

and extremely pure material can be grown using Ga-rich solution. It is possible, in principle, to grow from As-rich solution but here one runs into the problem of a high arsenic vapor pressure and the concept has not attracted significant research.

There are a variety of ways of growing epitaxial layers by LPE. These are indicated in Fig. 1. They differ principally in the manner in which the Ga-rich solution is made to contact the GaAs substrate in order to grow the epitaxial film. In the case of GaAs, the Ga-rich solution can be made to contact the GaAs substrate either (a) by tipping the saturated melt over the substrate, (b) by simply dipping the substrate into a saturated solution, or (c) by means of a horizontal sliding growth system. The choice of technology depends on the system undergoing epitaxy and the degree of control required. The tipping technique is relatively simple and can be used in closed systems which can readily be designed to avoid vapor loss. It generally lacks thickness control and there can be problems with dewetting. The dipping technique again can be used with volatile solvents, and with suitably designed multiple boats can be used to grow more than one layer, as for example in the growth of heterostructures. The horizontal sliding technique represents the most advanced technology and is used where maximum control accuracy and reproducibility is required. The sliding technology has been developed particularly for the growth of multilayers. The evolution of more advanced designs of semiconductor lasers, involving complex heterostructures, has been instrumental in the development of LPE and advanced sliding boat technology and other related systems.

The solution used for epitaxial growth is contained in a well or hole in the sliding bar A (Fig. 1), which is usually made from graphite. The initiation and termination of growth (wipe-off) is achieved by appropriate positioning of the solution relative to the substrates through movement of the slider A across the base B. Carefully programmed temperature cycles are used to establish equilibrium, etch-back of the substrate for surface cleaning, controlled growth and, finally, wipe-off. The use of multiple wells containing different solutions permits the growth of multiple layers. Intermediate wells containing different solutions permits the growth of multiple layers. Intermediate wells of solution can be incorporated for washing, where appropriate, so as to minimize the effect of carry-over and cross contamination between wells. In this way differently doped layers, involving homoepitaxial or heteroepitaxial growth, can be achived. In a double heterostructure laser device, for example, a sequence of layers may be required that involves growth of a heavily-doped p-type GaAs contact layer, a p-type GaAlAs layer for optical confinement, followed by a p- or n-type GaAs layer to define the laser and form the electrical p–n junction and, finally, an n-type GaAlAs layer also for optical confinement. LPE technology has been instrumental in the development and production of high-quality semiconductor lasers. Laser development involving complex structures tends now to be carried out now by MOVPE or MBE.

4.2. Vapor Phase Epitaxy: Conventional Inorganic Epitaxy

Vapor phase epitaxy normally refers to any technique that involves epitaxial deposition from the vapor. Here we consider specific situations involving the use of inorganic precursors in flow systems. Epitaxy is the result of chemical reactions that occur either directly or, following prereactions in the vapor, on the substrate where the transport is by diffusion.

By way of example, one may consider the growth of InP. Essentially one requires volatile compounds of the component elements that can be reduced on the substrate to form InP. The most convenient transporting agents for In are the halides. While phosphorus can be transported as P_4 this element is highly reactive and difficult to handle in a pure state. Following the original pioneering work of Knight et al.[2] on the growth of GaAs a standard technique has evolved that makes use of the chlorides of group V, PCl_3 in the case of InP. The process involves two stages. In the first stage, source saturation, H_2 is bubbled through PCl_3 and flowed over a heated boat of In. The PCl_3 decomposes in the H_2, releasing P, which dissolves in the In until the solution is saturated and a skin of InP forms across the surface of the In. The second stage involves flowing PCl_3 plus H_2 mixture over this same saturated In/InP source in an apparatus as illustrated diagrammatically in Fig. 2a. The reaction mechanism is a little complicated, but, essentially, the flow of the reagents over the source boat at $\sim700°C$ generates a near-equilibrium mixture of $InCl + P_4$ in the vapor stream. This mixture is transported down a heated silica reaction tube and over a substrate held at $\sim650°C$ where the supersaturated mixture reacts with the substrate to form an epitaxial layer of InP. In the source region in stage 1 the overall reaction is given by

$$PCl_3 + \tfrac{3}{2}H_2 + In_L \rightleftharpoons (1 - x)(In_L + P) + xInP + 3HCl \tag{1}$$

In stage 2 the pick-up and deposition reaction is given by

$$2InP + 2HCl \rightleftharpoons 2InCl + \tfrac{1}{2}P_4 + H_2 \tag{2}$$

A similar process for GaAs involving $Ga:AsCl_3:H_2$ was the pioneering process[2] for the production of high-purity GaAs. Alternative techniques involving HCl and solid GaAs can be used as reagents but purity and contamination can be

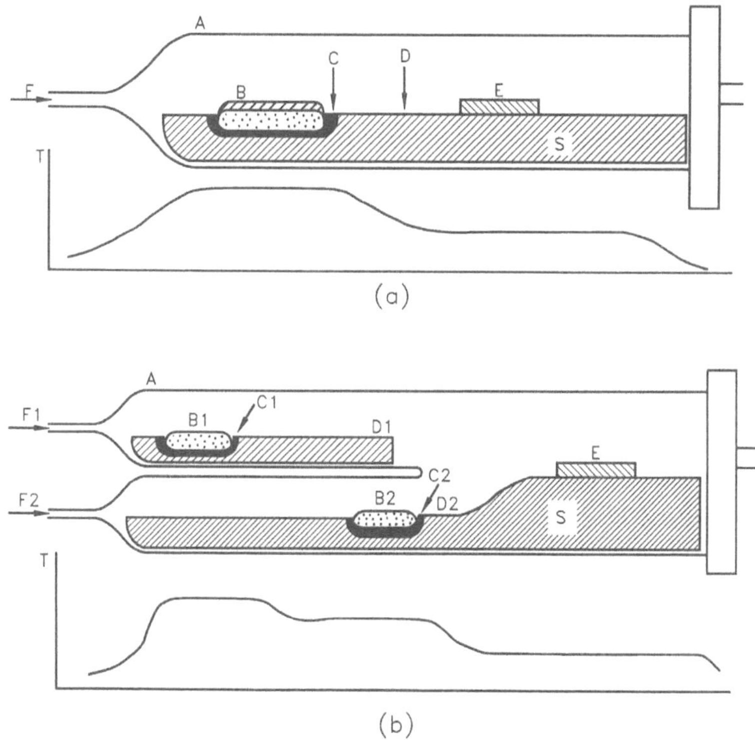

FIGURE 2.   Conventional inorganic VPE: Schematic illustrations showing the arrangement for the epitaxial growth of (a) a III–V compound; (b) a III–V alloy. In (a) the source boat contains the group III element over which a group V halide can be flowed. Two different source boats, c_1 and c_2, are used for different group III elements for the growth of $III_1\ III_2\ V$.

problems. Mixed alloys, including most of the alloys of the III-V compounds, can be grown in this way as illustrated in Fig. 2b. Separate source regions are used for the different matrix elements and their halides. A dual tube system is often used. This permits convenient mixing of the two reactant gas streams; a separate mixing region (not shown) is often used in order to achieve good uniformity of the epitaxially deposited alloy on the substrate E.

In the case of silicon, one can use halides of the element such as $SiCl_4$, $SiHCl_3$ as a transporting agent. These compounds can be reduced in hydrogen around ≈ 1200 and $\approx 1150°C$ respectively, as given by the following overall reactions:

$$SiCl_4 + 2H_2 \ \rightleftharpoons \ Si + 4HCl \tag{3}$$

$$SiHCl_3 + H_2 \ \rightleftharpoons \ Si + 3HCl \tag{4}$$

The reaction noted in Eq. (4) is also the basic process used for obtaining polysilicon rods used in float zoning. While $SiCl_4$ and $SiHCl_3$ were originally used in the growth of epitaxial layers for the preparation of device structures, the relatively high temperatures needed for reduction, coupled with the presence of HCl in the system, often caused memory problems in epitaxy. Epitaxial Si is normally grown using SiH_4 now that the latter compound is more readily available. This reaction occurs at significantly lower temperatures ($\approx 1000°C$) and eliminates the above problems. It is now the standard process for Si epitaxy. The overall reaction is as follows although the intermediate compound SiH_2 is involved in the reaction:

$$SiH_4 = Si + 2H_2 \tag{5}$$

The major difference in principle between the use of chlorides in the growth of III–V compounds and the use of hydrides with silicon is that the former requires a heated chamber as well as a heated substrate in order to prevent the semiconductor depositing throughout the system as well as on the substrate. In the case of SiH_4 the process can be carried out in a "cold wall system" using an RF heated susceptor plus substrate. It is effectively a very clean system. The silane process is used industrially for the growth of epitaxial Si device structures. (Silicon cannot be grown by MOVPE because it reacts readily with carbon-containing compounds to give SiC.) Silane in H_2 and doped silane are available in cylinders commercially, thus making the growth process relatively straightforward. However, it should be mentioned that silane needs very careful handling and stringent safety measures are needed since it can form explosive mixtures with air.

4.3. Molecular Beam Epitaxy (MBE): Metalorganic Molecular Beam Epitaxy (MOMBE)

This extremely important technique has rapidly assumed a position of premier importance in device fabrication since it can be used to grow epitaxial structures to atomic layer resolution.

The principle of the technique is illustrated in Fig. 3. The component elements of the compound or alloy are evaporated from heated Knudsen or effusion cells in an ultrahigh vacuum chamber, so arranged that the individual beams of atoms impinge on the heated substrate, where they combine to form a compound or alloy. The evaporation occurs under molecular flow conditions where the mean free path between atoms is larger than the system dimensions. This reduces to an absolute minimum the probability of gas collisions and premature reaction. It is a very clean, conceptually simple technology. In practice, however, it is extremely demanding, requiring enormous amounts of back-up technology and a very high level of experimental competence.

The flux F of atoms from a Knudsen cell of area A, at a distance d, is given by Eq. (6), where P is the equilibrium vapor pressure of the effusant, mass m, at the cell temperature T (K):

$$F = PA[\pi d^2 (2\pi mkT)^{1/2}]^{-1} \tag{6}$$

Equation (6) allows one to predict the maximum growth rate possible. Actual growth rates, however, require a knowledge of the sticking coefficients of the atoms.

Perhaps the most important characteristic of MBE is the capability of crystal growth at very low temperatures under ultraclean conditions. The low temperature can virtually eliminate significant interdiffusion between adjacent atomic layers,

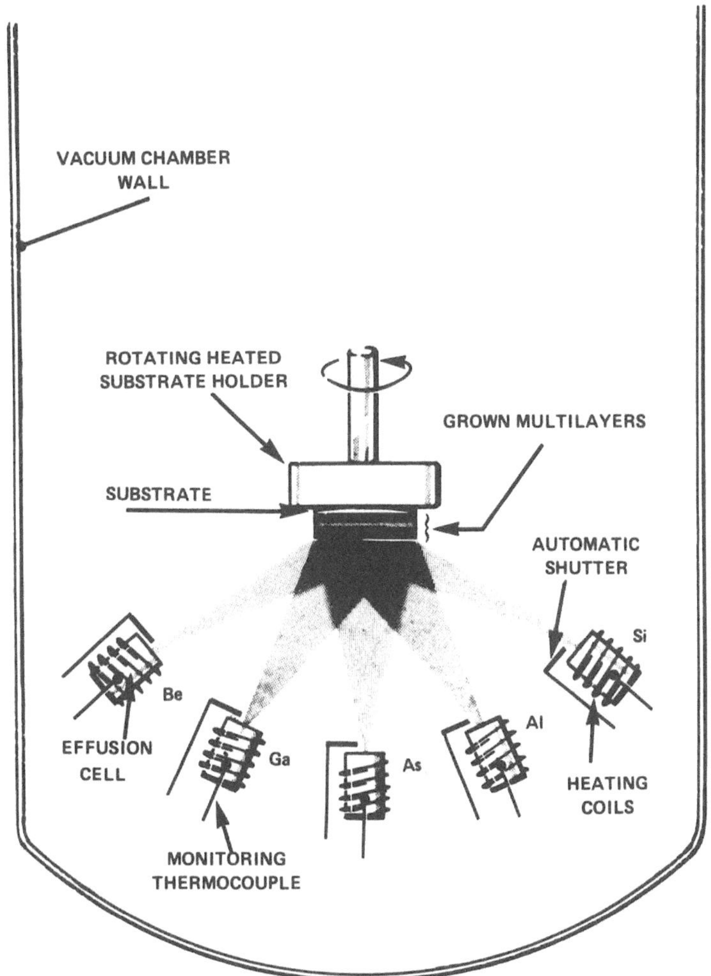

FIGURE 3.   MBE: Schematic illustrating the arrangement of effusion cells and rotating substrate/holder in the vacuum chamber.

thus enabling the growth of quantum wells, heterostructures, superlattices, etc. involving layers of different composition and thicknesses down to a few atomic layers. This permits the preparation of structures having engineered energy gaps. Figure 4 shows a photomicrograph in which alternate layers of GaAs and GaAlAs were grown in sequence, with periodically reducing thickness. The minimum thickness visible represents an intended single molecular layer. Crystal growth control to atomic dimensions is an astonishing achievement, which deserves wider recognition outside the field of crystal growth. The control achievable with MBE has enabled the growth of complex material structures and established the technology in the forefront of technologies for the exploitation of new devices based on new concepts in device physics and in solid state physics.

The high vacuum needed for MBE confers unique diagnostic capabilities for analyzing the growing or grown layer. A range of diagnostic equipment, including

FIGURE 4. Photomicrograph showing alternate layers of GaAlAs (white) and GaAs (black) grown by MBE. By reducing the thickness by a factor of 2 successively, one can discern near molecular layer growth. (Courtesy C. R. Whitehouse and A. Cullis, RSRE.)

RHEED, Auger, SIMS, etc., can be bolted onto the growth chamber to provide *in situ* analysis.

MBE does have disadvantages, apart from its size (see Fig. 5) and cost, which for a research machine may be upwards of £250k. As mentioned earlier, the high vacuum affects sticking coefficients and surface coverage, which can restrict its capability in growing certain materials or using particular dopants. It is not user-friendly with phosphorus compounds. It requires a very high degree of technical control and is not as well suited to production technology as MOVPE, but it is a research tool *par excellence* which has powered much new thinking in crystal growth and solid state physics. It is important to note that many of the MBE-initiated structures are subsequently developed by MOVPE techniques for production.

A technique related to MBE that is currently gaining considerable research interest is metalorganic molecular beam epitaxy (MOMBE). All vapor growth techniques require accurate control of the fluxes of the volatile components used in growth. It is particularly critical in the alloy growth of ternaries of specified composition, where the composition determines some physical property such as, for example, lasing wavelength. The production of accurate fluxes of the elements may require temperature control of the Knudsen ovens to better than 0.1°C, a difficult technical problem in a vacuum, especially if the temperatures need to be varied.

FIGURE 5. Photograph showing a typical MBE system.

Gas flows, however, can be accurately controlled using mass flow controllers to better than ±1.0%. The natural solution, then, is to use metalorganic sources for MBE. The gas flows of metalorganics and AsH_3 can both be cracked on passage through small separate heated furnace tubes. This permits the generation of accurate flows of the elements. The development of this hybrid technical approach, which is sometimes referred to as chemical beam epitaxy (CBE), is in an early stage of development but clearly has a promising future. There is every evidence that MBE and MOMBE will develop hybrid technologies, although the early demise of the parent basic technologies is unlikely.

4.4. Metalorganic Vapor Phase Epitaxy (MOVPE)

Metalorganic vapor phase epitaxy is often considered a rival technology to MBE, and, insofar as they can grow similar structures, this is true. The importance, however, of the two technologies lies in their ability to stimulate new knowledge on crystal growth, surface structure, adsorption characteristics, reaction mechanisms, etc., quite apart from the creation of new knowledge on materials properties.

In many ways MOVPE looked like a very unlikely technology, even some nine years ago in 1981. The use of organic compounds as transporting agents for the preparation of pure semiconductor compounds appeared distinctly unlikely, if only from the point of view of carbon contamination. However, results presented at the first international conference on MOVPE[3] by Dapkus[4] and Nakanisi,[5] who produced very pure, state-of-the-art high-mobility GaAs by MOVPE, eliminated that concern and the technology has developed from strength to strength with remarkable rapidity. MOVPE is now a central technology in the industrial production of lasers and photocathodes. It looks poised to take over the microwave field. It is interesting to examine why it has assumed such significance so rapidly.

The essence of the technique is as Manasevit[6] assessed it originally. Metalorganic precursors which can transport at room temperature are decomposed on a heated substrate supported within a cold wall environment. It is as simple and elegant as the decomposition of SiH_4 to produce Si. The growth of epitaxial GaAs provides an important example of the technology. The procedure involves flowing H_2 through a mass flow controller and then a cooled bubbler of Me_3Ga, which is a liquid that has a vapor pressure of 65 mm at 0°C. The diluted alkyl is flowed into, for example, a horizontal reactor (Fig. 6b) along with AsH_3 over an rf heated substrate of GaAs maintained at ~600°C. The overall reaction that takes place on the substrate can be represented by

$$(CH_3)_3Ga + AsH_3 \rightleftharpoons GaAs + 3CH_4 \tag{7}$$

It is interesting that the purest InP grown to date[7] has been prepared by MOVPE. Material having a peak mobility in excess of $400{,}000 \text{ cm}^2 \text{ V}^{-1} \text{ s}^{-1}$ at 50 K with a carrier concentration $<10^{14} \text{ cm}^{-3}$ has been grown using adduct purified Me_3In. The actual growth mechanisms in the growth of GaAs, for example, are complicated and the subject of intense study.[8] The loss of the first two methyl groups from a Me_3Ga molecule occurs more readily than the third and probably in the heated gas stream. Hydrogen from the AsH_3 is an important step in the process since it removes the methyl groups as methane and prevents C incorporation via, for example,

$CH_3Ga^{\cdot}$. It is important in this connection to maintain a high $[AsH_3]/[Me_3Ga]$ ratio in order to achieve good growth and prevent C incorporation.

The quality of MOVPE layers is also critically dependent on the flow conditions in growth reactors. The ideal situation is to achieve laminar flow as illustrated in Fig. 6b. The effect of the cold gas being rapidly heated when it meets the hot susceptor can cause a back pressure which may give rise to convective flow patterns near the input, as illustrated in Fig. 6a. Working at reduced pressure, typically ~ 0.1 atm or less can significantly reduce these convective instabilities, although they are best avoided by good design. Similar convection patterns to those at the input can also arise near the outlet. One of the effects of the initial convection pattern is often to cause premature reactions and semiconductor dust formation in the vapor, which results in poor quality surfaces in epitaxy. Premature reaction is an important phenomenon in MOVPE which should be eliminated. Giling[9] has reviewed the significance of the fundamental parameters that control fluid motion and highlighted the essence of good reactor design.

The key to the production of high-quality semiconductors lies in the preparation of high-purity precursors. An important innovation in this regard is the use of adducts, organic compounds that can release the base-free material on heating but are sufficiently stable to allow the removal of many unwanted organics by distillation. A second-stage distillation of the pure base-free alkyl allows the preparation of semiconductor grade material.[10] Much of the future of MOVPE lies with the interaction between the crystal grower and the organic chemist, in the design and production of tailor-made precursors having the right volatility and decomposition behavior. An example of this can be found in the growth of $Cd_xHg_{1-x}Te$, an important infrared detector material. This material can be prepared using Me_2Cd, Et_2Te, and Hg vapor from a liquid source using an MOVPE growth system such as that illustrated in Fig. 7. The epitaxial growth temperature depends on the decomposition temperature of the Te alkyl. Current research is aimed at reducing the growth

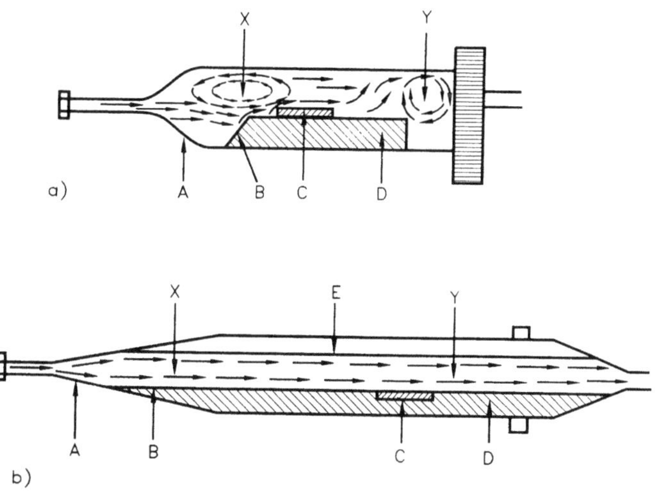

FIGURE 6. MOVPE: Schematic diagrams showing (a) convection patterns that can give rise to premature reaction; (b) system design aimed at achieving laminar flow.

FIGURE 7. Photograph showing an MOVPE growth system used for the preparation of epitaxial $Cd_xHg_{1-x}Te$.

temperature from $\approx 410°C$ set by the decomposition of Et_2Te to lower temperatures. The use of iPr_2Te can reduce the growth temperature to 350°C and methyllalyltelluride to 300°C. The growth of CdTe can be achieved at $\approx 200°C$ using diallyltelluride or dimethylditelluride.[10]

Lower temperature growth is not the only requirement in MOVPE. Chemical decomposition characteristics are also important. The search for alternative to arsine has led to the use of alkyl-substituted arsines. Here the use of fully substituted arsines like Me_3As resulted in the growth of highly *p*-type GaAs, a result of carbon contamination. There is a need to include H_2 on the As group in order to minimize C incorporation. One can expect future research to employ more cleverly designed precursors in order to achieve optimum growth results.

4.5. Photolytic Metalorganic Vapor Phase Epitaxy (Photo-MOVPE)

The thermal decomposition temperature of the most stable precursor is frequently the parameter that sets the lower limit of epitaxial growth of a compound by MOVPE. Thus, in the case of the growth of HgTe and $Cd_xHg_{1-x}Te$ using Hg(v), Me_2Cd, and Et_2Te it is the decomposition temperature $\approx 410°C$ of Et_2Te in H_2. In order to achieve lower temperature growth one can use less stable precursors, as noted previously, or alternatively use ultraviolet photons that are sufficiently energetic to bring about the decomposition of the organometallic. One of the first uses of this concept, namely, to bring about the epitaxial growth of a semiconductor on a substrate in contrast to the deposition of a semiconductor in a polycrystalline form, was reported by Irvine and Mullin[11] for the epitaxial growth of high quality HgTe on CdTe. The apparatus used is illustrated schematically in Fig. 8. In this work a broad-band ultraviolet (uv) lamp was used to bring about the decomposition of Et_2Te in a stream of flowing H_2 in the presence of an ambient of Hg(v). Epitaxial growth was achieved at $\approx 200°C$ some 200°C below the pyrolytic growth temperature using Et_2Te, a very significant result.

The importance of this work is not just in the ability to achieve a lower temperature growth but in the potential that it offers for local area depositon and

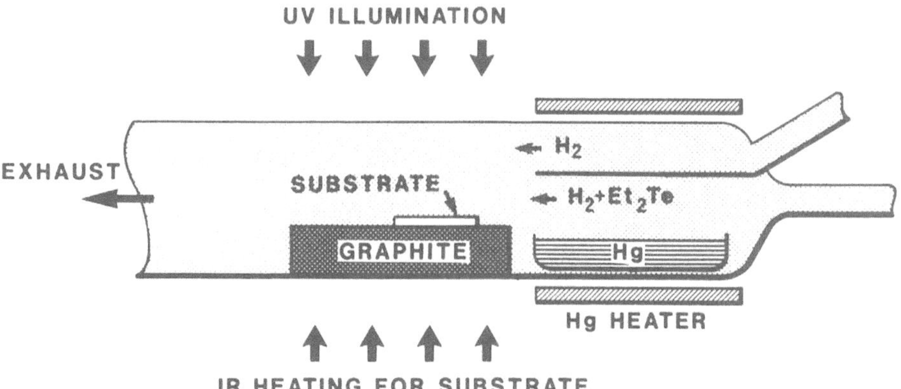

FIGURE 8. Photo-MOVPE: Schematic diagram illustrating the photolytic growth of epitaxial HgTe on a CdTe substrate using a broad band uv lamp.

epitaxial photopatterning. The use of uv light, especially using lasers, is now gaining widespread acceptance as a tool in order to achieve local area deposition, not just of semiconductors but also of metals and insulators. Photochemical processing is fast developing a branch of crystal growth that should meet many of the challenges and problems posed by device engineers. It could well, in the future, be used in an important role in automated *in situ* fabrication technologies, although much more research will be needed before wider engineering acceptance of this embryo technology can be achieved.

REFERENCES

1. G. B. Stringfellow, LPE of III–V semiconductors, in *Crystal Growth: A Tutorial Approach*, pp. 217–239 (W. Bardsley, D. T. J. Hurle, and J. B. Mullin, eds.), North-Holland, Amsterdam (1979).
2. J. R. Knight, D. Effer, and P. R. Evans, The preparation of high purity GaAs by vapor phase epitaxial growth, *Solid State Electron.* **8**, 178–180 (1965).
3. Metalorganic vapor phase epitaxy, *Proc. 1st Int. Conf. on Metalorganic Vapor Phase Epitaxy, Ajaccio, France, 1981*, in *J. Crystal Growth* **55**, 1–262 (1981).
4. P. D. Dapkus, H. M. Manasevit, K. L. Hess, T. S. Low, and G. E. Stillman, High purity GaAs prepared from trimethylgallium and arsine, *J. Crystal Growth* **55**, 10–23 (1981).
5. T. Nakanisi, T. Udagawa, A. Tanaka, and K. Kamei, Growth of high purity GaAs epilayers by MOCVD and their applications to microwave MESFETs, *J. Crystal Growth* **55**, 255–262 (1981).
6. H. M. Manasevit, Recollections and reflections of MOCVD, *J. Crystal Growth* **55**, 1–9 (1981).
7. E. J. Thrush, C. G. Cureton, and A. T. R. Briggs, MOCVD grown InP/InGaAs structures for optical receivers, *J. Crystal Growth* **93**, 870–876 (1988).
8. *NATO Workshop on Reaction Mechanisms of Organometallic Compounds with Surfaces* (D. J. Cole-Hamilton and J. O. Williams, eds.), University of St. Andrews, Scotland, 1988; NATO AS1 Series B Physics, Vol. 198, Plenum, New York (1989).
9. L. J. Giling, Principles of flow behavior: Applications to CVD reactors, in *5th Int. Summer School on Crystal Growth and Materials Research*, Davos, Switzerland.
10. A. C. Jones, A. K. Holliday, D. J. Cole-Hamilton, M. Munir Ahmad, and N. D. Gerrard, Routes to ultrapure alkyls of indium and gallium and their adducts with ethers, phosphines, and amines, *J. Crystal Growth* **68**, 1–9 (1984).
11. S. J. C. Irvine, J. B. Mullin, J. Giess, J. S. Gough, and A. Royle, Recent developments in the pyrolytic and photolytic deposition of (Cd, Hg)Te and related II-VI materials, *J. Crystal Growth* **93**, 732–743 (1988).

11

Amorphous Silicon-Electronics into the 21st Century

P. G. LeComber

1. INTRODUCTION

The last few decades have seen a revolution in the field of electronics and a tremendous increase in the number of applications of crystalline semiconductors such as silicon. These developments were possible because the electrical conductivity of crystalline silicon can be controlled over many orders of magnitude by doping, that is, by the addition during growth of the crystal of small concentrations of impurities. However, there are a number of areas where the expense of preparing these crystals and where the limited size to which they can be grown (at present about 25 cm in diameter) have prevented any very large-area applications. For example, crystalline silicon solar cells are widely used in space vehicles for converting sunlight into electrical power, but the economics of their production is such that their use here on earth is relatively limited.

Silicon can be prepared very cheaply in large areas by vacuum evaporation or by sputtering, but the material is then amorphous rather than crystalline. Until just over a decade ago it was believed that the amorphous material so prepared could not be doped and was not therefore suitable for use in modern electronics and in solar cells. This chapter describes how this limitation was overcome following the initial advances by Spear and LeComber.[1] In the 14 years since their work on doping amorphous silicon (a-Si) was published, there has been considerable research into and development of this material, leading to a number of commercial products. The products currently available are listed in Table 1, and these were estimated to have a turnover in 1988 of approximately \$500 M for the a-Si parts only! The annual turnover for the solar cells alone is estimated to already exceed \$50 M. It is also highly likely that within the next five years some of the other applications proposed for this material (Table 2) will appear commercially. By any standards the development from the early pioneering fundamental work to the present stage of industrialization is most remarkable.

P. G. LeComber ● Department of Applied Physics and Electronic and Manufacturing Engineering, University of Dundee, Dundee DD1 4HE, Scotland

TABLE 1.   Amorphous Silicon Products Commercially Available (1989)

Device	Product
Photovoltaic cell	Calculators, watches, battery chargers, etc.
Photoreceptor	Electrophotography LED printers

Photoconductors,
  image sensors, and
  position sensitive
  detectors
	Color sensors, light sensors,
	  contact-type image sensors, electronic white
	  boards, spatial light modulators, computer
	  paint-brush table, etc.

Heat control layer	Heat-reflecting float glass

Thin-film field-effect
  transistors (FETs)
	Displays, televisions
	  logic circuits for image sensors

High-voltage thin-film transistors	Printers

TABLE 2.   Other Proposed Applications of Amorphous Silicon

Image pick-up tubes	Optical recording
	LEDs
FETs for ambient sensors	Passivation layers
Fast detectors and modulators	Charge-coupled devices
Diodes	Memories
DiFETs	Strain gauges
Bipolar transistors	Photolithographic masks
Optical waveguides	X-ray detectors
Optically modulated neural networks	Charge particle detectors

2. AMORPHOUS MATERIALS

To see what is meant by the term "amorphous," note that in a crystal the atoms
or molecules are arranged in a regular structure and the periodicity of the arrangement can extend over a distance of centimeters with remarkably few structural
defects, as indicated in the model (Fig. 1a). On the other hand, in an amorphous
solid there is no ordered structure, although the basic properties of the chemical
bonds that bind the solid together define a certain amount of short-range order, as
shown in the model in Fig. 1b. The short-range order merely extends over a few
atomic spacings, that is, over a distance of about 1 nm from any given atom. In an
amorphous solid, the periodic structure of the crystal is replaced by a random
network of atoms or molecules. This difference in the atomic arrangement in the
crystalline and amorphous phases can influence many of the properties of the
material.

Within the present context the most important difference between crystalline
Si and an amorphous semiconductor such as a-Si is that in the latter there is a
continuous distribution of localized states within the forbidden energy gap. This is
illustrated in Fig. 2, where the density of states function $g(E)$ is plotted for the
crystal (Fig. 2a), and for the a-material (Fig. 2b). The density of states $g(E)$ is
simply the number of electron states, per unit volume of the material, per unit energy

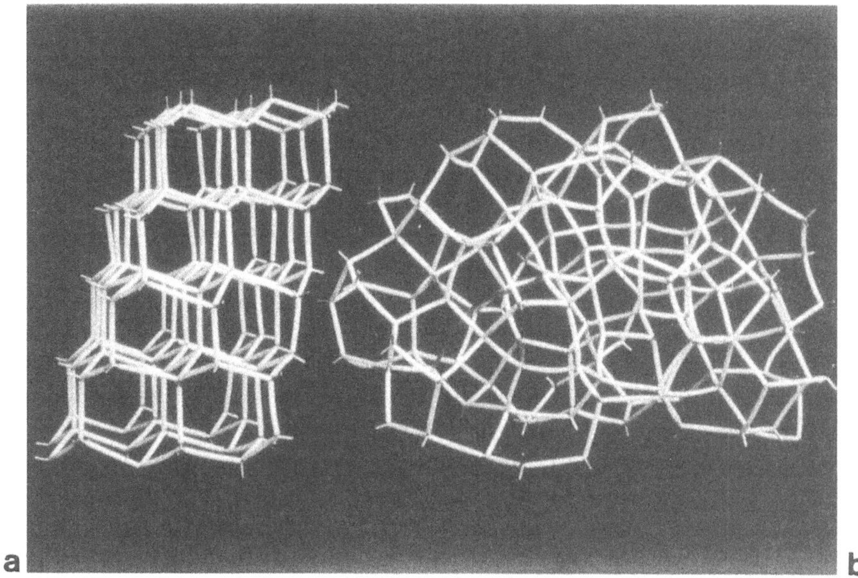

FIGURE 1.   Models illustrating the structure of (a) crystalline and (b) amorphous materials.

range in a narrow energy interval centered around the energy E, and is discussed in Chapter 3 by Barr. In the crystalline material $g(E)$ is finite in the allowed energy bands and, in the absence of impurities and crystalline defects, zero in the forbidden energy ranges. Although the details may vary, the form of $g(E)$ shown in Fig. 2b is believed typical of elemental amorphous solids such as a-Si and a-Ge. This curve

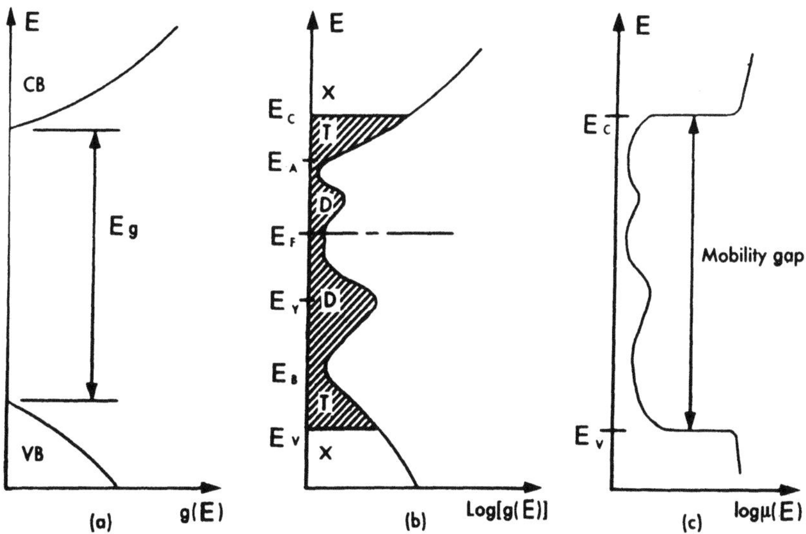

FIGURE 2.   (a) Density of states function $g(\varepsilon)$ in a crystalline semiconductor; CB and VB denote the conduction and valence bands, respectively. (b) Density of states in an amorphous semiconductor: E, extended states; T, localized tail states; D, localized defect states. (c) The mobility as a function of energy for the amorphous sample shown in (b).

is based on data determined experimentally for a-Si and will be used as an example in the following discussions.

Above E_C and below E_V the electron wave functions will extend throughout the crystal so that these states, denoted by X, are extended states. However, the states between E_C and E_V, shown shaded in the figure, are localized states. Those marked T getween E_C and E_A and between E_B and E_V are thought to arise because of the lack of long-range order and are called tail states. The states denoted by D arise from defects in the material and the density of these depends critically on the method used to prepare the amorphous film.

Another important difference between amorphous and crystalline materials concerns the mobility of the electrons or holes. In a crystal, an electron or hole in the extended band states can move many atomic distances before it is scattered by something that perturbs the periodicity of the lattice. In an amorphous material the periodicity of the lattice only extends over a few atomic spacings. Under these conditions the electron transport may no longer be considered as band motion with occasional scattering, as in crystalline theory. In this case the electron motion is essentially a diffusive process which can be considered similar to the Brownian motion of small particles in liquids. The analysis of electron motion in the extended states of amorphous Si typically leads to values of mobility of 10^{-3} m^2 V^{-1} s^{-1}, a factor of over 100 smaller than in the crystal. The fairly sudden transition from extended to localized states at E_C and at E_V is of fundamental importance in the understanding of amorphous materials. Since all the states between E_C and E_V, i.e., the tail states T and the defect states D in Fig. 2b, are localized, we would not expect the same transport process to occur in these levels as in the extended states just above E_C and below E_V. What is the nature of the transport through these localized gap states?

The process, normally called hopping, involves the electron picking up energy from the lattice vibrations in order to overcome any potential barrier between it and another localized site that is not too far away. In this case the mobility depends critically on the density of the localized sites, and on temperature, the latter since this provides the thermal energy for the lattice. Typically the hopping mobility will be less than 10^{-5} m^2 V^{-1} s^{-1}.

Hopping conduction in crystalline silicon and germanium is only observed at very low temperatures. However, it is probably the mechanism by which current flows in the majority of nonmetallic solids and plays an important role in many amorphous materials.

The electron mobility in the localized states shown in Fig. 2b is therefore significantly lower than in the extended states. The energies E_C and E_V identify the energies at which the states change from extended to localized in nature, and are therefore associated with a large change in mobility as shown in Fig. 2c. E_C and E_V are usually called mobility edges. Notice that in an amorphous material there is a mobility gap, defined by $E_C - E_V$ as in Fig. 2c, rather than the forbidden energy gap E_g in the crystalline density of states (Fig. 2a).

Even in the extended states of e.g., a-Si, the mobilities are more than two orders of magnitude smaller than in crystalline Si. It is therefore very unlikely that a-Si will ever replace crystalline Si in applications where very high speed is of importance. As we shall see shortly, a-Si has made the biggest impact in applications where cheapness and large area are important. Before this was possible, however, it was first necessary to dope the material, and this will be considered in the following section.

3. DOPING OF a-Si

Earlier we mentioned that the room temperature conductivity of crystalline semiconductors could be controlled by doping the material with suitable impurities. Many attempts were made to do the same with amorphous semiconductors prepared by vacuum evaporation and by rf sputtering but they were unsuccessful. Consequently, it was suggested that doping might even be impossible in an amorphous material, since the random structure might allow all bonds to be satisfied. Experiments at Dundee, and subsequently in many other laboratories, have shown, however, that if the material is prepared with sufficiently low density of localized gap states then it is possible to control the room temperature conductivity σ_{RT} of a-Si over many orders of magnitude. To understand how this is done it is necessary to discuss how a-Si films are prepared.

The technique, originally developed in the laboratories of STL in Harlow, Essex[2] is shown schematically in Fig. 3. Silane gas, SiH_4, flows via a flowmeter F into a reactor over a substrate S which is heated to about 300°C. The pressure inside the reactor is maintained at about 0.1 torr by vacuum pumps. Using a few watts of power from a radio-frequency generator usually operating at 13.56 MHz, a weak glow discharge is excited between the two electrodes E_1 and E_2, one of which contains the heated substrate S. In this discharge the silane breaks down and a reddish-brown a-Si film is deposited onto the substrate to a thickness typically of the order of 1 μm. The major advantage of this process is that it deposits a-Si films with some three orders of magnitude lower density of localized states (the shaded areas marked D in Fig. 2b) than in films prepared by thermal evaporation or sputtering. The reason this is so important is that those localized states would cancel out (compensate) the effects of the doping impurities and if present in too large a number effectively prevent doping.

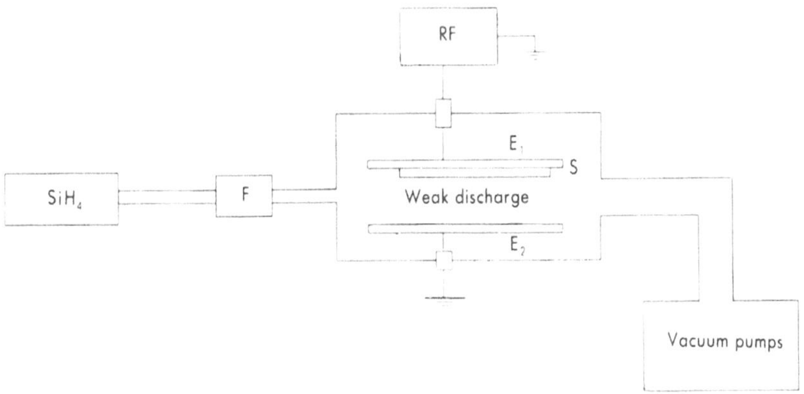

FIGURE 3.   Schematic diagram of systems used for plasma-enhanced deposition of a-Si from silane.

By adding small concentrations of phosphine (PH_3) or diborane (B_2H_6) to the silane gas used to prepare the film, Spear and LeComber demonstrated in 1975[1] that doping is possible in these glow discharge films with their much lower density of localized states. The changes in the magnitude of the room temperature electrical conductivity σ_{RT} that can be achieved in this way are shown in Fig. 4, where log σ_{RT} is plotted vs. the impurity gas concentration. On the right-hand side this is the ratio $N(PH_3)/N(SiH_4)$ of the number of phosphine to silane molecules in the gaseous mixture used to prepare the film, whereas on the left the corresponding diborane-to-silane ratio is shown. In the center of the graph, the conductivities of 10^{-8}–10^{-9} $(\Omega \text{ cm})^{-1}$ are typical of undoped glow discharge specimens.

Consider first phosphine doping where measurements show the samples to be n-type. Adding even a minute quantity of phosphine increases σ_{RT} dramatically.

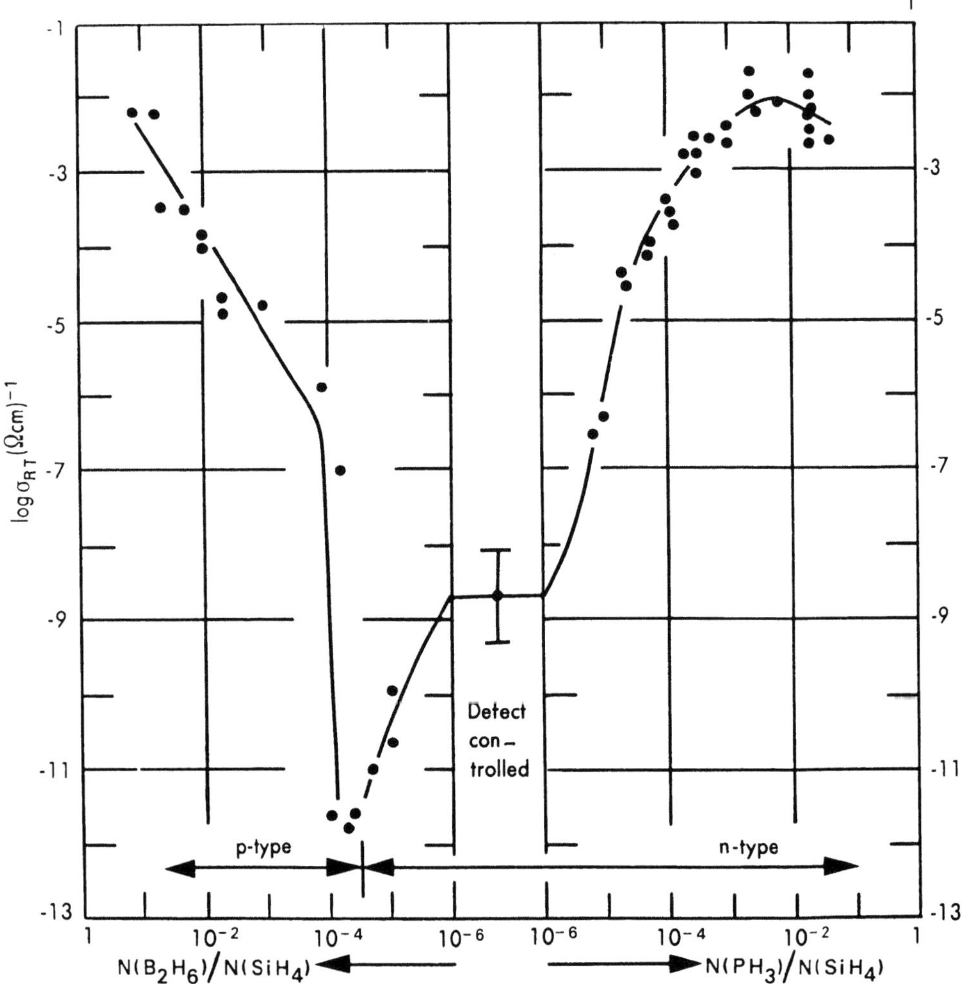

FIGURE 4.   The log of the room-temperature electrical conductivity of a-Si as a function of doping gas concentration.

With diborane doping σ_{RT} initially decreases because the B atoms compensate the normal n-type nature of the films. Once the concentration of B_2H_6 in the SiH_4 exceeds a certain concentration, however, the sample becomes p-type. Additional B doping increases the conductivity rapidly. Notice that the conductivity can be increased from a value of 10^{-12} $(\Omega\ cm)^{-1}$ by ten orders of magnitude to 10^{-2} $(\Omega\ cm)^{-1}$, by making the material either n-type with P doping or p-type with B doping. The ability to dope a-Si (and a-Ge) in a controlled manner over such a large range is of interest from both a fundamental and an applied point of view.

From the applied point of view the ability to dope the material was of crucial importance. Shortly after the publication of the doping results early in 1975, results for the first all-amorphous p–n junction were reported.[3] Preliminary measurements of the photoresponse of these devices showed the typical characteristics of photovoltaic cells. Shortly afterwards, Carlson and Wronski[4] of the RCA Laboratories in the United States reported more detailed photovoltaic characteristics of a-Si p–i–n structures. Although these early devices were relatively inefficient, these papers aroused considerable interest in the possible applications of the material to cheap, large-area photovoltaic devices and many industrial laboratories entered the field. Since that time progress has been very rapid, not only in solar cell efficiency but, as listed in Tables 1 and 2, in other applications.

4. APPLICATIONS OF a-Si

a-Si prepared by the glow discharge decomposition of silane has a number of properties, shown in Table 3, which make it suitable for many applications.[5,6] Of particular importance is the fact that it can be deposited relatively cheaply over large areas and can therefore be used in just those applications where crystalline Si would be too expensive. The first application where a-Si was produced commercially was in photovoltaic cells for converting light into electrical power and this will be described first.

4.1. Photovoltaic Cells

Photovoltaic cells are basically p–n junctions which when illuminated with light convert this energy into electrical energy. Figure 5 shows a typical I–V curve

TABLE 3. Properties of a-Si Films of Importance for Applications

• Thin film (1 μm thick)	• Room temperature electrical conductivity can be controlled over ten orders of magnitude by doping for both n-type and p-type material
• Low deposition temperature	
• Large area growth on many substrates such as glass, metals, flexible plastics	
• Mechanically very hard	• Easy to sequentially produce p-type and n-type material by switching from one gas mixture to another
• Chemically very stable	
• Inert material	
• Extremely photoconductive	• Easy to pattern arrays of devices using conventional photolithographic techniques developed for crystalline Si

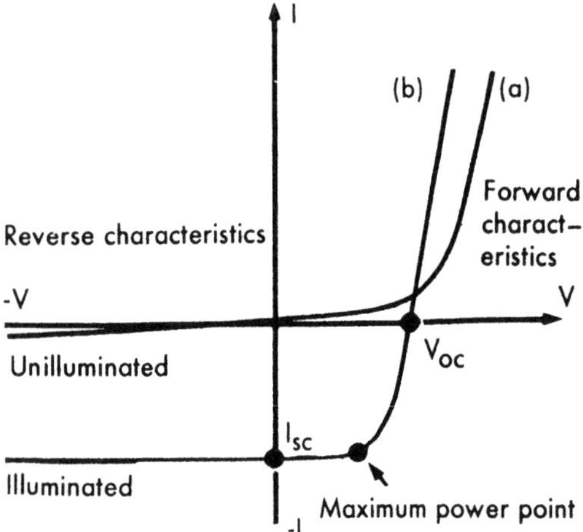

FIGURE 5. Current–voltage (I–V) characteristics of a p–n junction (a) in the dark and (b) illuminated.

of a p–n junction with its familiar dark characteristic shown as (a). The characteristics are usually generated by applying a potential to the p–n junction and measuring the current that flows. The part of the illuminated characteristic, shown as (b) in Fig. 5, between the points marked I_{SC} and V_{OC} can be generated simply by placing a load resistor in series with the p–n junction. If this load resistance is very small then the illuminated cell generates the maximum amount of current, called the short-circuit current I_{SC}, but zero voltage. If the load resistance is very large then the illuminated p–n junction generates the maximum voltage, called the open-circuit voltage V_{OC}, but a very small current. Somewhere in between these two extremes the cell generates the maximum output power given by the maximum product of current and voltage. This is the point where the cell has the highest efficiency for converting light energy into electrical energy. The solar cells of 1976 had a relatively low efficiency of about 2%–3% (compared to 15% for a good crystalline cell). Progress since then has been rapid, with conversion efficiencies increasing to 12% or 13% for small-area ($1\,\mathrm{cm}^2$) cells. For larger-area cells, today's technological problems decrease the efficiency, although more than 7% is now commonly obtained for 10 cm × 10 cm cells even in industrial production. Although further improvements will be required before a-Si p–n junctions can be used for large-scale electricity production, the efficiency of the cells is more than adequate for many consumer applications. The Sanyo Electric Company were the first to put a-Si cells into industrial production and they alone now fabricate over 5 million per month which would have a combined output of 5 MW under illumination by the sun when directly overhead! A familiar use is to power watches and calculators.

In spite of the present problems that arise from using a-Si solar cells for large-scale energy production, a number of pilot schemes are on trial. Sanyo Electric Company have had a-Si solar cells providing approximately 5 kW of power to a house in Osaka for a number of years. A few years ago Chronar Corporation installed a 100-kW a-Si solar cell micropower station for Alabama Power in the USA. Recently a start has been made on a 10-MW a-Si power plant in India and plans are far advanced by the Chronar Corporation for a 50-MW a-Si power station. Hopefully, the step required for a breakthrough into the important market of economic power production may not be far away.

It would be incorrect, however, to suggest that all our energy needs will eventually be provided by solar cells, a-Si or otherwise. An EEC report has estimated that, even by the year 2000, less than 10% of Europe's energy needs will be generated by solar cells. They will obviously continue to have a big impact as cheap power supplies for consumer products but perhaps more importantly they are likely to find a bigger application in third-world countries where national grids are not established. One such application is where the a-Si solar cells drive a water pump which can irrigate a small field.

4.2. Flat Screen Televisions and Displays

The flat liquid crystal displays used in watches and calculators are very familiar. Why do we not commonly have televisions making use of this technology? The answer is that the normal technology used to turn each picture point on and off will not work on a display with too many lines and certainly not with the 625 lines required by UK televisions. There is a way around this problem, which requires a small electronic switch at each picture point on the display. In the 1970s LeComber and Spear developed with their colleagues an a-Si field effect transistor (FET), also called a thin-film transistor, which has all the properties required by the electronic switch in this application.[7-15] (See Fig. 6 in Chapter 27 by Day, who discusses such displays in more detail.) They were able to demonstrate that the a-Si FETs could be used on 1000 line displays, more than enough for television. Since that time there has been a considerable development effort by many industrial companies throughout the world to put large area displays on the market.

Color is achieved simply by using at each picture point three liquid crystal elements each with a different primary color filter. Prototype resolution is typically half that of a normal TV because even that requires around a quarter of a million a-Si FETs, and commercial production with sufficient yield of 100% functional devices on a screen presents great difficulties. In particular, the large area of the product requires the development of new deposition systems and even new photo-lithography equipment, since display sizes are likely to soon exceed those conventionally used for VLSI fabrication. In spite of these difficulties, LC color displays using an active matrix of a-Si FETs are presently sold by Matsushita (3-in. display in Japan and Europe), Sharp (3 in. and 4 in. in Japan), Toshiba (4 in. in Japan), Hitachi (5 in. and 10 in.), and Hosiden (10 in.). In addition, Philips (6 in.) and CNET/Sagem have established pilot production plants and both Sharp and IBM Japan/Toshiba have stated that 14 in. color displays with more than 10^6 FETs will be commercially available sometime in 1990.

4.3. High-Voltage Thin-Film Transistors

In the last year or so, a very novel development of the a-Si FET has been developed by Tuan[16] and colleagues. The device, shown in Fig. 6, resembles the a-Si FET except that the gate electrode is displaced. Instead of being aligned with the gap between the source and drain electrodes, it is aligned directly opposite the

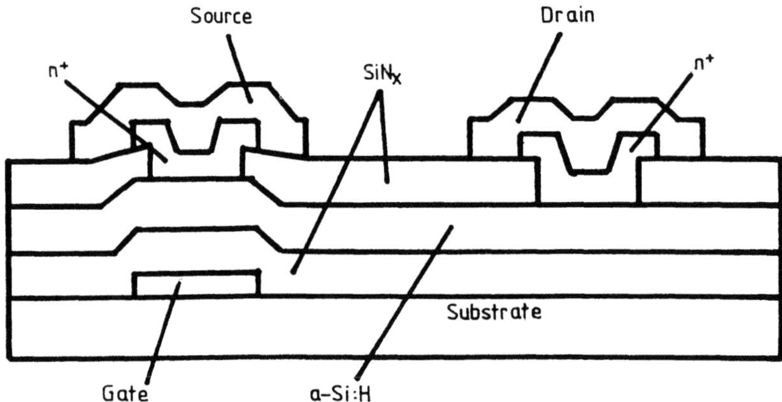

FIGURE 6. Structure of an a-Si high-voltage thin-film transistor (after Ref. 16).

source electrode. The device operates in the following way. With the gate voltage $V_G = 0$ the small conductivity ensures that only a small current flows between the source and drain electrodes. If, however, V_G is increased to only +6 V or so, then within a microsecond strong electron injection occurs from the source electrode into the a-Si. A large voltage (up to 400 V) applied between source and drain sweeps the injected carriers to the drain electrode resulting in source–drain currents of typically 5 μA for the devices investigated. Removal of the gate voltage causes the source–drain current to rapidly decrease to its OFF-value. The importance of this device is that small gate voltages, less than 10 V, can be used to control current flow in a high-voltage (400 V) circuit. Coupled with the fact that linear arrays of these devices can be produced, this means that they are likely to have a major impact on office equipment[17] in the form of electronic printers, such as ionographic and electrographic printers, and input scanners. In addition, a linear array of a-Si image sensors have been incorporated directly onto an a-Si print head to form an a-Si electronic copier.[17] These products are likely to form a major growth area for a-Si.

4.4. Electrophotography

In the last twenty years or so the photocopier has become an essential part of an office. The part of the machine that actually copies the document is called the photoreceptor and is typically contained on the outer surface of a cylindrical drum. The properties required for an ideal photoreceptor are listed in Table 4. Most of the materials that have been used over the years for this purpose have not met anywhere near all of these requirements. For example, the initial photoreceptor was

TABLE 4. Requirements of the Photoreceptor in a Photocopier

• High dark resistivity	• High photoconductivity
• Stable and chemically inert	• Not too sensitive to heat
• Mechanically hard	• Nontoxic

amorphous Se, which is soft, and therefore easily scratched, and also has the disadvantage of being toxic, thus causing problems when the photoreceptor drum has to be discarded.

Amorphous silicon[18] appears to satisfy all the requirements in Table 4, and Canon are now using it as a photoreceptor in three of their machines, which are on sale worldwide. The a-Si photoreceptor is very durable and produces images of excellent resolution and contrast. The lifetime of the a-Si photoreceptor is sufficient for many more copies ($>10^6$) that can be obtained with other materials. Sales have exceeded one million machines containing a-Si photoreceptor drums and each of these drums represents approximately $100 of business! This is therefore an important application of a-Si which will no doubt grow even further in the years ahead.

4.5. Image Sensors

Fuji–Xerox have made use of the photoconductive properties of a-Si to produce a linear image sensor with 12 lines per millimeter resolution, which is incorporated into facsimile machines. The fact that the image sensor is 25 cm wide, a full page width, means that no costly optics are required to reduce the image. Three-foot-wide versions of the a-Si image sensors have been incorporated into "white boards" (the modern equivalent of a blackboard) so that copies of lecture notes, etc. can be produced during a lecture!

5. CONCLUSIONS

Many other applications of a-Si have been proposed, and no doubt many more will be forthcoming. It is this large number of wide-ranging applications that makes the material so exciting commercially. If one application does not succeed then there are many others that will.

However, I would not like the reader to believe that a-Si is going to replace crystalline Si in all its applications. There are many applications where the cheapness and large-area advantage of a-Si will not offset, for example, its slower operating speed. What will happen is that a-Si will open up new applications not accessible for one reason or other to its crystalline counterpart. The Japanese have long been convinced that thin-film materials will play a major role in electronics in the next century. I hope that this chapter has made it clear that a-Si is already making many contributions in this area and will form an important part of the development of electronics into the twenty-first century.

REFERENCES

General

1. W. E. Spear and P. G. LeComber, *Solid State Commun.* **17**, 1193–1196 (1975); *Phil. Mag.* **33**, 935–949 (1976).
2. H. F. Sterling and R. C. G. Swann, *Solid State Electron.* **8**, 653–654 (1965); R. C. Chittick, J. H. Alexander, and H. F. Sterling, *J. Electrochem. Soc.* **116**, 77–91 (1969).

3. W. E. Spear, P. G. LeComber, S. Kimmond, and M. H. Brodky, *Appl. Phys. Lett.* **28**, 105–107 (1976).
4. D. E. Carlson and C. R. Wronski, *Appl. Phys. Lett.* **28**, 671–673 (1976).
5. *Semiconductors and Semimetals*, Vol. 21D (J. I. Pankove, ed.), Academic, New York (1984).
6. Review articles in recent Proceedings of the International Conferences on Amorphous and Liquid Semiconductors, viz., *J. Non-Crystal. Solids* **77/78** (1985), Rome Conference; *J. Non-Crystal. Solids* **97/98** (1987), Prague Conference; P. G. Le Comber, *J. Non-Crystal. Solids* **115**, 1–13 (1989).

a-Si FETs and Displays

7. P. G. LeComber, W. E. Spear, and A. Ghaith, *Electron. Lett.* **15**, 179 (1979).
8. A. J. Snell, K. D. Mackenzie, W. E. Spear, P. G. LeComber, and A. J. Hughes, *Appl. Phys.* **24**, 357 (1981).
9. P. G. LeComber, A. J. Snell, K. D. Mackenzie, and W. E. Spear, *J. Phys. (Paris)* **42**, Supp. C4, 423 (1981).
10. K. D. Mackenzie, A. J. Snell, I. French, P. G. LeComber, and W. E. Spear, *Appl. Phys.* **A31**, 87 (1983).
11. P. G. LeComber and W. E. Spear, Chap. 6 of *Semiconductors and Semimetals* (J. Pankove, ed.), Vol. 21D, 89, Academic, New York (1984).
12. D. G. Ast, Chap. 7 of *Semiconductors and Semimetals* (J. Pankove, ed.), Vol. 21D, pp. 115–138, Academic, New York (1984).
13. P. G. LeComber, *Materials Research Society Symposia Proceedings*, Vol. 49, pp. 341–351 (1985), and references cited therein.
14. M. J. Powell *et al.* *Proc. Int. Display Research Conference*, pp. 63–66 (1987).
15. T. Chikamura, S. Hotta, and S. Nagata, *Materials Research Society* 1987 Spring Meeting; S. Hotta *et al.*, *SID 1986 Digest*, pp. 296–297.

High-Voltage TFTs

16. H. C. Tuan, *Materials Research Society Symposia Proceedings*, Vol. 70, pp. 651–656 (1986).
17. Chuang *et al.*, *J. Non-Crystal. Solids* **97/98**, 301–304 (1987).

Electrophotography

18. I. Shimizu, *Semiconductors and Semimetals* (J. Pankove, ed.), **21D**, Vol. 55–73 (1984).

12

Control of Semiconductor Conductivity by Doping

E. D. Jones

1. INTRODUCTION

Doping is a procedure used for controlling the carrier concentration and hence the conductivity of semiconductors. It can be achieved by introducing into the semiconductor impurity atoms possessing a different number of valency electrons from those of the component elements of the semiconductor. In the case of compound semiconductors, an alternative procedure to control the carrier concentration is to introduce appropriate defects into the material, often by imposing small deviations from stoichiometry. The latter process, although analogous to doping, is not doping in the original sense. The available methods for obtaining conductivity control are listed below along with some introductory references to each of the subject areas:

1. Planar diffusion[1-3];
2. Ion implantation[4,5];
3. Transmutation doping[6];
4. Crystal growth and zone processes[7,8];
5. Epitaxial processing including chemical vapour deposition[9,10];
6. Stoichiometric changes in binary compounds.[11-13]

Whereas methods 3, 4, and 6 are normally used for the doping of bulk semiconducting materials, methods 1, 2, and 5 are used for the doping of thin films in, for example, the production of integrated circuits. This chapter discusses only 1–3 in detail; some discussion of 4–6 will be found elsewhere in this book and only a brief description of their application to carrier concentration control will be given here.

2. DIFFUSION

The diffusion of dopant atoms into crystalline materials can be studied from two different viewpoints. The first approach is to treat the solid as a continuous

---

E. D. Jones ● Department of Applied Physical Sciences, Coventry Polytechnic, Priory Street, Coventry CV1 5FB, U.K.

medium; the diffusion equations due to Fick can then be set up and solved to give the concentration of the dopant as a function of time and distance in the semiconductor. The second approach is based on the atomic structure of a crystalline solid where diffusion occurs by atoms jumping from one crystal site to another, usually by vacancy or interstitial movements.

Fick's first law of diffusion states that the flux of dopant atoms across a given plane in a crystal is proportional to the concentration gradient of such atoms across that plane. In mathematical terms this can be written in the form

$$\mathbf{J} = -D\nabla C \tag{1}$$

and if diffusion is constrained to take place in one direction only, the above equation can be simplified to

$$J = -D\frac{dC}{dx} \tag{2}$$

Here x is the distance parallel to the concentration gradient, J is the flux of atoms taking part in the diffusion, D is the diffusion coefficient sometimes referred to as the diffusivity, and C is the concentration of the diffusing species.

If a steady state does exist, then the above equation can be used to determine D, but in the majority of cases this is not the case. For example, if the concentration of the diffusing species is changing with time t the above equation is still valid but it is necessary to use Fick's second law, which is discussed below.

2.1. Constant Diffusivity

If it is assumed that D is constant and that diffusion takes place in one dimension only, then for a small element of volume of the crystal this may be written

$$\frac{\partial C}{\partial t} = D\frac{\partial^2 C}{\partial x^2} \tag{3}$$

Solutions to Eq. (3) are dependent upon the boundary conditions of the particular diffusion problem and can be found in the literature. The two standard solutions are given below.

2.1.1. Limited Source Diffusion

If a fixed total surface density of dopant Q is presented to the surface of a specimen for diffusion at time $t = 0$, the solution is a Gaussian function of the form

$$C(x, t) = \frac{Q}{(\pi D t)^{1/2}} \exp\left(\frac{-x^2}{4Dt}\right) \tag{4}$$

Plots of the above expression are shown in Fig. 1. The most important property of the above equation is that the total area under the curve of concentration versus depth will remain constant. This means that as the duration of the diffusion anneal increases, the surface concentration will decrease at a rate proportional to $1/\sqrt{t}$ and, at the same time, any point on the distribution where the ratio $C(x, t)/C_0(t)$ is constant will proceed into the crystal at a rate proportional to $\sqrt{t}$.

2.1.2. Infinite Source Diffusion

If the surface concentration C_0 of dopant remains constant throughout the diffusion anneal then the solution to the diffusion equation is a complementary error function of the form

$$C = C_0\left\{1 - \mathrm{erf}\left[\frac{x}{(4Dt)^{1/2}}\right]\right\} \tag{5}$$

which is normally written in the form

$$C = C_0\,\mathrm{erfc}\left[\frac{x}{(4Dt)^{1/2}}\right] \tag{6}$$

Diffusion profiles that exhibit this type of behavior are also shown in Fig. 1.

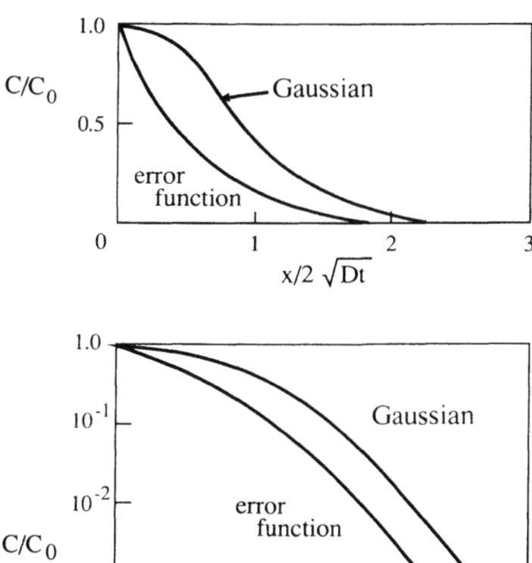

FIGURE 1. Normalized error function and Gaussian distributions.

2.2. Nonconstant Diffusivity

In many diffusion systems D is not a constant, and when this occurs the type of mathematical dependency that D does possess must be established and it is then substituted into the basic diffusion equation, which is

$$\frac{\partial C}{\partial t} = \frac{\partial}{\partial x}\left(D\frac{\partial C}{\partial x}\right) \tag{7}$$

Solutions with a variable D are usually involved and specialist techniques have to be used. One example is where D varies as the first, second, or third power of the concentration of the diffusing dopant (i.e., $D \propto C^n$, where $n = 1, 2, 3$) and a recognized method for solving the diffusion equation in this case is to use the Boltzmann–Matano technique (1) for analyzing the profiles. This can be used if D is a function only of dopant concentration and the boundary conditions can be described in terms of one variable which combines both depth and time, which is expressed in terms of x/t. The diffusivity at any other concentration C' on the concentration dependent profile is given by

$$D(C') = -\frac{1}{2t}\left(\frac{dx}{dC}\right)_{C'}\int_0^{C'} x\, dC \tag{8}$$

At any point on the profile where the concentration is C', the diffusivity can be determined from the diffusion time t, the reciprocal of the slope, and the area under the profile at that point.

A typical experimental profile for the diffusion of In in CdS[14] which has been analyzed using the above formula is shown in Fig. 2. If an infinite source of diffusion dopant vapor is maintained around the semiconductor the diffusivity at the surface

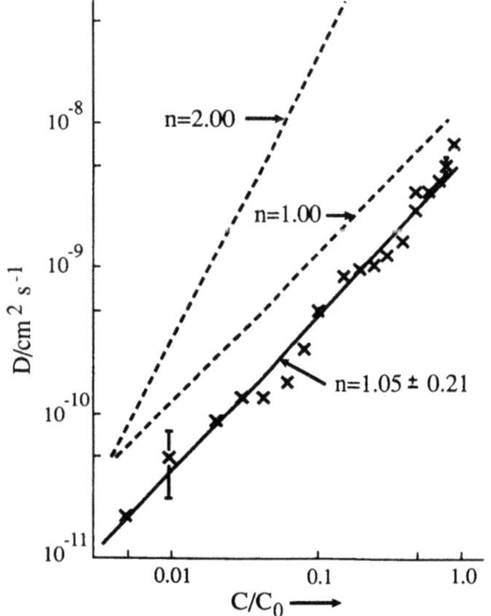

FIGURE 2.   A graph showing a typical experimental concentration profile for the diffusion of In in CdS which has been analyzed using the Boltzmann–Matano formula, where the diffusivity D is linearly proportional to the concentration of the diffusing species.[14]

of the slice, D_0, is given by

$$D_0 = \frac{1}{4t}\left(\frac{x}{0.81}\right)^2 \tag{9}$$

where x is the depth of the diffusion front. The value of D for any other concentration on the same profile is given by

$$D(C) = D_0\left(\frac{C}{C_0}\right) \tag{10}$$

where C_0 is the dopant concentration at the surface of the slice.

Concentration profiles in which $D \propto C^n$ are shown by Fig. 3[15] along with a solution for an invariant D. Solutions of this type have been used for the diffusion of Zn in GaAs[16] where the diffusion is assumed to take place by a substitutional–interstitial mechanism, and also for the diffusion of In in CdS.[14] In the latter case, charge neutrality is maintained on the Cd sublattice during the diffusion by the replacement of three Cd^{2+} ions on the Cd substrate by two In^{3+} ions from the diffusion source and the mobile defect is assumed to be a charged vacancy associated with an In^{3+} ion.

The diffusivity is related to the absolute temperature T by the Arrhenius relationship:

$$D = D_0 \exp\left(-\frac{E}{kT}\right) \tag{11}$$

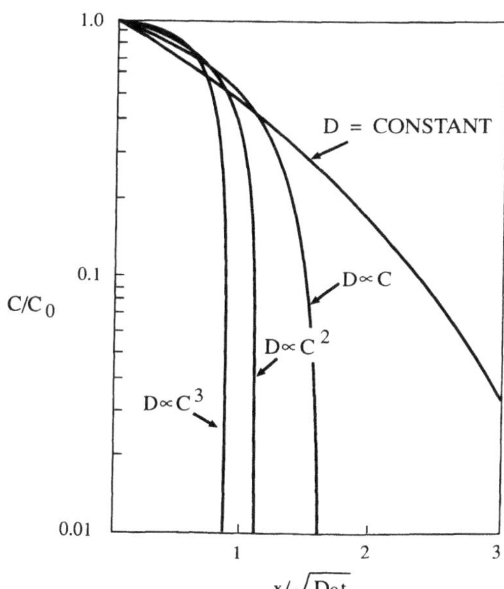

FIGURE 3. Normalized concentration profiles where the diffusivity D is proportional to the first, second, or third power of the concentration of the diffusing dopant,[15] i.e., $D \propto C^n$, where $n = 1, 2,$ or 3.

where E is the activation energy for a dopant atom to jump from one site to a neighboring one, k is Boltmann's constant, and D_0 is a constant which involves many factors.

2.3. Diffusion Techniques

The diffusion doping of semiconductors is often carried out in a tube furnace under continuous flow conditions by passing the dopant vapor over slices of the crystal. Diffusions are also carried out in closed silica capsules. The former method is generally used in batch production processes involving multiple slices, whereas the latter conditions are used when "one-off" measurements are required, usually in research and development. If the boundary conditions used in the doping process are not the same as either of the sets used in the standard solutions to the diffusion equation given earlier for constant D, then it may be necessary to determine the actual dopant profile for the conditions used. This involves either solving the diffusion equation using new boundary conditions, or using computer fitting to determine the best mathematical function to represent the concentration profile (or a combination of both).

In, for example, the self diffusion of Cd in CdTe,[17] carried out in a closed silica capsule under Cd saturated vapor conditions, the resulting profiles can best be represented by the sum of two complementary error functions. In addition, the plot of D versus the reciprocal of temperature can be represented by the sum of two Arrhenius relationships. This indicates that two diffusion processes are occurring simultaneously, each represented by different distinct activation energies. This is probably due to Cd atoms diffusing through the crystal interstitially and then being trapped onto substitutional sites to give a slower substitutional diffusion process. These profiles are shown in Fig. 4.

2.3.1. The Planar Diffusion Process

In practice, the doping of semiconductor materials using this process to form diffused layers and junctions is normally a two stage operation.

2.3.1a. Dopant Deposition. In this operation, the required amount of dopant is introduced into the semiconductor surface using, for example, a flow process similar to the one described earlier, or ion implantation (discussed in Section 3).

2.3.1b. Drive-in Diffusion. After the dopant deposition has been completed a further heat-treatment is carried out. During this drive-in diffusion, the total concentration of dopant in the slice remains constant. It is essential that the surface is sealed at the start of the heat treatment to prevent any loss of dopant from the semiconductor by out diffusion from the slice. In the case of silicon, the drive in diffusion is carried out in an oxidizing gas stream or with an oxide capping layer.

Most dopants diffuse much more slowly through the oxide than they do through pure silicon and this is why silicon dioxide SiO_2 is used as a mask in silicon technology. It is not possible to use oxide masks in other semiconducting materials because no similar correlation on diffusivity exists between the element and its

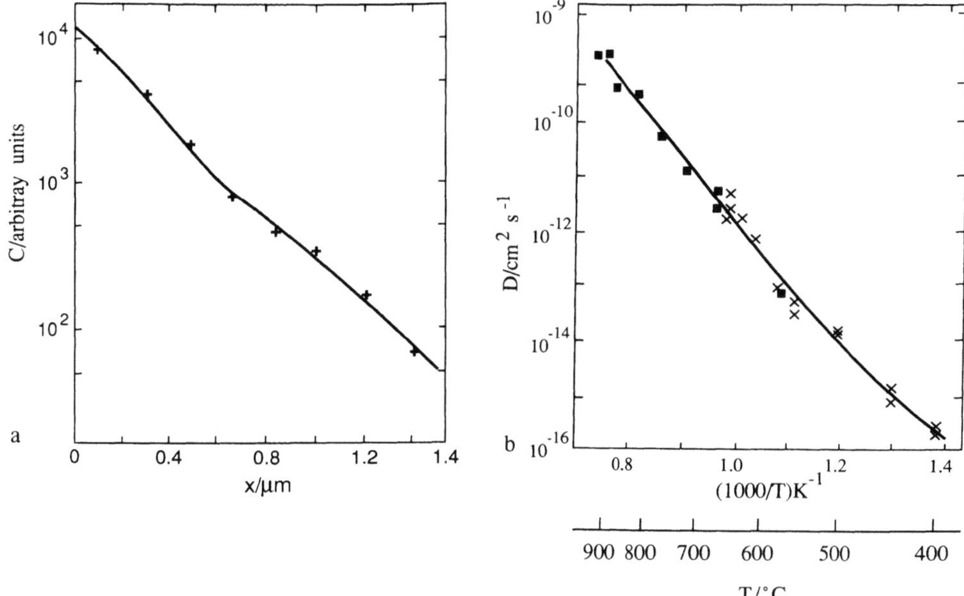

FIGURE 4.   Graphs illustrating the use of computer fitting to determine the best mathematical function or functions to represent experimental data[17]: (a) shows a concentration profile which can best be represented by the sum of two complementary error functions and (b) shows an Arrhenius plot illustrating the presence of two diffusion mechanisms. The experimental points have been measured by mechanical sectioning (■) and by anodic oxidation (×).

oxide, similar to that which occurs in silicon. As a consequence of this, deposited SiO_2 and silicon nitride are frequently used as diffusion masks in other semiconducting materials.

2.4. Deviations from Simple Diffusion Theory

This does occur in practice, as a result of many causes, and some of them will be outlined briefly below:

2.4.1. Lateral Diffusion

This occurs at the edge of a mask window where diffusion takes place along the semiconductor surface under the mask, as well as perpendicular to the surface.

2.4.2. The Emitter–Push Effect

This is observed in transistor structures that employ two consecutive diffusions to form a device. For example, in an $n-p-n$ device boron is used in the first diffusion to form the base region in an n-type substrate. Then during the second diffusion, which uses phosphorous to form the emitter region, the boron is pushed deeper into the device. Similar effects have been observed with doped buried layers, and in addition, an opposite effect to the "emitter-push" which is referred to as "base-retardation" has also been observed.

2.4.3. Field-Aided Diffusion

This occurs when a neutral donor or acceptor atom enters a semiconductor during diffusion and becomes ionized. As the two charged species formed will have different mobilities they will move through the semiconductor at different rates. This will result in an internal electric field being set up which will either aid the motion of the slower member, slow down the motion of the faster member, or possibly affect the motion of both types of particles.

2.4.4. External Rate Limitations

During the dopant deposition stage, the surface concentration of dopant does not reach an equilibrium value instantly at the start of the diffusion, but takes a finite time to do so. This is because a period of adjustment is required, at the start of the diffusion, for the surface concentration of dopant to increase from zero up to the equilibrium value, and in certain cases it takes a significant time to do so.

2.4.5. Dopant Redistribution During Oxidation

In silicon technology, dopant atoms located near the silicon/silicon dioxide interface will redistribute themselves as the interface advances into the silicon slice. The reasons for this are as follows:

a. The partition coefficient k, which is defined as the ratio of the equilibrium concentration of dopant in Si to that in SiO_2 is not unity. Typical values will vary from 0.3 for boron up to 20 for gallium.
b. The rate of diffusion of dopant through the SiO_2.
c. The interface between the Si and the SiO_2 is advancing into the silicon as the oxidation is proceeding.
d. The thickness of the growing SiO_2 layer is approximately twice as great as that of the silicon layer it replaces.

All the above factors, either taken individually or together, can have a significant effect on the planar diffusion process and their possible effects should be taken into consideration when designing production processes to manufacture devices routinely. These topics are discussed at length in the literature[18] and they will not be described in detail in this chapter.

3. ION IMPLANTATION

Ion implantation involves bombarding the surface of a semiconductor with ions that have been accelerated in an electric field to enable them to penetrate the crystal lattice. The areas to be implanted are defined by a window of the required shape cut in a masking layer and the ions penetrate a short distance through the window into the lattice of the semiconductor. The masking layer must be sufficiently thick to prevent the dopant ions penetrating through the mask and entering the semiconductor and it does not need to resist high temperatures. Sometimes, therefore, the photoresist itself is used for masking in ion implantation.

One of the main advantages of the ion implantation technique is the dopant uniformity and reproducibility that can be obtained, which cannot be achieved by any other method; in particular, it is a very important tool in the production of metal oxide semiconductor devices.

A large variety of masking materials are possible in ion implantation, but the two that are used most frequently are silicon dioxide (SiO_2) and silicon nitride (Si_3N_4); they can be used for both silicon and compound semiconductors. The minimum thickness of masking material required to stop a given fraction of incident dopant ions, which will depend on the applied voltage, can be estimated from the range parameters for the ions in that material.

The lattice damage caused by the dopant atoms colliding with the lattice can be annealed out after the implantation has been carried out. This can be achieved either by thermal heat treatment using a conventional anneal or by using pulsed laser power to bring about rapid lattice heating with the bulk of the specimen remaining at room temperature. The implanted dopant atoms diffuse into lattice sites during this anneal.

Ion implantation is a very important tool in the production of semiconductor devices.

In a typical experimental system, similar to the one shown in Fig. 5, ions of the dopant atoms first pass through an electromagnetic separator in order to eliminate unwanted ions, and then they are accelerated by an electric field in high vacuum to the required implantation energy. Lateral dopant uniformity beneath the semiconductor surface is achieved by passing the high-energy beam through vertical and horizontal electric fields so as to impart a "raster" scan to the beam.

The ion implantation technique, unlike other methods, produces highly reproducible laterally uniform, and precisely controlled dopant profiles. Moreover, the dopant concentration is not limited by the equilibrium solid solubility of the dopant in the host material. As the implantation is performed at room temperature, the spread under the edges of the mask is limited to that caused by scattering.

Another advantage of the implantation technique is that the method is independent of the physical condition at the surface of the semiconductor and the depth of penetration is dependent only on the accelerating voltage. This contrasts with diffusion, which is very dependent on the physical condition at the surface of the semiconductor. The variable that will determine the final concentration in the

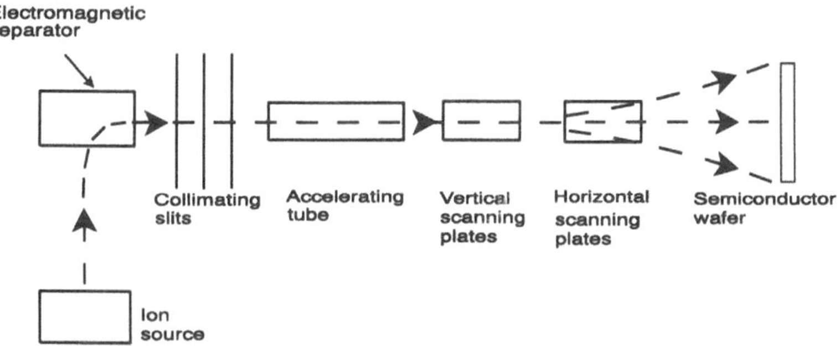

FIGURE 5. A block diagram of a typical ion implantation system.

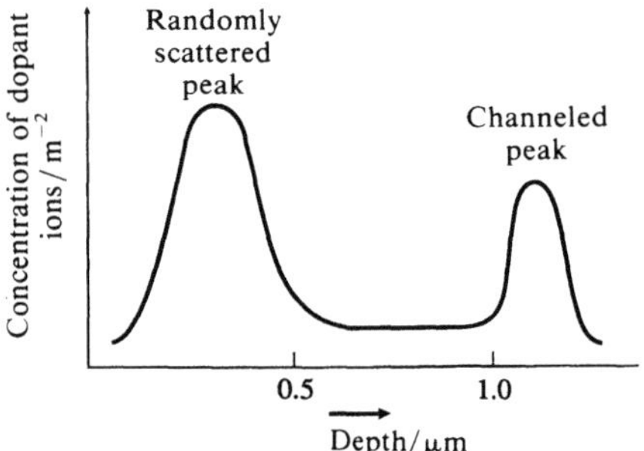

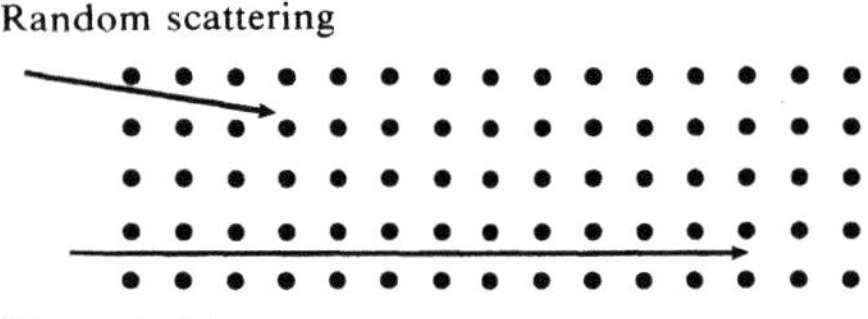

FIGURE 6. A schematic diagram showing a concentration profile of dopant atoms that have been fired into a semiconductor by ion implantation. The diagram shows two peaks; the first is formed by random scattering between the ions of the beam and atoms of the lattice and the second is due to channeling.

semiconductor is the total dose received by the semiconductor slice, and this is dependent on the energy and the intensity of the ion beam.

The disadvantages of ion implantation include the following: (a) only shallow profiles (less than about 1 μm) can be implanted; (b) there is considerable damage caused to the lattice by the high-energy particles; and (c) the equipment used for the implantation is very expensive.

A major problem with ion implantation is due to an effect referred to as "channeling." This can occur when the ion beam is aligned along certain principal crystal directions of the lattice. Under these circumstances, ions penetrate more freely into the lattice along atom-free channels as shown in Fig. 6. Channeling can be reduced if a slight misalignment is imposed between the crystal directions concerned and the direction of the ion beam. A typical concentration profile possessing two peaks is shown in Fig. 6; the first is formed by random scattering between the ions of the beam and atoms of the lattice, and the second peak, which is much deeper, is due to channeling.

When a beam of ions possessing a single energy enters a semiconductor, the randomness of the collisions gives a profile of dopant concentration as a function of depth that is best represented by a Gaussian function of the form

$$C(x) = \frac{\phi_t}{(2\pi)^{1/2}\,\Delta R}\exp\left[-\frac{(x-R)^2}{\Delta R^2}\right] \tag{12}$$

where C is the dopant concentration, x is the depth into the semiconductor, ϕ_t is the ion dose received by the silicon slice, R is the mean range of all particles in the Gaussian distribution, and ΔR is the standard deviation of the distribution. In fact a low concentration tail does exist on the side of the distribution between it and the surface of the slice, which makes the distribution asymmetric. Consequently concentration distribution fittings using more advance theories are required.[5]

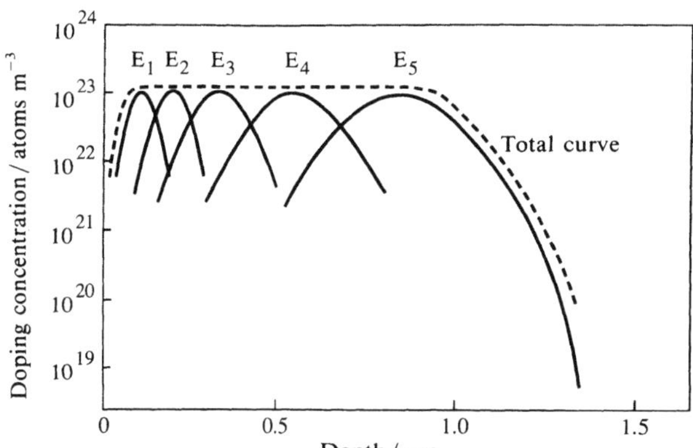

FIGURE 7. A schematic diagram showing how a reasonably flat dopant profile in a semiconductor can be obtained using five implants where each implant was carried out at a predetermined voltage and for a predetermined time.

In many applications, complex doping profiles are required, and these can be obtained using multiple implantations in which appropriate total doses are used at various accelerating voltages. As an illustration, a reasonably flat profile is shown in Fig. 7, which is composed of five dopant implants, with each implant carried out at a predetermined voltage and for a predetermined time. A profile of this shape cannot be obtained using planar diffusion, because in diffusion the shape of the concentration profile is determine by the geometry of the system and the boundary conditions used in the solution of the diffusion equation.

It may be seen that the ion implantation technique, producing shallow precisely controlled profiles, is complementary to the diffusion technique, which can be used to produce much deeper profiles.

4. TRANSMUTATION DOPING

It is possible to produce dopant atoms in a semiconductor material by bombarding it with nuclear particles, usually neutrons, so as to transmute a host element into a different, dopant, element. The products of such a reaction are radioactive and will decay by one of the standard decay processes—either by K-electron capture, or by the emission of an alpha or beta particle, which is accompanied by one or more gamma photons. It is possible that a series of decays could be involved that would involve the emission of several nuclear particles with different half-lives. The end product of the decay will give n-type doping if there is an overall increase of one in the atomic number of the target nuclei, and p-type doping if there is a decrease of one in the atomic number.

The feasibility of such a reaction occurring will depend on the nuclear parameters of the target nucleus, and these are the isotopic abundance, the nuclear cross section for the particular reaction, the branching ratios, and the mode of decay of the product nucleus. In a reaction of the type $A(n, x)B$, the concentration of nucleus B that is formed by irradiating nucleus A with neutrons is given by

$$N_B = \frac{\Sigma_A \phi}{\lambda_B}[1 - \exp(-\lambda_B t)]\exp[-\lambda_B(t - t')] \tag{13}$$

where N is the number of nuclei per unit volume, t is the time variable whose origin occurs at the start of the irradiation, $\sum_A$ is the macroscopic absorption cross section for the reaction, λ_B is the decay constant of the radioactive isotope formed, ϕ is the neutron flux intensity, and t' is the duration of the neutron irradiation.

A plot of the radiation emitted by B will produce a similar curve to Eq. (13) and is shown as a function of t and t' in Fig. 8. If isotope B decays to produce C, which is stable, then the final concentration of C is

$$N_C = \frac{\sum_A \phi}{\lambda_B} (\exp \lambda_B t' - 1) \tag{14}$$

It is desirable that the required dopant concentration should be obtained without having to use an excessive neutron dose in order to keep the radiation damage to a minimum. Normally, after the neutron irradiation has been completed the irradiated samples are annealed at high temperatures for a short period of time to reduce any damage that may have been caused. In addition, it is advantageous if the half-lives of the products are short so that the induced radioactivity will decay away in a short period of time; a period of a few hours is satisfactory, but it should not extend beyond a few days. If this did occur it would create problems with handling the material after it had been irradiated.

If several types of reactions occur in a single target material then it is desirable that all the reactions produce dopants of the same conductivity type. If dopants of both types are produced, some electrical compensation will occur, which could decrease the effectiveness of the process.

The transmutation doping reaction that has been used on a commercial scale to a significant extent is the thermal neutron irradiation of silicon to produce phosphorous, which acts as an n-type dopant. In the United Kingdom, there is an extensive program being carried out using the Harwell reactors where silicon boules are irradiated in bulk.

Naturally occurring silicon is composed of three isotopes, which are listed below along with their percentage abundances:

$$^{28}\text{Si}(92,2\%); \qquad ^{29}\text{Si}(4.7\%); \qquad ^{30}\text{Si}(3.1\%)$$

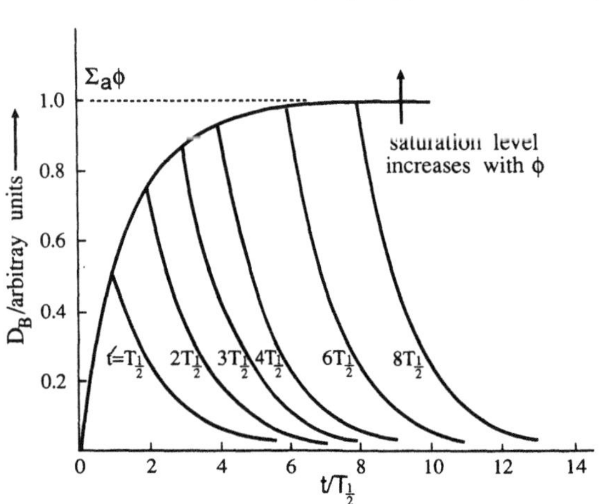

FIGURE 8. A graph showing how the induced activity in a target nucleus that has been irradiated with neutrons varies with time t and also with the time of neutron irradiation t'.

When silicon is irradiated with thermal neutrons, both the isotopes ^{28}Si and ^{29}Si produce stable products, and, theoretically, only ^{30}Si produces a radioactive product, ^{31}Si, which decays to ^{31}P with a half-life of 2.6 h as shown below:

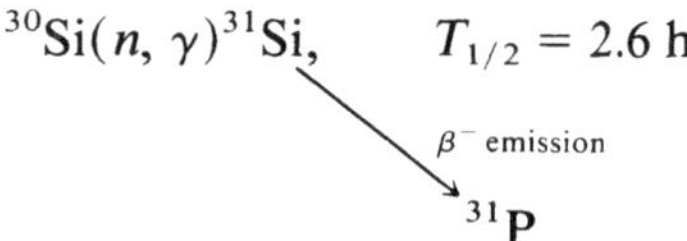

The 2.6 h half-life means that the radioactivity will decay away to insignificant levels within a few days after the neutron exposure has terminated.

The main advantage of the method is that, as a result of low neutron flux attenuation in the samples, bulk material possessing a more uniform resistivity is produced when compared with the doped material produced by the other techniques that are available. For example, in the float zone and crystal-pulling techniques large changes in dopant concentration occur because phosphorous has a partition coefficient of 0.31 (see Fig. 9). This leads to a varying distribution of phosphorus in the material. The reasons for this are discussed below.

Theoretically, it is possible to dope other semiconductors using the irradiation technique, but no known commercial development has taken place to date. In the case of germanium, when it is irradiated with thermal neutrons, both n-type and p-type dopants are produced and so some compensation occurs. In the case of compound semiconductors, the situation is even more complex since possibly two, three, or even four target nuclei could be involved. When such compounds are irradiated with thermal neutrons, many more reactions could arise and the resultant activity is often intense and long-lived. In these cases, unless the doping can be achieved by a more conventional method, it is very unlikely that it would be economic to pursue trying to dope these materials using transmutation doping, but the doping of gallium arsenide has been investigated using this technique.[19]

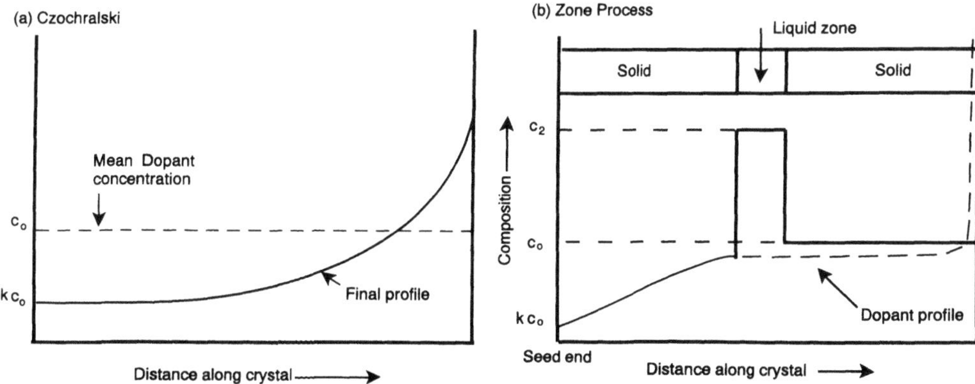

FIGURE 9. Graphs showing how the concentration of dopant atoms will increase along the length of an ingot when the segregation coefficient $k < 1$ in (a) the Czochralski technique and (b) the zone process. This occurs because dopant rejection occurs at an advancing solid–liquid interface under progressive solidification.

5. CRYSTAL GROWTH AND ZONE PROCESSES

In this section the doping of bulk semiconductors using the Czochralski crystal-pulling method, zone processes, and the Bridgman–Stockbarger process is considered. All these techniques are used for growing single crystals, but the float zone process can also be used for purification of the semiconducting ingot. The Bridgman–Stockbarger technique is used mainly for growing compound semiconductors. All three techniques are described in detail in Chapter 9 and consequently they will not be described further in this chapter; only their application to the doping of semiconductor materials in bulk will be referred to in this section. In these techniques, doping of the bulk material is carried out by adding dopant to the melt in the Czochralski technique and by using a highly doped seed crystal in the float zone technique, whereas in the Bridgman–Stockbarger technique either dopant can be added to the melt or a highly doped seed crystal can be used.

In all these techniques, uneven dopant distributions are produced (see Fig. 9) and the reason for this is the relevant differences in the solubility of the dopant in the solid and liquid phases. Under equilibrium conditions, the ratio of the solubilities in the solid and liquid phases is referred to as the equilibrium partition coefficient or distribution coefficient, and at any temperature it is a constant for a given dopant–semiconductor combination. Values of the partition coefficient, for different elements, vary considerably and the shape of the dopant concentration profile along the length of a boule is very dependent on the value of k_0. In particular, it is important if it is less or greater than unity.

A schematic representation of a typical binary phase diagram at low solute concentrations for the case where $k_0 < 1$ is shown in Fig. 10, where it can be seen that freezing will occur over a range of temperatures starting at T_1 and finishing at T_2. This will produce considerable local variations in the solid solubility ranging from C_1 at temperature T_1 to C_2 at temperature T_2. When $k_0 < 1$, dopant rejection will occur at an advancing solid–liquid interface under progressive solidification.

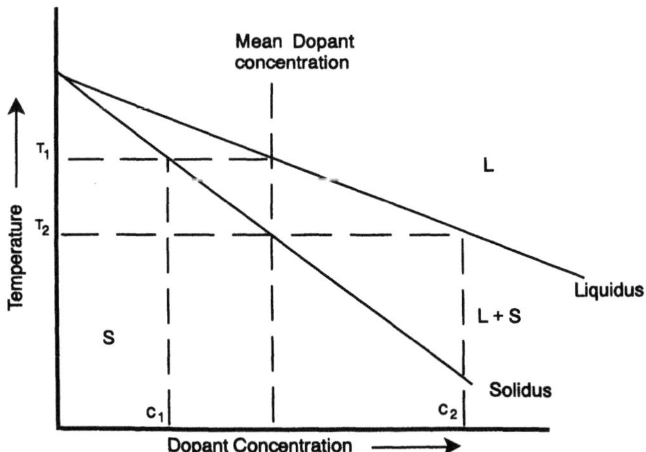

FIGURE 10. A typical binary phase diagram representing the case of a dopant in an elemental semiconductor at low dopant concentrations and where the partition coefficient $k < 1$.

The distribution of dopant concentration, C_s, in Czochralski growth or in one pass of the molten zone through a uniformly doped bar in the zone process is given by the Pfann equation:

$$\frac{C_s}{C_0} = k_e(1 - f_s)^{(k_e - 1)}$$

where k_e is the effective partition coefficient, f_s is the fraction of the charge that has solidified, and C_0 is the equilibrium concentration of the dopant at the start of the process.

The situation is more complicated in processes involving the passage of zones of liquid and dopant through an ingot of material which originally contained no dopant and also in the cases where $k_0 > 1$; the reader is referred to the literature.

6. EPITAXIAL PROCESSES

Epitaxy is a crystal growth process involving the deposition of material on top of a crystalline substrate in a way that preserves the overall single-crystal structure. (See Chapter 10 by Mullin.) Within the growth constraints, each new layer is quite independent of the last and it may be differently doped. The total impurity concentration of successive layers can be maintained at accurately controlled predetermined levels and the only limit to the number of layers that can be produced is the time involved. The main advantage of this technique, when compared with planar diffusion doping, is that here abrupt junctions can be produced. If the epitaxial layers are grown at high temperatures, the method may suffer from the disadvantage that some interdiffusion may take place across junction interfaces while the device is being grown. This is why successive attempts are being made, with success, to grow the devices at lower temperatures and temperatures as low as 300°C are currently being considered.

There are many ways in which epitaxial layers are grown; these include liquid phase epitaxy, vapor epitaxy, molecular beam epitaxy, metal organic vapor phase epitaxy, and plasma deposition.[9] In all cases, the doping process is carried out by introducing the dopant to the surface of the semiconductor in the appropriate form and at the appropriate time.

7. STOICHIOMETRIC CHANGES IN COMPOUND SEMICONDUCTORS

A truly stoichiometric compound can be represented by a single point at a particular temperature on the phase diagram, and this corresponds to that partial pressure at which the number of cations and anions in the solid phase are equal. Perfect crystals with true stoichiometric composition are normally intrinsic. At any other equilibrium vapor pressure a departure from true stoichiometry must exist, and if this occurs over a range of vapor pressures the semiconductor must have a phase field that possesses a finite width on the phase diagram, but usually the departure from true stoichiometry is very small, perhaps as small as 0.01% or 0.001%. Normally a spread in stoichiometry of this magnitude is so small that in a temperature composition phase diagram such a phase field would be represented by a vertical straight line, and this is illustrated in Fig. 11a.

When a departure from true stoichiometry does occur in the equilibrium state in a crystalline binary compound semiconductor, the excess of one type of ion must be incorporated in the crystal lattice in some way. If the anion sublattice is complete, excess cations can be incorporated interstitially or, if the anion sublattice is so closely spaced to prevent cations occupying interstitial sites, then an excess of cations can alternatively be obtained by creating additional vacancies on anion sites. An excess of anions can be incorporated into the crystal using similar methods, but in all cases charge neutrality must hold when excess atoms are incorporated into any crystal to create defects.

When such defects are created in a crystal, it is possible for them to be ionized and to contribute either electrons, or holes, or both to the conductivity, and the extent to which this will occur will depend on the value of the Fermi–Dirac distribution function for the energy of the particular defect state that exists. In general, interstitial metal ions, anion vacancies, and anions placed on cation sites will act as donors, whereas interstitial anions, cation vacancies, and cations on anion sites will act as acceptors. In theory, the stoichiometry of pure compound semiconductors can be changed so that either type of pure compound semiconductors can be changed so that either type of conductivity can be obtained, but in general the II–VI family of semiconductors can only accommodate one type of conductivity with the exception of CdTe. This can be made either n-type or p-type by subjecting it to a heat treatment in an overpressure of Cd or Te, respectively. It would be

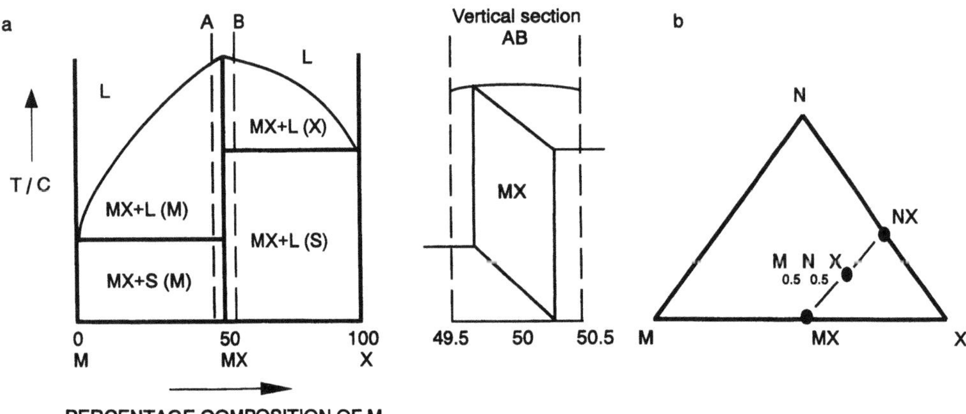

FIGURE 11. Phase diagrams illustrating how the doping of compound semiconductors occur: (a) changing the composition of the compound within the limits set by the boundary of the phase field will change the defect structure within the compound as shown in this binary phase diagram; (b) by the addition of impurities of the appropriate valency will create a ternary compound and is represented on a ternary phase diagram; in this diagram an isothermal section through such a diagram is shown.

logical to assume that as the Te ion is large and immobile in CdTe, when it is n-type, Cd ions will occupy interstitial sites, whereas when it is p-type additional charged vacancies exist on the Cd sublattice. In fact, similarly to all other II–VI semiconductors, the defect structure in these materials has proved to be very much more complex than that which has been proposed above[19].

It is also possible to dope compound semiconductors by adding traces of impurities of the appropriate valency in a manner similar to that which is used for elemental semiconductors, but in this case there are a greater number of possibilities. In both cases, in elemental and compound semiconductors the conditions inside the semiconductor are influenced by the physical conditions that exist around the semiconductor during the doping operation. In the case of elemental semiconductors the state inside the semiconductor can be represented on a binary phase diagram similar to the ones shown in Figs. 10 and 11a, whereas in the case of a binary compound the state inside the semiconductor is represented on a ternary phase diagram. An isothermal section through such a diagram is shown in Fig. 11b. In both systems, the interpretation of each situation is complex, and one way to control this is to dope the semiconductor by carrying out diffusion anneals in closed capsules where the overpressure conditions can be controlled more easily using the phase rule. This method of using impurities to dope compound semiconductors will not be discussed any further in this chapter and the reader is advised to refer to the literature.[11,12]

REFERENCES

1. B. Tuck, *Atomic Diffusion in III–V Semiconductors*, Adam Hilger, Bristol (1988).
2. J. Crank, *The Mathematics of Diffusion*, Oxford U.P., Oxford (1975).
3. S. Wolf and R. N. Tauber, *Silicon Processing for the VLSI Era*, Vol. 1, Lattice Press, Sunset Beach, California (1986).
4. J. L. Stone and J. C. Plunkett, *Impurity Doping Processes in Silicon* (F. F. Y. Wang, ed.), North-Holland, Amsterdam (1981).
5. H. Ryssell and I. Ruge, *Ion Implantation*, Wiley, Chichester (1986).
6. B. D. Stone, *Impurity Doping Processes in Silicon* (F. F. Y. Wang, ed.), North-Holland, Amsterdam (1981).
7. A. J. R. de Kock, *Handbook on Semiconductors*, Vol. 3 (S. P. Keller, ed.), North-Holland, Amsterdam (1980).
8. J. C. Brice, *Crystal Growth Processes*, Blackie, London (1986).
9. S. P. Keller (ed.), *Handbook on Semiconductors*, Vol. 3, North-Holland, Amsterdam (1980).
10. D. K. Ferry (ed.), *Gallium Arsenide Technology*, Howard W. Sams and Co., Indianapolis (1985).
11. D. Shaw (ed.), *Atomic Diffusion in Semiconductors*, Plenum Press, London (1973).
12. W. Van Gool, *Principles of the Defect Chemistry of Crystalline Solids*, Academic, London (1966).
13. M. Aven and J. S. Prener (eds.), *Physics and Chemistry of II–VI Compounds*, North-Holland, Amsterdam (1967).
14. E. D. Jones and H. Mykura, *J. Phys. Chem. Solids* **39**, 11–18 (1978).
15. L. R. Weisberg and J. Blanc, *Phys. Rev.* **131**, 1548 (1963).
16. B. Tuck and M. A. Kadhim, *J. Mater. Sci.* **7**, 585 (1972).
17. N. M. Stewart and E. D. Jones (private communication).
18. S. M. Sze (ed.), *VLS Technology*, McGraw-Hill, New York (1985).
19. J. H. M. Stoelinga, D. M. Larsen, W. Walukiewicz, R. L. Aggarwal, and C. O. Bozler, *Neutron Transmutation Doping on Semiconductors* (J. M. Meese, ed.), Plenum Press, New York (1978).
20. M. Brown and A. F. W. Willoughby, *J. Crystal Growth* **59**, 27 (1982).

13

Silicon Processing: CMOS Technology

D. J. Foster

1. INTRODUCTION

CMOS is the dominant commercial process technology for the fabrication of integrated circuits and will be the topic of this chapter. Originally, CMOS processes were developed in the 1960s and employed metal as the gate conductor; hence the term "CMOS" for complementary metal–oxide–silicon defined from a cross section through the gate electrode and transistor channel region. Minimum feature sizes achievable by the photolithographic systems of the day were approximately 10 μm. Circuit sizes were very small by present day standards and only a limited number of digital logic functions could be integrated within a chip. Subsequently, CMOS technology has evolved by process technique innovations supported by significant equipment developments and modern production processes have 1-μm minimum feature sizes that permit a million transistors or more to be made in silicon chips of 1 cm^2. Although the name CMOS has been retained the gates are no longer made from metal but from polysilicon.

The end of the evolutionary line for CMOS has still not been reached. The trend is for a continuing reduction in minimum feature sizes to around 0.25 μm, which is about the limit of resolution for optical lithography. These smaller minimum features sizes will lead to higher circuit complexities, lower gate delays, and a lower power dissipated by each logic gate. Additional circuit features such as analog and nonvolatile memory will exist alongside digital functions on the chip to enhance circuit performance and extend the applications of CMOS still further; already logic circuits and microprocessors make far more use of on-chip memory for improved data processing speeds.

The main technology driver for the development of CMOS is the DRAM. DRAMs permit easy failure analysis, are sensitive to process flaws, and are the basis for the next generation of process technology. The minimum feature size used in DRAMs is currently 0.8 μm for a 4-Mbit device with the 16-Mbit device expected to employ 0.5-μm features and be in full production in 1992. DRAM process technology becomes adapted for general CMOS use about two years after the

---

D. J. Foster ● Plessey Research Caswell Limited, Caswell, Towcester, Northamptonshire NN12 8EQ, U.K.

173

introduction of a new generation DRAM. The new technology CMOS has added to it analog components (capacitors and bipolar transistors), nonvolatile memory cells (EPROMs and E^2PROMs), and multiple layers of metal to satisfy a wide range of applications.

Currently, CMOS logic circuits with 1-μm minimum feature sizes are beginning to be fabricated on production lines for general digital applications. The best gate delays achievable for CMOS logic are around 350 ps for NAND gates operating from the standard 5-V supply and the best packing density is 100k logic gates in 1 cm^2 of silicon (a logic gate is considered to be a two-input NAND gate consisting of four transistors). Two examples of commercially available circuits are 100k logic gate arrays and a 32-bit microprocessor operating from a 25-MHz clock.

As CMOS processes are scaled to 0.5-μm feature sizes, logic gate delays should fall to around 200 ps, operate from a new standard supply of 3.3 V, and enable CMOS with up to 500k logic gates to be fabricated in 1 cm^2 of silicon. Speed–power products for logic gates should fall from around 700 pJ to 200 pJ, thereby allowing lower power per gate performance. Examples of new products for the 1990s are real time continuous speech recognition circuits and a 32-bit microprocessor operating from a 40-MHz clock for 30 MIPS.

2. CIRCUIT DESIGNS, MASKS, AND PATTERN PRINTING

A CMOS circuit is initially defined in terms of logic equations in order to implement a particular function such as data analysis of telecommunication signals. These equations are then transformed into circuit symbols which represent electronic components such as transistors, resistors, and capacitors that will implement the logic function. Once the circuit configuration is defined, the individual circuit symbols of the design are converted into rectangular patterns which will after fabrication become the electronic components themselves. The conversion to physical components is made by reference to the process design rules which determine the order in which layers of insulators and conductors are formed and patterned on silicon wafers. The coordinates of each individual component are fed into a computer and stored. The use of a computer in circuit design permits common elements to be permanently retained as logic cells (such as a NOR gate) and subsequently recalled and reused at any point in circuit design.

On completion of the circuit's design, a data tape is made and subsequently used in mask-making to regenerate the same patterns on photographic masks. There are about 14 such masking layers in a modern CMOS process. Masks are used to build components into the silicon in a sequential manner. The common method for converting the coordinate data into masks is by writing either with a laser or electron beam in a machine called a mask pattern generator.

In a mask pattern generator, circuit layout data from the tape are fed into the machine's computer, which then controls the direction of a beam of electrons or a laser beam which is scanned across a photographic plate covered with a photosensitive film. Such plates are made from high-quality glass with a low coefficient of expansion, covered on one side by a thin film of chrome. A layer of photoresist on the chrome side of the plate is selectively exposed by the beam which defines a circuit feature. Exposed resist is dissolved in developer to reveal the chrome underneath. The plate is then immersed in an etch solution, where the exposed chrome is removed. Finally, the photoresist is removed from the plate and inspected to assess pattern quality. The complexity of modern-day circuits make it impossible to visually inspect masks, and computers with the pattern data stored in memory are used to check for correctness; such a system is known as a die-to-database checker.

It is common for the circuit pattern on the plate to be either five or ten times the actual chip size, which is optically reduced when printed onto wafers. The equipment used to print circuit patterns onto wafers is called an optical wafer stepper; it sequentially prints an array of circuit patterns across the wafer surface. The wafer stepper projects a chip pattern on a wafer to within $\pm 0.25\ \mu$m from its ideal position relative to an existing circuit layer on the wafer. This slight misalignment is allowed for in the design rules of the process.

A mask set is a master copy for all the same circuits and must be physically perfect. Those defects generated during the mask-making process can be corrected before the plate is used; small scratches in the chrome layer can be repaired by localized chrome deposition and unwanted spots of chrome removed by laser evaporation. The presence of dust particles on a mask surface will lower the circuit chip yield since they are printed into resist layers on wafers and appear as permanent defects (holes and spots) after pattern etching. Mask surfaces are kept dust free by thin transparent membranes called pellicles, which are positioned several millimeters away from the glass surface to ensure than any dust particle is defocused during wafer printing.

3. SILICON WAFER QUALITY

As device sizes are reduced and packing densities rise, the demands placed on the quality of silicon wafers are severe. Metallic contaminants (such as Ni, Cu, Fe), crystal imperfections (such as silicon atoms interstitially positioned in the lattice or missing from crystal sites), must be reduced or eliminated since in transistors the presence of defects in depletion regions results in currents being generated that dissipate off-state power and discharge capacitor nodes. The increased complexity of CMOS circuits means there will be a greater chance of defects occurring in the transistor structure. Typical wafer diameters are currently 125–150 mm, but 200–250 mm wafers are predicted for the future. Such wafers will be made from a perfect crystal boule by sawing it into thin slices and polishing the surface to remove abrasions and scratches.

Poor control of wafer doping leads to problems in maintaining the uniformity of electrical parameters. Poor doping uniformity arises from high-temperature gradients across the molten silicon during crystal growth; the dopant is contained in the melt. Volumes of molten silicon are now very large since the wafer dimensions are also large and the liquid must be maintained uniformly above the melting point of silicon >1414°C. In addition, impurities from the crucible can be incorporated into the melt to reappear later in wafers during device processing. The most troublesome impurities are oxygen and carbon. Oxygen comes from the walls of the quartz crucible, which slowly dissolves at the high temperatures and carbon appears from the CO_2 content in the surrounding air. Oxygen will later precipitate in wafers to give n-type electrical donors whose activity is related to the process schedule temperatures. Carbon acts in conjunction with oxygen in many instances to alter its doping behavior.

The beneficial effect of oxygen in wafers is to harden the wafer and reduce crinkling and warp that can occur during room temperature transfer to and from the high furnace temperatures. In addition, there are techniques for precipitating oxygen within the crystal lattice to produce a local defect deep within a wafer. These induced defects are aids in the removal of metallic ions that generate high leakage currents if located in transistor depletion regions. Such induced defects are known as gettering sites because they remove or getter impurities from the silicon wafer. As circuit dimensions become smaller such sites have to be positioned closer to the wafer surface where they are more effective.

4. EPITAXIAL SILICON SUBSTRATES

The use of epitaxial silicon is common in CMOS to suppress the action of active parasitic devices that occur from unintentional combinations of n-well, p-substrate, and source–drain diffusions. A later section will describe the parasitic silicon controlled rectifier (SCR) in more detail. Epitaxial silicon permits a layer of doped silicon to be grown crystallographically on another of different doping. The doping and thickness of the epitaxial layer are chosen for good electrical performance from transistors. The heavily doped substrate acts as a shunting resistor for high-current pulses and prevents latchup from destroying circuits. The epitaxial layer is of (100) surface crystal orientation, boron doped with 10 Ω cm resistivity and 8 μm thick. The underlying substrate is heavily doped with boron to achieve 25-mΩ cm resistivity and of the same crystal orientation.

5. DEVICE SCALING

The idea of device scaling has been invoked to predict which processing factors should be changed for fabricating smaller devices to maintain electrical performance. Table 1 shows how a scaling factor k can be applied to an existing process in order to generate one of smaller geometry. In additional studies, however, transistor electrical performance has been found to peak for geometries of about 1 μm and fall for smaller dimensions. This is because not all device parameters scale and the second-order effects in larger transistors become increasingly troublesome in smaller devices. Junction breakdown voltages fall due to the higher doping in silicon, solid solubility is reached in source and drain diffusions leading to high resistance, and velocity saturation of carriers between the source and drain limits transistor output currents. Processing techniques are used to fix many of the failings in MOS transistors and restore some performance, but these fixes are expensive and will not be effective for all small geometry processes.

TABLE 1. Simple Scaling Theory Identifies the Process Change Needed for a Performance Improvement; the Scaling Factor is k

Device parameter	Scaled value	Change from scaling
Channel length, L	L/k	Shorter polysilicon tracks required
Channel width, W	W/k	Narrower active areas needed
Drain depletion width, x_d	x_d/k	Higher well dopings required
Gate oxide, d_{ox}	d_{ox}/k	Thinner gate oxides to be grown
Transistor gain, β	$k\beta$	Improved by thinner gate oxides
Transistor drive, I_D	kI_D	Improved by gain increase
Track capacitance, C	C/k	Reduced by smaller areas
Gate delay, t_g	t_g/k^2	Improved by C and I_D
Speed–power product, S	S/k	Improved by I_D and t_g

Although it is possible to make significantly denser CMOS circuits the improvements in electrical performance are not as great. Table 2 illustrates the performance of CMOS relative to 1-μm technology as device dimensions are shrunk in size; the larger geometry processes indicate the improvements that have been made in moving to 1-μm features.

6. CMOS PROCESS FLOW

In formulating a CMOS process schedule the isolation pattern is the first section to be built into the substrate; that is followed by the well doping schedules which involve high-temperature diffusions and determine the electrical parameters of transistors and other components; well schedules are followed by the gate oxide growth and definition of the gate conductor; the sources and drains are fabricated next followed by the incorporation of the metallization layers.

Figure 1 shows a typical process flow for a 1-μm CMOS process. The starting material is p-type epitaxial silicon on a p^+ substrate. Fabrication begins with an

TABLE 2. A Comparison of CMOS Performance as a Function of Minimum Feature Size (Often the Channel Length)[a]

Feature size (μm)	2.0	1.5	1.0	0.8	0.5
Inverter gate delay (350 ps)	2.2	1.2	1.0	0.8	0.7
Packing density (4400 gates/mm^2)	0.3	0.6	1.0	2.5	4.0
SRAM cell size (126 μm^2)	4.0	2.0	1.0	0.7	0.3
Speed–power product (700 pJ)	2.6	1.6	1.0	0.8	0.3
Power supply (V)	5.0	5.0	5.0	5.0	3.3

[a] Actual 1-μm CMOS technology values are shown in brackets. Packing density improvements are much greater than for gate delays.

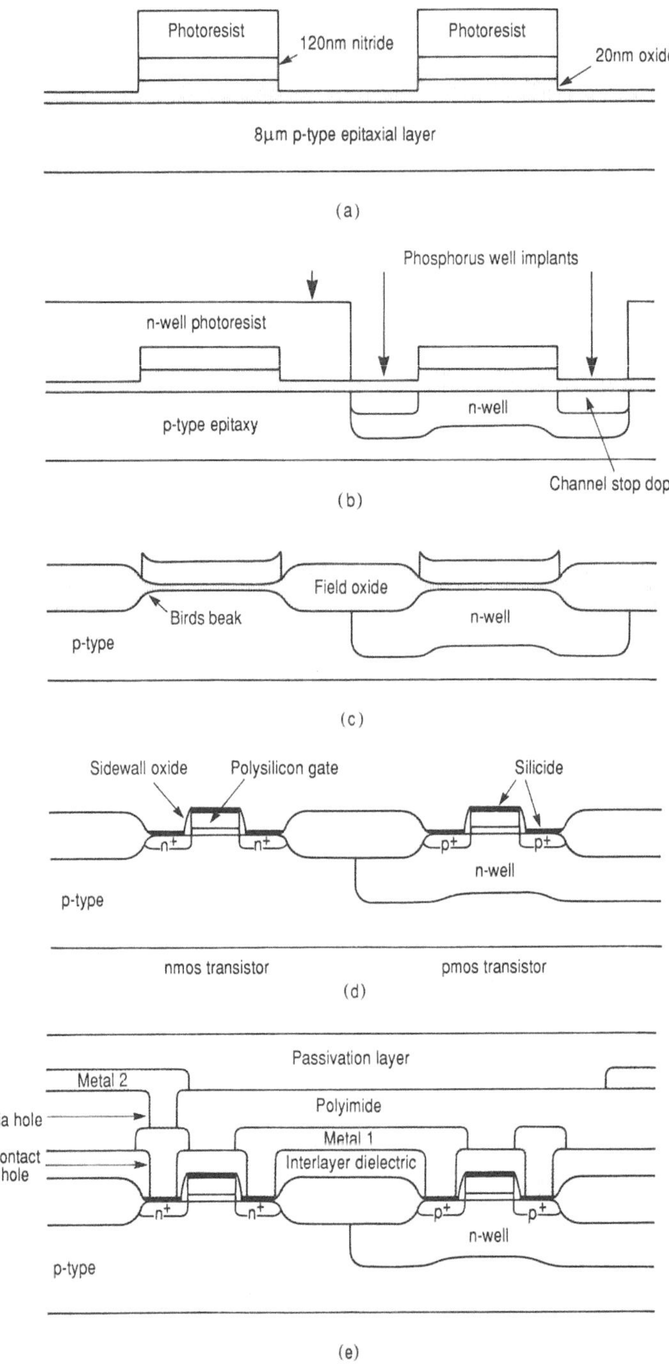

FIGURE 1.   Cross-sectional diagrams illustrating the CMOS process sequence. (a) Active area etched. (b) *N*-well implanted. (c) *N*-well diffused and field oxide grown. (d) Gate conductor and source–drains formed. (e) Interlayer dielectric and metal layers.

initial oxide of 20 nm followed by a nitride deposition. The oxide is grown at 950°C in oxygen for around 1 h. The nitride is deposited by low-pressure chemical vapor deposition (LPCVD) until 120 nm thick. LPCVD has the advantage of being performed at low temperatures (about 600°C), giving good uniformity, and easily covers any surface steps.

The nitride is patterned so that transistors and other components will be formed where it remains after etching. In areas where the nitride is removed an isolating oxide is grown, which is known as the field oxide. The second photolithographic step involves the fabrication of the n-well by printing and implantation. Phosphorus is implanted beneath the surface with a dose of 2×10^{12} cm^{-2} atoms at an energy of 120 kV; a second dose of lower energy is then given to raise the surface concentration. Often the surface of the p-type material is raised by implantation of boron. The dopant is driven in by a heat treatment in a furnace, 1150°C for 8 h in nitrogen being typical. The field oxide is then grown in a steam ambient at 950°C for about 3 h to grow 600 nm of oxide. The original nitride and oxide are stripped and the gate oxide grown using the same schedule as the initial oxidation. A threshold adjust implant of boron is performed to fix the threshold voltages of the $n$mos and $p$mos transistors. Boron raises the surface concentration of the $n$mos device and counter-dopes some of the phosphorus in the $p$mos. Thus the transistor positions have been defined.

Polysilicon, 450-nm-thick film, is then deposited and doped n-type with a POCl$_3$ gas in a furnace for a resistivity of 25 Ω/sq. Subsequent patterning defines transistor gates. The formation of sidewall spacers to off-set the source and drain diffusions and permit unpatterned silicidation are defined next. The sidewall oxide prevents the sides of polysilicon tracks from being silicided and therefore shorting to the sources and drains. Sources and drains are then implanted followed by their silicidation. Implants for sources and drains involve heavy doses, typically 2×10^{15} cm^{-2}. A combination of arsenic and phosphorus is used for the $n$mos device and boron for the $p$mos; energies of around 25–50 kV are typical. The sources and drains are silicided, along with the gate, to reduce their resistance from >50 Ω/sq. to about 2 Ω/sq. The surface of the wafer is then smoothed out by an oxide layer before contact holes are cut to the source–drain diffusions and the polysilicon gates. This interlayer dielectric is important in forming reliable metal tracks which can crack or thin down where they lie over sharp steps, such as polysilicon edges.

Wafers are then metallized to wire up the individual components. Typically two or three layers of metal are used to allow tracks to cross and improve the packing density of components. Each metal layer is isolated vertically from another by a high-temperature plastic called polyimide. Vias are etched through the polymide to make vertical connections. Finally the circuit is protected from the atmosphere by another layer of polyimide. The chip is then tested for functionality before being packaged.

7. DEVICE ISOLATION AND n-WELL DOPING

The conventional way of isolating neighboring devices is by the LOCOS technique (local oxidation of silicon), which is used in both MOS and bipolar technologies. LOCOS requires a thin silicon dioxide layer to be grown on a wafer surface followed by a deposited layer of silicon nitride. The active area mask of the circuit is then printed in a layer of photoresist spun on the wafer and the pattern transferred from it to the nitride by plasma etching; the resist protects the nitride from the plasma. Etching is commonly achieved by exposing the layers to a CF_4 plasma, which removes the nitride at a much faster rate than the photoresist.

A further masking stage is given to define where *n*mos and *p*mos transistors are to be located. When the wafer material is *p*-type, *n*mos transistors can be made directly into it but in order to make *p*mos transistors an *n*-type doped region is required. This *n*-type region is known as the *n*-well and is created by the implantation of phosphorus. Implantation involves directing a beam of ions at the wafer when situated in an ion implanter in order to dope it in a highly controlled manner. The ions are selected in a magnetic ion separator, which operates using the same principle as a mass spectrometer, and are accelerated to the desired energy. The energy determines their depth in the silicon. A mask of photoresist, usually 1 μm thick, blocks the ions from reaching the silicon surface in areas where doping is not required. In the silicon itself the dopant is situated just below the surface, and in order to move it deeper it is diffused in a furnace. Furnaces are used because of the high temperatures (900–1200°C) required to move dopants. Typically the phosphorus diffuses to a depth of 4 μm. The surface dopant is supplemented by a channel stop to prevent current carriers from spilling across from one *n*-well to another. Following *n*-well formation the wafer is furnaced at 950°C in a steam ambient to grow a field oxide between the nitrided regions. The nitride now acts as a mask against the oxidizing ambient.

During the growth of a field oxide the edges of the nitride pad curl up and oxide extends underneath into the active device area as shown in Fig. 2. Since the patterned nitride is used to define critical feature sizes of circuit components the sideways encroachment of the isolation oxide and channel stop diffusion introduce a dimensional loss. This loss is overcome by making the nitride pad larger to allow for the size change, but then components must be further apart to maintain isolation, and packing density is consequently reduced. For large geometry processes the loss per edge can be tolerated since the overall size change is small. However, this is not the case for small geometry processes where the size change can be an appreciable fraction of the device dimension. As can be seen in Fig. 3, deep submicron technologies employing LOCOS will find it impossible to achieve dimensional control of small features.

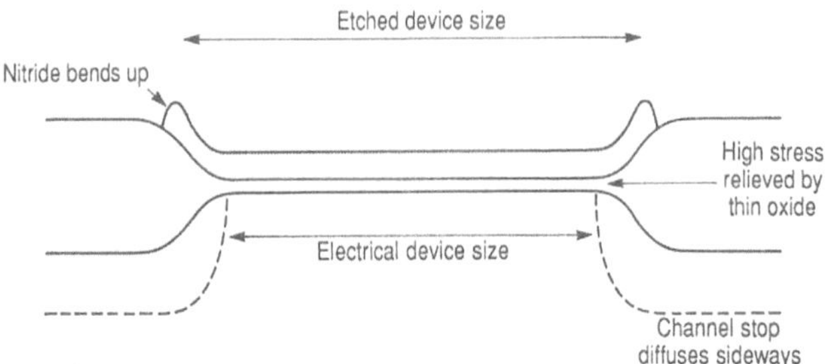

FIGURE 2.   Size reduction from LOCOS.

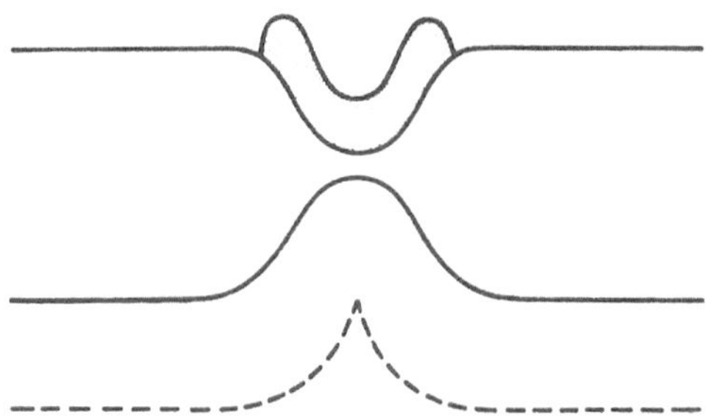

FIGURE 3.   LOCOS fails for deep submicron dimensions.

Size control in LOCOS has been investigated by assessing nitride and suboxide thicknesses. If the wrong combination is chosen then oxidation stresses due to volume expansion will generate dislocations in the silicon. If the nitride is made thick it is more rigid and bends less easily, but if the pad oxide is too thin the growing field oxide pushes against the nitride and causes cracks in the underlying silicon. The minimum limit for birds beak penetration is approximately 0.25 μm per edge for a 0.3-μm field oxide thickness with no defects generated. The use of a sealed interface using oxynitride instead of oxide can reduce the birds beak to 0.1 μm per edge but involves additional processing, and steep field oxide steps occur on the wafer surface.

Many alternative isolation approaches have been investigated with the aim of reducing component size change after oxidation but they all involve complicated fabrication stages. Of the alternatives, trench isolation has been assessed by many industrial research groups; Fig. 4 outlines the technique for CMOS. Trench widths are defined as minimum photolithographic distances and edge loss is controlled by anisotropic plasma etching. The process is more complicated than for LOCOS but uses established fabrication equipment. Trenches extend just beyond the n-well into the substrate, which is diffused after their formation. After oxidizing the trench sidewall and depositing a thin layer of nitride the trenches are filled with polysilicon. The excess polysilicon on the surface is etched back and when the field oxide is grown the thin nitride layer prevents all sideways oxidation. Thus dimensions are maintained during field oxide growth. Peripheral junction capacitance of source and drain diffusions in trench isolated CMOS is reduced since junctions are bounded on three sides by dielectric; device speed is consequently improved by the lower parasitic capacitance of the junctions. In addition, the technique permits very high packing densities since device separation is at the minimum photolithographic distance. An isolation change away from LOCOS is likely to be made for CMOS with around 0.5 μm dimensions, but no universally agreed technique presently exists among circuit manufacturers.

8. GATE OXIDATION AND GATE CONDUCTOR FORMATION

After field oxide growth, the nitride and oxide used to define the active device areas are stripped and the gate oxide is then grown. A furnace is used at a temperature of 950°C for an oxide thickness of 20 nm (typical for a 1-μm CMOS technology).

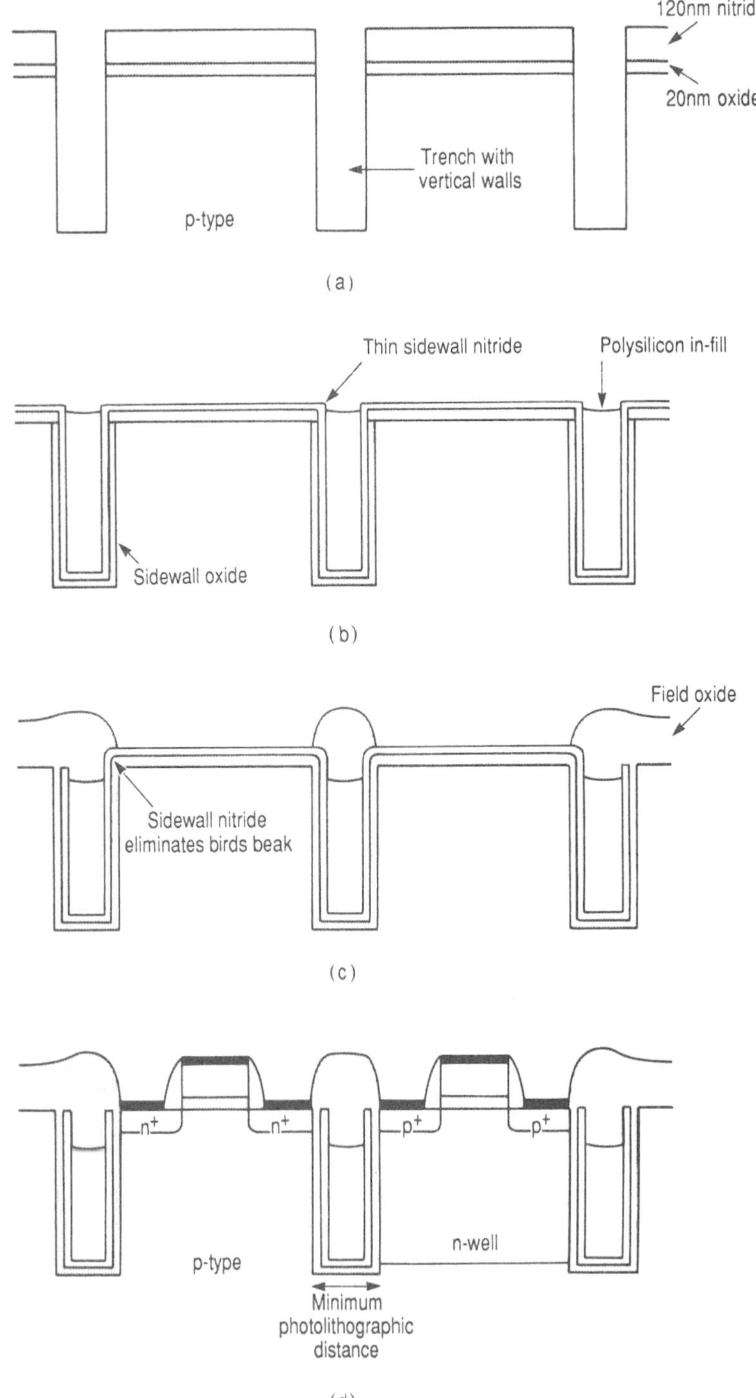

FIGURE 4.   Process sequence for trench isolation. (a) Print trenches and etch. (b) Fill trenches and etch back surface polysilicon. (c) Pattern active area and grow field oxide. (d) Processing as for LOCOS.

The gas ambient in addition to oxygen contains 3% HCl to render any metallic contaminants into the volatile chloride form or to lock them in a fixed position within the oxide. Growth conditions are capable of achieving highly stable and reliable oxides with 5×10^{10} cm^{-2} bulk trap densities and interface trap densities of around 3×10^{10} cm^{-2} eV^{-1}.

In a circuit the supply voltage (5 V) will be dropped across the gate oxide (20 nm) and the field will therefore be 2.5 MV/cm. As devices are scaled it could be expected that a 0.5-μm process would have a field of 5 MV/cm (10 nm oxide thickness and 5 V supply). These are high fields for an oxide to sustain during the lifetime of a circuit. During transistor operation charge will be injected through the oxide and will eventually lead to dielectric wearout causing device failure. The oxide will actually break down completely with a 10 MV/cm field across it, which could occur from an electrical spike from handling or incorrect assembly. In order to prevent a 0.5-μm circuit failing from a gate oxide fault the power supply should be reduced. A suitable JEDEC standard of 3.3 V has been proposed but has yet to be agreed upon and implemented.

Alternatively for submicron CMOS the oxide can be replaced by oxynitride. Oxynitride is created by first growing a layer of oxide and then exposing it to NH_3 in a rapid thermal annealing kit. The rapid thermal kit could also be used to grow the oxide in an oxygen ambient at high temperatures, around 1100°C for a few minutes. The kit consists of a quartz chamber surrounded by infrared lamps. The wafers are kept inside for only a short time since only thin films are grown. The oxynitride has a better resistance to charge injected from electric fields and a much higher breakdown field strength of 16 MV/cm. Thus submicron CMOS may not employ oxides as a gate dielectric.

Polysilicon is deposited on top of the gate oxide by a chemical vapor deposition technique at 610°C, doped, and patterned. A thin layer of oxide (100 nm) is then deposited to cover the entire wafer surface and is etched back in an anisotropic etcher to leave fillets or sidewalls along the polysilicon edges; Fig. 5 shows their location. The sidewalls off-set the source and drain implants to reduce the gate-to-drain feedback capacitance (Miller capacitance), which is an important feature for obtaining fast switching speeds in logic gates.

9. SOURCE–DRAIN FORMATION AND SILICIDATION

As CMOS technology evolved and the depth of the sources and drains was reduced by the appropriate scale factor, transistor output currents increased as predicted by the junction scaling theory and circuit performance improvements occurred due to the increased drive current. However, shallow diffusions then reached the solid solubility limit for dopant in silicon and subsequently produced an increased resistance when made shallower; the resistance worsened the electrical characteristics of transistors by reducing the level of output current from the theoretical value. A new source–drain technology was therefore developed to restore a loss of performance. The present approach throughout the industry is to have shallow highly resistive source and drains created by implantation followed by the deposition of a layer of low resistivity silicide on top to short out the high-resistance diffusions. The silicide is usually $TiSi_2$ or $CoSi_2$. Figure 5 illustrates a modern source–drain technology $n$mos structure.

Shallow n^+ regions are obtained using a heavy element such as arsenic which has a low ion range in silicon resulting in 0.2-μm junction depths. Heavy ions such as arsenic come to rest in the silicon after nuclear collisions with the silicon lattice atoms. Shallow p^+ junctions are not as easy to make because the element boron is universally used in their fabrication. The problem occurs because boron is a light ion and is stopped only by electron interaction with the silicon lattice atoms; consequently, boron is able to channel along crystallographic directions to reach large depths. The heavier dopant BF_2 can be implanted (the implant energy shared between the B and F_2), which breaks up on impact enabling very low-energy boron to be implanted; but even then only depths of around 0.4 μm are possible since channeling still occurs. One way to reduce the boron depth is to implant the crystal with silicon or germanium in order to render the surface amorphous. A light dopant ion can then be implanted into the amorphous region, which is about 100 nm thick; no channeling is possible in amorphous material because it is noncrystalline. A rapid thermal annealer can then out-diffuse the boron in a highly controlled manner to produce 0.2-μm junction depths that are at the same depth as the n^+ junctions.

Shallow junctions improve electrical performance by having low peripheral junction capacitance but are highly resistive. By using a thin silicide layer the resistance of the diffusion can be lowered by over an order of magnitude. This is achieved by sputtering a 50-nm layer of Ti over the surface of the wafer and creating TiSi in a rapid thermal kit with nitrogen as gas ambient; the reaction temperature is 625°C. The unreacted Ti is stripped off in a hydrogen peroxide solution and the TiSi converted to $TiSi_2$ by a second anneal at a higher temperature of 800–1000°C. Annealing times are about 30 s each, which results in a final resistivity of around 2 Ω/sq. The Ti consumes Si in the reactions to produce a 100-nm layer of $TiSi_2$.

Modern CMOS processes must have a special n-type doping profile to spread out the electric field around the $n$mos drain. The reason for reducing the field concerns the generation of hot-carriers generated within the drain depletion region. For larger geometry processes hot-carrier injection was not a problem because electric fields in depletion regions were small. As device sizes were reduced by scaling, channel lengths became shorter and depletion regions became thinner due to the higher doping in the silicon. However, the industry standard 5-V supply voltage was not scaled and consequently devices with channel lengths of 1 μm and smaller contained very high electric fields around $n$mos transistor drains. Such high fields impart high energies to carriers in the vicinity of the drain, as Fig. 6 illustrates. Some hot-carriers become injected into the gate oxide where they can be trapped;

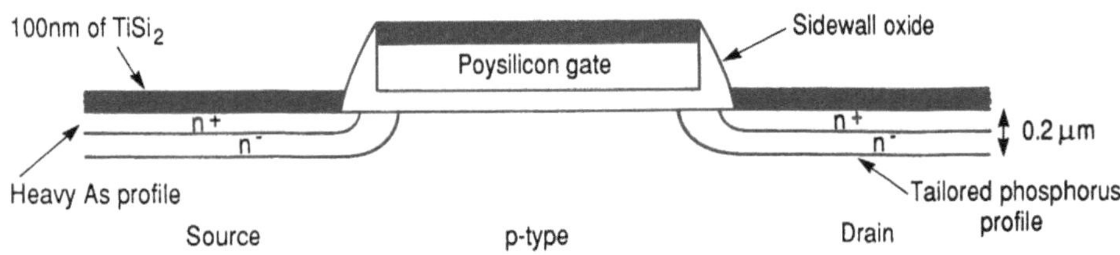

FIGURE 5. Double diffused $n$mos and self-aligned silicide.

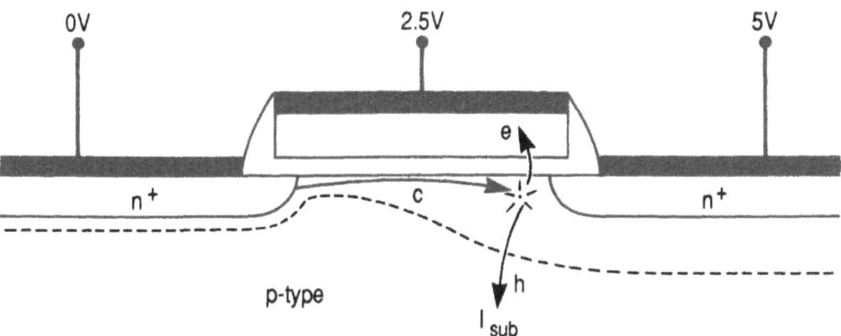

FIGURE 6.   Hot-carrier injection; electron hole generation at high field point.

trapped negative charge raises the threshold voltage and lowers the gain, due to the damage caused at the oxide–silicon interface during impact with the oxide. The consequence is that the device output current will reduce with time, especially in the linear region of device operation. For an arsenic n^+ junction operating from a 5-V power supply the transistor output current will be reduced by around 5% after a single day of operation; longer periods of time reduce it further affecting circuit performance and ultimately cause failure. By surrounding the As n^+ diffusion with a second one of phosphorus to produce an n^- region, the electric field is spread out by the n^- dopant. The peak field is considerably reduced and is insufficient to provide the carriers with sufficient energy to mount the oxide–silicon interface barrier. These $n$mos drains are known as double-diffused junctions; with a double-diffused junction the device will operate untroubled for many decades. An alternative approach is to build a lightly doped drain (LDD) junction into the transistor structure. Because holes are heavier than electrons the problem only appears in $p$mos transistors at around 0.5 μm channel lengths.

10. MULTILAYER METALLIZATION

VLSI CMOS circuits are complex and contain hundreds of thousands of contact holes etched for metallization links; in a logic cell such as a two-input NAND gate there is approximately one contact hole per transistor. As contact hole sizes are reduced to below 1 μm^2 the contact resistance rises exponentially. Ohmic contacts between the metal and silicon are formed at temperatures around 425°C where some silicon within the contact hole is consumed by the metal; the amount increases as furnace temperature, time, and volume of aluminum in contact with the silicon increases. The conventional means of suppressing this reaction is to introduce silicon (typically 1.5% by volume) into the aluminum film to saturate it at the annealing temperature. In large area contacts the silicon has no detrimental effects, in fact it must be present to enable contacts to be reliably produced without the aluminum pitting through the underlying p–n junction. In contacts around 1 μm^2, the silicon segregates during sintering and grows as an aluminum-doped, epitaxial layer to cover the contact hole and to produce a large resistive barrier. The smaller the contact hole area the higher the resistance and the lower the output current from transistors. However, silicon can be eliminated if a TiW barrier is used.

An alternative approach for CMOS is not to use aluminum at all but to use tungsten; tungsten is used as an interconnect layer because of the superior reliability obtained for thin tracks. Figure 7 illustrates the advantages of using tungsten not only to eliminate high contact resistances but to obtain better step coverage of the contact hole sidewalls. Tungsten is deposited to completely fill the contact hole by a chemical vapor phase reaction and is then patterned for the first metal layer. The current density capacity for tungsten (its resistance to electromigration) is about five times greater than for aluminum, which always has copper (1%–4%) added to it in order to improve its performance. Consequently tungsten is more reliable as a current-carrying conductor but has the disadvantage of being slightly more resistive.

If only a single layer of metal were to be used for connecting components in a circuit a huge chip area would be required; in a circuit using a single metal layer the interconnect would consume 75% of the chip area, falling to 50% for two metal layers and zero for three. CMOS now uses two or three layers of metal to improve the packing density of components. Each layer is electrically isolated from the others by a dielectric. The dielectric can be a hard material such as oxide or a soft material such as polyimide. Polyimide, which is spun on the wafer as a liquid before being cured, has the advantage of planarizing the wafer surface after the first layer of metal has been formed. Vias are subsequently etched through the polyimide dielectric to allow connection between the metal layers. For a 1 μm CMOS process, track widths are typically 2 μm wide separated by gaps of 1 μm giving a metal pitch of 3 μm. These dimensions are reduced for submicron processes making high demands on the metallization techniques employed both for electrical reliability and chip yield.

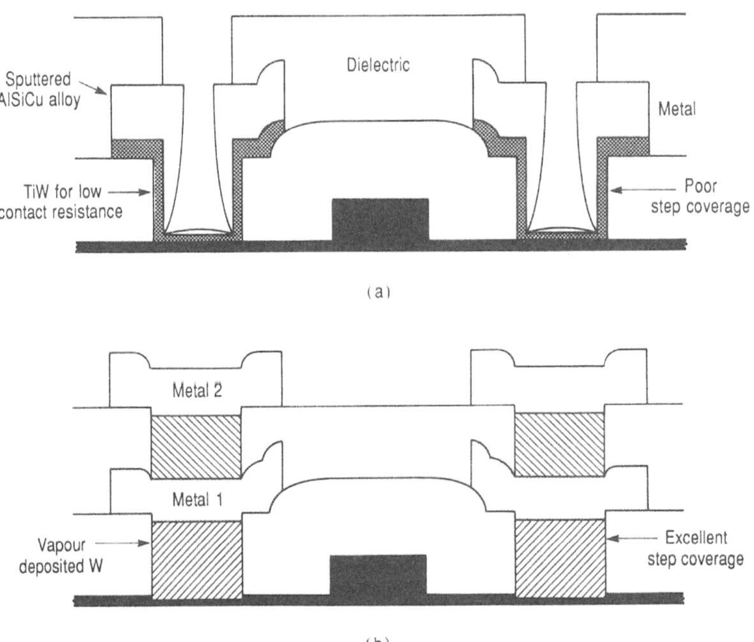

FIGURE 7.   Metallization schemes. (a) Aluminum metal in submicron contact holes. (b) Tungsten in submicron contact holes.

11. BICMOS

BICMOS technology combines bipolar and CMOS transistors on the same chip by a merging of process techniques. The advantage of a BICMOS process is that capacitive loads can be driven faster by bipolar transistors (usually configured as ECL gates) than by MOS transistors. Also ECL compatibility can be made for CMOS chips to interface with bipolars on a circuit board. The main application for BICMOS is found in logic circuits, SRAMs, and in some microprocessors where long signal tracks are to be driven and improved chip interfaces required. A further advantage of BICMOS is the ease of realizing analog circuit functions for special applications. The two main disadvantages of BICMOS are that the process technology is very complicated and the bipolar cells consume large chip space. Therefore BICMOS finds niche application areas where a performance gain, primarily speed, is required.

12. SILICON ON INSULATOR

Silicon on insulator (SOI) substrates permit neighboring components to be isolated by dielectric, as Fig. 8 shows. Neighboring components are free from the depletion layer distances that must be provided for conventionally isolated LOCOS devices and can therefore be placed at minimum photolithographic distances apart as in trench isolation. In principle, SOI offers the best packing densities for CMOS. Traditionally SOI substrates have been silicon on sapphire (SOS) but SOS cannot achieve the high levels of integration required for commercial CMOS owing to a poor silicon to sapphire interface that causes high levels of off-state leakage current.

Superior device performance can be offered by substrates created by oxygen implantation into silicon to produce a buried isolating layer. Large doses of oxygen, 2×10^{18} cm^{-2}, are implanted at 200 kV into silicon wafers to build up a buried oxide layer about 300 nm thick. The implantation is followed by a high-temperature anneal at 1300°C to restore the crystal quality damaged by the implant and to diffuse excess oxygen out of the top silicon layer. Processing is then similar for bulk silicon but much simpler for device isolation, which is achieved by a single silicon etch. By careful control of the top silicon island thickness, below 100 nm, device output currents can be greater than for equivalent bulk devices by up to 50%. Lower junction capacitances also are achieved by the dielectric isolation, which gives an improvement in inverter speed; 17-ps delays have been achieved in ring oscillators. Continuing process developments could lead to SOI by oxygen implantation becoming a mainstream commercial technology in the 1990s.

13. LATCH-UP IN CMOS

Latch-up in CMOS refers to parasitic bipolar transistor action formed when unrelated diffusions link electrically. Latch-up has been a major concern for the reliability of CMOS but has now been effectively eliminated. Figure 9 shows a cross-sectional diagram of CMOS inverter. Superimposed on the inverter are two

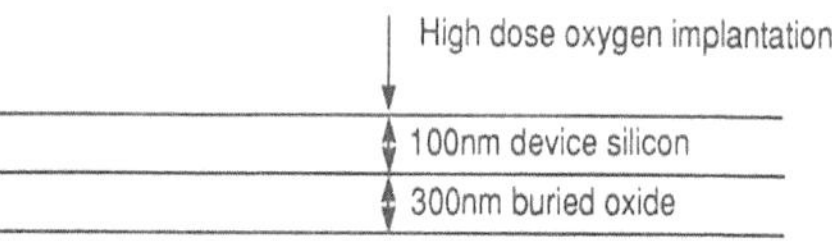

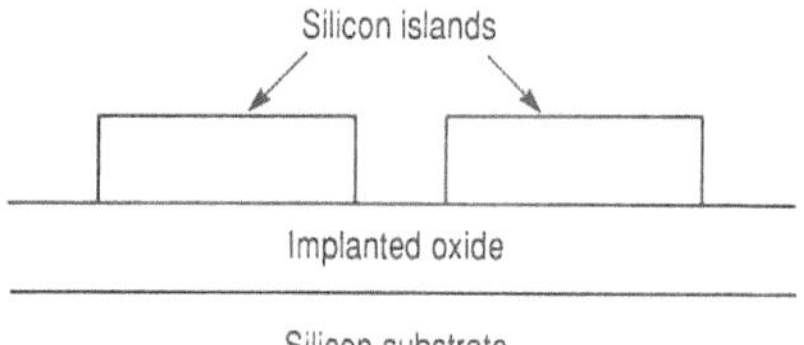

Anistropically etch silcon islands 82-90° sidewalls

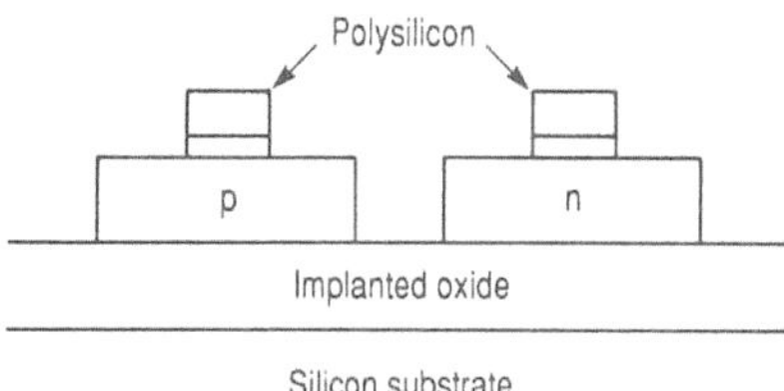

Form n- and p-type regions, gate oxide and form polysilicon pattern

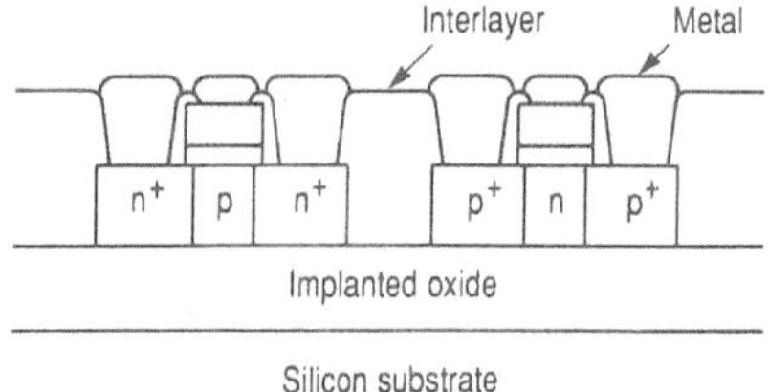

Deposit interlayer, etch contacts, metallize

FIGURE 8.   Silicon on insulator CMOS.

parasitic bipolar transistors formed from sources and well diffusions; one transistor is a lateral *npn* and the other a vertical *pnp*. These transistors form a *pnpn* path or silicon controlled rectifier (SCR) between the supply terminals. Other *pnpn* paths in CMOS circuits exist in transmission gates, gate protection structures, and logic blocks, but this path in the inverter is the most common.

Figure 10 shows a typical SCR characteristic. In normal operation the SCR is "off" in an unlatched or a high-resistance state but it can be triggered "on" into a latched or low-resistance state by a current flowing laterially across the emitter–base junctions of the parasitic bipolar transistors. When a parasitic bipolar transistor is "on" its collector current acts as the base current of the other transistor thus creating a state of positive feedback between them. Trigger currents are shown in Fig. 9 as

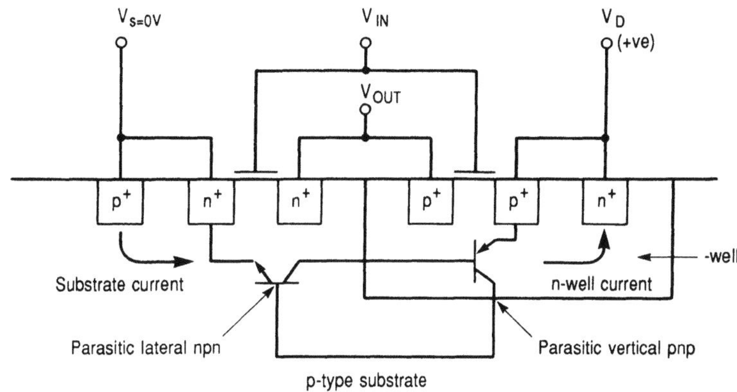

FIGURE 9.   Cross section of CMOS inverter.

n-well and substrate currents and can be generated by voltage supply fluctuations, electromechanical pulses, and ionizing radiation.

In the low-resistance state, the 5-V terminal voltage across the inverter will have fallen to about 1.5 V and very high currents of 10–100 mA will flow across the power supply terminals. A loss of logic states and possible permanent damage will be experienced by the circuit. Once latched the SCR can only be turned off by reducing the current flowing through it to below the threshold level of the holding current. Often this is only possible by disconnecting the power supply. To reduce latch-up susceptability, holding currents and voltages should be as high as possible to shift the low-resistance region away from all reasonable operating conditions.

The use of epitaxial silicon grown on heavily doped silicon substrates has been shown to eliminate latch-up entirely. The heavily doped substrate introduces a shunting resistor below the emitter–base junction of the lateral bipolar transistor. The low substrate resistance prevents the emitter to base voltage from becoming

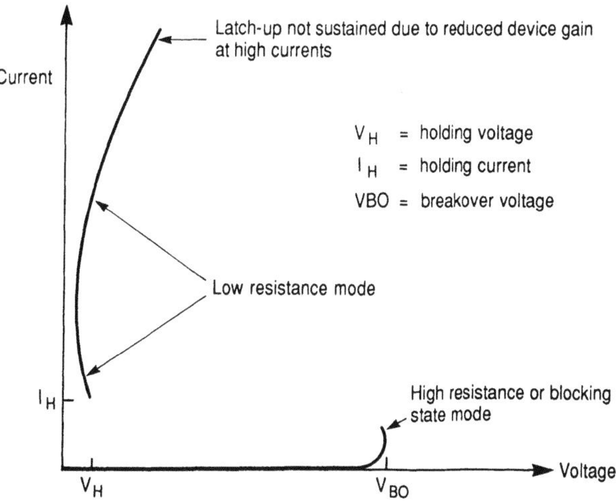

FIGURE 10.   SCR characteristic.

forward biased by greater than 0.7 V. By proper choice of substrate resistivity and epitaxial thickness only very high currents above those normally encountered in CMOS structures are able to forward bias the lateral junctions.

14. CHIP PACKAGING

Individual circuits on a wafer are separated from each other by unpatterned tracks of bare silicon known as scribe channels. During photolithography, scribe channels are othogonally aligned relative to a major crystallographic flat machined on the wafer edge so that each chip is bounded by major crystallographic axes. A laser beam or a narrow bladed diamond saw is positioned over each scribe channel and a deep cut made across the wafer from edge to edge. The wafer is subsequently pressed by a roller, which initiates cleaves along each scribe channel and the wafer fractures into individual circuits. After scribing and dicing, chips are encapsulated in chip carriers or packages before being mounted onto printed circuit boards. These package types fall into two main categories, plastic and ceramic. For each type a chip is firmly attached to the metal base of the package. Thin wires are bonded to aluminum pads located on the outside of the chip and then to the package terminals. In this way connection between the chip and external electrical equipment is possible through the package leads.

Ceramic carriers have a recess into which a chip is placed. After bonding, a lid hermetically seals the chip into a moisture-free cavity to avoid metal bond pad corrosion. The plastic moulding used in packaging contains a mixture of novolac epoxy resins containing fused silica with high chemical stability to prevent any shrinkage, which would stress and crack the package. Some moulded plastic packages have been identified as causing soft errors in DRAM circuits by alpha particle emission and some particle emission has also been identified from the ceramic material. A technique to give further chip protection is to cover the chip with a thick layer of polyimide to absorb the energy of the alpha particle.

In high-volume circuit manufacture where costs are reduced, plastic packages are cheaper to use than ceramic packages and many standard CMOS parts such as DRAMs, SRAMs, and microprocessors are sold commercially in moulded plastic. Despite the increase in the scale of CMOS integration, the majority of integrated circuits produced have lead counts in the range 8–40. However, 32-bit microprocessors and other specialist professional or military systems require lead counts of 100 or more. These advanced packages require leads that are placed very close together and can be expensive because of the skills needed in making them.

BIBLIOGRAPHY

There are many excellent publications where detailed information can be found about silicon process technology and device performance. The following books will provide the student with such information:

A. B. Glaser and G. E. Subak-Sharpe, *Integrated Circuit Engineering: Design, Fabrication and Applications*, Addison-Wesley, Reading, Massachusetts (1979).

A. S. Grove, *Physics and Technology of Semiconductor Devices*, Wiley, New York (1967).

J. Mavor, M. A. Jack, and P. B. Denyer, *Introduction to MOS LSI Design*, Addison-Wesley, Reading, Massachusetts (1983).

S. M. Sze, *Physics of Semiconductor Devices*, Wiley, New York (1981).

S. M. Sze (ed.), *VLSI Technology*, McGraw-Hill, New York (1983).

S. Wolf and R. N. Tauber, *Silicon Processing for the VLSI Era*, Vol. 1, *Process Technology*, Lattice Press, Sunset Beach, California (1986).

14

Technologies for High-Speed Compound Semiconductor ICs

J. A. Turner

1. INTRODUCTION

The development of process technologies for high-speed integrated circuits in compound semiconductor materials presents a number of wide-ranging challenges for the process engineer. For analog applications the diversity of frequencies from about 1 GHz to 100 GHz means that a standard technology cannot be adopted for the whole frequency spectrum as the circuit topologies required at each end of this frequency range differ widely. At low frequencies, ~1 GHz, traditional distributed (microstrip) style matching elements consume too much surface area of GaAs and so rf techniques using lumped inductors, capacitors, and resistors are used. It is only at frequencies in excess of 6–7 GHz that fully distributed circuits are small enough to enable utilization of this type of impedance matching. In digital circuits the challenge is in yield improvement, the close spacing of components and signal lines, and the level of integration.

The component and technology requirements for digital and analog circuits are given in Table 1. The process engineer must first develop technologies for manufacturing all the components given in this table and then devise ways of integrating them into a reproducible, reliable IC process consuming as little GaAs surface area as possible.

The circuit manufacturing procedures can be described by four basic processing technologies:

1. Lithography;
2. Etching of GaAs;
3. Metal deposition and removal;
4. Dielectric deposition and removal.

These are used to generate all the components required in the IC and will be described in Section 4.

J. A. Turner • Plessey Research Caswell Limited, Allen Clark Research Centre, Caswell, Towcester, Northamptonshire NN12 8EQ, U.K.

TABLE 1. Components and Technologies for Analog and Digital ICs

	Analog	Digital
Transistors	√	√
Resistors	√	—
Capacitors	√	√
Inductors	√	—
Transmission/interconnect lines	√	√
Multilevel metal	√	√
Air bridges	√	√
Via hole technology	√	—
Material uniformity	√	√√
Number of gates per chip	≤ 20	Large

2. THE MATERIAL CHOICE FOR ACTIVE DEVICES

Presently three major choices of material technology are available to the process technologist:

1. Vapor phase epitaxy[1] and molecular beam epitaxy[2] for heterostructure FET and bipolar devices;
2. Ion implantation[3] for these MESFET.

The operation of the devices mentioned here are described by other authors in this book and will not be covered here.

The choice of material technology obviously depends on the particular device being used. There is no doubt, however, that MESFET technology is benefiting from the introduction of ion implantation on the grounds of cost and parameter control. An ion implantation machine is capable of implanting ≥ 350 GaAs wafers per hour, compared to less than ten wafers per hour from the most sophisticated vapor phase and molecular beam epitaxy equipments. Table 2 compares the three material technologies in terms of their useful and less desirable attributes.

TABLE 2. General Comparison of Material Processes

Material technology	Profile control	Electrical quality	Cost per layer
Vapor phase epitaxy (VPE)	Least control of the three techniques	Highest electron mobility	High—scaling up possible
Molecular beam epitaxy	Monolayer accuracy and very versatile	Similar to VPE	Highest—not readily adapted for high throughput
Ion implantation (II)	Few percent variation across whole wafer	Acceptable for analog and digital IC fabrication	Low—true "mass production" process

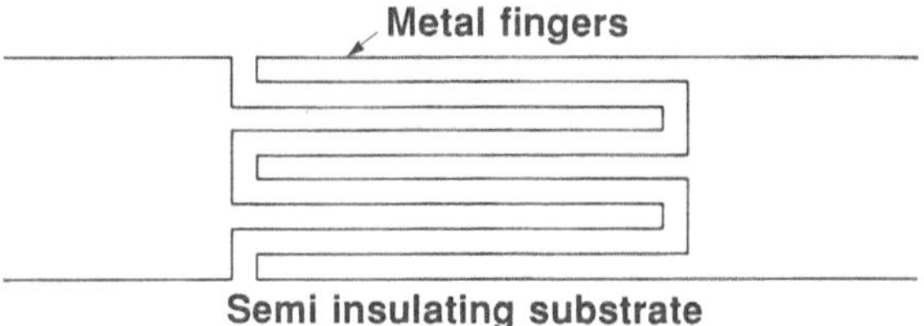

FIGURE 1.   An interdigitated capacitor structure.

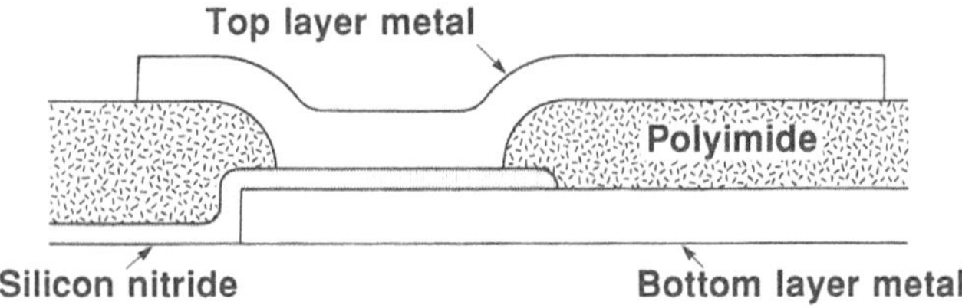

FIGURE 2.   A metal—insulator—metal (MIM) capacitor.

3. CIRCUIT-BUILDING ELEMENTS

3.1. Capacitors

Capacitors are used for tuning elements, for interstage isolation, and for bypass. Two types are currently in use:

a. Interdigitated for low capacitance (less than 1 pF)[4];
b. Parallel plate, metal insulator metal for 1–20 pF.[5]

The interdigitated capacitor shown in Fig. 1 utilizes the capacitative coupling between adjacent fingers. To fabricate these elements, sputtered films of a gold-based alloy are deposited on to the semi-insulating substrate, annealed, and the finger pattern defined by ion beam milling. The milling process removes metal not protected by a photoresist layer at the rate of a few micrometers per hour. The metal–insulator–metal capacitor (MIM) is formed by two metallized areas separated by a dielectric material. Figure 2 shows this type of capacitor. The dielectric used is chosen to give a reasonable capacitance/unit area, is easy to deposit, and stable in use. Table 3 gives three dielectrics in common use, the most commonly used being Si_3N_4 or SiO_2.

The dielectric thicknesses used have a major outcome on the final yield of the circuit. In analog circuits there are generally many more capacitors than MESFETs

TABLE 3.   Capacitor Dielectrics

Material	Dielectric constant	C/A (pF/mm^2)	Deposition technique
Silicon nitride	6.5	480	Plasma-assisted CVD
Silicon dioxide	4.0	340	Low-pressure CVD
Polyimide	3–4.5	30	Spun and cured film

and the capacitor yields determine the yield of the circuit. Generally dielectric thicknesses of 1000–2000 Å are chosen in order to minimize the number of pinholes in the film that cause dc short circuits while maintaining a reasonably high capacitance per unit area.

3.2. Inductors

Inductors are used particularly in lower-frequency analog circuits as tuning elements between transistors.[5,6] Figure 3 shows schematically two common configurations of inductor. They pose two major fabrication challenges—definition of the metallization and the connection between the center of the spiral and the rest of the circuit.

To maintain a high circuit quality factor (Q) at low microwave frequencies the metal thickness must be of the order of 3 μm and the separation of the spirals is typically 12 μm.

An alternative technology for achieving a thick metallization is to plate up the spiral using a process similar to that described for the air bridge process in Section 3.4. One disadvantage of this process is that the edge definition of the plated spiral is somewhat worse than the milled metallization; however, it is easier to build up the metal thickness to values considerably greater than 3 μm, which can be an advantage in some lower-frequency circuits.

The second challenge, that of connecting the center of the spiral to the rest of the circuit can be met either by the use of the airbridge technology or by a metal underpass connection.

The airbridge technology offers a very good microwave solution with a low parasitic capacitance but is prone to mechanical deformation. The span is necessarily large and any downward pressure on the bridge will cause it to be distorted and to short to the spirals. A mechanically more rigid solution is one in which the connection is made to run between two dielectric layers underneath the spiral. Figure 4 shows a cross section through a completed inductor with an underpass connection.

3.3. Resistors

Resistors are used for determining the bias applied to the transistors, for feedback networks, and for terminations in power combiners and Lange couplers.[5] Two families of resistor are commonly used, those fabricated in the GaAs active layer and those produced in deposited resistive films.

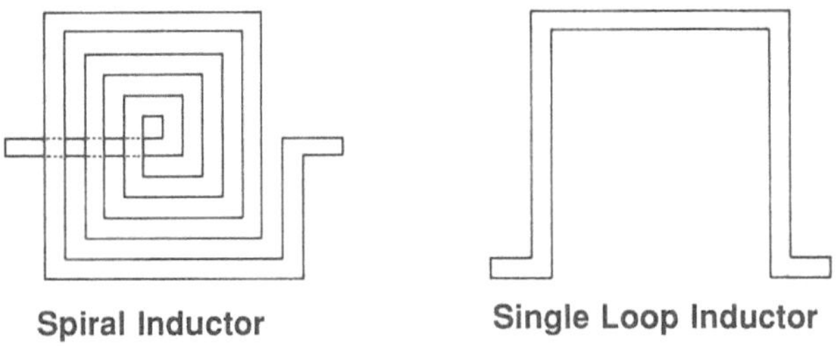

FIGURE 3.   Inductor configurations.

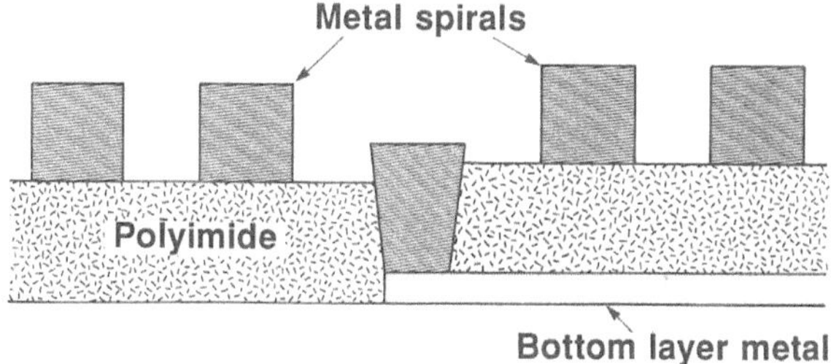

FIGURE 4.   Inductor with "underpass" metal.

3.3.1. GaAs Resistors

The simplest way to produce a resistor is to use the GaAs material in which the active device is fabricated. In this approach the resistor is fabricated in a manner similar to the FET itself apart from the fact that the gate electrode is omitted. Two types of GaAs resistor are shown in Fig. 5. Figure 5a shows a mesa resistor where isolation is achieved by etching. In Fig. 5b isolation is achieved by selectively implanting resistive areas in the semi-insulating GaAs substrate.

The use of GaAs resistors is not without its problems, as an ungated resistor can act, under certain bias conditions, as a planar transverse electron oscillator (Gunn oscillator). It is therefore important to ensure that the bias voltage across the terminals of the resistor is below the threshold voltage for the onset of Gunn oscillations.

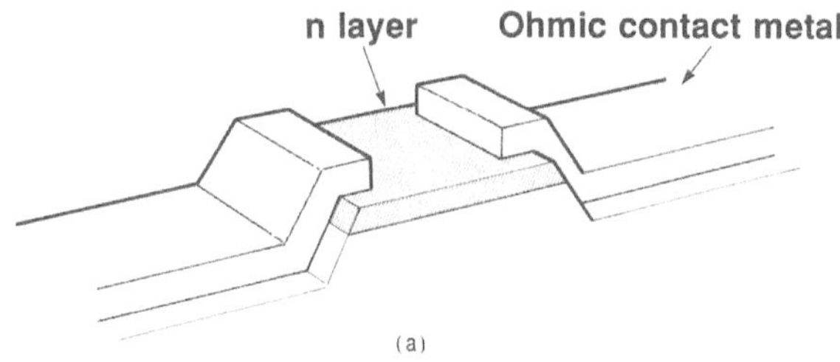

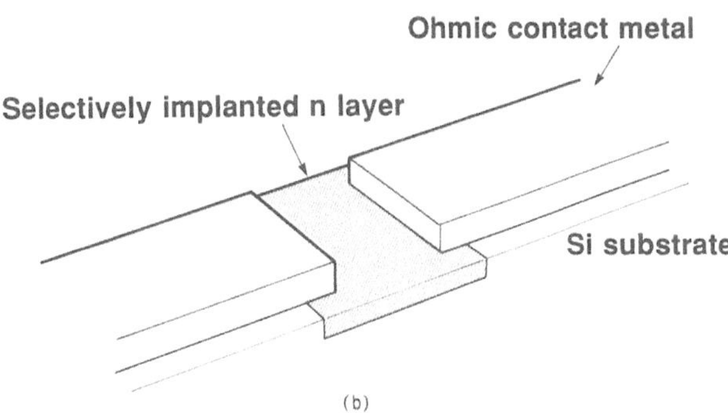

FIGURE 5.   Two available resistor technologies. (a) mesa resistor; (b) implanted resistor.

3.3.2. Thin-Film Resistors

In Table 4 are listed a number of metallizations that are being used for resistors. Cermets are potentially a very good resistor material as changes in composition of the silicon monoxide-chromium mix can alter the sheet resistivity of the deposited layer. However, control of the composition is not easy and more work is necessary on the deposition technology if cermets are to be widely accepted as a resistor material. Generally, whichever material from Table 4 is chosen, additional process steps are required in order to incorporate it into an MMIC process—consequently wherever possible GaAs resistors are used.

3.4. Airbridges and Dielectrically Isolated Interconnections

In many sections of an MMIC a connection is required between nonadjacent metallized areas. This can be accomplished using either airbridges or dielectric interconnections.

An airbridge is a bridge of metal running above the surface of the circuit between metal areas to be interconnected; they are formed by a photoresist and plating process outlined in Fig. 6. By careful control of the resist thicknesses and plating conditions, a very strong well-defined bridge can be formed.

The scanning electron microscope picture in Fig. 7 shows an FET with airbridge interconnections to the source contact areas. Some manufacturers rely solely on this form of interconnection using it for passive component integration as well as for interconnecting the MESFET and in digital as well as analog circuits. In digital circuits many hundreds of such interconnections are made and the process must therefore be very reliable.

An alternative interconnection process is to make use of multilevels of dielectric films. This is shown in Fig. 8. One advantage of this approach is the fact that it is essentially a planar process and overcomes the possibility that airbridges can be damaged during die separation and mounting. Three dielectric materials for this purpose are in common use: silicon dioxide, silicon nitride, and polyimide. For fabrication simplicity the interlayer dielectric can also be the film that forms the capacitor dielectric.

In some multilevel schemes polyimide is a preferred dielectric material as it has the ability to "planarize" the surface of the MMIC making subsequent metallization over sharp underlying irregularities a relatively easy task.

TABLE 4. Some Resistor Materials Used in MMICs

Material	TCR (ppm/°C)	Ω/sq	Manufacturing technique
CrSiO (cermet)	−300 to +100	50–500	Sputtering from composite target
Bulk GaAs	+3200	50 for 2×10^{17} material	Epitaxial or ion implanted
NiCr	200	90 typically	Sputtering

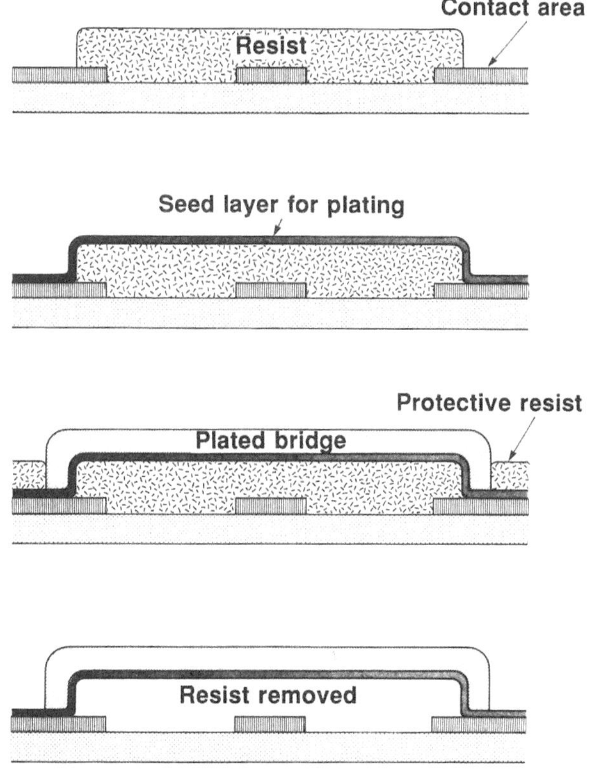

FIGURE 6.   Air bridge fabrication process.

3.5. Via Hole Technology

In more complex multifunction analogue ICs (Fig. 9) it is desirable to remove the constraint of rf grounding the circuit chip around its periphery, and processes have been developed to directly ground individual components. One way that this can be achieved is by a "via hole" process, in which holes are made from the back of the wafer to contact grounding pads on the front side.[7] This is illustrated in Fig. 10. Once the holes have been made they can be filled with the appropriate metallization, which makes a direct contact to the front side metal areas requiring to be grounded. The process of mounting the chip onto a suitable substrate automatically grounds the components on the front side of the wafer.

3.6. The Integration Process

Defining the components that were described in the preceding sections of this chapter is not sufficient for the realization of a full integrated circuit process. Its development is an iterative exercise where the requirements of the technologies to produce the components are combined with those needed to interconnect them. Generally many compromises have to be made before a manufacturable process can be finally realized not only involving aspects of the technology, but also satisfying the requirements of circuit design, assembly, and packaging. The interaction

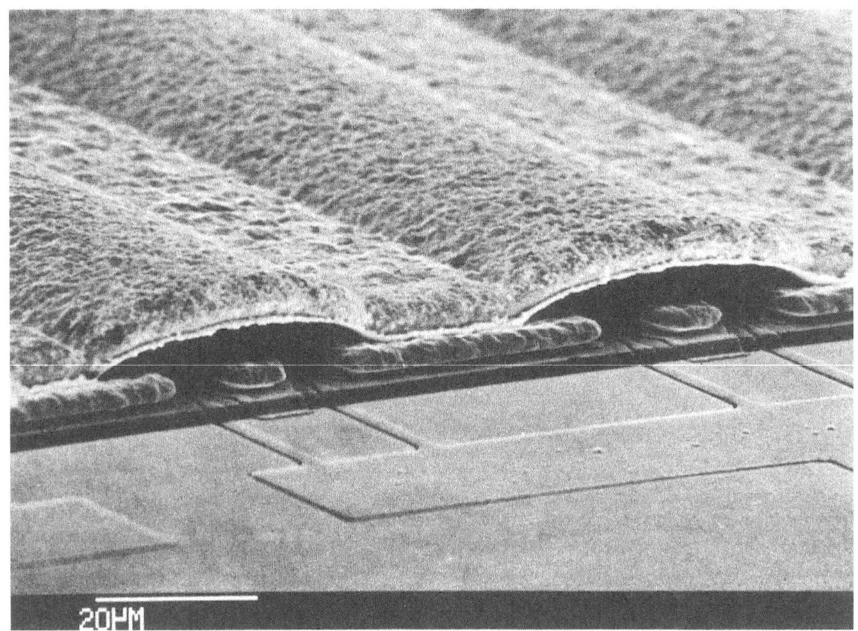

FIGURE 7.   SEM picture of an airbridged power MESFET.

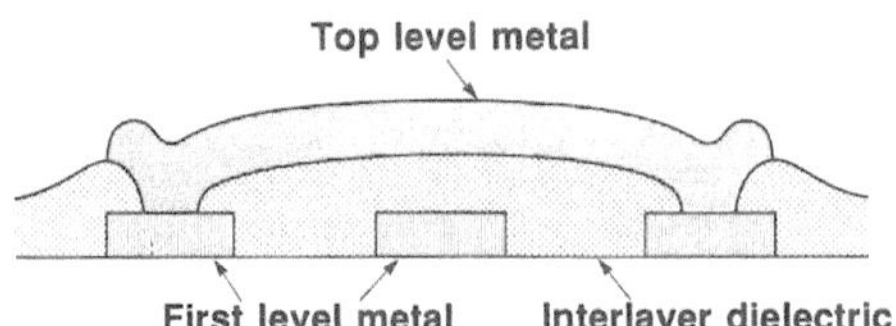

FIGURE 8.   Dielectric interconnection.

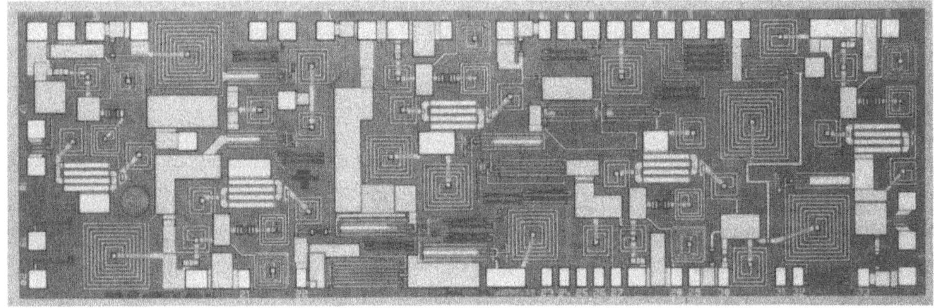

FIGURE 9.   Phased array radar Tx/Tx module.

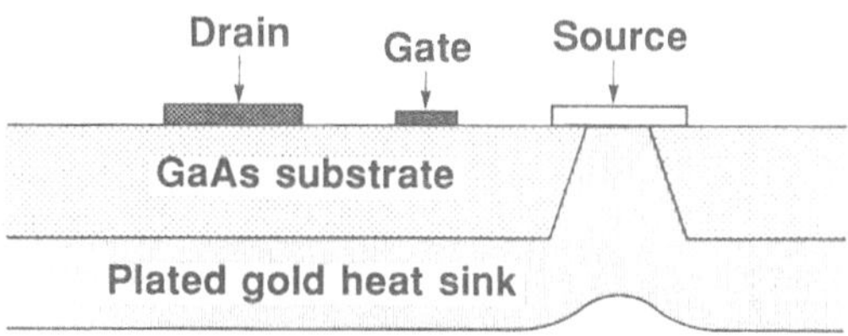

FIGURE 10.   Component grounding using via holes.

necessary between circuit fabricators, circuit designers, and users cannot be over-stressed, and all successful processes are the result of constant consultation between *all* interested parties.

One example of a GaAs integrated circuit process that is equally applicable to analog and digital ICs is shown schematically in Fig. 11. In this figure it is seen how the layers of the circuit are built up sequentially starting from the initial defining

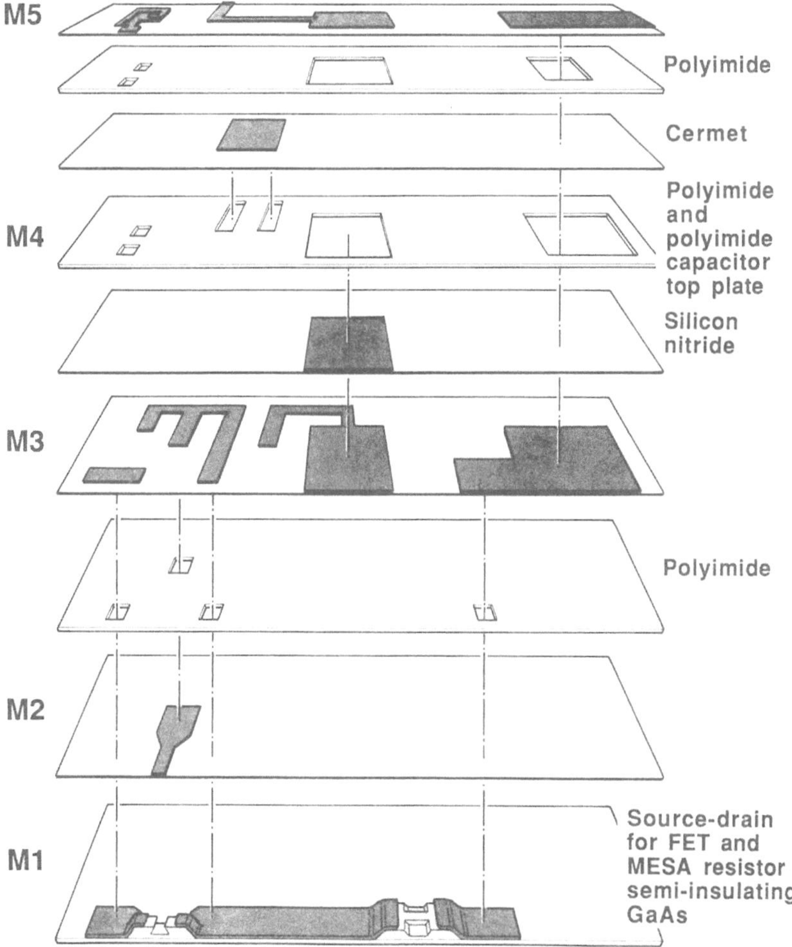

FIGURE 11.   Schematic of a GaAs IC process.

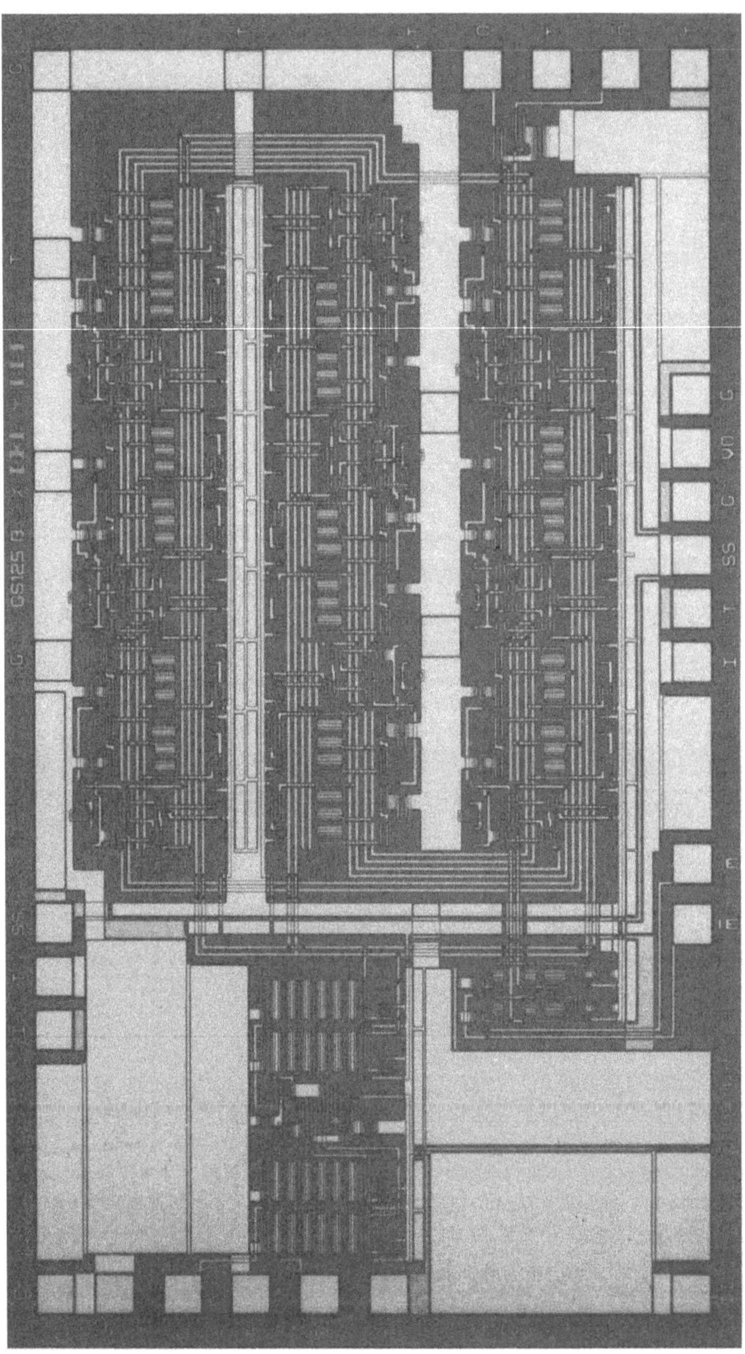

FIGURE 12. A $\div 4/5$ modulus divider.

of the MESFET and ending up with the passivation of the complete circuit. This is representative of the many process technologies in use worldwide; however, there are many variations. It is probably true to say that there are as many IC processes as there are laboratories fabricating them, each one subtly different from the other. Figure 11 shows final metal interconnections being made using interlevel dielectric isolation; another and perhaps more popular approach is to use airbridge interconnections for this process.

Whereas Fig. 9 in Section 3.5 showed a complex analog circuit fabricated using this process, the technology is also capable of producing the equally complex digital circuit shown in Fig. 12.

4. IC FABRICATION PROCESSES

In this section we describe the process techniques that are used to produce the structures described in the preceding sections of this chapter.

Once the active n layer has been formed, the whole integrated circuit can be processed using three basic fabrication techniques. These are as follows:

a. Lithography;
b. Deposition—metal and dielectric;
c. Etching—metal, dielectric, and semiconductor.

Various sequences of these three techniques are needed to allow complete circuits to be fabricated.[8]

4.1. Lithography

Lithography forms a vital part of any process technology, for all process stages are preceded by a lithography step that defines the pattern for etching or for metal deposition. Basically the lithography process involves defining a pattern in a thin photosensitive film (photoresist) by a process of optical exposure through a photographic mask and subsequent differential removal by immersion in a resist developer solution. Figure 13 shows two different methods of photolithography: by direct contact and by optical stepper. In the direct contact method, a mask containing the pattern to be defined is placed in hard contact with the photosensitive resist film that has previously been spun onto the GaAs wafer as a uniform thickness film. The resist not protected by the dense areas in the mask is irradiated with ultraviolet light. The irradiated film of resist is placed in a developer solution, and, if the resist is a "positive" one, those resist areas exposed by the radiation will be removed to expose the substrate surface. If the resist is a negative one then those resist areas not exposed will be removed in the developer. The resist film left after development can then be used as an etch resistance film or as a film for the metal lift-off process.

The optical stepper process is similar to the in-contact one in terms of exposure and development of the resist except that the mask is held some way away from the resist coated sample and is exposed to projecting an image of the reticle (mask) on to the surface of the GaAs slice.

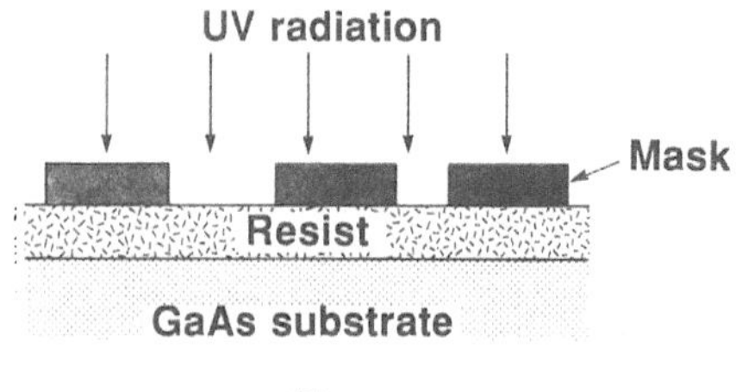

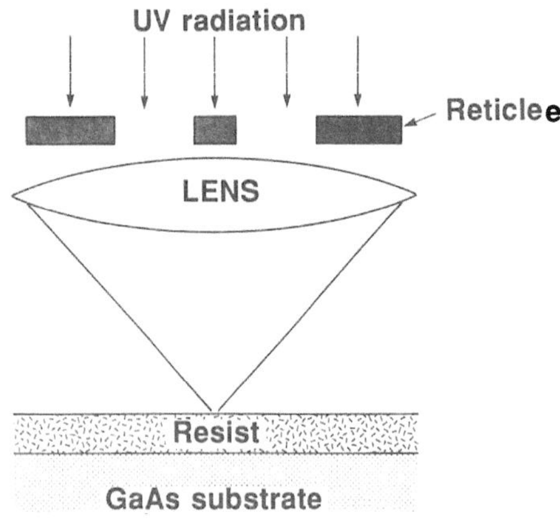

FIGURE 13.   Contact printing and optical stepper schematic, (a) contact printing; (b) optical stepper.

Each process has advantages and disadvantages over the other, and these are listed in Table 5.

As far as producing high-performance MESFETs is concerned, the preferred method of lithography is contact printing, for at present it gives the shortest gate lengths.

However, neither approach is capable of defining dimensions less than 0.5 μm. For sub-0.5-μm geometries electron beam lithography is used. Here a very narrow beam of electrons is used to expose electron sensitive resists. This type of lithography overcomes the problems of present optical processes and enables geometries down

TABLE 5.   Comparison of Contact Lithography and DSW

Photolith method	Advantages	Disadvantages
Contact lithography	0.5-μm resolution on existing equipment	Mask procurement difficult
		Mask wear due to contact with resist
Direct step on	No mask wear 5:1 or 10:1 projection system so mask procurement less difficult	Resolution guaranteed only to ~1 μm Process conditions critical

to 0.1 μm or less to be defined. Table 6 shows the factors affecting the choice of electron beam lithography, and from this it can be seen that the main disadvantages are cost of the equipment and the time taken to expose the patterns. It is, however, the most practical way of defining sub-0.5-μm gate stripes and therefore is widely used for fabricated discrete MESFETs and integrated circuits.

Another real advantage of the electron beam process is its high yield, giving over 80% properly defined 0.5-μm gate lines on each wafer. For this reason many companies are using the electron beam technology for gate dimensions up to 1μm despite its throughput limitations.

4.2. Deposition Processes

The fabrication of an MMIC requires the deposition of both metal and dielectric films, and this is achieved using some of the different techniques explained in the following sections of this chapter.

4.2.1. Metal Deposition Techniques

Metal films are applied to GaAs wafers either by evaporation or sputtering dependent on the required final metal thickness. In the evaporation process the metal is heated to a temperature sufficient to cause vaporization. This vaporized metal then deposits onto the GaAs slice situated inside the vacuum chamber held at a pressure of around 10^{-7} torr. Vaporization of the metal is achieved either by heating the metal in an electrically heated coil or by directing an electron beam on to the charge of the metal to be deposited. This method of metal deposition is a somewhat directive one and is not recommended for metallizing over steep edges but is good for lift-off.

The deposition of thick films (≥ 1 μm) and of alloys is best carried out by sputtering.

In this process high-energy ions are directed onto a metal target knocking metal atoms from the target, which redeposit onto the GaAs slice. This is a low-temperature process and relatively nondirective and hence good for step coverage. Step coverage can be enhanced by biasing the substrate as this causes deposited metal atoms to be redeposited onto steep steps. Sputtering is particularly good for alloy deposition for the nature of the process causes the deposited film to retain the composition of the target.

TABLE 6. Electron Beam Lithography

Resolution to at least 0.2 μm
No masks required
Good yield over a large area
Throughput time slow but can mix electron beam and photolithography to
    partially surmount this problem
Machine cost high

4.2.2. Plasma Enhanced Chemical Vapor Deposition (PECVD)

None of the above processes can effectively be used for depositing dielectric films. The standard process for dielectric deposition is chemical vapor deposition in which the relevant gases are reacted to form the nitride or oxide which deposits onto the substrate at a high temperature. Such processes cannot, however, be used in GaAs technologies as the temperatures required will adversely affect prior process steps (e.g., ohmic contacts) and may cause the GaAs to decompose. Consequently PECVD processing is used. Here the temperature of the reaction required to form the dielectric film is lowered by the formation of a plasma in which the energetic electrons increase the reaction temperature allowing substrate temperatures below 300°C to be maintained. Silicon nitride is by far the most popular dielectric film used in GaAs processing and this is formed by reacting silane and ammonia in a closed system under a plasma.[9]

4.3. Etching Processes

Once the metal and dielectric films have been deposited they need to be selectively removed to form the various patterns that build up the integrated circuit. This is carried out by a combination of lithography and etching. There are many etching processes available to the GaAs fabrication technologist. The major ones are listed below and will be described in some detail:

a. Ion beam milling[10];
b. Reactive ion etching and plasma etching[11];
c. Wet chemical etching.

Of these processes only (c) is not a "dry" process and the move away from wet chemical etching is becoming an important aspect of GaAs technology. In many instances the dry process technology is being adopted because of its controllability and cost effectiveness. In others it is proving to be the only practical method of removing many metals and dielectrics.

4.3.1. Ion Beam Milling

In this process an energetic ion beam is directed onto the material to be etched and metal removal is purely by the impact of these ions onto its surface. Inert gases are used and no chemical reaction takes place. There is provision within the milling equipment to rotate and tilt the substrate material and this is important as experience has shown that etch rate is a function of the angle that the substrate presents to the ion beam. Rotation of the substrate is intended to unify the etch rate across the wafer. Ion beam milling is a rather slow process with etch rates around 1–2 μm per hour.

4.3.2. Reactive Ion Etching (RIE) and Plasma Etching

These processes are similar to ion beam milling except that chemical compounds are introduced into the equipment to improve the etch rate. It uses relatively high-energy directive ions giving only a small amount of undercutting of the masking

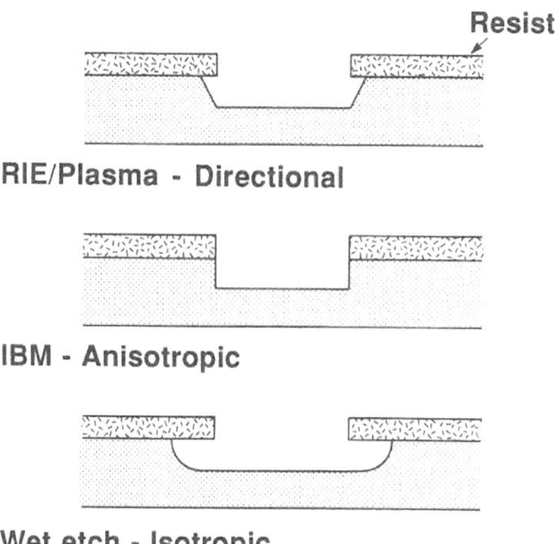

FIGURE 14. Etching profiles in GaAs.

medium. It therefore has advantages over wet etch processes and is used for instance for the etching of via holes in GaAs slices where a minimum amount of undercutting is highly desirable. For this purpose chlorine-based compounds such as carbon tetrachloride are introduced into the Rie equipment. RIE is also being utilized more and more on multilevel resist processes because of its high anisotropic etching behavior. In this process fluorine-based chemicals are used.

The plasma etching process is very similar to reactive ion etching and they are distinguished from each other only by the operational pressures and voltages at which the equipments are operated. RIE operates at lower pressures ($\approx 10^{-2}$ torr) than plasma etching and is therefore somewhat more directional.

Figure 14 summarizes these techniques with regard to the directionality of the etching process.

It is important to note that although wet processes are rapidly being superseded by dry ones they are not without their problems. A major concern of the GaAs technologist is the damage caused by the dry process techniques, which generally are high-energy processes. It appears to be relatively easy to cause damage to the surface layers of crystalline GaAs, which can cause serious degradation in electrical properties of the material and hence cause changes in the circuit performance.

5. FUTURE TRENDS

Predicting the future in such a rapidly advancing area as microwave IC technology is extremely difficult, but it is certain that improvements in the technology are necessary in order to satisfy the higher frequency requirements of the systems designer.

In materials technology one of the major concerns is to improve the reproducibility of the semi-insulating substrate. Whole boule annealing improves the uniformity of the threshold voltage of ion-implanted MESFETs and is particularly advantageous for digital technologies; however, this has opened up a new area of research where considerable optimization can be achieved.

The need for improved performance, low noise figures, higher frequency, and higher power has focused research onto new device structures based on heterojunction technologies. In these devices the material structures have to be epitaxially grown and developments are underway to improve the uniformity and throughput of conventional epitaxy technologies such as MOCVD and MBE.

Lithography developments for $<0.5\ \mu$m feature sizes appear to be centered around the use of electron beam lithography. This technique is capable of very fine geometries but at present limited in its use by a poor throughput, lack of equipment reliability, and through extremely variable quality and nonstandard electron-sensitive resists. Increased throughput is one of the main topics for the electron beam machine manufacturer and many machines are currently being designed to turn them from laboratory research tools to high-throughput manufacturing equipments. Resist developments are also proceeding rapidly; here adhesion to the GaAs surface, resistance to plasma processing, and increased sensitivity are top priority items for improvement.

Other lithographic processes are also in development for reducing dimensional tolerances with minimal effect on yield. X-ray lithography is a theoretically attractive method for producing extremely fine-line geometries. The short wavelength (2–10 Å) makes diffraction effects nonexistent as is backscattering from the substrate. These properties mean that "proximity" lithography in very thick resist is possible with very high definition. There would also be no problems with dust on the mask as it would be transparent to x-rays. It would appear to be an ideal solution to a difficult problem. However, it also has a number of distinct disadvantages. The first is in the manufacture of a suitable x-ray source. In the main these are big and very expensive, though work is under way to develop lower-cost desk-top synchrotrons capable of producing the necessary flux or soft x-rays. The second is the difficulty in manufacturing the mask. At present very thin membranes are used coated with a metal film (gold) that is opaque to x-rays. This film is patterned to allow the x-rays to penetrate to the underlying resist-coated sample. The manufacture of the masking membranes is not a trivial matter and is generally carried out by first depositing the membrane onto a silicon slice (used as a dispensible substrate), coated with gold and then patterned. The silicon substrate is subsequently etched away to leave a very flimsy x-ray mask.

The optical stepper until the present day has been of somewhat limited use to the GaAs technologist mainly because of its resolution limits of around 1 μm. However, recent dramatic improvements in lens design have pushed the resolution limits to a reproducible 0.5 μm and the prospects of reducing this further are very real. These recent developments have caused the GaAs technologist to reappraise the situation regarding the type of lithographic process he will use for $\leq$0.5-μm feature sizes, and we shall find more and more use being made of the submicrometer capabilities of the optical stepper for the same reasons as those advanced by the silicon circuit fabricator at the 1-μm resolution level.

The move, in GaAs technology, away from wet etching processes has had tremendous gains with regard to fabrication yield, dimensional control, etc., but not without a number of problems. Dry processing techniques based on energetic plasmas have introduced a challenge to the fabricator: that of damage to the semiconductor surface. This damage can cause severe degradation of circuit performance and lead to premature operational failure. New plasma techniques involving microwave frequency sources and large magnetic fields have been shown in laboratory experiments to cause far less surface damage than conventional dry process techniques. This technique based on electron cyclotron resonance (ECR) utilizes a powerful magnetic field into which a microwave (2.45 GHz) signal is propagated. Early experiments have shown that etch rates can be achieved that are some five times higher than for normal plasma etching.

ECR ion beam etching is also being researched at present. In this technique a collimated beam of high-energy ions can be directed at the substrate at a variety of predetermined angles giving the fabrication technologist an option to adjust the etched wall angle, which can be beneficial in a number of process stages.

The use of lasers in the GaAs processing technology is opening up new areas for research. Already they are being used to condition the surface of the substrate prior to epitaxial growth by removing traces of unwanted oxide. They are also being used for the selective heat treatment of contacts. Here a finely focused steered laser beam is used to individually sinter contact areas without causing overall heating of the GaAs slice. In this way the stoichiometry of the non-heat-treated areas of the slice can be preserved, leading to improved device and circuit performance.

There is also a move away from cleanrooms as we know them today. The smaller geometries called for when increasing frequency performance have led to a rethinking of cleanroom design. A technology being widely evaluated at present is the standard mechanical interface (SMIF) isolation concept. This approach keeps wafers sealed in protective boxes except when processing is carried out. The wafers can, therefore, be transported between processes through relatively "dirty" environments without being contaminated. Robotics are being examined to remove the wafers from the box into the clean process areas. Eventually an "operator-free" processing facility is envisaged with human intervention only being necessary when problems arise. This approach is believed to be very cost effective as it removes the need for large cleanrooms and their attendant air conditioning facilities as fabrication is carried out in clean air bays needing only modest air processing.

Although this chapter has dealt basically with a GaAs technology, the technologies and components described can, with only minor modifications, successfully be applied to heterostructure FET and bipolar transistor analog and digital circuits.

REFERENCES

1. J. R. Knight, D. Effer, and P. R. Evans, The preparation of high purity gallium arsenide by vapor phase epitaxial growth, *Solid State Electron.* **8**, 178 (1965).
2. B. R. Pamplin, *Molecular Beam Epitaxy*, Pergamon Press, Oxford, (1980).
3. J. F. Gibbons, W. S. Johnson, and S. W. Mylroie, *Projected Range Statistics, Semiconductors and Related Materials*, 2nd Edition Dowden, Hutchinson, and Ross, Stroudsburg, Pennsylvania (1975).
4. R. Esfandiari, D. W. Maki, and M. Siracusa, Design of interdigitated capacitors and their application to GaAs monolithic filters, *IEEE Trans. Microwave Theory Tech.* **31**, 57 (1983).
5. R. S. Pengelly, *Microwave Field Effect Transistors—Theory, Design and Applications* Wiley, New York (1982).

6. R. A. Pucel, Design considerations for monolithic microwave circuits, *IEEE Trans. Microwave Theory Tech.* **29**, 513 (1981).

7. L. A. D'Asaro, J. V. DiLorenzo, and H. Fukui, Improved performance of GaAs microwave field effect transistors with low inductance via-connections through the substrate, *IEEE Trans. Electron Devices* **25**, 1218 (1978).

8. R. E. Williams, *Gallium Arsenide Processing Techniques*, Artech House, Dedham, Massachusetts (1984).

9. J. R. Hollahan and R. S. Rosler, Ion-beam techniques for device fabrication, in *Thin Film Processes*, Academic, New York, (1978).

10. E. G. Spencer and P. H. Schmidt, *J. Vac. Sci. Technol.* **8**, S52 (1971).

11. J. W. Coburn, *Plasma Chem. Plasma Proc.* **2**, 1 (1982).

15

Phosphors and Luminescence

Brian Ray

1. BACKGROUND

Phosphors are defined classically as materials that exhibit emission of visible radiation for significant periods of time following the removal of the excitation means (radiation, electron beam, electric field, etc.); such delayed emission of light is called phosphorescence. One of the first recorded observations of natural phosphorescence was that by Cellini in diamond in 1568, while the Bolognan cobbler, Vincenzo Cascariolo, who was an exponent of alchemy, synthesized in 1604 the first artificial phosphor by calcining sulfur-rich barium sulfate to give it a golden appearance in daylight.

The use of the word "phosphor" has been extended to describe materials exhibiting not only limited phosphorescence but also massive fluorescence in the infrared, visible, and ultraviolet spectral regions. "Fluorescence" is the near instantaneous emission of radiation following excitation. The inability to discriminate between phosphorescence and fluorescence at the time when the term "phosphor" came into use was at the root of it taking on a broader working definition. It was not in fact until the time of Stokes in 1853 that time resolutions of 100 μs and less became possible and the distinct emission characteristics of phosphorescence and fluorescence could start to be observed. Today an arbitrary time distinction between fluorescence and phosphorescence of 10 ns is used often.

The systematic design of phosphors for a wide range of optoelectronic applications has taxed the minds and ingenuities of researchers over many decades. This has encompassed both fluorescence and phosphorescence within the description "luminescence," which is the nonincandescent radiation emitted by such materials. Steady and progressive development has occurred in a number of fields, of which perhaps the two most striking that have been around for many years are fluorescent lamps and color television; here, enhanced emission efficiency, better balanced color, more ready processing, and extended product lifetimes are improved features that have resulted from targeted research and increased understanding of the phosphors concerned.

Brian Ray • Faculty of Applied Science, Coventry Polytechnic, Priory Street, Coventry CV1 5FB, U.K.

A major area of recent development has been in light-emitting solid state devices and displays; this has evolved through advances in semiconductor and integrated circuit technologies. p-n junction device arrays using III–V compounds, e.g., GaAs, GaP, $Ga_{1-x}In_xP$, etc., have provided solid state emitters in the infrared, red, yellow, and green spectral regions. However, the indirect character and magnitude of the band gaps in some III–V compounds have led to low intrinsic luminescence efficiencies and limitations in providing effective emitters across the whole of the visible spectrum. II–VI compounds, with their direct and larger band gaps as key qualifications to overcome the limitations of the III–V compounds, have been the focus of sustained research programs in this application area over the past two decades. The major problems associated with their development have been the following:

i. Creating appropriate physical structures whereby efficient minority carrier injection into basically n-type materials can occur;
ii. Extending the performance lifetime through materials selection and preparation, and optimization of operational conditions;
iii. Optimizing the effectiveness of particular impurity centers in different host matrices for a given color emission requirement.

The sections that follow focus on three characteristic areas of the luminescence behavior of phosphors, providing some basic theory and direct applications of particular phosphors as appropriate. These are (i) photoluminescence analyzing both phosphorescence decay characteristics and fluorescence behavior; (ii) concepts and applications of thermoluminescence; and (iii) electroluminescence, of both powder and thin-film structures under a range of excitation conditions. The emphasis of the presentation will be on luminescence emission in the visible spectral region.

2. PHOTOLUMINESCENCE

2.1. Phosphorescence

Consideration is given to phosphorescence and its decay using both configurational energy diagram and continuum energy band models to reflect differing conducting characteristics of phosphors. Two principal decay modes are observed set against these two different energy diagram approaches.

2.1.1. Exponential Decay*

Figure 1 illustrates a configurational energy diagram of ground and excited states in which no trapping of excited electrons can occur. The time spent in the excited state is dependent then upon the probability p for the radiant transition BB'. Thus, we have

$$I = -dn/dt = pn \qquad (1)$$

* The literature on phosphorescence and thermoluminescence uses n to define the number of centers in the excited state or of ionized centers and p as the radiant transition probability from the excited to the ground state; this nomenclature has been retained in the text here on phosphorence and thermoluminescence.

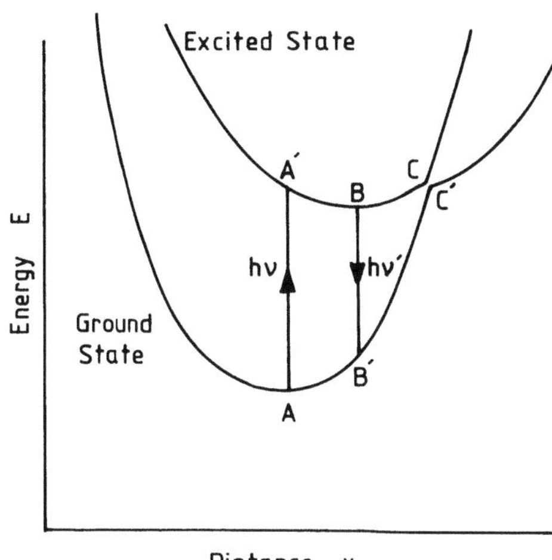

FIGURE 1.   Configurational energy diagram for ground and excited states.

where there are n centers in the excited state at time t. This leads to the expression

$$I = I_0 \exp(-pt) \tag{2}$$

Low transition probabilities are observed in phosphors exhibiting such exponential decays, e.g., $SrSO_4$:Eu has $p = 0.5\ s^{-1}$.

2.1.2. Hyperbolic Decay

The long afterglows associated with ZnS and alkaline earth sulfides (CaS, SrS, MgS) are sometimes linked to a hyperbolic form of decay, when free carrier concentrations and mobilities are sufficient to use the energy band model, Fig. 2, particularly in heavily doped materials.

If there are n free electrons in the conduction band, n ionized centers (at which radiative recombination can occur) uniformly distributed throughout the phosphor, and a recombination probability δ then

$$I = -dn/dt = \delta n^2 \tag{3}$$

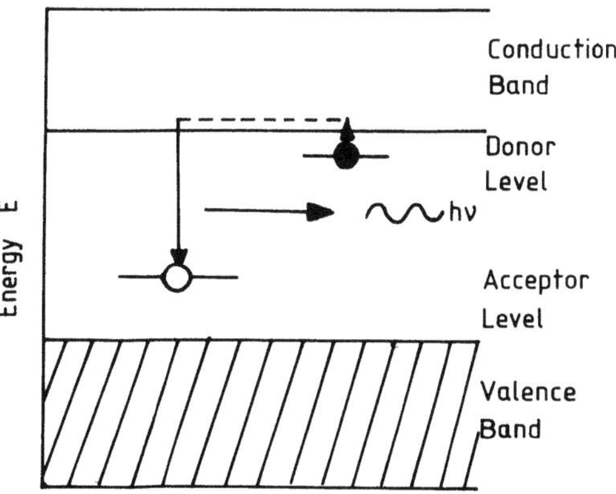

FIGURE 2.   Energy band diagram for phosphorescence and thermoluminescence.[5]

leading to

$$I = \delta n_0^2/(1 + \delta n_0 t)^2 = I_0/[1 + (I_0\delta)^{1/2}t]^2 \tag{4}$$

Since δ is a product of the capture cross section of the center and the mean velocity of the electron, it is dependent on $T^{1/2}$, where T is the absolute temperature. Thus, the intensity of the decay is dependent on the initial excitation intensity and temperature.

2.2 Fluorescence

2.2.1. Basic Principles

At high excitation intensities, the number of electron–hole pairs generated, or excited states created, is so great that the transitions associated with phosphorescence decay are insignificant. The example of Fig. 3 with single donor and acceptor levels will be used. Essentially for free electrons and holes generated, the holes are rapidly captured at luminescence centers (acceptors) while the mobile and longer-lived electrons recombine radiatively with them at the centers [all available electron trapping states (relatively deep donors) would be filled at high excitation levels]. The exciting energy is greater than the emitted energy (Franck Condon shift/Stokes emission). Fluorescent lamps afford an illustration of these principles in that an electrically excited gas discharge generates high-energy ultraviolet photons, which stimulate fluorescence emission in the phosphor mixture on the wall of the lamp. The phosphorescence characteristics of the phosphor are chosen so as to damp out any significant low-frequency fllicker originating from the exciting source. In order to achieve color balance and appropriate half-life, lamp phosphors are complex blends of hosts such as $CaCO_3$, $CaHPO_4$, CaF_2, NH_4Cl, Sb_2O_3, and $MnCO_3$ doped with rare earth and other emitting chemical groups. Butler[1] gives a very full and detailed account of the development of lamp phosphors.

2.2.2. Characterization of Fluorescence

It is interest in the fluorescence characteristics of large band gap semiconductors that has led to significant advances in the techniques available for their study.

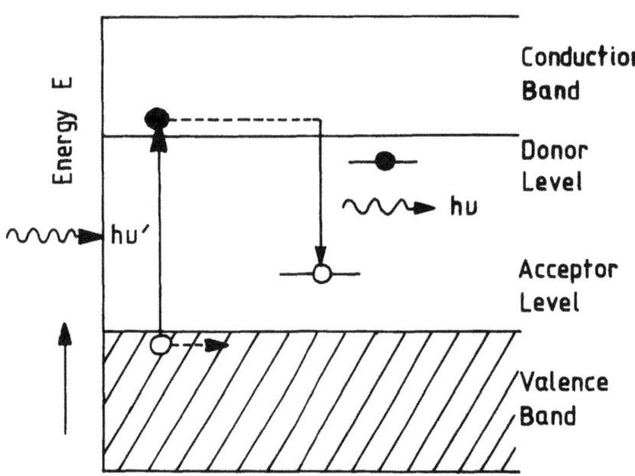

FIGURE 3. Energy band diagram for fluorescence with donor states filled and direct radiative recombination of free electrons with holes in acceptor states.

Excitation and fluorescence emission spectroscopy as a function of temperature have been most widely used. Isolation of overlapping radiative transitions has been achieved through time-resolved spectra, and correlations between the structure of excitation and emission spectra have provided insights into the recombination processes occurring.[2,3] Optically detected magnetic resonance spectroscopy has assumed special significance in the identification of deep radiative recombination centers permitting determination of both the chemical and structural character of radiating centers.[4]

2.2.3. Radiative Recombination Processes

It is the II–VI compounds with large direct band gaps that have provoked much of the recent work on fluorescence behavior under different forms of excitation. The most significant features of these materials are the ionic component in the bond encouraging exciton formation and the tendency to be only *n*-type. Thus of the major radiative recombination means band-to-band transitions take on a very low probability, and the competing processes are reduced to free exciton, bound exciton, free electron to acceptor, and donor–acceptor pair recombinations. In the unexcited condition, the phosphor will have no free holes while there will be free electrons in the conduction band. Hole minority carriers generated by optical excitation generally have a short lifetime in II–VI compounds and a hole will become bound to an electron to form a free exciton, to a charged defect acceptor level, or to a neutral (isoelectronic) defect level. All other factors being equal, the most probable event will be the process with the highest capture cross section. However, in the design of phosphors high concentration levels can be used to accentuate the impact of one or more centers, and provided that thermal quenching is not possible because of the shallowness of the center, tailoring of color is possible. ZnS:Cu, Al is a classic example where the concentration of the acceptor (Cu) relative to that of the donor (Al) can be used to shift emitted color from green to blue (see Fig. 4), while the level of doping determines the intensity achievable for a given particle size.[5] The copper green emission referred to above is an example of a donor–acceptor pair emission, which is a common occurrence not only in II–VI compounds but in

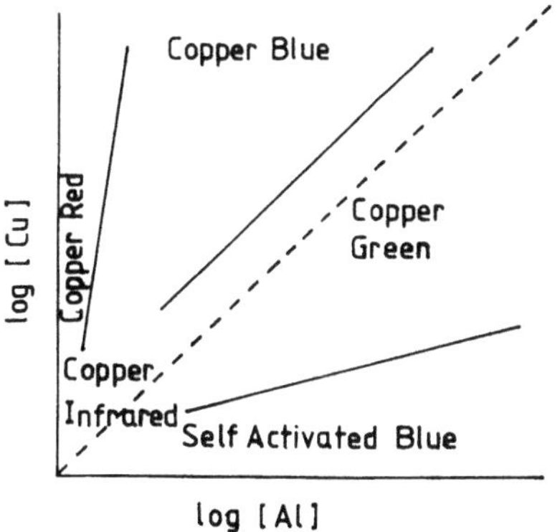

FIGURE 4. Color dependence of luminescence on copper (activator) and aluminum (coactivator) concentrations[5]; the broken line defines equal activator and coactivator concentrations, while the solid lines are the approximate concentration boundaries for the particular characteristic color emission indicated.

any large band gap semiconductor where there is tendency towards self-compensation.

2.2.4. Transition Metal and Rare Earth Elements

These groups of elements are, perhaps, worth singling out for their distinctive behavior as emitting centers in a wide variety of host matrices. This is due to their unique outer electron configurations involving unfilled $3d$, $4d$, and $5d$ shells in the transition metals and unfilled $4f$ shells in the rare earths. In the latter case the electrons within the $4f$ shell are screened against the crystal field by $5s$ and $5p$ electrons, and this results in extremely sharp lines similar in form to those for the free ions with small shifts and splittings of the order of 10 meV linked to the crystalline environment; the rare earths are generally incorporated in II–VI compounds as trivalent ions. The d-electron levels in the transition metals are more sensitive to the crystalline environment than those of the rare earths with crystal field splittings of up to 1 eV. Manganese normally goes into 11–VI compound in a divalent state taking up the tetrahedral coordination of the zinc blende structure; with d–d transitions forbidden in the free ion, exchange interactions with the p-states of the neighboring ions gives rise to characteristic yellow emission at 2.12 eV at 300 K in ZnS.

3. THERMOLUMINESCENCE

Important phosphor parameters such as trap depth, transition probability, and concentration may be studied using thermoluminescence.

3.1. Phenomenon and History

Robert Boyle in 1663 is claimed to have been the first person to observe thermoluminescence, although the understanding of the phenomenon has had to await the extensive studies during the middle part of the present century. Thermoluminescence is the emission stimulated thermally by the release of carriers (electrons or holes) from traps to recombine radiatively with ionized centers or carriers of opposite sign; for thermoluminescence to be observed, it is necessary for the phosphor to have been excited previously by some other means, most commonly electromagnetic radiation.

3.2. Basic Theory

The basic theory outlined below is for phosphors with electron trapping states at a single energy level in the forbidden energy gap, and for which an exponential phosphorescence decay at constant temperature is observed. However, this approach can be adapted to a wide range of conditions. Taking then the conditions cited in Section 2.1.1,

$$I = -dn/dt = pn$$

the transition probability, p, may be assumed to have a temperature dependence on the activation energy, E, of the carrier in the trapping state given by

$$p = s \exp(-E/kT) \tag{5}$$

If the temperature of the excited phosphor is raised linearly at a rate

$$\beta = dT/dt \tag{6}$$

then the emission intensity, I, can be shown[5] to be given by

$$I = I_0 \exp(-E/kT) \exp\left[-\int_0^T s \exp(-E/kT)\, dT/\beta \right] \tag{7}$$

It is possible to determine the energy of the trapping level relevant to the appropriate conduction or valence band edge and thereby characterize the phosphor; two immediate possibilities exist for this from: (i) the initial exponential rise of intensity with reciprocal temperature, when the second exponential approximates to 1, and (ii) the temperature at which the maximum emission occurs (Fig. 5):

$$(i) \quad I = I_0 \exp(-E/kT) \tag{8}$$

$$(ii) \quad \ln(\beta/T_m^2) = \ln(ks/E) - E/kT_m \tag{9}$$

3.3. Interpretation of More Complex Structures

In practice, there are many variations in the practical conditions prevailing in phosphors, which include (i) a range of trapping states at different energies with widely varying escape probabilities, (ii) trapping states related to particular emitting

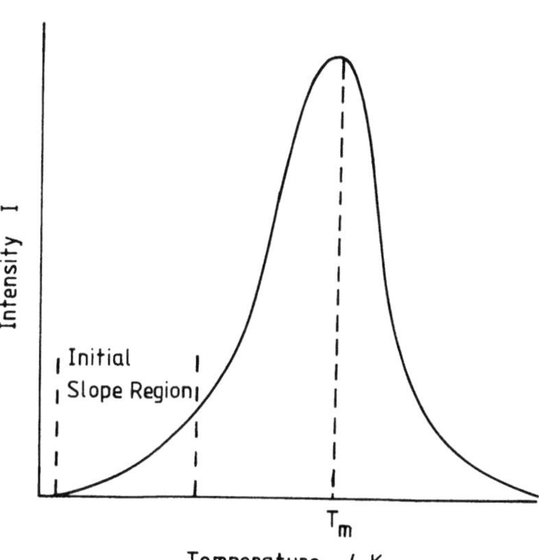

FIGURE 5. Thermoluminescence emission curve for trapping states of a single energy depth.

centers, and (iii) retrapping of the carriers released competing with the radiative recombination process. In these circumstances, alternative approaches to determining trapping energy levels have to be adopted, and these are summarized by McKeever.[6] Of such approaches, the following have proved particularly significant: (iv) the fractional glow method, involving a cyclic heating/cooling process that delineates a distribution of trapping energy states[7]; (v) the peak shape method to identify the kinetic order relating the balance between retrapping and recombination processes[8]; and (vi) the fractional area method at different heating rates to discriminate emission originating from overlapping trapping energy states.[9]

Figure 6 provides an illustration of the thermoluminescence behavior of a CaS phosphor, in which there are trapping states at different energies giving rise to a mixture of recombination emission energies.[10]

3.4. Applications

While thermoluminescence can provide fundamental information on the electron and hole trapping levels in large band gap semiconductors, it has also been applied to the measurement of radiation levels and to the dating of objects.[6,11,12] In dosimetry, these encompass selective measurement of ultraviolet, x, γ, β, and neutron radiation in personal, environmental, and medical contexts. The materials used have thermoluminescence emission peaks well above room temperature and generally above 200°C. Typical materials employed are LiF doped with Mg, Ti or Mg, C, P, $Li_2B_4O_7$ with Mn or Cu, $CaSO_4$ with Dy or Tm, Mg_2SiO_4 with Tb, and CaS with Dy. Dating of pottery, sediments, rocks, and geological specimens draws on the history of exposure of stable oxygen-based thermoluminescence emitters in them with calibrations being made against appropriately doped host material of the type identified above.

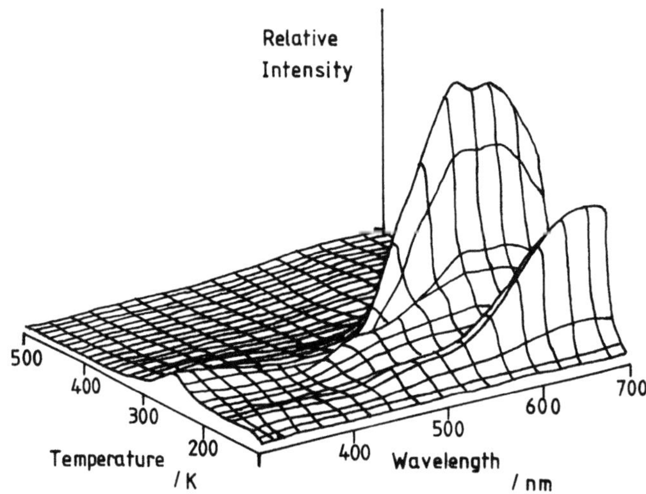

FIGURE 6.   Thermoluminescence emission intensity as a function of wavelength and temperature for single-crystal CaS heated at 0.1 K s^{-1}.[10]

4. ELECTROLUMINESCENCE

4.1. Overview

The earliest significant developments of electroluminescence (EL) in polycrystalline semiconductors can be traced back to Destriau,[13] who observed first luminescence in electrically excited copper-doped zinc sulfide in 1936. In the 1950s, much effort was devoted to developing large area, powder-based, electroluminescent emitting panels using zinc sulfide embedded in a dielectric binder, but lifetime problems caused this to fade as a commercial prospect.[14,15] p-n junction electroluminescent devices based on single-crystal III-V compounds took the center stage of commercial development and exploitation in the next two decades. However, such devices based on single-crystal material and high-quality epitaxial layer fabrication techniques were too expensive for extension to large-area applications. Thus, a revival of interest in the development of polycrystalline II-VI compounds for large-area emitters occurred in the 1970s and 1980s; this was stimulated by the advances made with ZnS:Mn in dc EL powder devices by Vecht.[16] The range of II-VI materials encompassed has been extended to the alkaline earth chalcogenides with their larger though indirect band gaps to increase the possibilities of obtaining effective emitters across the visible spectral range and into the near ultraviolet. Both powder and thin-film large area devices have been developed for small to medium volume production levels using ZnS:Mn. 1000×1000 pixel definition in up to 1 m^2 display areas in thin-film devices have been disclosed.[17]

4.2. Basic Theory

A generally observed feature of large-area EL under any excitation conditions is the pointlike character of the emission in both continuous and particulate thin layers; this suggests that localized electric field conditions are giving rise to radiative recombination of carriers. Two principal models considered usually to describe EL emission, (i) impact ionization, and (ii) carrier injection. Impact ionization requires the acceleration of free carriers under high local electric field conditions, so that carriers have sufficient energy to impact-ionize lattice atoms and generate further carriers and give rise to radiative recombination:

$$I = I_0 \exp(-a/V^{1/2}) \tag{10}$$

describes the EL emission as a function of voltage V for impact ionization and this is modified at high electric field values to

$$I = I_0 \exp(-b/V) \tag{11}$$

The inverse square root relationship is taken as evidence of a Mott–Schottky barrier providing a local high-field condition for the injection of carriers. With impact ionization, the existence of a threshold voltage, such that $qV_0 > h\nu$, the optical excitation energy for emission to occur, is observed. With carrier injection, such as occurs at p-n junctions, there would be no threshold for emission. In II-VI compounds, because of charge compensation effects, doping to achieve both p- and

n-type material does not readily occur, although it is possible to form p-n heterojunctions within particles through concentration gradients. The earliest explanations of EL in ZnS:Cu were based on such heterojunctions.[5]

The frequency dependence of ac electroluminescence, where impact ionization applies, usually follows broadly a relationship of the form[14]

$$I = Bf[1 - \exp(-A/f)] \tag{12}$$

with A being a constant at a given voltage, B also a constant, and f the ac frequency. In practice, the fit observed to this relationship is less than good and reflects an oversimplification in the assumptions applying to the derivation of this equation to describe the operational configurations used to observe electroluminescence.

4.3. Powder EL Devices

Much effort has been devoted in trying to exploit the EL characteristics of ZnS in powder-based devices because of the potentially much cheaper production costs compared to devices requiring large-area vacuum deposition techniques. While successes have been achieved in both dc and ac excited EL powder devices, the combination of performance stability and production yields have not been as yet sufficiently high to capitalize on the process cost advantages. One particular strength of powder devices is the good optical extraction of emitted radiation from them, although in bright light conditions only poor constrast is achievable owing to diffuse reflection and back scattering of ambient radiation by the particles.

The basic structures of powder EL devices are as in Fig. 7; careful control of the phosphor particle size is essential, particularly for dc devices where diameters of around 1 μm are required. Figure 7a relating to the dc EL powder structure identifies that the presence of a Cu_xS phase around the ZnS:Cu, Mn particles sandwiched between the two electrodes is required for satisfactory performance; an initial forming process results in a copper depletion of the ZnS:Mn layer adjacent to the transparent electrode in order to achieve effective EL emission. Copper migration in dc EL device does set limitations on lifetimes, although pulsed operation

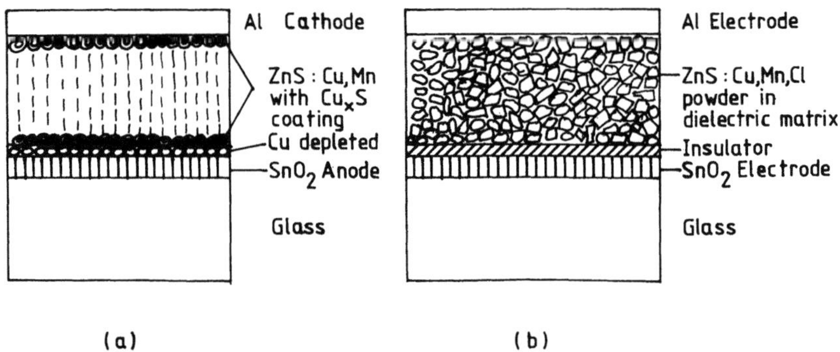

FIGURE 7.   Structures of cells used for powder electroluminescence (a) dc and (b) ac.[20]

at brightnesses of $100 \, \text{cd m}^{-2}$ have given working lives exceeding 10,000 hours Vecht.[18] More recently rare earth doped CaS dc powder EL devices have been reported with brightnesses exceeding $1000 \, \text{cd m}^{-2}$.

The ac EL powder devices have a structure similar to that used by Destriau and shown in Fig. 7b. Their performance is dependent upon heavy doping of the ZnS layer with copper, which in the manufacturing process becomes inhomogeneously distributed as partially conducting Cu_xS phases. An inhomogeneous field distribution results and at high field points hot carrier injection occurs and impact ionization of centers occurs. A range of ZnS-based EL phosphors are used in commercial applications of powder ac EL devices particularly linked to background lighting.

4.4. Thin-Film Devices

Thin-film ac EL devices have been the focus of much recent work, and as indicated in Section 4.1 have resulted in significant advances in large-area commercially available devices in the medium cost range. Limited work has been done on thin film dc EL devices, although they do offer scope for operation at low voltages, provided that migration effects can be controlled; brightnesses up to $3000 \, \text{cd m}^{-2}$ have been obtained for ZnS doped with Mn, Er, and Tb, although efficiencies have generally only been in the region of $0.1 \, \text{lm W}^{-1}$. The work on ac EL thin-film devices has thus overshadowed that on their dc counterparts. Mach[20] provides an extensive review of electroluminescence in thin film polycrystalline semiconductors.

The structure adopted almost universally for ac EL thin film emitters is of five layers:

   i. Transparent conducting/metal electrode, M;
   ii. Insulator, I;
   iii. Doped semiconductor, S;
   iv. Insulator, I; and
   v. Metal electrode, M.

Figure 8 illustrates this MISIM structure, in which the thickness of the active semiconductor layer can range in practice from 0.4–$1.5 \, \mu\text{m}$. The semiconductors employed as the active medium include ZnS, ZnSe, ZnF_2, CdF_2, CaS, and CaSe. Transition and rare earth elements, because of the possibility of their respective

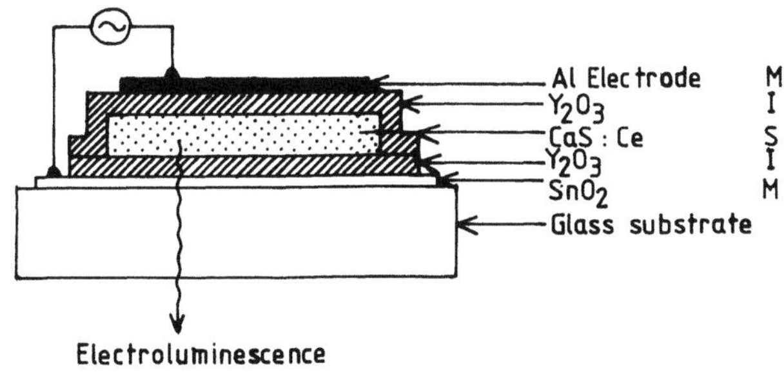

FIGURE 8. Metal–insulator–semiconductor–insulator–metal (MISIM) structure used for the study of thin-film electroluminescence in CaS : Ce.[21]

inner shell $(d\text{-}d)$ and $(f\text{-}f)$ transitions providing the appropriate electronic screening to permit effective impact ionization, have been used as the emitting centers in these base matrices.

The solubility of the centers in the host lattice is essential to allow the effective capture of the hot carriers injected, although optimization of the center's concentration is necessary to avoid concentration quenching effects. The solubility of Mn in ZnS is extensive, and, in consequence, the most significant development of this combination as a thin film has occurred in ac EL devices; the crystal field splitting of Mn^{2+} in ZnS leads to a broadened emission spectrum in the yellow-orange at the optimum doping level of 1 mol % Mn. Of the rare earths, Tb^{3+} inserted as TbF_3 has resulted in brightnesses of 1.6×10^3 cd m^{-2} in the green with a much sharper spectral structure. However, the limited solubilities of some of the rare earths in ZnS have prompted alternative host lattices to be considered and, within the II–VIs, the alkaline earth sulfides have given brightnesses up to 1.5×10^3 cd m^{-2} in CaS : Ce, Cl; however, the indirect band gap in CaS makes it an intrinsically less efficient luminescent emitter, typically peak values of 0.1 lm W^{-1} being obtained. Figure 9 illustrates the brightness dependence on voltage of a CaS : Ce MISIM structure excited at 5 kHz.[21]

The insulating layer has a critical role to play and requires a material of high dielectric strength to maximize the capacitance per unit area $(\varepsilon\varepsilon_0/d)$ for a given thickness (d) and to have high breakdown resistance. Y_2O_3 and $PbTiO_3$ have been extensively investigated as insulators having these qualities with ZnS : Mn as the active layer; some interesting hysteresis effects have been observed that permit the device to be operated with a large offset voltage and small oscillating voltage. Atomic layer epitaxy, sputtering, and electron beam evaporation are techniques that have been adopted for the deposition of a range of insulating and semiconducting materials in these thin-film devices.

Much of the future development of EL displays will center on thin-film ac structures, although the potential cost advantage of powder ac and dc structures will provide an impetus there also. A greater understanding of the memory effect

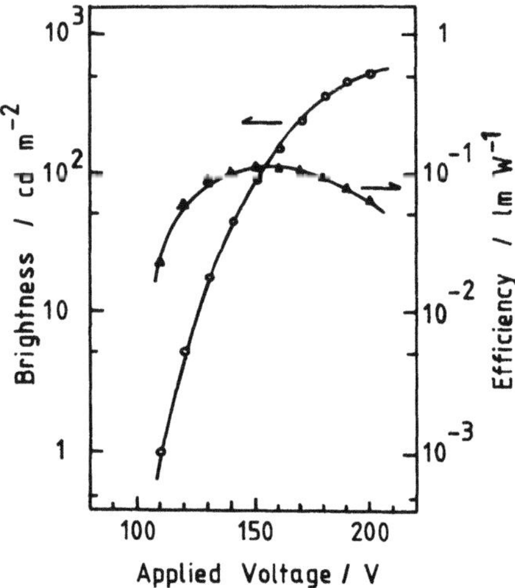

FIGURE 9. Voltage dependences of brightness and electroluminescence emission efficiency in CaS : Ce excited at 5 kHz.[21]

in thin-film devices is needed to promote its further exploitation, while on the materials front some opportunities exist for the investigation of admixtures of conventional and alkaline earth II–VI compounds to optimize their individual strengths.

REFERENCES

1. K. H. Butler, *Fluorescent Lamp Phosphors* Pennsylvania State University Press, University Park, Pennsylvania (1980).
2. H. Hartmann, R. Mach, and B. Selle, *Current Topics in Materials Science*, Vol. 9, North-Holland, Amsterdam (1982).
3. T. Taguchi and B. Ray, *Prog. Crystal Growth Characterization* **6**, 103–162 (1983).
4. J. J. Davies, *J. Crystal Growth* **72**, 317–325 (1985).
5. B. Ray, *II-VI Compounds*, Pergamon, Oxford (1969).
6. S. W. S. McKeever, *Thermoluminescence of Solids*, Cambridge U.P., Cambridge (1985).
7. H. Gobrecht and D. Hoffman, *J. Phys. Chem. Solids* **27**, 509–522 (1966).
8. R. Chen, *J. Appl. Phys.* **40**, 570–585 (1969).
9. M. A. S. Sweet and D. Urquhart, *J. Phys. C.* **14**, 773–781 (1981).
10. A. G. J. Green, I. V. F. Viney, J. W. Brightwell, and B. Ray, *Materials Lett.* **7**, 19–21 (1988).
11. I. Bailiff, *SERC Bull.* **3**(1), 27–28 (1988).
12. J. A. Perry, *RPL Dosimetry*, Adam Hilger, London (1987).
13. G. Destriau, *J. Chem. Phys.* **33**, 587–625 (1936).
14. H. K. Henisch, *Electroluminescence*, Pergamon, New York (1962).
15. J. I. Pankove, *Optical Processes in Semiconductors*, Prentice-Hall, New York (1977).
16. A. Vecht, N. J. Werring, R. Ellis, and P. J. F. Smith, *Proc. IEEE* **61**, 902–907 (1973).
17. R. Mach and G. O. Muller, *J. Crystal Growth* **86**, 866–872 (1988).
18. A. Vecht, *J. Crystal Growth* **59**, 81–97 (1982).
19. B. Ray, *Proceedings 4th International EL Workshop in Tottori, Japan* Springer, Heidelberg (1989).
20. R. Mach, *Solid State Sci.* **57**, 186–208 (1985).
21. S. Tanaka, V. Shanker, M. Shiiki, H. Deguchi, and H. Kobayashi, *Proc. Soc. Inf. Display Conf.* **26**, 255–258 (1986).

16

Microstructural and Compositional Characterization of Thin-Film Semiconductor Materials by Transmission Electron Microscopy (TEM)

A. C. Wright and J. O. Williams

1. INTRODUCTION

The field of semiconductor devices has been transformed in recent years by the availability of preparative techniques capable of routinely providing materials in thin-film form and of extremely high purity. With advances in the field of metalorganic vapor phase epitaxy (MOVPE), molecular beam epitaxy (MBE), and metalorganic molecular beam epitaxy (MOMBE) has come the need to characterize materials using methods with higher sensitivities and much improved spacial resolution. It is no exaggeration to say that single atoms need to be seen and their influences recorded. These requirements have placed additional demands on characterization techniques. Table 1 summarizes a selection of the important techniques currently available for the chemical, structural, optical, and electrical characterization of epitaxial materials. Most of these can only be used to determine macroscopic (average) properties. Some, such as secondary ion mass spectrometry (SIMS) and the electrical techniques, can be combined with profiling to probe properties as a function of depth normal to the epitaxial layers, but at distances $\leqslant 10$ nm (100 Å) resolution limits begin to be approached.

In this chapter we shall focus on microscopical techniques for the characterization of the structural and compositional properties of epitaxial semiconductors. The power of the electron microscope stems from its ability to observe directly the defective nature of crystalline materials and also from the wide range of operating modes possible based on electron imaging, diffraction, analysis, and a hybrid of imaging and diffraction. The basic operation of the transmission electron microscope (TEM) or the detailed underlying theory will not be covered. This has been

A. C. Wright and J. O. Williams • Solid State Chemistry Group, Chemistry Department, UMIST, P.O. Box 88, Manchester M60 1QD, U.K.

TABLE 1. Common Characterization Techniques

Method	Abbreviation	Property measured
Chemical		
Electron spectroscopy	ESCA, PES, XPS	Elemental analysis
Auger electron spectroscopy	AES	
Secondary ion mass spectrometry	SIMS	
Energy dispersive spectrometry	EDS, EDAX	
Synchrotron radiation	SRS (EXAFS)	Local environment
Structural		
X-ray diffraction	XRD	a_0, a/a_0, composition
Electron diffraction	SAD	a_0, a/a_0, segregation
	LEED, RHEED	Surface structure
Electron microscopy	TEM. HRTEM	Lattice structure, interfaces, dislocations, twins, etc.
	SEM	Morphology
Synchrotron radiation	SRS	Dislocation structure, strains
Scanning tunneling microscopy	STM	Surface structure
Optical		
Photoluminescence	PL	Impurities, defects, electronic states
Absorption, excitation	—	Band gaps, electronic states
Electrical		
Hall effect	—	Carrier concentration, carrier mobility
Capacitance–voltage	CV	Carrier profiles, barrier heights
Deep level spectroscopy	DLTS, TSC, TL	Trapping centers

adequately documented elsewhere[1,2] and it will be assumed that the reader has a basic working knowledge of both elementary experimental and theoretical electron microscopy. It is also assumed that the structure and basic crystallography of tetrahedral semiconductors is understood.

Rather than structure the review around different materials systems or simply catalog the modes of operation of the TEM, it will be based on the type of information that is required about the material system under study. Most researchers know what information they require and the problem is often one of finding out what techniques are available. In the context of semiconductor layers and device structures, the problem may include accurate assessments of the layer widths, composition profiles, and interface sharpness. It is also possible that information concerning the state of strain near an interface and a measure of very small variations in the lattice parameter are required. As compound semiconductor materials lack a centre of symmetry, then knowledge of the polarity of the grown layer may be required, particularly if it has been grown on a nonpolar substrate.

Most materials contain defects. Semiconductors are no exception, and so a means of characterizing the detailed structure of these is required. It is also possible to study the structure of semiconductor surfaces using a TEM, although this is really classed as a reflection rather than a purely transmission mode of operation. New techniques continue to be introduced into the repertoire of the microscopist for the analysis of these materials.

None of this is possible without the ability to prepare suitable specimens for TEM and information is included on this topic at the end of the review. Applications of TEM for the development of semiconductor devices have been presented by Oppolzer[3] and in memory chip development by Oppolzer *et al.*[4]

2. THE ANALYSIS OF STRUCTURE

2.1. Basic Structure, Symmetry, and Lattice Parameter Measurement

In structural analysis, the researcher is primarily interested in the basic (perfect) structure of the material under study. Only when this is known can the defect structure be examined. In semiconductor systems, the basic structure is usually either cubic (sphalerite) or hexagonal (wurtzite) and one is generally interested in viewing along a specific direction. When new structural phases occur, e.g., precipitates, it may be possible to obtain diffraction patterns which can be used for identification. Electron diffraction gives data in a coded form on the basic structure type, and unlike x-ray data, can give information from regions as small as 1 nm across when using very fine electron probes, e.g., a dedicated scanning transmission electron microscope (STEM) or a TEM fitted with a field emission gun (FEG). The resulting pattern of spots can be compared with standard spot patterns for a given structure type or phase (see, e.g., Andrews *et al.*[5]) or if not available, the spots can be converted into d values, which can then be compared with tabulations of possible structures.

A useful facility to have is a computer program which can generate spot patterns along a given projection when provided with the unit cell and positions and types of atoms. In this case, not only can the positions of the spots be predicted but also their intensity. Caution must be used as such programs do not usually take into account the possibility of dynamically allowed reflections occurring. It is also possible to find the direction of projection, $\langle uvw \rangle$ and hence the crystallographic orientation of specimens, given measured d spacings and angles between the Bragg spots (together with the lattice parameters) using computer methods. Determination of the orientation of the crystal with respect to the electron beam is often desired, and while this can be achieved using ordinary spot patterns in conjunction with a stereographic projection, it is far easier to use a Kikuchi map for the structure in question. Here, the diffraction pattern is created from a region of the specimen thick enough to give rise to Kikucki lines through diffuse scattering. Kikuchi maps are available for cubic and hexagonal materials such as sphalerite and wurtzite.

More detailed information on the structure can be achieved from convergent beam electron diffraction (CBED). In this technique, convergent illumination is used to spread the diffraction points out into disks. While conventional spot patterns give only a projection of the symmetry of the crystal, CBED patterns reflect the symmetry of the whole crystal and a careful analysis of all the symmetry elements present in these patterns can yield the space group. A good description of this procedure can be found in the article by Eades.[6] Such patterns are best obtained from defect-free crystals since the presence of defects can lead to a lowering of symmetry. The availability of microscopes with a facility to produce small electron probes without serious contamination of the specimen has meant that this problem can usually be avoided. The quality of convergent beam patterns are often much improved at low specimen temperatures as high-angle scattering into higher-order Laue zones is increased. Thus a liquid-nitrogen-cooled specimen stage is advantageous for this kind of work.

A feature known as higher-order Laue zone lines (HOLZ lines) obtained in the zero-order disk of convergent beam patterns for certain directions $\langle uvw \rangle$ can be used for lattice parameter measurements as well as for symmetry determination as shown by Ecob *et al.*[7] and for semiconductor structures by Rackham *et al.*[8] The positions of these lines are extremely sensitive to the lattice parameter and the resulting pattern can be simulated using a computer. By matching the simulations to the experimental pattern an accurate value for the lattice parameter can be obtained. Originally, the simulations were performed using a purely geometrical construction and it was found that the input voltage for the incident beam required to obtain a match between theory and experiment was different by several hundred volts from that expected. This voltage difference also varies between different zone axes of the same crystal, and so while lattice parameter variations between two given points on the specimen can be detected with great sensitivity (1 part in 1000), absolute measurements are not possible.

This problem can be overcome by performing a full dynamical calculation for the zone of interest as shown by Blom and Schapink.[9] While in principle this approach gives great accuracy, it is not suitable for routine use owing to the massive computational demands required even to obtain relatively low-resolution plots of the zero-order disk. Calculations performed in the case of silicon along [111] taking into account 22 beams required over one hour of computer time on a large mainframe [10], giving a resultant image of only 64 by 64 point resolution, which is hardly adequate for matching purposes. An alternative approach taking into account the effects of the crystal potential on these patterns has been described by Lin *et al.*[11] and by Bithell and Stobbs.[12] The magnitude of the voltage correction can be calculated from a simple summation of Fourier coefficients taken from the zero-order Laue zone (ZOLZ). These corrections automatically take into account the effect of changing composition (i.e., when studying superlattice structures) and permit absolute lattice parameter measurements to be performed. In this case, the microscope voltage has to be determined first by matching patterns from a material for which the lattice parameter is accurately known. Patterns can then be obtained from the material of interest without changing any of the microscope operating conditions (i.e., filament bias)[11] and, taking into account the voltage correction, can be matched to a suitable value for the lattice parameter.

Careful measurements on the positions of the HOLZ lines (i.e., by pattern matching) can reveal distortions from the ideal structure, say a change from cubic to tetragonal at or near an interface. While such measurements can be made at high sensitivity from very small regions of crystal using a focused convergent beam, it should be pointed out that significant relaxation of the crystal can occur when prepared as a thin foil for TEM and that the observed symmetry or lattice parameter may not be representative of the bulk specimen. Relaxations of this type have been observed by Gibson *et al.*[13] in superlattice structures of silicon and germanium using imaging methods.

The basic structure can also be studied using high-resolution electron microscopy or lattice imaging. In this technique, the crystal is oriented along some suitable zone axis, often $\langle 110 \rangle$ in the case of sphalerite, to produce a large number of diffracted beams. A large objective aperture permits interference between the diffracted beams and the main beam (and also interference between the diffracted beams themselves) so that a fringe pattern is seen in the image. Under certain stringent conditions of defocus and crystal thickness, the resultant interference pattern can resemble the projected charge density of the crystal to some limiting resolution. The interaction between the electrons in the illuminating beam and the crystal potential is complex, and intuitive interpretations of the image contrast cannot be made. Analysis of the contrast in terms of the structure is done by performing image simulations usually by the well-known multislice method. Computer packages are available commercially. Input to such programs includes details of the crystal structure, its orientation and electron-optical data such as objective lens defocus, its spherical abberation coefficient, accelerating voltage, and parameters describing the coherence of the illuminating electrons. Images are computed as a function of crystal thickness and lens defocus to build up a library of simulated images.

In the case of sphalerite projected along $\langle 110 \rangle$, which is the projection most useful for the general study of defects using HREM, dark "blobs" in the image usually correspond to either pairs of unresolved atoms or tunnels in the structure, depending on the defocus and the crystal thickness. Current electron microscopes do not have sufficient resolving power to be able to directly image the atomic arrangement of sphalerite when projected along $\langle 110 \rangle$. Direct resolution can now be achieved when the structure is projected along $\langle 100 \rangle$ providing the point resolution exceeds about 2 Å and identification of particular atomic species may be possible if the scattering difference between metal and nonmetal is large enough.[90] This direction is not suitable for the study of defects, however. Under certain conditions, "dumbbell-like" images may be obtained indicating that the atomic structure has been resolved. This is not so and such images are created as a result of dynamical diffraction effects. It has been shown that the spacing of the dumbbell-like atom pairs varies depending on the electron optical conditions and is generally not equal to the correct value of $a_0/4$.

Such basic studies are a necessary first step if lattice imaging methods are to be applied to the study of defects or interfaces. More information on the technique can be found in the work of Spence[14] and on computational procedures in the paper of Self *et al.*[15]

2.2. Determination of Crystal Polarity

When attempting to correlate structural features such as defects with the growth of a thin film of a compound semiconductor, it is useful to have knowledge of the crystal polarity. Because the sphalerite lattice lacks a center of symmetry, crystallographic planes such as {111} are populated by either the metal or nonmetal atoms, and defects terminating at these planes show different charge scattering behavior. This leads to an anisotropic structure in the resultant film when mismatch has been relieved by the formation of dislocations as shown by Brown *et al.*[16] in the case of ZnS grown onto GaAs(100). Thus for cross-sectional studies it is necessary to image along mutually perpendicular [110] directions and to be able to distinguish between them so that the (111)A and (111)B planes can be identified. Also, compound semiconductors such as GaAs grown onto nonpolar substrates such as Si or Ge can exhibit a domainlike structure created by the presence of $a_0/4$ steps on the substrate surface. Polarity determination is likely to be of use in studies of these features in such films.

When projected along [110], the sphalerite structure shows pairs of atoms or dumbells aligned along $\langle 100 \rangle$. One end of the dumbell is taken by the metal, e.g., Ga and the other by the nonmetal, e.g., As. The projection thus lacks twofold rotation symmetry and several techniques have been evolved to determine the orientation of the structure. Conventional selected area diffraction patterns can be employed to determine the polarity as shown in the case of CdTe by DiCioccio *et al.*[17] In such polar structures, dynamical diffraction gives rise to visible differences in the intensities of symmetry-related beams such as the $\{004\}$, $\{00\bar{4}\}$ pair. The method requires careful alignment of the crystal along the [110] axis, however. Dynamical calculations of the beam intensities are required (summed over a given thickness) and can be achieved using standard multislice programs.[18]

Polarity has also been determined from the asymmetries present in Kikuchi line patterns by Bird and Wright[19] for the case of GaAs. The method involves tilting off [110] to the [116] axis and looking for the relative positioning of light and dark lines along the (331) Kikuchi band.

Convergent beam electron diffraction has been successfully applied to GaAs by Preston and Spellward.[20] Under conditions of strong dynamical diffraction along [110], the $\{002\}$ and $\{00\bar{2}\}$ diffraction disks appear recognizably different and can be used to orient the crystal. The method requires a facility for the simulation of the CBED disks as a function of specimen thickness so that a match can be obtained and the disks identified. Computer time for this procedure is much the same as for calculating HOLZ lines described above, and although it can be lengthy, is only required once for a given material. An alternative CBED method has been outlined by Tafto and Spence[21] in which the crystal is tilted to excite odd order reflection in HOLZ (e.g., 911) which double diffract into the $\{002\}$ or $\{00\bar{2}\}$ disks. Phase differences between these paths lead to recognizable differences on the (002) type disks which do not require extensive calculations. The present authors have found that this method is less easy to implement in practice than diffracting along [110]. Both of these methods require that the probe be placed on strain-free material. This is not always easy when the material is heavily dislocated such as in GaAs grown onto Si.

High-resolution lattice imaging offers another means of polarity determination providing sufficient resolution is available. It has been applied successfully to GaSb by Wright *et al.*[22] by matching the experimental images to theoretical simulations. In this case the dark, round "blobs" normally seen in the image when the defocus is set of maximum contrast (Fig. 1a), can be distorted into elongated chevrons by setting the defocus nearer to Gaussian (Fig. 1b). The orientation of the chevrons is opposite to that expected intuitively from the strength of the scattering factors of Ga and Sb, however (see Fig. 1).

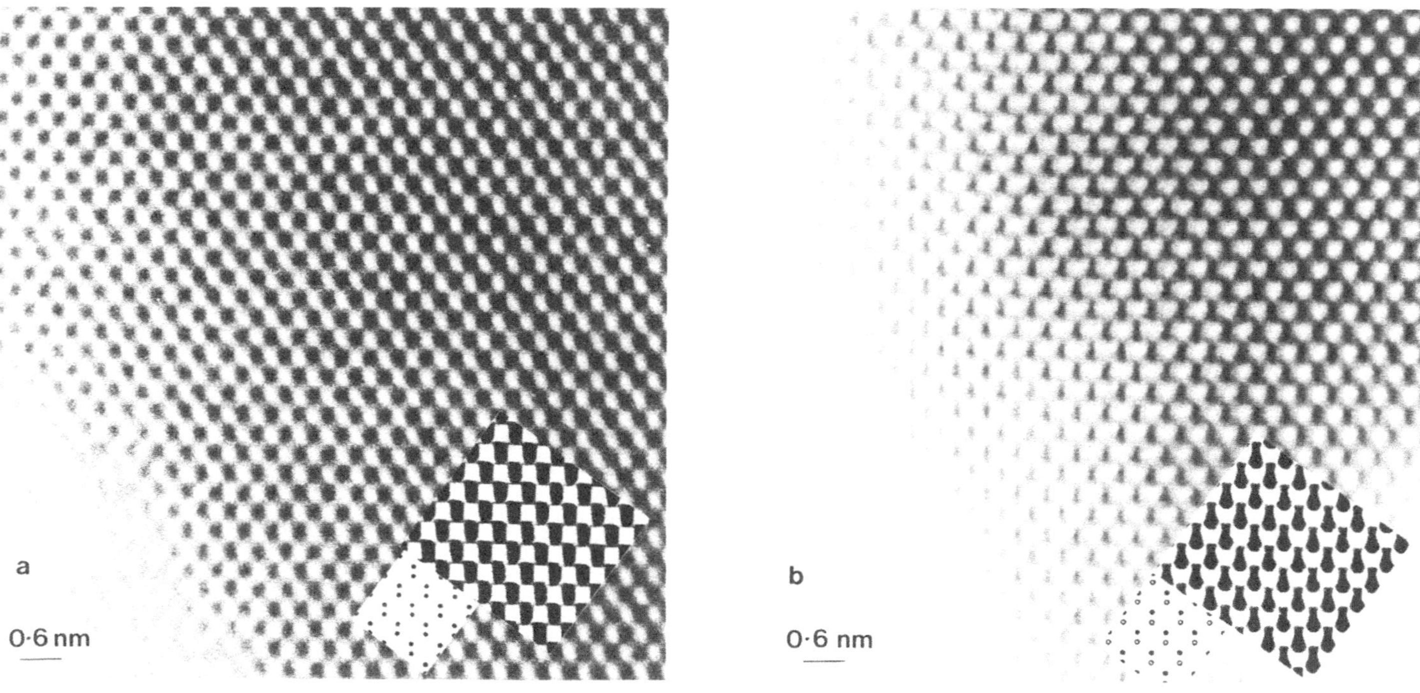

FIGURE 1.   (a) High-resolution lattice image of GaSb $\langle 110 \rangle$ recorded at approximately Scherzer defocus (-583 Å). Simulation and projected structure, shown as insets, show that dark "blobs" in the image correspond to unresolved atom pairs as seen in this projection of the structure. Note that it is impossible to determine the relative orientation of the structure (polarity) with respect to the image under these conditions. In the projected structure no distinction between Ga and Sb can be made. (b) High-resolution lattice image of the same area of GaSb as shown in Fig. 1a but recorded with a defocus closer to Gaussian. Note that dark blobs have become chevron or arrow shaped. Simulation (-110 Å) and projected structure, shown as insets, now reveal the true orientation or polarity of the crystal with respect to the image, but show that chevrons do not correspond exactly with atom pairs. In the projected structure solid circles are Sb and open circles are Ga.

There are other features in high-resolution images that can be used for polarity determination as shown by Smith *et al.*[23] In this case, a sideways shift in the (002) fringes can be seen towards the lower atomic number end of the dumbbell. High-quality images are required for this type of approach and specimens with as little surface damage or amorphous layers as possible are necessary.

2.3. Defect Analysis

Semiconductor layers often contain defects of one kind or another. While they are generally detrimental to the electrical and/or optical properties of a device structure, the effect on the properties by a specific defect type varies (see, e.g., Stutius and Ponce[24]). Slip systems are thermally activated and so the defect structures observed depend on the growth temperature of the film. It is thus desirable to characterize these defects when they arise and try to correlate them to both the resultant properties and other factors such as the growth technique and state of the substrate surface.

Defects generally come in two varieties, line and point. Line defects include dislocations, stacking faults, twin, and (in polar semiconductors) antiphase boundaries. These are relatively easy to image using TEM, but their analysis can be nontrivial. Single point defects such as vacancies or atoms on the wrong site are virtually impossible to image directly except where they occur in some form of ordered array giving a kind of superlattice structure.

Line defects in semiconductors, particularly dislocations, have been extensively studied. Examples of their structures have been given in Hirth and Lothie[25] and, with particular respect to the core structure in doped material, by Burle-Durbec *et al.*[26] Dislocations are often dissociated, giving rise to stacking faults and partials. While it is possible to obtain information on the Burgers vector of a dislocation by TEM fairly readily, the details of their core structure are harder to obtain. Several theoretical studies of the cores of dislocations in silicon have been performed, e.g., Heggie and Jones,[27] Lapiccirella and Lodge,[28] and Jones.[29] Core structures in GaAs have been obtained by Myung.[30] The structural models given by these workers provide useful data when attempting to determine the core structure using high-resolution lattice imaging. Calculations of any core structure should ideally take into account the relaxation of the atoms at the surface and is likely to be greater when the core is located in a very thin part of the specimen, i.e., near the edge. The presence of kinks along the dislocation core cannot be discounted, particularly in images from thicker regions. Decoration of dislocation cores can occur when the impurity concentration is high, and lattice imaging has been used to study this as shown by Bourret.[31]

Burgers vector analysis is readily performed using two-beam diffraction contrast and the $\mathbf{g} \cdot \mathbf{b} = 0$ criterion. Information on this type can also be found in the works of Thomas and Goringe[2] or Loretto and Smallman.[32] Determination of whether a dislocation is of the shuffle or glide type has to be performed by high-resolution lattice imaging, and examples of this type of work for silicon include publications by Bourret *et al.*,[33] Anstis *et al.*,[34] Hutchison *et al.*,[35,36] and Olsen and Spence.[37] Burgers vector analysis is also possible using HREM as shown by Qin *et al.*[38] The studies show that it is possible using HREM to distinguish between shuffle and glide type cores but the match between the experimental image and the simulated image is not always perfect. The presence of amorphous surface layers often makes such comparisons difficult, as shown by Kilaas and Gronsky.[39] Lattice images of kinks in CdSe (hexagonal) have been obtained and studied by Glaisher et al.[40] This involves imaging with the dislocation line normal to the electron beam and leads to difficulties in image interpretation when applied to cubic (sphalerite) semiconductors. Methods for overcoming this problem have been investigated by Shindo et al.[41]

The number of different dislocation types in the sphalerite structure is doubled in comparison with elemental semiconductors because dislocations in GaAs can end with a row of gallium or arsenic atoms. Arsenic dislocations in the glide set and gallium dislocations in the shuffle set are denoted as alpha dislocations. Gallium dislocations in the glide set and arsenic dislocations in the shuffle set are beta dislocations. Studies of dislocation cores in GaAs by HREM have been performed by Tanaka and Jouffrey[42] and by Gerthsen et al.[43] In the latter study, the effect of changing from the alpha set to the beta set on simulated images was found to be too small to detect, at least under the conditions used, although it was possible to distinguish between glide and shuffle types. Identification of whether the core is of the alpha or beta type using HREM is likely to require that the core is situated at a particular specimen thickness where dynamical effects give rise to unique features in the image. Stacking faults in semiconductors have also been extensively studied by HREM. While the separation of the partials can be achieved using two-beam contrast under weak beam conditions, it is most readily performed using HREM. Analysis of the pattern of white spots determines whether the white spots correspond to atom pairs or tunnels in the structure as shown by Olsen and Spence.[37] Extrinsic and intrinsic faults have characteristic patterns which enables identification as shown by Ponce et al.[44]

2.3.1. Grain Boundaries

While they are less common in device structures, grain boundaries in semiconductors have also been the subject of study. Where they occur, they may be electrically active and it is important to characterize their structure. Diffraction contrast (two-beam) can be used to obtain information on the rigid body shift across the boundary by the analysis of the one-dimensional fringe patterns produced (so-called alpha fringes) as shown by Pond et al.[45] More information can be obtained using lattice imaging in conjunction with image simulation. Grain boundaries in germanium have been studied using lattice imaging by Krivanek et al.[46] and the contrast was modeled using an optical out-of-focus image of a line drawing of the structure. Image simulation methods have been applied to boundaries in germanium by d'Anterroches et al.[47] in which the complex atomic arrangement and also the rigid body shift at the boundary was determined. All of these studies only give the projected structure (along $\langle 110 \rangle$), and to determine the full three-dimensional nature of the boundary more than one projection is required as shown by Bourret et al.[48] in the case of the twinlike {211} boundary in germanium. As with dislocations, the analysis of grain boundaries in compound semiconductors is more complicated because of the polarity of the matrix leading to a greater number of possible models.

An analyis of boundaries in GaAs by Cho *et al.*[49] revealed the presence of antisite bonding at the interface. Low-angle boundaries were also found to exist and determined as an array of beta dislocations which are likely to be electrically active.

2.3.2. Twin Boundaries

Twin boundaries in sphalerite usually occur on (111) planes and can, in principle, exist in three different forms (Fig. 2a). Two of these subtypes have antiphase-type interfaces and should exhibit electrical activity. An image simulation study by Williams and Wright[50] has shown that the three types can be distinguished using lattice imaging (Fig. 2b). The expected, low-energy, inversion twin shows an asymmetry about the twin plane (Fig. 2c), and this has been confirmed by Qin *et al.*[38] Such asymmetries were also observed by Shiojiri *et al.*[51] and interpreted in terms of a nontetrahedral boundary structure. However, these asymmetries can be ascribed to the fact that, for the inversion twin, while the lattice is in twin orientation, the distribution of atom types with respect to the boundary is not. Williams and Wright also showed that asymmetries in the image of twin boundaries can also be caused by crystal misalignment and that extreme care must be taken in accurately orienting the crystal along the $\langle 110 \rangle$ pole.

2.3.3. Antiphase Boundaries (APBs)

The presence of antiphase boundaries must be considered when polar semiconductors are grown onto nonpolar substrates, e.g., GaAs on Si. They can be readily revealed using diffraction contrast with the (002) spot under the same diffraction conditions as used for polarity determination (see Ref. 21) as shown by Eaglesham *et al.*[52] Intensity differences between the (002) and (00$\bar{2}$) spots then reveal APBs as the boundary between brighter and darker regions. Conventional bright field diffraction contrast imaging can also reveal APBs as shown by Cho *et al.*,[49] but the contrast, originating from the rigid-body displacement across the boundary, can be superficially similar to that obtained from stacking faults. High-resolution lattice imaging does not reveal APBs readily.[53] Some attempts have been made to image point defects using lattice imaging. Point defects such as vacancies and interstitials can aggregate into intermediate defect configurations (IDCs), which may later evolve into dislocations. Various types of IDCs have been discussed by Tan *et al.*[54] and high-resolution lattice imaging has been applied by Krakow *et al.*[55] to study these using image simulation methods for identification. It was found that in a given column of Si atom pairs as seen along $\langle 110 \rangle$, at least 50% must be defective in order to be visible. The visibility of these defects will be reduced further when the effects of amorphous surface layers are added to the image contrast. Point (and line) defects can also be imaged using diffuse scattering as described by Kuan,[56] although no experimental data were presented. In this method, a small objective aperture is used to collect diffusely scattered electrons between the elastic peaks and form the image. Larger objective apertures can be used, and hence brighter images obtained, if the $\langle 111 \rangle$ direction is chosen in preference to $\langle 110 \rangle$.

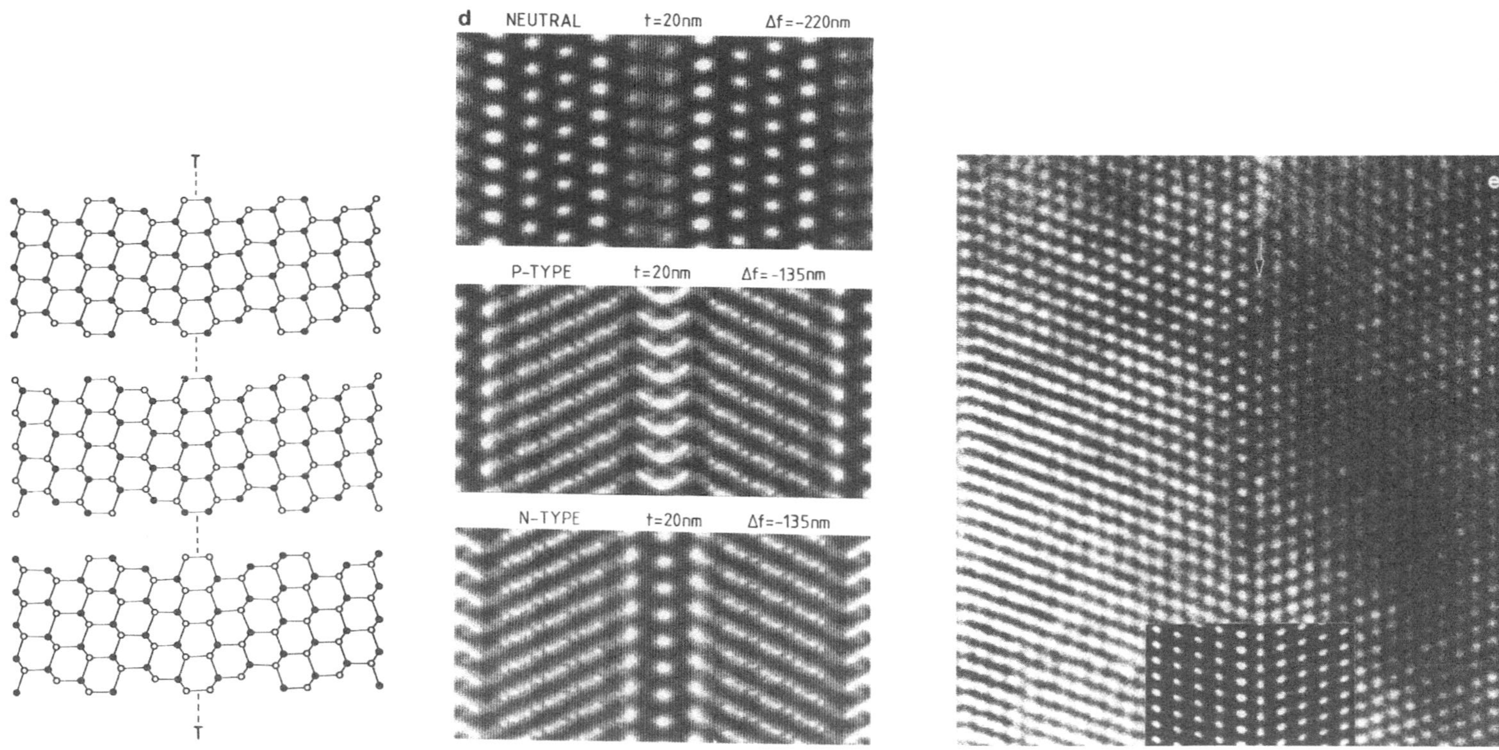

FIGURE 2.   Structural models of (a) the upright or "neutral" {111} twin boundary and (b), (c) inversion {111} twins called "*P*-type" and "*N*-type," respectively, in ZnSe. The ZnSe sphalerite lattice is seen here along [110] and the position of the twin plane is marked by the dotted line T. Zinc atoms are shown as solid circles and selenium atoms as open circles. The upright twins show both layers of Zn and Se atoms adjacent to the mirror plane, in contrast to the inversion twins, which show an antiphase behavior. (d) Simulation of high-resolution lattice images using the structural models of Figs. 2a, 2b, and 2c showing that unique images can be formed at similar crystal thickness (but not defocus). (e) High-resolution lattice image of {111} neutral twin plane in ZnSe viewed along [110] showing asymmetry in image via offset bright view of spots. Simulated image is shown as an inset and position of twin plane is marked by arrow.

2.4. Detection of Long- and Short-Range Order in Alloy Systems

When there are more than two elemental constituents present in a semiconductor layer, there exists the possibility of more ordered structures occurring. The nature of the ordered structure often depends on the choice of substrate orientation and deposition conditions. In addition, certain alloy systems exhibit a phenomenon called spinodal decomposition, in which the material separates out into regions of two different composition, usually on a very fine scale.

Evidence of ordering is usually best found in the diffraction pattern and is often revealed as extra spots (perfect order) or streaks along certain directions (imperfect order). Diffraction patterns provide the best information on the details of the ordered structure, and it is desirable to obtain patterns from more than one direction so that a three-dimensional model of reciprocal space can be built up. Usually, one plan view and one cross-sectional specimen will suffice. Ordered structures have been observed in the $Al_xGa_{1-x}As/GaAs(110)$ system by Kuan *et al.*[57] in which alternate monolayers of Al and Ga occurred. High-resolution imaging was able to show that the new ordered structure was considerably faulted in the growth direction.

Similar ordered structures have been observed in $Ga_{1-x}In_xP$ grown onto (100) oriented InP substrates[58] and enabled an explanation of why the energy band gap varied by as much as 50 meV between material grown under different conditions/substrate orientation. No ordering was observed, however, when the substrate orientation was either (111) or (110).[59] In this case, the ordering took place on (111) planes. The degree of ordering was found to depend also upon the deposition conditions, with diffraction patterns varying from sharp superlattice spots to either straight or wavy streaks. This could be interpreted in terms of antiphase boundaries lying coplanar with the substrate surface. The substrate misorientation [e.g., from (100)] was also found to influence the ordered superlattice structure. High-resolution imaging was able to image the ordered structures directly and so show the extent of the ordered regions. Similar effects have also been observed in $Al_xIn_{1-x}As$ layers.[60]

The system $Ga_{1-x}In_xAs_{1-y}P_y$ shows regions that undergo a phase separation or spinodal decomposition into GaP-rich and InAs-rich zones. Such compositional modulations could be detected by EDX methods[61] but is readily observable using diffraction contrast. A "basket-weave"-like contrast, obtained using the (220) spot, may arise due to the surface relaxation of misfit strains between the zones of different composition.[62] A very fine-scale "speckle" contrast is also visible under the same conditions, and its origin is still unclear at present.

2.5. Characterization of Interfaces

The fabrication of electronic devices naturally involves the creation of interfaces between various semiconductor, metal, or insulating layers. The structure of these interfaces at the atomic level often determines the properties of the device, and so it is important to characterize both structural and compositional changes at the interface.

Depending on the thickness of the grown layer, mismatches between the lattice parameters of layer and substrate can lead to defects in the layer. When the dislocation density is great, as in the case of large mismatch systems such as GaAs on Si, weak beam imaging is useful for revealing defect densities along the interface as shown by Eaglesham *et al.*[52] High-resolution lattice imaging along the $\langle 110 \rangle$ direction can also reveal readily the interfacial dislocations and their type as shown by Sharan and Narayan.[63] Lattice imaging is particularly useful for determining the roughness of interfaces and whether this is responsible for nucleating defects such as twin boundaries.

2.5.1. Position of the Interface

When quantum well structures are being studied, the position of the interface must be known so that an assessment of the well width and the interface roughness can be determined. The visibility of the interface depends largely on the difference in electron scattering between the materials on either side of the interface, and considerable progress has been made in this area of study. In sphalerite, the intensity of the (002) reflection is sensitive to the composition, being greater when the scattering difference between the metal and nonmetal components is large. Thus in the case of the $GaAs/Ga_{1-x}Al_xAs$ system, Al-rich areas appear brighter when this reflection is used in dark field mode to image the specimen. The resolution of this approach is limited by the size of the objective aperture to around 5 Å.

Lattice imaging gives much better spatial resolution, better than 2 Å point to point with modern medium voltage machines. However, this does not mean that the interfaces will be any easier to see. When projected along $\langle 110 \rangle$, the sphalerite diffraction pattern produces two (002)-type beams, their intensity being sensitive to the composition, and a great many more beams that are not. Thus contrast difference between two layers is not easy to see unless differences in scattering are large. By tilting about an axis normal to the interface [for (100)-type surfaces], many noncomposition-sensitive beams are reduced in intensity while retaining the (002)-type beams so that interlayer contrast is enhanced.

Under these conditions, the lattice image consists mainly of (200)-type fringes. In addition, theoretical simulations for the $\langle 110 \rangle$ disection show that the actual position of the interface is displaced from that indicated from the lattice image.[64] A better approach is to image along the $\langle 100 \rangle$ direction. There are now four (002)-type beams and a fewer overall number of diffracted beams giving much improved contrast. Also, the indicated interface position is nearer the true value. This direction is less suitable for defect analysis at interfaces, however, and may require a special specimen to be made up for this purpose, particularly if only low tilt specimen stages are available, as on many high-resolution machines. High contrast differences between layers in the $Al_xGa_{1-x}As$ system may be ascribed to dynamical effects such that the intensity of the (002) reflections is at a peak for a particular crystal thickness. At high Al concentrations, oxidation of the Al-rich layers at the specimen surface can lead to high contrast. Preferential ion-milling of one layer type, e.g., the well, can also lead to contrast enhancement via thickness changes across the interface.

Image processing can be used to good advantage for contrast enhancement. On-line processing is particularly useful for noise reduction and contrast boosting so that the optimum focus setting or crystal thickness can be found. Processing methods are becoming more common for this type of work as shown by Furuta *et al.*[65]

2.5.2. Structure of the Interface

In systems in which the materials on either side of the interface have the same basic structure, e.g., sphalerite, the problem of interfacial structure usually centers around such parameters as interfacial roughness and simply requires that the position of the interface be determined as described above. Steps on the interface, often the cause of defects such as twins and antiphase boundaries, can then be imaged. In the case of superlattice structures, imaging under convergent beam conditions using superlattice reflections with plan view specimens can reveal fluctuations in the interface position on monolayer dimensions as shown by Vincent *et al.*[66]

When the grown layer does not have the same structure as that of the substrate, e.g., silicides on silicon or aluminum on GaAs, then a more complicated interfacial structure results. Parameters such as the relative displacement of the layer over the substrate and the rigid body displacement normal to the plane of the interface need to be determined. For this purpose, both convergent beam electron diffraction and lattice imaging methods have been used successfully. Examples are $NiSi_2$ or $CoSi_2$ on Si in which the layer has the CaF_2 structure and there are two possible orientations for the layer with respect to the substrate. High-resolution lattice imaging cannot only determine which model is appropriate[67] but also obtain information on the rigid body shift normal to the plane of the interface.[68] Convergent beam electron diffraction using plan-view specimens has also been used to determine the in-plane displacement for this system[69] and also that of Al on GaAs(100). The structure of the Al/GaAs interface has been studied by Wright *et al.*[70] using HREM as shown in Fig. 3 and has confirmed the results obtained from CBED studies.

HREM (in cross-sectional specimens) and CBED methods are complimentary in studies of the interfaces of these systems and both should be considered for any serious study. Only HREM can give information on the rigid body displacement normal to the interface. In both cases, access to computational methods is required if the experimental results are to be interpreted at all.

2.5.3. Investigation of States of Strain Near Interfaces

Only homoepitaxial systems are perfectly lattice matched at the interface, e.g., Si on Si. All other systems will involve some degree of mismatch and, therefore, a degree of strain at the interface. When the strain exceeds a critical value, defects will be nucleated and so it is desirable to be able to measure this strain at some point in the specimen. The effects of strain can often be imaged in conventional diffraction contrast and high-resolution modes of operation. Quantification is performed by measuring the lattice parameter at a given point on the specimen. Thus the CBED method described in Section 2.1. can be used. This method can be extended into a hybrid imaging/diffraction mode as described by Humphries *et al.*[71] In this technique, the electron probe is brought to a focus above the specimen thus forming an image of the specimen superimposed into the HOLZ line pattern. Variations in the lattice parameter across the specimen are reflected in variations in the positions of the HOLZ lines.

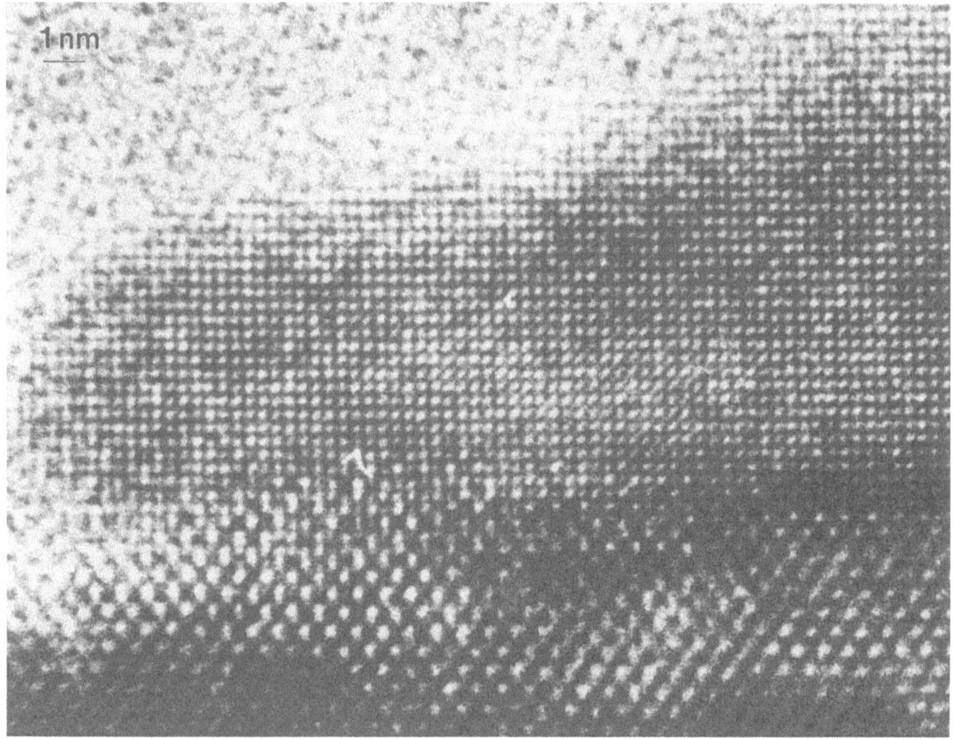

FIGURE 3. High-resolution lattice image of cross-section of the Al/GaAs (100) interface. The GaAs is viewed along [110] and the Al along [100]. White spots correspond to tunnels in both structures. Computer analysis of the image confirms results obtained from earlier diffraction studies. Square lattice is Al and oblique lattice is GaAs.

However, it should be noted that when the specimens are prepared in the form of a thin foil, extensive relaxation of the material will take place as found by the present authors in the case of the strained layer system $GaSb_xAs_{1-x}/GaAs$. The wedge shape nature of such foils gave rise in this case to excessive relaxation near the specimen edge. Thus only data acquired from parts of the specimen that are both thick and of constant thickness should be used. Even then, the data are not likely to be representative of the true state of strain as occurs in bulk material. By way of example, Fig. 4a shows a convergent beam image of a strained-layer superlattice structure. Considerable variations in the lattice parameter across the layers are evident, which increase as the specimen edge is approached. However, such wide variations are not observed from a thicker area which is also more uniform in thickness (Fig. 4b).

3. DETERMINATION OF COMPOSITION CHANGES ACROSS INTERFACES

The determination of composition variations is an important part of the characterization of ultrathin layers such as quantum well structures, and there are several

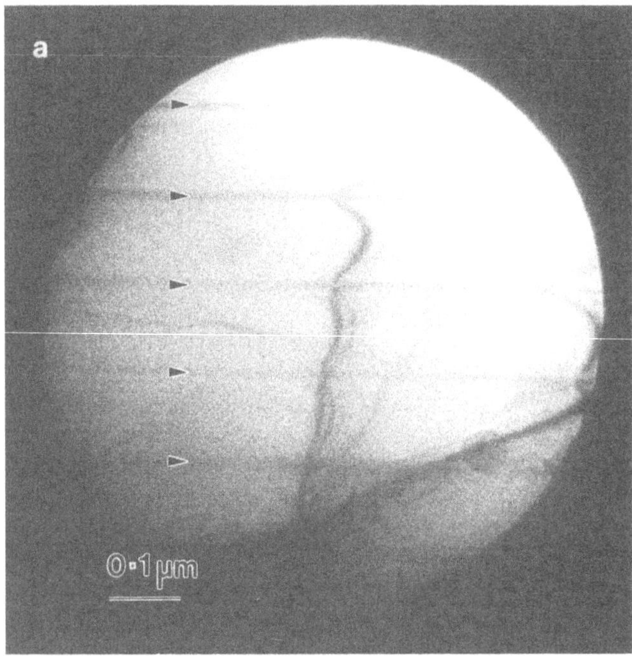

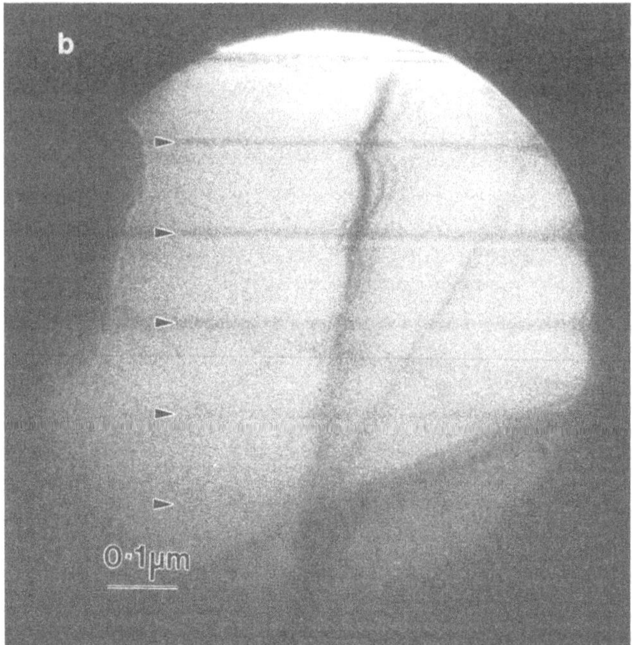

FIGURE 4.   (a) Convergent beam image of a GaSbAs/GaAs strained layer superlattice. This wedge-shaped specimen shows increasing variation in the state of strain between Sb-rich wells (arrowed) as the specimen edge is approached. (b) An image recorded from a thicker region shows far less variation, proving that considerable relaxation of the structure occurs in thin specimens.

techniques now available to the microscopist for this purpose. Compositional analysis in the TEM has usually been performed using spectroscopic methods such as energy or wavelength dispersion of x-rays (EDX or WDX) using a fine electron probe. The spacial resolution is set by a combination of probe size, typically 1–5 nm in modern instruments, and beam broadening as the probe spreads out into the specimen.

Beam broadening can be a great problem unless the specimens are very thin and so the effects of probe size and its shape (i.e., Gaussian profile) mean that any compositional profiles have to be subjected to deconvolution procedures if the true composition profile is to be obtained. Nevertheless, useful data have been obtained using EDX as shown by McGibbon et al.[72] Other problems arise when EDX peaks overlap substantially, making quantification difficult. Fluorescence from nearby layers can also bias the result such as attempting to monitor the Ga and As levels in ZnSe grown on GaAs, giving spurious peaks in the EDX spectrum. The only solution here is to remove the substrate entirely as shown by Wright et al.[73] It is now possible to obtain two-dimensional chemical maps using EDX, but acquisition times are long and thus specimen drift can limit the resolution. Field emission gun (FEG) equipped TEMs are better suited to elemental mapping at high spatial resolution because the greater brightness of the electron source gives much better statistics. Electron energy loss spectroscopy (EELS) offers another approach to microanalysis, but there are problems with quantification unless the specimens are very thin so that multiple scattering is limited. It has been used to interpret HREM data on $Ga_{1-x}In_xAs/InP$ multilayers as shown by Petford-Long and Long.[74]

A very useful technique for monitoring composition changes is to use annular dark field imaging (ADF), which gives a form of Z or atomic number contrast.[72] This requires the use of a STEM and the conditions are set up for a short camera length so that only high-angle diffusely scattered electrons are captured by the STEM dark field detector. Because heavier atoms scatter more at higher angles, these regions show up as brighter areas giving Z-contrast. An advantage of this technique is the ability to use smaller probes as the cross section for high angle scattering is many orders of magnitude greater for electrons than for x-ray generation. Consequently, data acquisition times are much reduced (15 s as opposed to 1 h on an EM430 system fitted with both EDX and STEM units[75]) so that specimen drift is almost eliminated. Serious disadvantages include the difficulty of quantification, meaning that EDX data will also be required and systems with more than one variable, e.g., quaternaries, cannot be unambiguously interpreted.

An example of this technique is shown in Fig. 5, in which compositional variations can be detected in quantum wells of $GaSb_xAs_{1-x}$. Compositional analysis can also be performed using diffraction contrast imaging. As described earlier, the (002) reflections are sensitive to difference in scattering between the metal and nonmetal components in sphalerite. Thus the intensity in the image obtained using this reflection should be proportional to the compositional changes in a system such as $Ga_{1-x}Al_xAs$ as shown originally by Petroff.[76] However, the intensity of the (002) reflection also varies with specimen thickness (which is generally not constant) and so a simple interpretation of the contrast in terms of the composition cannot be made. Indeed, in the case of the $Ga_{1-x}In_xAs/GaAs$ system anomalous contrast effects have been observed by Baxter et al.[77] in which In-rich layers were seen to be darker than the surrounding GaAs.

FIGURE 5. Annular dark field (ADF) STEM image of a GaSbAs/GaAs strained layers superlattice. Atomic number contrast clearly shows the Sb-rich wells as bright lines and that fluctuations in the well width or composition occur over their length.

The anomalous contrast was ascribed to the effects of multiple elastic and inelastic scattering, but it is clear that further study is required in this area. Nevertheless, it remains a useful technique for qualitative characterization of multilayer structures and an example of its application to $Al_xGa_{1-x}As/GaAs$ multilayers grown by MOCVD is shown in Fig. 6. It is clear that the variations in the Al concentration along the growth direction do not follow an exact square wave. Peaks in the Al concentration occur when trimethyl aluminum is switched in and out of the system.

In the case of the well-studied $Al_xGa_{1-x}As/GaAs$ system, it is now possible to use the (002) dark field method to obtain compositional information to a level of ±0.02 as shown by Bithell and Stobbs.[78] Potential sources of error such as dynamical scattering and inelastic effects can be taken into account in a simple way using a ratiometric method. For the method to work with any degree of accuracy, orientations for which multiple scattering can occur must be avoided. In practice this means that the crystal should be oriented within 10–15° of [100]. The method also requires the use of a microdensitometer to quantify image contrast. Problems with thickness

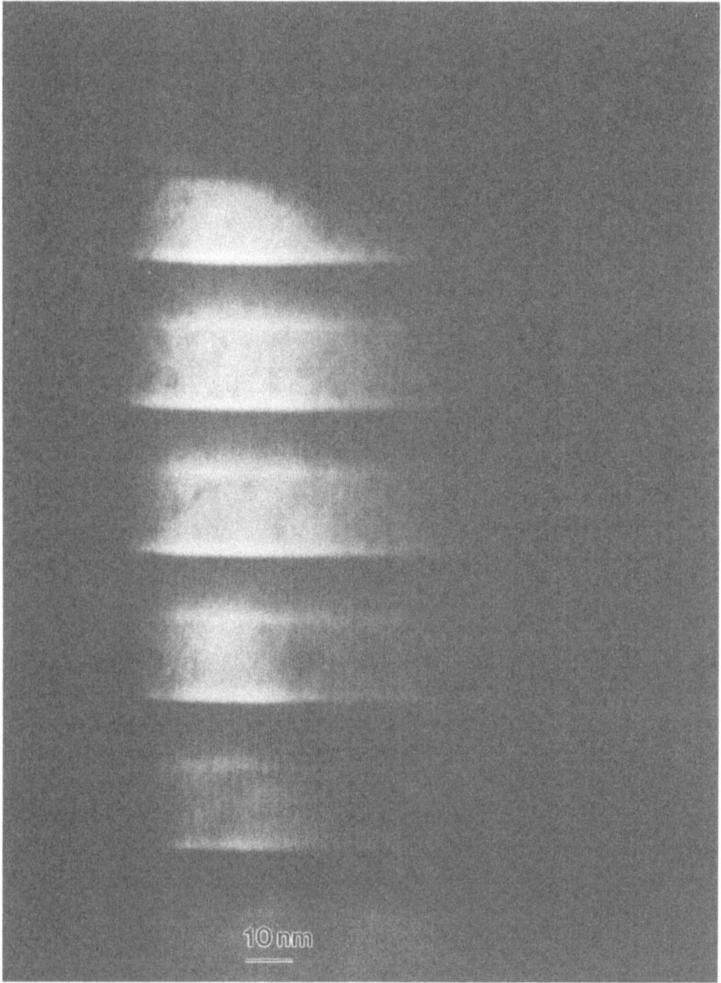

FIGURE 6. Dark field (002) image of a AlGaAs/GaAs quantum well structure grown by MOCVD using a 90° wedge specimen. Brighter areas are richer in Al and show clear peaks occurring during switching cycles of reactor system. Specimen edge is at left and is viewed near ⟨100⟩ pole.

variations can be avoided by obtaining the image from 90° wedge specimens (see Section 4 and also below) as shown in Fig. 6.

When the layers are very thin, i.e., a few nanometers, Fresnel fringe effects can appear and be mistaken for compositional variations. In this case, it is better not to use dark field methods but to employ bright field Fresnel fringe profiles as a means of determining the composition profile. By matching simulations of the fringe profiles to the compositional profile, the latter may be deduced as shown by Ross et al.[79] The problem of determining absolute values for the concentration is, however, compounded by the effects of inelastic scattering.

A novel method for assessing compositional variations in multilayer systems was invented by Kakibayashi and Nagata.[80] This technique uses a special type of specimen in which the wafer is cleaved along two orthogonal ⟨110⟩ directions to

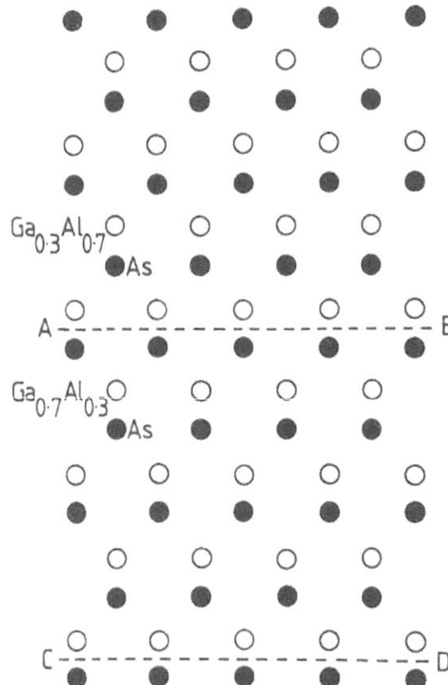

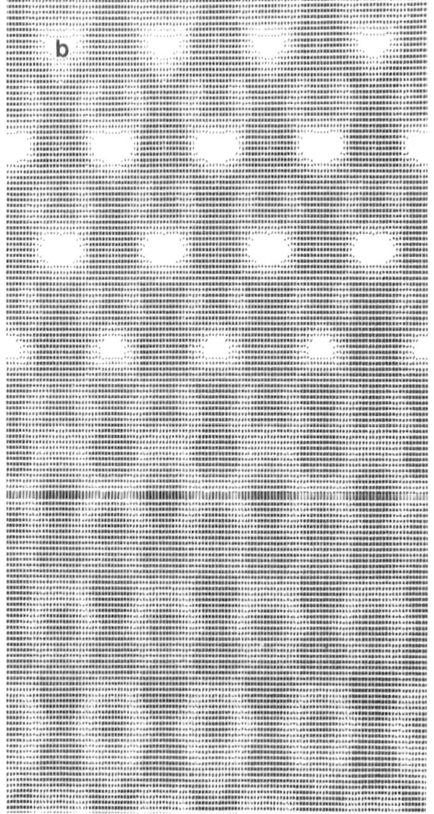

FIGURE 7. (a) Simulations of high-resolution image contrast of a Al–Ga–As quantum well structure. For certain crystal thicknesses, images showing (b) abrupt appearance, or (c) diffuse can be obtained, depending on the degree of focus. Direction is along $\langle 110 \rangle$.

give a 90° wedge. The specimen is then imaged along $\langle 100 \rangle$ using bright field diffraction contrast and shows fringes of constant thickness running along the specimen edge. The position of these fringes relative to the specimen edge is also dependent on the composition and can be used to obtain compositional profiles across the layers. The position of the fringes can be calculated as a function of the composition using standard theoretical methods[81] and so obtain absolute values for the composition at a given point. However, neglecting the effects of absorption in such calculations will lead to errors. This can be overcome by using calibration data obtained from specimens of known composition as shown by De Jong and Janssen.[82] The precise position of the fringes is not easy to determine as they are rather broad, and this will also limit the accuracy to which composition can be determined.

Convergent beam electron diffraction can also be applied to the study of compositional variations as shown by Hetherington *et al.*[83] This technique uses the separation of lines seen in the higher-order Laue zone disks to monitor compositional changes. Several problems arise in attempting to quantify these line shifts in terms of the composition, however.[83] Also the method does not give directly a map of the composition superimposed on the image of the specimen as does the thickness fringe method. Lattice imaging methods have also been applied to the study of compositional variations across interfaces. The use of asymmetry in dumbbell-like images seen along $\langle 110 \rangle$ has been investigated by Olsen *et al.*[84] for the $Ga_{1-x}Al_xAs$

FIGURE 8.   High-resolution lattice image of $Al_xGa_{1-x}As/GaAs$ superlattice structure viewed along $\langle 110 \rangle$. Clear contrast between the GaAs and Al-rich layers is seen. Al-rich areas appear less sharp. Image gives an indication of interfacial roughness.

system. While it may be possible in principle to use this approach for compositional analysis, the images are sensitive functions of both objective lens defocus and crystal thickness and require extensive simulations of the image for a given accelerating voltage. A theoretical study by Wright and Williams[85] indicates that lattice imaging methods can show up compositionally sharp interfaces as diffuse and vice versa, depending on the degree of defocus and crystal thickness. Figure 7 shows simulations of an abrupt AlGaAs layer in which images revealing the appearance of both abrupt and diffuse interface depending on the degree of defocus. Nevertheless, high-resolution lattice images can give useful information on interface roughness as shown for the case of $Al_xGa_{1-x}As/GaAs$ multilayers in Fig. 8.

Another method with the potential for determining the degree of interfacial abruptness in well-ordered superlattice structures has been shown by Kuan et $al.$[86] and uses the intensities of superlattice spots seen in the electron diffraction pattern. This involves calculating the average relative intensities of these spots (using dynamical methods) for a range of compositional profiles so that these can be compared with the experimental data. The method has the advantage over lattice imaging in avoiding the complicating effects of focus dependence. A disadvantage of this technique is that to obtain spots in the pattern, the structure under study must have a sufficiently large number of periods, i.e., 100 or so. Any disorder in the layer sequence will result in streaking in the diffraction pattern, thus making comparisons difficult.

4. STUDIES OF SEMICONDUCTOR SURFACES IN THE TEM

Knowledge of the state of the semiconductor surface is important. The surface is where growth of layers takes place and so it is desirable to understand its structure and how this affects the structure of the layer. By using the TEM in a glancing reflection mode, both diffraction patterns and images can be obtained, which give useful information on the structure of the surface. Diffraction patterns are useful for determining the nature of surface order. By imaging with only one beam, diffraction contrast can be used to reveal the presence of surface steps and also determine the coverage of domains of particular states of surface order (i.e., surface reconstructions). The method has been shown to have considerable use for the characterization of surface roughness on silicon, which is an important parameter in the production of very thin oxide layers for memory chips.[87]

Specialized instrumentation has been developed in several laboratories in which the specimen environment can be carefully controlled enabling the state of the surface to be modified by heating or cleaning operations using additional attachments such as ion-guns. Dynamic processes such as adsorption, sublimation, oxidation, and phase changes can be studied in this way using the above reflection mode. A review of reflection electron microscopy and its applications has been given by Yagi.[88]

Another possibility being explored is to examine TEM diffraction patterns for the very weak spots expected from surface order in thin foils.[89] This approach involves being able to clean and anneal the thin foil surface under UHV conditions. A low-level CCD camera (which has a very high dynamic range) is then used to record the diffraction pattern.

5. SPECIMEN PREPARATION

Specimen preparation from semiconductor samples is generally performed by one of three methods.[100] For cross-sectional studies, this usually involves mechanical prethinning (e.g., using a dimpling instrument) followed by ion-milling and has been well described by Chew and Cullis[91] and by Bravman and Sinclair.[92] Problems with this approach are differential thinning of layer and substrate and ion-beam-induced defects. Such damage is prone to occur in the more ionic materials, e.g., ZnS and can be reduced by using iodine vapor instead of the more usual argon gas.[93] This technique is also useful for materials containing indium which forms elemental In on the specimen when thinned with Ar.[91]

Plan-view specimens are often best produced using chemical removal of the substrate (i.e., jet polishing). GaAs substrates can be removed successfully using bromine or chlorine in methanol. Jet vapor methods have been developed for thinning III–V materials.[94] A chemical polishing method for cross sections was presented by Chu and Sheng.[95] Silicon substrates require etches containing HF in order to remove the native oxide. Alternatively, the substrate can be mechanically removed using a dimpling tool and thinned to electron transparency using ion-milling from the substrate side only.

More recently, cross-sectional specimens have been prepared very quickly by simple cleavage along {110} planes to yield a 90° wedge. This is useful if only layer thicknesses are required or if compositional measurements are of interest. A simple method for producing these wedges is given by Graham.[96]

Further useful details on semiconductor specimen preparation can be found in the reviews of Newcomb,[97,98] and for device structures in the paper by Garulli et al.[99]

6. CONCLUSION

In this chapter we have attempted to describe the power of transmission electron microscopy through its many and varied modes of operation to yield detailed information on the microstructure and spatially resolved compositon of semiconductor thin-film materials. The advantages and limitations of the various approaches have been described and it is hoped that the reader can now select the technique or combination of techniques most appropriate to the problem at hand. The combined power of both experimental and theoretical electron microscopy is underlined and examples have been given that illustrate the need for the application of a combination of techniques in order to obtain an unequivocal quantitative answer. Even though the TEM can reveal atomic structure, the conditions under which this is achieved must be fully appreciated.

ACKNOWLEDGMENTS

The authors acknowledge the support of the SERC and UMIST in this work. Stimulating discussions and experimental assistance from other members of the Manchester HVEM facility are greatly appreciated.

REFERENCES

1. P. B. Hirsch, A. Howie, R. B. Nicholson, D. W. Pashley, and M. J. Whelan, *Electron Microscopy of Thin Crystals*, Krieger, New York (1977).
2. G. Thomas and M. J. Goringe, *Transmission Electron Microscopy of Materials*, Wiley, New York (1979).
3. H. Oppolzer, TEM for support of VLSI technology, in *Microscopy of Semiconducting Materials*, Inst. Phys. Conf. Ser. No. 76, pp. 461–470, IOP Publishing, Bristol (1985).
4. H. Oppolzer, H. Cerva, C. Fruth, V. Huber, and S. Schild, TEM studies during development of a 4-megabit D-RAM, in *Microscopy of Semiconducting Materials*, Inst. Phys. Conf. Ser., No. 87, pp. 433–438, IOP Publishing, Bristol (1987).
5. K. W. Andrews, D. J. Dyson, and S. R. Keown, *Interpretation of Electron Diffraction Patterns*, Hilger, London (1971).
6. J. E. Eades, Symmetry determination by convergent beam diffraction, in *EUREM '88*, Inst. Phys. Conf. Ser., No. 93, Vol. 1, pp. 3–12, IOP Publishing, Bristol (1988).
7. R. C. Ecob, M. P. Shaw, A. J. Porter, and B. Ralph, Small symmetry changes accompanying phase transformations, *Phil. Mag.* **A44**(5), 1117–1133 (1981).
8. G. M. Rackham and J. W. Steeds, Convergent beam observations near boundaries and interfaces, in *EMAG '75*, Inst. Phys. Conf. Ser., pp. 457–460, IOP Publishing, Bristol (1975).
9. N. S. Blom and F. W. Schapink, Computer simulation of convergent beam electron diffraction patterns from bi-crystal, *J. Appl. Crystallogr.* **18**, 126–130 (1985).
10. A. C. Wright, unpublished results.
11. Y. P. Lin, D. M. Bird, and R. Vincent, Errors and correction term for HOLZ line simulations, *Ultramicroscopy* **27**, 233–240 (1989).
12. E. G. Bithell and W. M. Stobbs, The simulation of HOLZ line positions in electron diffraction patterns, *J. Microsc.* **153**(1), 39–49 (1989).
13. J. M. Gibson, R. Hull, J. C. Bean, and M. M. J. Treacy, Elastic relaxation in transmission electron microscopy of strained-layer superlattices, *Appl. Phys. Lett.* **46**(7), 649–651 (1985).
14. J. C. H. Spence, *Experimental High Resolution Electron Microscopy*, Clarendon Press, Oxford (1981).
15. P. G. Self, M. A. O'Keefe, P. R. Buseck, and A. E. C. Spargo, Practical computation of amplitudes and phases in electron diffraction, *Ultramicroscopy* **11**, 35–52 (1983).
16. P. D. Brown, A. P. C. Jones, G. J. Russel, J. Woods, B. Cockayne, and P. J. Wright, In-plane anisotropy of the defect distribution in ZnSe, ZnS and ZnSe/ZnS epilayers grown onto (001) GaAs by MOCVD, in *Microscopy of Semiconducting Materials*, Inst. Phys. Conf. Ser., No. 87, pp. 123–128, IOP Publishing, Bristol (1987).
17. L. DiCioccio, E. A. Hewat, A. Million, J. P. Gaillard, and M. Dupuy, High resolution transmission electron microscopy of CdTe-HgTe superlattice cross-sections, in *Microscopy of Semiconducting Materials*, Inst. Phys. Conf. Ser., No. 87, pp. 243–248, IOP Publishing, Bristol (1987).
18. M. A. O'Keefe, P. R. Buseck, and I. Iijima, *Nature* **274**, 322–324 (1978).
19. D. M. Bird and A. G. Wright, Polarity determination from Kikuchi patterns, in *EUREM '88*, Inst. Phys. Conf. Ser., No. 93, Vol. 2, pp. 39–40, IOP Publishing, Bristol (1988).
20. A. R. Preston and P. Spellward, Polarity determination in GaAs by matching of [110] CBED patterns and simulations, in *EUREM '88*, Inst. Phys. Conf. Ser., No. 93, Vol. 2, pp. 27 and 28, IOP Publishing, Bristol (1988).
21. J. Tafto and J. C. H. Spence, A simple method for the determination of structure factor phase relationships and crystal polarity using electron diffraction, *J. Appl. Crystallogr.* **15**, 60–64 (1982).
22. A. C. Wright, T. L. Ng, and J. O. Williams, Polarity determination in ⟨110⟩-oriented GaSb by high-resolution transmission electron microscopy at 300 kV, *Phil. Mag. Lett.* **57**(2), 107–111 (1988).
23. D. J. Smith, R. W. Glaisher, and P. Lu, Surface polarity determination in ⟨110⟩-oriented compound semiconductors by high resolution electron microscopy, *Phil. Mag. Lett.* **59**(2), 69–75 (1989).
24. W. Stutius and F. A. Ponce, Crystal orientation dependance of the electrical transport and lattice structure of zinc selenide films grown by MOCVD, *J. Appl. Phys.* **58**(4), 1548–1553 (1985).
25. J. P. Hirth and J. Lothe, *Theory of Dislocations*, McGraw-Hill, New York (1968).
26. N. Burle-Durbec, B. Pichard, and F. Minari, Interaction of In atoms with partial dislocation cores in GaAs: 0.3% In, *Phil. Mag. Lett.* **59**(3), 121–129 (1989).
27. M. Heggie and R. Jones, Atomic structure of dislocations and kinks in silicon, in *Microscopy of Semiconducting Materials*, Inst. Phys. Conf. Ser., No. 87, pp. 367–374, IOP Publishing, Bristol (1987).

28. A. Lapiccirella and K. W. Lodge, Dislocation core structure in silicon, in *Microscopy of Semiconducting Materials*, Inst. Phys. Conf. Ser., No. 60, pp. 51-55, IOP Publishing, Bristol (1981).

29. R. Jones, Reconstructed dislocations in covalently bonded semiconductors, in *Microscopy of Semiconducting Materials*, Inst. Phys. Conf. Ser., No. 60, pp. 45-50, IOP Publishing, Bristol (1981).

30. N. Maung, Ph.D. thesis, Gottingen, F.R.G. (1987).

31. A. Bourret, Direct observation of impurity decoration in dislocation core, *Electron Microsc.* 1, 306-307 (1980).

32. M. H. Loretto and R. E. Smallman, *Defect Analysis in Electron Microscopy*, Chapman and Hall, London (1975).

33. A. Bourret, J. Desseaux, and C. D'Anterroches, Defect structure in CZ silicon and germanium studied by high-resolution electron microscopy, in *Microscopy of Semiconducting Materials*, Inst. Phys. Conf. Ser., No. 60, pp. 9-14, IOP Publishing, Bristol (1981).

34. G. R. Anstis, P. B. Hirsch, C. J. Humphries, J. L. Hutchison, and A. Ourmazd, Lattice images of the 30° partials in silicon, in *Microscopy of Semiconducting Materials*, Inst. Phys. Conf. Ser., No. 60, pp. 15-22 (1981).

35. J. L. Hutchison, G. R. Anstis, and P. Pirouz, Lattice images of undissociated 60° dislocations in silicon, in *Microscopy of Semiconducting Materials*, Inst. Phys. Conf. Ser., N. 67, pp. 21-26, IOP Publishing, Bristol (1983).

36. J. L. Hutchison, C. J. Humphries, A. Ourmazd, and P. B. Hirsch, Structure images of dislocations in silicon, *Electron Microsc.* 1, 304-305 (1980).

37. A. Olsen and J. C. H. Spence, Distinguishing dissociated glide and shuffle set dislocation by high resolution electron microscopy, *Phil. Mag.* A43(4), 945-965 (1981).

38. L. C. Qin, D. X. Li, and K. H. Kuo, A HREM study of the defects in ZnS, *Phil. Mag.* A53(4), 543-555 (1986).

39. R. Kilaas and R. Gronsky, The effect of amorphous surface layers on images of crystals in high resolution transmission electron microscopy, *Ultramicroscopy* 16, 193-202 (1985).

40. R. W. Glaisher, M. Kuwabara, J. C. H. Spence, and M. J. McKelvy, On the lattice imaging of kinks in wurtzite semiconductors, in *Microscopy of Semiconducting Materials*, Inst. Phys. Conf. Ser., No. 87, pp. 349-354 IOP Publishing, Bristol (1987).

41. D. Shindo, J. C. H. Spence, H. Alexander, N. Long, and Vanderschaever, *Electron Microsc. (Kyoto)* 1, 785 (1986).

42. M. Tanaka and B. Jouffrey, Dissociated dislocations in GaAs observed in high resolution electron microscopy, *Phil. Mag.* A50(6), 733-743 (1984).

43. D. Gerthsen, F. A. Ponce, and G. B. Anderson, High resolution transmission electron microscopy of 60° dislocations in Si-GaAs *Phil. Mag.* A59(5), 1045-1058 (1989).

44. F. A. Ponce, T. Yamashita, R. H. Bube, and R. Sinclair, Imaging of defects in cadmium telluride using high resolution transmission electron microscopy, in *Defects in Semiconductors*, North-Holland, Amsterdam (1981).

45. R. C. Pond, D. A. Smith, and V. Vitek, Inst. Metals Conf., Jersey (1961).

46. O. Krivanek, S. Isoda, and K. Kobayashi, Lattice imaging of grain boundary in crystalline Ge, *Phil. Mag.* 36, 931 (1977).

47. C. D'Anterroches, G. Silvestre, A. M. Papon, J. J. Bacmann, and A. Bourret, Atomic structure of $E = 9$ grain boundary in germanium, *Electron Microsc.* 1, 316-317 (1980).

48. A. Bourret, L. Billard, and M. Petit, HREM determination of the structure of the $\{211\}$ $E = 3$ twin in Ge, in *Microscopy of Semiconducting Materials*, Inst. Phys. Conf. Ser. N. 76, pp. 23-28, IOP Publishing, Bristol (1985).

49. N-H. Cho, C. B. Carter, D. K. Wagner, and S. McKernan, Grain boundaries and APB's in GaAs, in *Microscopy of Semiconducting Materials*, Inst. Phys. Conf. Ser., No. 87, pp. 281-286, IOP Publishing, Bristol (1985).

50. J. O. Williams and A. C. Wright, High resolution lattice imaging study of twin boundaries in epitaxial films of ZnSe grown by MOVPE, *Phil. Mag.* A55(1), 99-110 (1987).

51. M. Shiojiri, C. Kaito, S. Sekimoto, and N. Nakamura, Polarity and inversion twins in ZnSe crystals observed by HREM, *Phil. Mag.* A46(3), 495-505 (1982).

52. D. J. Eaglesham, R. Devenish, R. T. Fan, C. J. Humphries, H. Morkoc, R. R. Bradley, and P. D. Augustus, Defects in MBE and MOCVD-grown GaAs on Si, in, *Microscopy of Semiconducting Materials*, Inst. Phys. Conf. Ser., No. 87, pp. 105-110, IOP Publishing, Bristol (1987).

53. T. S. Kuan and C. A Chang, Electron microscope studies of a Ge–GaAs superlattice grown by MBE, *J. Appl. Phys.* 58(4), 4408-4413 (1984).

54. T. Y. Tan, H. Foll, and W. Krakow, Intermediate defects in silicon and germanium, in, *Microscopy of Semiconducting Materials*, Inst. Phys. Conf. Ser., No. 60, pp. 1–7, IOP Publishing, Bristol (1981).

55. W. Krakow, T. Y. Tan, and H. Foll, The identification of atomic chain configurations in ion irradiated Si by HREM, in, *Microscopy of Semiconducting Materials*, Inst. Phys. Conf. Ser., No. 60, pp. 23–28 IOP Publishing, Bristol (1981).

56. T. S. Kuan, The imaging of point and line defects in silicon using diffuse scattering, *Electron Microsc.* **1**, 336–337 (1980).

57. T. S. Kuan, T. F. Kuech, W. I. Wang, and E. L. Wilkie, Long range order in $Al_xGa_{1-x}As$, *Appl. Phys. Lett.* **51** 51 (1987).

58. T. Suzuki, A. Gomyo, S. Iijima, K. Kobayahi, S. Kawata, I. Hino, and T. Yuasa, Band-gap energy anomaly and sublattice ordering in GaInP and AlGaInP grown by MOVPE *Jp. J. Appl. Phys.* **27**(11), 2098–2106 (1988).

59. A. Gomyo, T. Suzuki, S. Iijima, H. Hotta, H. Fujii, S. Kawata, K. Kobayashi, Y. Ueno, and I. Hino, Nonexistence of long range order in $Ga_{0.5}In_{0.5}P$ epitaxial layers grown on (111)B and (110)GaAs substrates, *Jpn. J. Appl. Phys.* **27**(12), L2370–L2372 (1988).

60. A. G. Norman, R. E. Mallard, I. J. Murgatroyd, G. R. Booker, A. H. Moore, and M. D. Scott, TED, TEM, and HREM studies of atomic ordering in $Al_xIn_{1-x}As(x \sim 0.5)$ epitaxial layers grown by OMVPE, in, *Microscopy of Semiconducting Materials*, Inst. Phys. Conf. Ser., No. 87, pp. 77–82, IOP Publishing, Bristol (1987).

61. P. Henoc, A. Izrael, and A. Langier, *J. Crystal Growth* **51** 387 (1981).

62. M. M. J. Treacy, J. M. Gibson, and A. Howie, On elastic relaxation and long wavelength microstructures in spinodally decomposed $In_xGa_{1-x}As_yP_{1-y}$ epitaxial layers, *Phil. Mag.* **A51**(3), 389–417 (1985).

63. S. Sharan and J. Narayan, Strain relief mechanisms and the nature of dislocations in GaAs/Si heterostructures, *J. Appl. Phys.* **66**(6), 2376–2380 (1989).

64. Y. Suzuki and H. Okamoto, Transmission electron microscope observation of lattice image of $Al_xGa_{1-x}As$–$Al_yGa_{1-y}As$ superlattices with high contrast, *J. Appl. Phys.* **58**(9), 3456–3462 (1985).

65. T. Furuta, H. Sakaki, H. Ichinose, Y. Ishida, M. Sone, and M. Onoe, Structural evaluation of GaAs/AlGaAs heterointerfaces by atomic-resolution electron microscopy with clear contrast, *Jpn. J. Appl. Phys.* **23**(5), L265–267 (1984).

66. R. Vincent, D. Cherns, S. J. Bailey, and H. Morkoc, Structure of AlGaAs/GaAs multilayers images in superlattices reflections, *Phil. Mag. Lett.* **56**(1), 1–6 (1987).

67. D. Cherns, G. R. Anstis, J. L. Hutchison, and J. C. H. Spence, *Phil. Mag.* **A46**, 849 (1982).

68. J. M. Gibson, R. T. Tung, C. A. Pimental, and D. C. Joy, Interfacial atomic structure and Schottky barrier height, in, *Microscopy of Semiconducting Materials*, Inst. Phys. Conf. Ser., No. 76, pp. 173–181, IOP Publishing, Bristol (1985).

69. D. J. Eaglesham, C. J. Kiely, D. Cherns, and M. Missous, *Phil. Mag.* **A60**(2), 161–175 (1989).

70. A. C. Wright, T. L. Ng, J. O. Williams, M. Missous, and E. H. Rhoderick, to be published.

71. C. J. Humphries, D. M. Maher, H. L. Fraser, and D. J. Eaglesham, Convergent-beam imaging—a transmission electron microscopy technique for investigating small localized distortions in crystals, *Phil. Mag.* **A58**(5), 787–798 (1988).

72. A. J. McGibbon, J. N. Chapman, A. G. Cullis, and N. G. Chew, Microanalysis of III–V semiconductor interfaces, *EMAG '87, Analytical Electron Microscopy Workshop,* (G. W. Lorimer, ed.), Inst. Metals, London, pp. 219–222 (1988).

73. T. L. Ng, M. E. Pemble, J. O. Williams, A. C. Wright, and Z. Zainal, The preparation of plan-view TEM samples of ZnSe epilayers on GaAs(100) substrates by selective photoelectrochemical etching, *J. Microsc.* **150**(1), 31–40 (1988).

74. A. K. Petford-Long and A. J. Long, Electron energy loss spectroscopy applied to III–V semiconductor quantum-well structures; correlated with atomic resolution imaging, *EMAG '87, Analytical Electron Microscopy Workshop,* (G. W. Lorimer, ed.), Inst. Metals, London, pp. 201–204 (1988).

75. A. C. Wright unpublished results.

76. P. M. Petroff, *J. Vac. Sci. Technol.* **14**, 974 (1977).

77. C. S. Baxter, W. M. Stobbs, K. J. Monserrat, and J. N. Tothill, Anomalous intensity in dark field images of (Ga, In)As/GaAs multilayers, *EMAG '87, Analytical Electron Microscopy Workshop,* (G. W. Lorimer, ed.) Inst. Metals, London, pp. 209–212 (1988).

78. E. G. Bithell and W. M. Stobbs, Composition determination in the GaAs/(Al, Ga)As system using contrast in dark field transmission electron microscope images, *Phil. Mag.* **A60**(1), 39–62 (1989).

79. F. M. Ross, E. G. Britton, and W. M. Stobbs, Application of Fresnel fringe contrast analysis to the measurement of composition profiles in GaAs/(Al, Ga)As heterostructures, *EMAG '87, Analytical Electron Microscopy Workshop,* (G. W. Lorimer, ed.), Inst. Metals, London, pp. 205–208 (1988).

80. H. Kakibayashi and F. Nagata, Composition dependence of equal thickness fringes in an electron microscope image of GaAs/Al_xGa_{1-x}As multilayer structures, *Jpn. J. Appl. Phys.* **24**(12), L905–L907 (1985).

81. H. Kakibayashi and F. Nagata, Simulation studies of a composition analysis by thickness fringe (CAT) in an electron microscope image of GaAs/Al_xGa_{1-x}As superstructure, *Jpn. J. Appl. Phys.* **25**(11), 1644–1649 (1986).

82. A. F. de Jong and K. T. F. Janssen, Compositional analysis of Al_xGa_{1-x}As layers with the thickness fringe method on calibrated samples, in, *EUREM '88,* Inst. Phys. Conf. Ser., No. 93, Vol. 2, pp. 153–154, IOP Publishing, Bristol (1988).

83. C. J. C. Hetherington, D. J. Eaglesham, C. J. Humphries, and G. J. Tatlock, TEM compositional microanalysis in III–V alloys, in, *Microscopy of Semiconducting Materials,* Inst. Phys. Conf. Ser., No. 87, pp. 655–658, IOP Publishing, Bristol (1987).

84. A. Olsen, J. C. H. Spence, and P. Petroff, Compositional analysis of III–V interface lattice images, 38th EMSA meeting, pp. 318–319.

85. A. C. Wright and J. O. Williams, Structural evaluation of hetero-epitaxial interfaces by high resolution transmission electron microscopy, *Mater. Lett.* **3**(3), 80–88 (Jan. 1985).

86. T. S. Kuan, W. I. Wang, and E. L. Wilkie, Electron diffraction studies of atomic abruptness of Al_xGa_{1-x}As/GaAs interfaces, *EUREM '84,* Budapest, pp. 113–118 (August (1984).

87. K. Honda, A. Ohsawa, and N. Toyokura, Silicon surface roughness—Structural observation by reflection electron microscopy, *Appl. Phys. Lett.* **48**(12), 770–781 (1986).

88. K. Yagi, Reflection electron microscopy, *J. Appl. Crystallogr.* **20**, 147–160 (1987).

89. R. W. Glaisher, Personal communication; work being carried out at Arizona State University.

90. A. Ourmazd, in, *Defects in Semiconductors* (H. J. Bardeleben, ed.) Vols. 10–12 of *Materials Science Forum,* pp. 735 ff., Trans. Tech. Pub., Switzerland (1986).

91. N. G. Chew and A. G. Cullis, The preparation of transmission electron microscope specimens from compound semiconductors by ion milling, *Ultramicroscopy* **23**(2), 175–198 (1987).

92. J. C. Bravman and R. Sinclair, The preparation of cross-section specimens for transmission electron microscopy, *J. Electron Microsc. Tech.* **1**, 53–61 (1984).

93. N. G. Chew and A. G. Cullis, TEM specimen preparation for semiconductors using iodine ion milling, in, *EMAG '85,* Inst. Phys. Conf. Ser., No. 78, IOP Publishing, Bristol (1985).

94. G. Wagner, A. Dreilich, and E. Butter, Chemical thinning of III–V compound semiconductors for transmission electron microscopy, *J. Matter. Sci.* **23**, 2761–2767 (1988).

95. Ṣ. N. G. Chu and T. T. Sheng, TEM cross section sample preparation technique for III–V compound semiconductor device materials by chemical etching, *J. Electrochem. Soc.* **11**, 2663–2666 (Nov. 1984).

96. R. J. Graham, A high-yield method for the preparation of cleaved wedge specimens through semiconductor devices, *Ultramicroscopy* **27**, 329–332 (1989).

97. S. B. Newcomb, C. B. Boothroyd, and W. M. Stobbs, Specimen preparation methods for the examination of surfaces and interfaces in the transmission electron microscope, *J. Microsc.* **140**(2), 195–207 (1985).

98. S. B. Newcomb, C. S. Baxter, and E. G. Bithell, The preparation of cross-sectional TEM specimens, in, *EUREM '88,* Inst. Phys. Conf. Ser., No. 93, Vol. 1, IOP Publishing, Bristol (1988).

99. A. Garulli, A. Armigliato, and M. Finetti, A preparation technique for TEM cross-sections of test structures with reduced feature size, *Ultramicroscopy* **26**, 295–300 (1988).

100. T. L. Ng, A. C. Wright, and J. O. Williams, to be published.

17

Dielectric Properties and Materials

Robert M. Hill

1. INTRODUCTION

Elsewhere it has been noted[1] that what might well be the first area of experimental physics has benefited least by the quantum revolution. The area in question, of course, was that opened up by Thales of Milerus around 500 BC, when he is reported to have rubbed some amber with fur and noticed that the amber then attracted particles over small, but finite, distances and that this attractive property decayed in time. Frictional electricity, as it came to be called, was investigated by Faraday, who deduced that the effect was due to the induction of charges on the surface of the amber by the action of rubbing and the retention of these surface charges over a period of time. Both these properties are still required of a di-electric, or dielectric— viz., the storage of surface charge and a low rate of dissipation of this charge within the bulk of the material. In more conventional terms we say that a good dielectric has to have a high capacitance and a low leakage current.

2. BACKGROUND

Consider the slab of dielectric shown in Fig. 1. We will assume that the opposing faces of the slab have been charged by amounts $+Q$ and $-Q$ over their areas A. The opposing surface charges give rise, from Gauss's law, to a constant internal

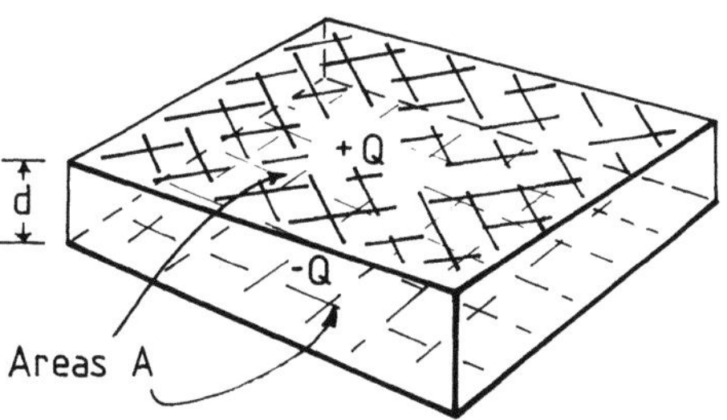

FIGURE 1. Schematic representation of a slab of dielectric of thickness *d* and surface area *A*.

Robert M. Hill ● Department of Physics, King's College London, The Strand, London WC2R 2LS, U.K.

field in the medium, and hence a voltage difference occurs across the thickness of magnitude

$$V = Qd/(\varepsilon\varepsilon_0 A) \tag{1}$$

where d is the thickness between the faces, ε_0 the absolute permittivity of vacuum $(8.854 \times 10^{-12}\ \mathrm{F\,m^{-1}})$ and ε is the relative permittivity, or simply the permittivity, of the material of the slab. At low surface charge densities the voltage induced is proportional to the charge and the capacitance of the slab is given as

$$\frac{dQ}{dV} = C = \varepsilon\varepsilon_0 A/d \tag{2}$$

In applications making direct use of surface charging, such as in photocopying, the quantity of interest is the density of charge per unit surface area. In the simplest terms this controls the amount of powder that can be retained on the surface. From Eq. (1) and (2) we have

$$Q/A = \varepsilon\varepsilon_0 V/d = \varepsilon\varepsilon_0 \mathscr{E} = P \tag{3}$$

where P is the polarization, $\mathscr{E}$ is the electric field, and signs are ignored. In all dielectrics, without exception, there is a maximum value of electric field that can be withstood without causing the material to physically break down and therefore there is a maximum value to this surface charge density.

Alternatively one may wish to use the dielectric to store electrical energy. The energy stored in a capacitor is

$$W = \int Q\,dV = \tfrac{1}{2}CV^2 = \tfrac{1}{2}\varepsilon\varepsilon_0 A\,d\mathscr{E}^2 \tag{4}$$

From which we have that the energy stored, per unit volume, is

$$W/\mathrm{volume} = \tfrac{1}{2}\varepsilon\varepsilon_0 \mathscr{E}^2 \tag{5}$$

From Eqs. (2), (3), and (5) we see that the usable properties of a dielectric are defined by the permittivity and the magnitude of the applied electric field, as long as this is less than the breakdown value for the material. In practice, nonlinearity in the response to high electric fields occurs before breakdown is reached and the specific forms of the above equations would not be expected to apply. The nonlinear properties are of interest in their own right and form the basis of a number of modern optical signal control devices as discussed in other chapters. Here we shall restrict our overview to the linear region of response, which normally applies up to fields of the order of $10^5\ \mathrm{V\,m^{-1}}$, and examine the nature and information contained in the (linear) permittivity.

3. BOUND CHARGE EFFECTS

Equation (2) gives a convenient method of measuring the dielectric permittivity. By definition that of vacuum is unity; typically polymeric materials have low values of permittivity, say from 2 to 4; glasses exhibit a larger response with the permittivity in the range from 5 to 6; while semiconductors show a strong response, with the permittivity of silicon being about 12. Two classes of materials have anomalously high values: one group comprises H_2O and D_2O, both in the solid and liquid states, and the other group is known collectively as ferroelectrics. With the water molecule the center of charge for the hydrogen (deuterium) atoms does not coincide with that of the oxygen atom. The molecule is therefore highly polar and every molecule will contribute to the net polarization. The resultant permittivity is about 80. A similar effect occurs in the solid semiconductors where with no field applied the polarizations induced by the electrons in the covalent bonds and the nucleii charges exactly balance. However, in this case the effect of a field is to induce a nett polarization, which is, however, small as it is due to the difference in opposing effects. The ferroelectrics, which are so termed in comparison with the ferromagnetics, can exhibit permittivities well into the tens or hundreds of thousands. In every unit cell of these crystalline materials there is a small natural polarization as for the H_2O molecule. In the absence of a field these add randomly and there will be no nett effect. However, in the presence of a field the elemental polarizations align and strong cooperative interactions then give the extraordinarily high permittivity values that are observed.

Permittivity is a macroscopic material property. It is the volume-averaged response to an applied external force field. One can, in specific cases, be certain that the observed response is due to a known number of responding elements and hence determine the microscopic individual dipole moment of the elements. If we continue with the assumption that our system is comprised of a number of identical dipolar elements that do not interact with themselves, and only by way of viscous drag forces with the matrix in which they are embedded, then the total polarization will be a measure of the excess number of dipoles that have been forced into alignment by the field. For this ideal, noninteracting, system the rate of decay of the polarization after the removal of the field can only be proportional to the residual polarization; that is

$$\frac{dP}{dt} = -P/\tau \tag{6}$$

from which we have that

$$P = P_0 \exp((-t/\tau) = P_0 f(t) \tag{7}$$

where the zero–time polarization P_0 is given by Eq. (3), τ is the relaxation time and can be associated with the viscous force, and $f(t)$ is a generalized time decay function. If more than one process of polarization relaxation occurs these will be characterized by their magnitude and by their relaxation time. Hence by choosing a suitable time scale in which to examine $f(t)$ we can select particular processes.

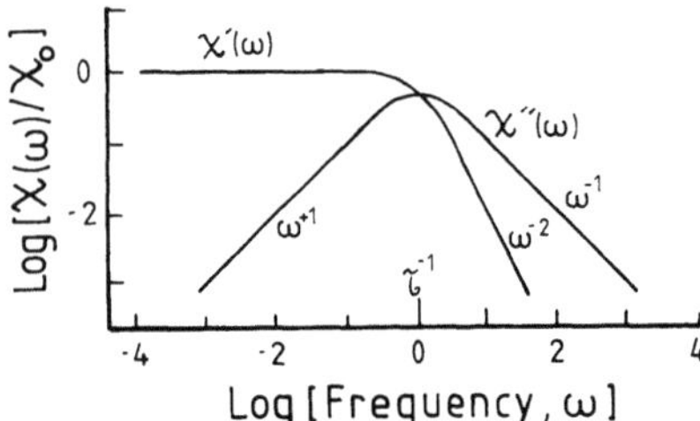

FIGURE 2.   The Debye spectral response of Eq. (8).

In practice it is more convenient to work in the frequency domain. A one-sided Fourier transformation of Eq. (7), after division by $\varepsilon_0 \mathscr{E}$, gives the dielectric susceptibility for one of these processes as

$$\chi(\omega) = \chi_0 (1 + i\omega\tau)^{-1}$$

$$= \chi_0 \left(\frac{1}{1 + (\omega\tau)^2} - i \frac{\omega\tau}{1 + (\omega\tau)^2} \right) \tag{8}$$

a response originally proposed by Paul Debye in 1912[2] and which is given in Fig. 2. Because of the nature of relaxation the frequency component is complex. In the figure the response has been plotted on log/log scales in order to emphasize the power laws inherent in the relationships.

The frequency-dependent permittivity is the summation of the individual susceptibility responses from the frequency being considered up to the fastest process possible and includes the relative permittivity of vacuum. Hence we can write

$$\varepsilon(\omega) = \sum_i \chi_i(\omega) + 1 = \chi(\omega) + \varepsilon(\infty) = \varepsilon'(\omega) - i\varepsilon''(\omega) \tag{9}$$

in which a single process has been singled out and all the others, at higher frequencies, summed into an effective "infinite frequency" permittivity value $\varepsilon(\infty)$.

3.1. Experimental Data

Unfortunately few of the experimental data on typical solid dielectric systems show the Debye behavior characterized by Eq. (8). Polymeric materials generally meet our requirement for low charge leakage and typical data is shown in Figs. 3

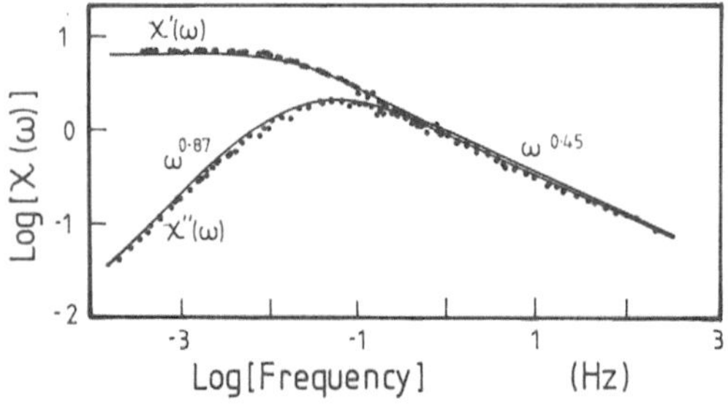

FIGURE 3. The spectral response of poly(vinyl acetate) from the data reported in Ref. 3. The fractional power law behaviors are indicated.

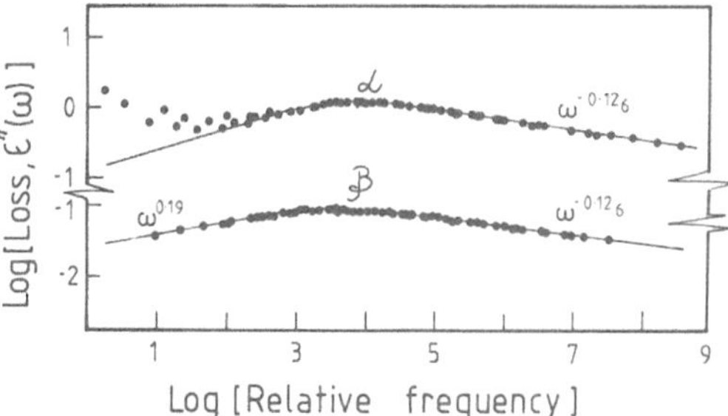

FIGURE 4.   The spectral responses of the loss components of the α and β relaxation processes in poly(vinyl chloride) from the data reported by Ishida.[5]

and 4. The first of these, Fig. 3, gives a temperature normalized response for poly(vinyl acetate) as reported by Johnson *et al.* in 1981.[3] In constructing this plot from the data a value for $\varepsilon(\infty)$ of 3.47 has been assumed and the data from the temperature range 306–349 K have been assembled into a single master plot.[4] It is obvious that the imaginary component of the permittivity, $\varepsilon''(\omega)$, which is commonly termed the dielectric loss, has a much wider response than that indicated in Fig. 2. Furthermore the frequency exponent values of the Debye response now appear as fractional. The two loss peaks observed by Ishida[5] in poly(vinyl chloride) are indicated in Fig. 4. As before these plots have been assembled to give extended master responses and these are of the same nature as the loss characteristic in Fig. 3. The higher-temperature process has been termed α and the lower-temperature process β following the conventional notation.[6] Taking the frequency at which the maximum in the loss occurs as a characteristic frequency for the relaxation process Fig. 5 shows the temperature dependence of relaxation in an Arrhenius plot. The low-temperature β response is activated with an activation energy of 0.6 eV and an infinite-temperature intercept with the frequency axis at 2×10^{14} Hz, typical of a lattice vibration frequency. The α plot, however, is strongly curved with a characteristic frequency which becomes zero at a finite temperature, T'_g, a true glass transition temperature[7] at which the molten polymer solidifies on cooling. Without, at present, looking at the details, it is apparent that the dielectric responses are, at least, related to the nature of the physical structure of the matrix in which the responding dipoles

FIGURE 5.   The temperature dependencies of the α and β relaxation peaks from poly(vinyl chloride) from the data reported in Ref. 5. The temperature T'_g defines that at which the relaxation rate for the glassy α process becomes zero.[7]

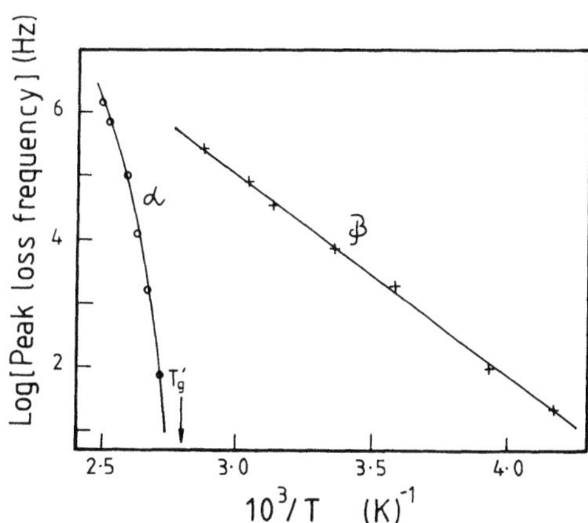

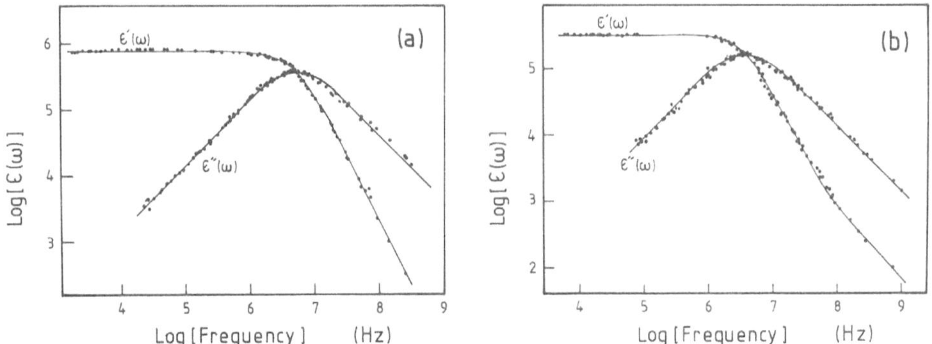

FIGURE 6. Ferroelectric responses. (a) The spectral response for caesium dihydrogen phosphate in the region of the Curie temperature at 156 K. A Debye plot (cf. Fig. 2) has been superimposed on the data that were reported by Deguchi et al.[8] (b) The spectral response for $CsH_{0.04}D_{1.96}PO_4$ in the region of its Curie temperature at 267 K. A distortion of the almost perfect Debye response can be seen at the highest, normalized, frequencies. From the data reported in Ref. 8.

exist and can be used to give information about the structural relaxation of the medium.

The second group of materials that showed anomalously high permittivity values were the ferroelectrics. In Fig. 6 two sets of data, both measured on ferroelectric samples, are presented from the data reported by Deguchi et al.[8] The first of these, Fig. 6a, is for a sample of CDP (caesium dihydrogen phosphate) and the second, Fig. 6b, from a similar sample where 98% of the hydrogen had been replaced with deuterium. The former shows an almost perfect Debye characteristic, following the Debye response superimposed on the data, while the latter exhibits a break from the Debye form only at the highest-frequency region. The parallel plots for the real and imaginary components of the permittivity are a necessary consequence of fractional power law behavior, as indicated in Fig. 3, and indicates that the high-frequency exponent is $-(1 - 0.035)$. It is only with exceptionally good experimental data that one is able to discern this small deviation between the data sets. The data were measured in temperature ranges around the ferroelectric Curie temperature, 156 K for CDP and 267 K for C(D/H)P and the difference in temperature indicates a difference in the dynamics of the two structures due to the heavier deuterium atoms. The plots have been scaled at the Curie temperatures and the magnitudes there achieve values between 10^5 and 10^6, with the more perfect CDP having the larger response. Of particular interest to us is the observation that it is in the spectral response of the two samples that we can see, most clearly, the small effect that composition imperfection has induced.

4. MOBILE CHARGE EFFECTS

Having examined, however briefly, the role of bound, dipolar, charges in dielectrics it is useful to return again to basics when we come to consider mobile charges. D. K. Davies, in a fascinating piece of work in the 1960s,[9] examined the nature of contact charging at the surface of a range of dielectrics. The work functions of a range of metal probes were determined relative to that of a standard gold surface and then repeated contacts were made to the dielectric under consideration by each of these in turn. When the work function of the metal contact is low, negative charge flows from the metal into the dielectric giving a negative surface charge on the latter. Conversely, a metal of high work function results in a positive surface charge on the dielectric after contact. Equilibrium of the surface charge could only be achieved after a number of contracts, and the results for two separate experiments are shown in Fig. 7. In these figures the range of equilibrium charge values gives an indication of the order of local inhomogeneities in either work function and/or contact area. Nevertheless, there are clear trends and the intercept of the best fit line through the data and the zero charge axis gives the measure of the work function value for the dielectric. With a metal the definition of the work function is clear; it is the energy required to extract an electron from the Fermi surface of the metal and place it at infinity. A similar definition applies for a semiconductor even when the Fermi surface lies well within the forbidden gap and hence we may assume that this experiment has shown that one may consider dielectrics to have Fermi surfaces as well.

It was found possible, however, to use the technique in a different manner. The charge induced onto the surface is attracted by its own field into the bulk of the material. As it moves the center of charge becomes displaced from the free surface and the surface potential decays. By following the decay the effective mobility of the charge cloud can be determined and, typically these were found to be very small, in the range 10^{-17}–10^{-13} m^2 V^{-1} s^{-1}. In the terminology of the introduction the retention of the surface charge is because it cannot move away with any speed. But, as Mott has shown,[10] for charges to move within bands there has to be a minimum mobility of the order of 10^{-4} m^2 V^{-1} s^{-1} or else the mean free paths are less than the interatomic distances. The charges in the dielectrics cannot be "free" in the semiconductor or metal sense but are constrained to move by localized hopping between trapping centers. This is in agreement with the observation that, on the whole, dielectrics are transparent to visible light, the normal white milky appearance being due to scattering by the crystalline/amorphous structure. Hence the forbidden gap would have to be in excess of at least 4 eV and the probability of obtaining intrinsic conduction would become negligible. The differences between

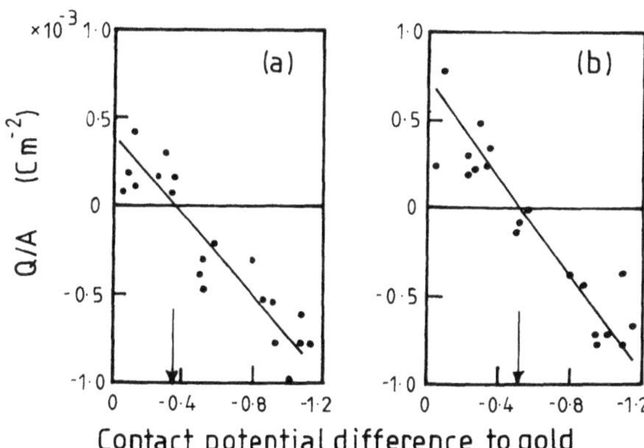

FIGURE 7. Plots of the measured surface charge as a function of relative work function (with respect to a gold standard) for (a) polycarbonate and (b) nylon. From Ref. 9.

dielectrics and semiconductors, and particularly amorphous semiconductors,[11] is quantitative rather than qualitative.

As an example of this we show in Fig. 8 two dielectric response plots of data that were measured from a p/n junction by Barsony and Jonscher.[12] Such a junction forms, in principle, a perfect dielectric sample once the charge carriers have been frozen out. The p and n regions make ideal contacts with no discontinuity in structure and the width of the depletion region, which forms our sample, can be varied by imposing a dc bias voltage on the probe signal. As shown in the figure, at zero bias the response is large, d is small, and with a negative bias the response decreases by almost two orders of magnitude. In neither case do we see the Debye loss characteristic. It appears in fact that the shape of the loss response may be a function of bias. From this we can deduce that the electrodes may not play a major role in determining the form of the spectral response.

4.1. Experimental Data

In principle there is no difficulty in considering a dielectric that contains mobile charges and hence supports a current. It may be convenient to consider the carriers as ionic, but this is not necessary. The conductivity can be represented in Eq. (9) by the addition of a term to the loss, i.e.,

$$\varepsilon(\omega) = \varepsilon'(\omega) - i\varepsilon''(\omega) - i\sigma/(\omega\varepsilon_0) \tag{10}$$

where σ is the conductivity. A convenient case to examine is that of a semiconductor in which the carriers have not been frozen out. Here the conductance will be in parallel with the capacitance and can be picked out by its inverse frequency signature. In Fig. 9a we give a slightly more complex sample configuration in which the low-frequency part of the response is dominated by a Schottky barrier, which is electrically in series with the InP semiconductor bulk. The Schottky barrier itself has a parallel conductance so that the complete electrical description is of the form given as an inset to Fig. 9b and the calculated response is the main feature of that figure. In dielectric terminology this is a Maxwell–Wagner element with series resistance capacitance.[13]

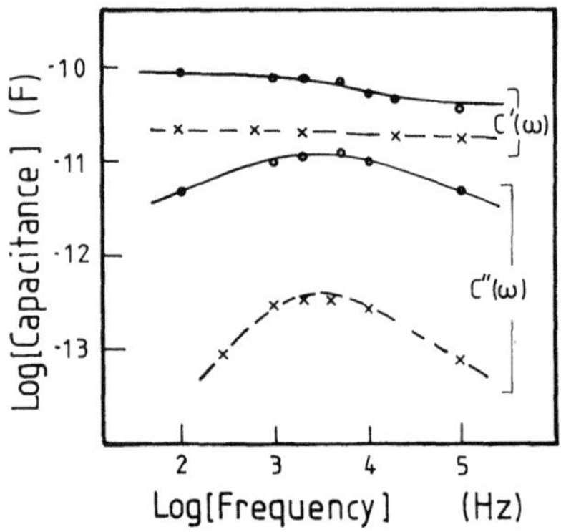

FIGURE 8. Spectral responses measured on a p–n junction in silicon. ——, zero bias; ---, −6 V bias. From the data reported by Barsony and Jonscher in Ref. 12.

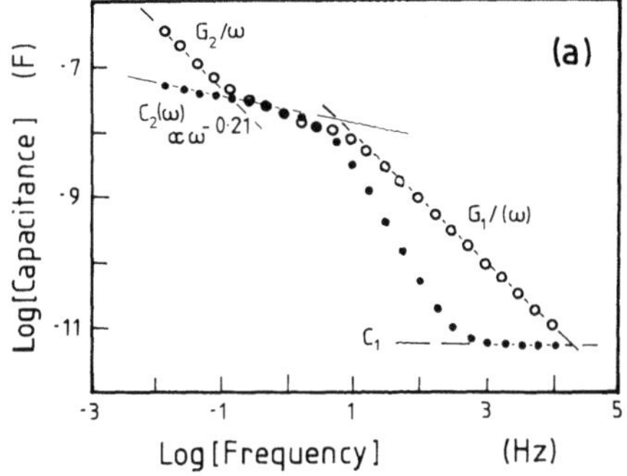
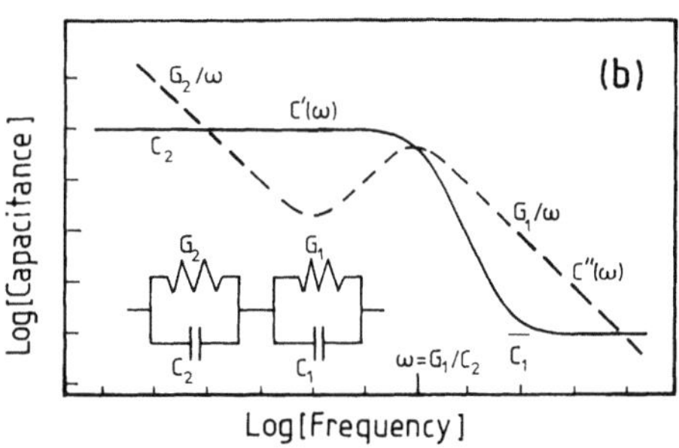

FIGURE 9.   (a) The measured dielectric response for a sample of semi-insulating InP and an aluminum contact that gave a Schottky barrier. The sample was 0.3 mm thick and the high doping gave a Schottky depletion width of only 30 nm. (b) Schematic representation of a series connected pair of parallel resistance–capacitance circuit elements and their frequency response. The plot shows how the response in (a) can be related to the individual components of the sample.

With the double logarithmic plots that we have been using the individual elements of the response in Fig. 9b can be related directly to the components that give rise to them. Returning to Fig. 9a it can be seen that the broad features of expected behavior are present but that the capacitance of the Schottky barrier is frequency dispersive. In the particular case investigated the doping level in the sample was high and the width of the Schottky barrier no more than 30 nm—about 10^{-4} of the bulk sample thickness—and this has given rise to the very large Schottky dispersion.

The ability to differentiate between "bulk" and "barrier" regions in a noninvasive manner and to be able to cope with large changes in the magnitude of capacitance are strengths of the dielectric technique and are, at present, being applied to such complex physical systems as gels and emulsions[14,15] in order to determine both the microscopic and macroscopic natures of these heterogeneous systems. In all such investigations one makes use of the series conductance–capacitance properties of the system in order to examine it at different levels of scale.

A different class of materials, which are also of interest because of their unusual charge transport properties, are the fast ion conductors. These are essentially insulating matrices within which large, on an atomic scale, channels exist in the structure. Ionic charges can travel along these channels relatively unimpeded by the structure and high conductivities have been reported. These have led to consideration of the structures as a basis for solid state batteries. As an example of the type of dielectric response exhibited by these materials, Fig. 10 presents data measured on a sample of β-alumina.[16] The perfect response for a battery would be the parallel capacitance and conductance as seen in the frequency range between 10^3 and 10^4 Hz in Fig. 9a, but here we see gigantic dispersions in both capacitance and loss at low frequencies. This has been termed the "anomalous low-frequency" dispersion by Jonscher[17] and arises from an imperfect charge transport within the bulk of the material.[18] The observation that the capacitance exhibits significant dispersion indicates a bulk polarization and consequently the rate of energy recovery will be low so that the current that such a material could supply would be limited.

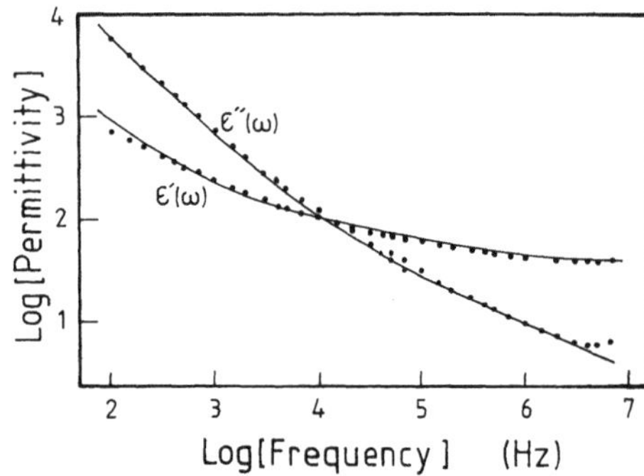

FIGURE 10.   An example of the anomalous low-frequency dispersion exhibited by Na β-alumina. From the data reported in Ref. 16.

These large dispersions are quite common at low frequencies and can be expected in materials in which ionic transport takes place. One would expect that at sufficiently low frequencies the effect of barriers might be dominant, as in Fig. 9, but if such barriers exist their capacitance is very high (which might indicate that their nature has to do with charge double layers) and the frequency below which they can be observed is very low.

5. THE CLUSTER MODEL OF RELAXATION RESPONSE

Even this brief overview of the spectral response of a range of dielectric materials has been sufficient to show that the Debye model is of limited applicability. A number of empirical spectral response functions have been proposed but most of these fail as they either consider symmetrical loss characteristics or loss characteristics with a low-frequency exponent of $+1$. Over the last few years a fully cooperative relaxation model has been proposed by Dissado and Hill,[18,10] and it is this relaxation function that has been used here to fit all the experimental data.

The cluster model of relaxation response has been derived from a purely quantum mechanical approach. It is considered that the action of a field is to supply energy to a responding element such as a dipole. As long as the field is present the dipole retains this energy. On removal of the field the dipole relaxes as an oscillator but sharing the energy with an increasing number of other, not necessarily dipolar, elements as time evolves. Making the assumption that the cooperation can be described as a resonance of the responding elements, the energy that is available will be shared equally among them, and this defines a cooperative cluster. Letting the original energy stored by E_0 and the fraction of energy transferred to the cluster be n, then when the cluster contains N elements each of these will have energy

$$E_N = nE_0/N \qquad (11)$$

and the density spectrum of energies is then

$$N(E)\, dE = (nE_0/E)\, dE \qquad (12)$$

which leads to an infrared divergence in the number of responders at small energy. Following Hopfield[20] it can be shown that once the cluster has begun to develop the time decay is of the form

$$f(t) \propto t^{-n}, \qquad 0 \leqq n \leqq 1 \tag{13}$$

which, on transformation, gives a high-frequency decay in the form

$$\chi(\omega) \propto \omega^{n-1} \tag{14}$$

However, it is not possible to consider only one single relaxation event. Other centers will also form clusters which develop in parallel as long as they do not lose energy to the heat bath. Once this does occur then there is a change in the nature of relaxation and cooperation is now between the clusters and occurs by way of phonons. The onset time, that is, the time taken to couple to the phonon bath, is likely to be thermally activated and is equivalent to the characteristic rate shown in Fig. 5. The nature of the cooperation between clusters is of the same form as that within the clusters, but we now have decay and restoration leading to a dynamic equilibrium.[19] Essentially the intracluster exchanges form a second energy bath. Letting the efficiency of the exchanges between clusters be m, the long-time development of relaxation takes the form

$$f(t) \propto t^{-(1+m)} \tag{15}$$

for which the equivalent frequency domain response is

$$\chi''(\omega) \propto \omega^{m} \propto \chi(0) - \chi'(\omega), \qquad 0 \leqslant m \leqslant 1 \tag{16}$$

as already indicated in the figures.

Equations (14) and (16) are only asymptotic approximations to the full spectral response, which can be written, most conveniently, in terms of the Gaussian hypergeometric function, $_1F_2(\ ,\ ;\ ;\ )$,[21] as

$$\chi(\omega) = \chi_0(1 + ix)^{n-1} {}_2F_1\left(1 - n, 1 - m; 2 - n; \frac{1}{1 + ix}\right) \tag{17}$$

where $x = \omega/\omega_c$ with ω_c the characteristic frequency, m and n are as defined above, and χ_0 is the magnitude of the susceptibility response. All the curves plotted through the experimental data here have been obtained using Eq. (17) and the values of m and n determined from the asymptotic relationships (14) and (16), except in the case of Fig. 10, where an equivalent relationship for the imperfect charge transport case to that of Eq. (17) was used.[18]

The importance of the cluster model approach is twofold. At the simplest level it gives excellent fits to the observed experimental data over wide ranges of the parameters m and n, and hence over a wide range of spectral forms, and reduces the data to four specific parameters, viz., χ_0, ω_c, m, and n. The ability to reduce data to quantitative numbers is of real advantage in experimental investigations of a comparative nature, such as the examination of the properties of complex systems. Of fundamental importance, however, is the ability to define both short-range and long-range coupling efficiency parameters by a noninvasive technique. This yields a completely new insight into the nature of the relaxation response of materials and, once fully understood, is capable of forming structural descriptions of real significance.

For example, it is already obvious that broad descriptions of classes of materials can already be made from the evidence that has been presented here in the figures. The classic Debye dipole has $n = 0$ and $m = 1$; that is, there are no local interactions, each particles exists by itself and cannot form a cluster, and $m = 1$ defines[22] an almost exponential distribution of intercluster events, i.e., a random noise process. This is precisely the terms in which the Debye model was proposed. In the polymers we have m small and n large. Hence energy can be transported locally along the polymer backbone, but each single, folded, molecule is poorly coupled to the others. In this context we note that the n values for the α and β processes in PVC were identical. This suggests that as the chains break up on melting the individual remnants have the same local structure as the original molecule. The perfect ferroelectric has each unit cell identical and hence no exchange can take place, n is zero, and with no cluster formation m is unity. The imperfect material appears locally different but macroscopically identical. Even the $p-n$ junction reveals that electrically there are imperfections and it is likely that these are due to the compensated centers, which will have a strong dipolar nature. It is interesting to observe that the nature of the response as well as its magnitude are functions of the bias voltage.

6. CONCLUSIONS

The application of modern physical concepts to dielectric relaxation theory has allowed description of the relaxation in terms of a strongly cooperative model. It has been shown that the nature of the cooperation is to weaken the conventional, single-particle, relaxation behavior so that fractional power law dispersions are observed. Conversely it has been shown that examination of the dielectric relaxation response as a function of frequency can be used to obtain information about both short- and long-range structural coherence of the medium in which the dielectrically active element resides.

REFERENCES

1. R. M. Hill and A. K. Jonscher, The dielectric behavior of condensed matter and its many-body interpretation, *Contemp. Phys.* **24**, 75–110 (1983).
2. P. Debye, *Polar Molecules*, Dover, New York (1945).
3. G. E. Johnson, E. W. Anderson, and T. Furukawa, Fourier transform dielectric spectroscopy utilizing a real time executive (RTE) system, *IEEE Trans. Conf. Elec. Insul. Diel. Phenom.* 258–263 (1981).
4. R. M. Hill, Thin film dielectrics—The future, *Thin Solid Films* **100**, 319–323 (1983).
5. Y. Ishida, Studies in dielectric behavior of high polymers, *Kolloid-Z.* **168**, 29–36 (1960).
6. G. P. Johari, Glass transitions and secondary relaxations in molecular crystals, *Ann. N.Y. Acad. Sci.* **279**, 117–140 (1976).

7. R. M. Hill and L. A. Dissado, The temperature dependence of relaxation processes, *J. Phys. C: Solid State Phys.* **15**, 5171–5193 (1982).

8. K. Deguchi, E. Okaue, and E. Nakamura, Effect of deuterization on the dielectric properties of the ferroelectric CsH_2PO_4, *J. Phys. Soc. Jpn.* **51**, 3569–3582 (1982).

9. D. K. Davies, Charge generation on dielectric surfaces, *Brit. J. Appl. Phys. (J. Phys. D.)* **2**, 1533–1537 (1969).

10. V. F. Mott, in *Electronic and Structural Properties of Amorphous Semiconductors* (P. G. Le Comber and J. Mort, eds.), pp. 1–54, Academic, London (1973).

11. N. F. Mott and E. A. Davis, *Electronic Processes in Noncrystalline Materials*, Oxford U.P., Oxford (1979).

12. I. Barsony and A. K. Jonscher, Dielectric properties of silicon p-n junctions, *Solid State Electron.* **21**, 471–473 (1978).

13. R. M. Hill and C. Pickup, Barrier effects in dispersive media, *J. Mater. Sci.* **20**, 4431–4444 (1985).

14. L. A. Dissado, R. C. Rowe, A. Haidar, and R. M. Hill, The characterization of heterogeneous gels by means of a dielectric technique. I. Theory and preliminary evaluation, *J. Colloid Interface Sci.* **117**, 310–324 (1987).

15. R. C. Rowe, L. A. Dissado, S. H. Zaidi, and R. M. Hill, The characterization of heterogeneous gels by means of a dielectric technique. II. Formulation and structural considerations, *J. Colloid Interface Sci.* **112**, 354–366 (1988).

16. D. P. Almond, G. K. Duncan, and A. R. West, The determination of hopping rates and carrier concentrations in ionic conduction by a new analysis of ac conductivity, *Solid State Ionics* **8**, 159–164 (1983).

17. A. K. Jonscher, Low frequency dispersion in carrier-dominated dielectrics, *Phil. Mag.* **B38**, 587–601 (1978).

18. L. A. Dissado and R. M. Hill, Anomalous low frequency dispersion—A quasi-dc response, *J. Chem. Soc. Trans. Faraday Soc. 2* **80**, 291–317 (1984).

19. L. A. Dissado and R. M. Hill, A cluster approach to the structure of imperfect materials and their relaxation spectroscopy, *Proc. R. Soc. London Ser. A* **390**, 131–180 (1983).

20. J. J. Hopfield, Infrared divergence, X-ray edges, and all that, *Commun. Solid State Phys.* **11**, 40–49 (1969).

21. M. Abramowitz and I. A. Stegun, *Handbook of Mathematical Functions*, Dover, New York (1970), Chap. 15.

22. R. M. Hill, L. A. Dissado, and R. Jackson, The examination of correlated noise, *J. Phys. C: Solid State Phys.* **14**, 3915–3925 (1981).

18

Xerographic Photoreceptors

Michael G. Carter

1. INTRODUCTION

1.1. Background

It is 50 years since the first "xerographic" copy was produced by Chester Carlson (October, 1938) in Astoria, New York. Since that time, there have developed several different generic types of xerographic photoreceptors, but all have employed the same empirical process of electrophotography.

The development pattern for those types that were commercially successful in the rapidly expanding world of the xerographic office copier, has a common thread— modification to the base chemical composition in order to meet the increasingly severe requirements of the modern business environment.

This chapter will introduce the principles of the xerographic process and discuss the characteristics of the "ideal" photoreceptor.

1.2. Generic Types

The generic photoreceptor types covered are selenium, cadmium sulfide, zinc oxide, amorphous silicon, and organic photoreceptors. Future developments, not yet fully realized, are outside the scope of this chapter.

Comparisons between the various devices are made and an assessment of industry trends is discussed, based on the anticipated needs of the "Office of the Future."

2. THE XEROGRAPHIC PROCESS

2.1. Schematic

The basic process is shown pictorially in Fig. 1.

Michael G. Carter • Rank Xerox Limited, Mitcheldean, Gloucestershire GL17 ODD, U.K.

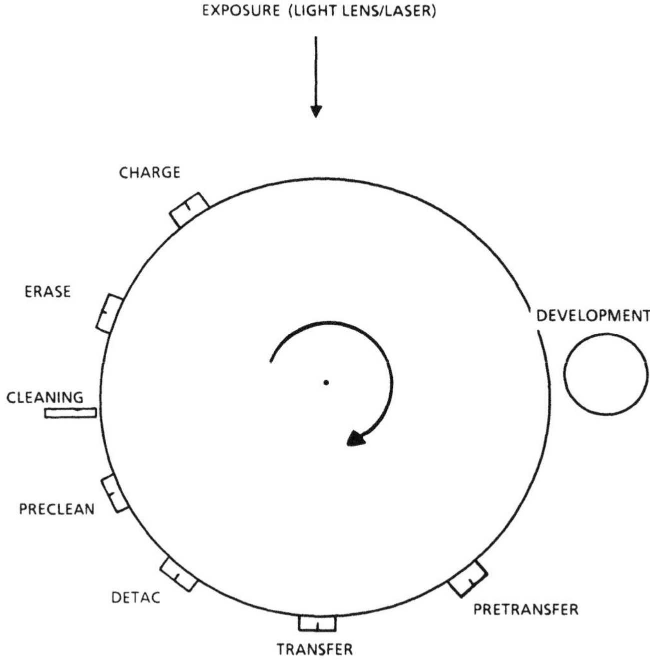

FIGURE 1.   The xerographic cycle, starting with the charge-up of the photoreceptor.

The function of each sequential step is as follows, with those that are refinements (and not necessarily present in the actual copier/printer) marked with an asterisk:

Charge	Use of corotron to generate corona discharge onto photoreceptor (up to 1000 V, ± polarity).
Exposure	Imagewise light exposure from light lens or laser source.
Development	Movement of toner/liquid ink from carrier material to photoreceptor by differential charging.
Pretransfer(*)	Corotron/lamp used to minimize electrostatic forces that cause developed toner to adhere to the photoreceptor surface.
Transfer	Movement of toner from photoreceptor surface to paper, using corotron to generate differential charging.
Detac(*)	Added charge to paper to aid removal from the photoreceptor where flexibility geometry does not facilitate self-stripping.
Preclean(*)	Balanced charge applied to loosen residual toner on photoreceptor surface.
Clean	Physical wipe on photoreceptor surface by brush or doctor blade to remove residual toner.
Erase (*)	Lamp exposure to remove residual voltage on photoreceptor surface before charge process recommences.

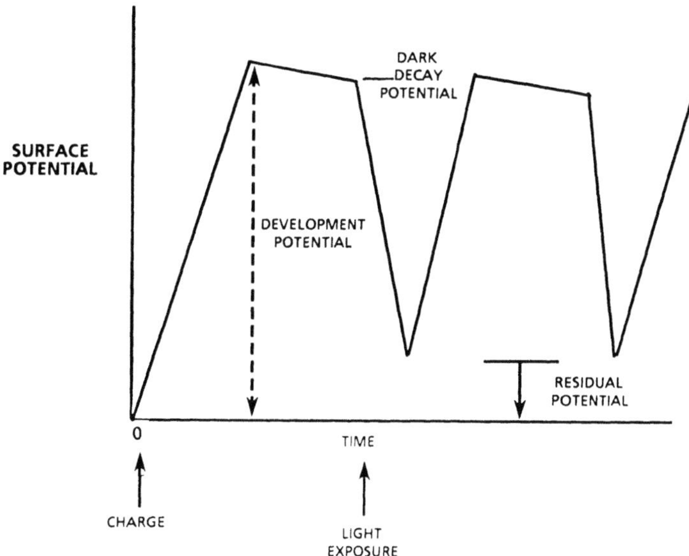

FIGURE 2.   Photoreceptor surface potential during the xerographic cycle.

2.2. Electrical Requirements

The electrical changes to the surface of the xerographic device are illustrated in Fig. 2. The basic charging requirements of the photoreceptor are as follows:

- Rapidly accept a high level of charge (sufficient surface traps/effective blocking substrate).
- Little dark decay (charge retention in the dark preexposure interval).
- Quick discharge of its potential when exposed to light.
- Little residual potential after exposure.

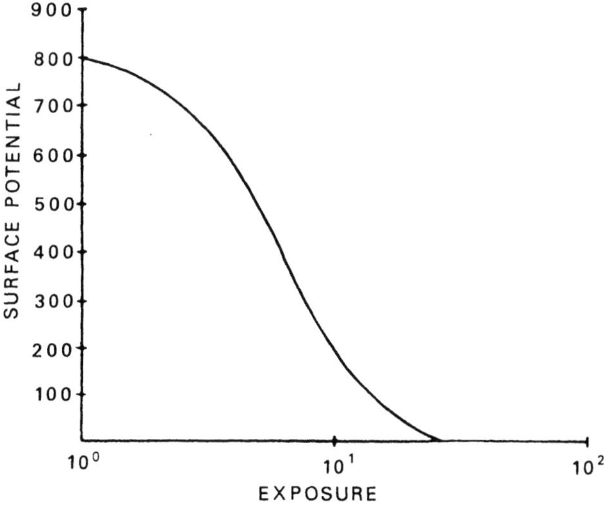

FIGURE 3.   Example plot of a PIDC for a xerographic photoreceptor.

In addition, cycle-to-cycle electrical stability is an important parameter. However, there are other equally important criteria for a photoreceptor to be a viable commercial entity, often contradictory in terms of processability.

A comprehensive description of the mechanisms in play during the xerographic process is given by Baker-Self,[1] in which is described how the various levels of the photoreceptor structure interact to provide the versatility and performance required. The photogenerator layer allows input photons to be absorbed in the photoreceptor and create electron–hole pairs, while the charge transport layer allows movement of these free holes and electrons through the photoconductor to neutralize surface charges of opposite polarity.

A plot of surface potential as a function of the magnitude of the input exposure is called the photoinduced discharge curve (PIDC), the functional form of which is one of the most important xerographic parameters (Fig. 3).

3. THE IDEAL PHOTORECEPTOR

3.1. Operating Characteristics

The current photoreceptor has to satisfy at least 10–12 operating criteria, to a greater or lesser degree, depending on the market segment it will be operating in (low, medium, or high, in relation to copies per minute).

These criteria range from electrical performance (i.e., light sensitivity, process speed, cycle stability), through cost effectiveness (cost/copy, life versus machine life) to configuration (drum–belt, geometrical flexibility) and environmental aspects (disposability).

The ultimate capability—that of all essential requirements maximized in one photoreceptor—can never be realized, especially when it is not uncommon for the same product to exhibit totally different operating characteristics (e.g., life, copy print defects) in different hardware systems.

Fatigue factors include light, heat, moisture, electric field, and chemical exposure. To ensure, however, that key features are maintained, only limited changes can be made in attempts to reduce performance variations and degradation by these fatigue factors.

Figure 4 illustrates the designers spider web matrix for the ideal photoreceptor, illustrating the complexity of these requirements.

4. SELENIUM ALLOY PHOTORECEPTORS

4.1. Amorphous Selenium

There was a gap of some 20 years between Carlson's initial discovery and the advent of commercially viable xerographic copying systems. These employed amorphous selenium photoreceptors (late 1950s).

While amorphous selenium has the photoelectric properties necessary for xerography, its stable crystalline state does not, and the gradual transition to this latter structure often aided by physical damage to the working surface, selenium being relatively soft, remains one of the product's inherent weaknesses—that of

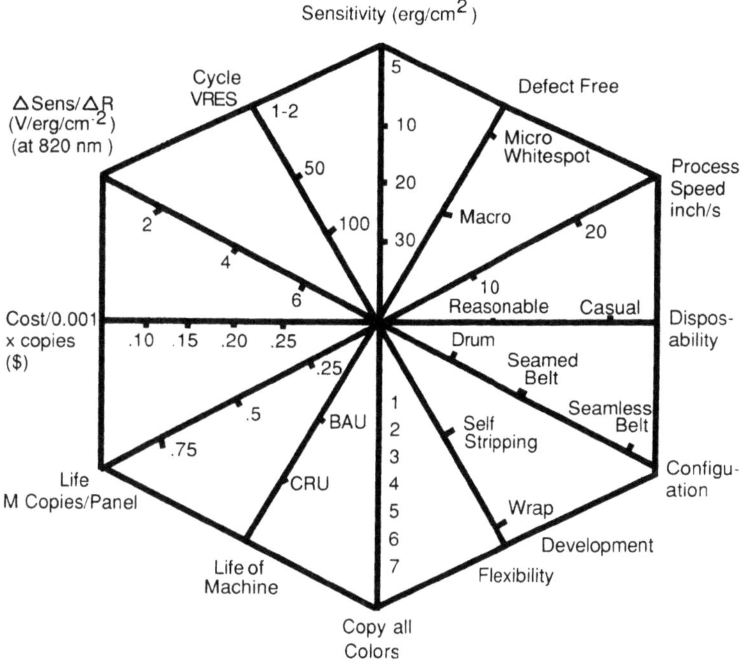

FIGURE 4.   The ideal photoreceptor.

short life. Depending on copying process speed, a life of only 30–500 k panels can be obtained.

Addition of up to 0.5% arsenic is made to reduce the rate of crystallization.

The photoreceptor consists of a single layer of Se, 50–70 μm thick, laid down by vacuum disposition in a planetary or in-line coating configuration, onto either a rigid aluminum or flexible nickel belt substrate.

Selenium has low panchromaticity, having effective spectral response restricted to 400–550 nm. This ensures poor response to blue/infrared light and diode/helium-neon laser emissions. However, it does have good xerographic cycling properties [process speed greater than 30 in./s (75 cm/s), low dark decay—50 V/s—low residual voltage coupled with correspondingly high contrast potential]. Disposability is an increasingly difficult subject, owing to the basic toxicity of selenium.

4.2. Arsenic Tri-Selenide

For more than 10 years, the basic selenium arsenic alloy remained the main source of xerographic photoreceptors. The early 1970s saw the introduction of arsenic triselenide (As_2, Se_3) as an addition to the range, aimed at greater panchromaticity and longer life (greater resistance to crystallinity). Lives in excess of one million copies can be obtained.

With arsenic contents as high as 50%, the effective spectral response ranges from uv to 800 nm (near infrared), bringing He–Ne laser capability. There are, however, xerographic property issues, namely, early life dark decay and light fatigue, overcome by aging and shielding, respectively. Process speed is improved by small additions of, e.g., Iodine (<0.3%) to improve hole mobility.

4.3. Selenium–Tellurium

The late 1970s saw the advent of selenium–tellurium alloys and multilayer structures. The substrates—rigid aluminum/flexible nickel belt—remained the same, but the working xerographic layer became split into two, separating out the charge transport layer (low arsenic selenium alloy at 50-70 μm) and generator Layer ($\sim$1 μm, up to 15% tellurium-selenium alloy). The tellurium content is varied to suit the photosensitivity requirements, on a product-by-product basis. The spectural response is effective from uv to 650 nm, thus enhancing the basic blue rendition of selenium. There can be an optional (top) layer of tellurium–selenium with a small amount of arsenic, designed to protect the photoreceptor from environmental impacts.

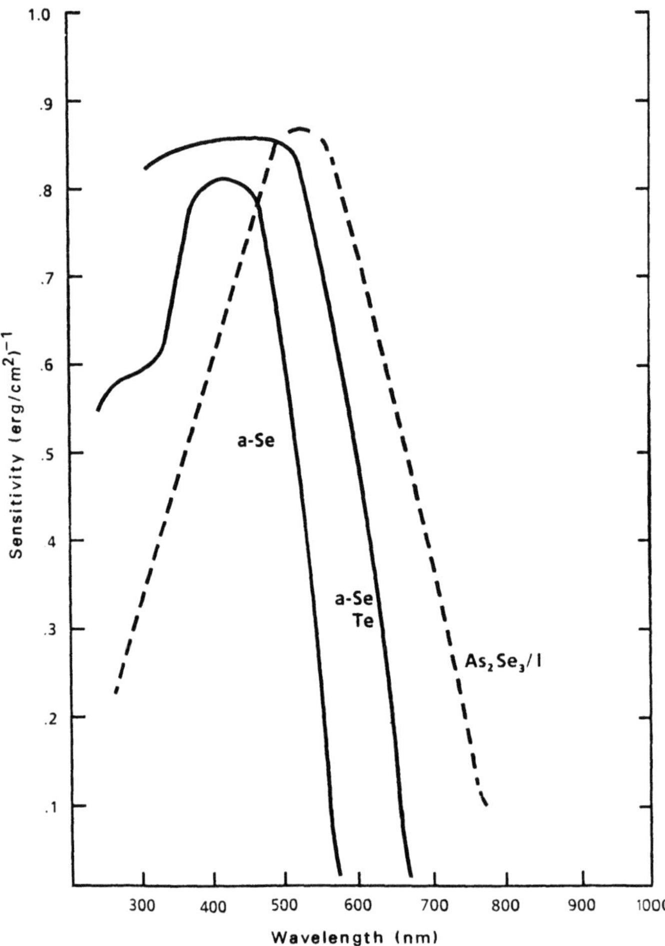

FIGURE 5. Spectral response for selenium family of photoreceptors.

In comparison with selenium, Te-Se devices have slightly less crystallization, better xerographic cycling stability (after break-in), but need more time between exposure and development. For a comparison of spectral response for the three types of selenium alloy, see Fig. 5.

5. CRYSTALLINE CADMIUM SULFIDE

5.1. Cadmium Sulfide Photoreceptor

In the early 1970s the Japanese entered the office copier market, basically at the low volume (copies per minute) end. The photoreceptor used was based on either selenium or cadmium sulfide. This device is significantly different from the selenium alloy device, the working thickness of cadmium sulfide being only 0.3 μm, obtained by dipping or spraying onto an aluminum substrate. This thin film leads to a very low contrast potential (charge potential achieved minus residual charge after exposure) in the order of 40 V, necessitating a very sensitive development system, such as liquid or powder cloud, as opposed to the more robust systems employed in the selenium alloy process.

Cadmium sulfide has a low relative spectral response, not normally effective beyond 600 nm. However, doping can extend this range to enable infrared or laser diode imaging (Fig. 6).

5.2. Market Share

Despite these inherent drawbacks, the worldwide population in 1984 was estimated to be 0.8 million, or some 14% of the market for photoreceptors,[2] production still being mainly by Japanese manufacturers.

Later in this chapter it will be seen how the Industry will be moving away from cadmium sulfide and indeed selenium alloy, into organic photoreceptors. The cadmium sulfide share is expected to reduce to 11% of the world's population, but to grow to 1.4M, in 1989, such being the overall growth of the office copying phenomenon.[2]

6. ZINC OXIDE PHOTORECEPTORS

Formulated for a very specialized segment of the market, the zinc oxide photoreceptor has failed to make any real impact when compared to the previous devices discussed, and with the advent of low-cost organic photoreceptors it will increasingly disappear. This device is effective only in the lowest-volume product range, where its very low cost offsets an inherent life of 1000/panel.

The substrate is aluminized polyester, the overall package having one other advantage, that of disposibility. Inherent problems with the effect of humidity, low xerographic process speed, and limited spectral response (see Fig. 6), further restrict its usefulness, though the latter problem is helped by dye sensitization to enhance spectral response at the higher wavelengths.

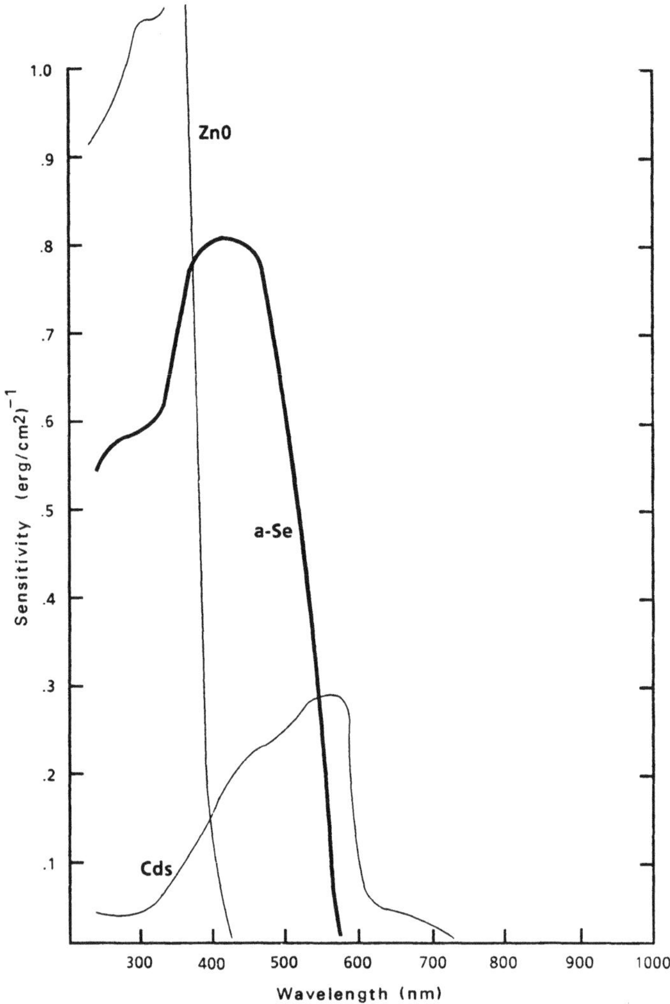

FIGURE 6.   Spectral response of cadmium sulfide and zinc oxide, compared with amorphous selenium.

7. AMORPHOUS SILICON PHOTORECEPTORS

7.1. Amorphous Silicon

This generic type of device is also discussed by LeComber in his chapter on amorphous silicon materials (Chapter 11), and this section should be read in conjunction with that chapter.

The development of amorphous silicon photoreceptors follows a similar pattern to that of selenium, the use of the basic chemical being followed by iterations aimed at enhancing the basic characteristic of the device. In this case a-Si (p type), first developed in the early 1970s, has been followed by the hydrogenated version (a-Si:H), and then its current multilayer structure involving doping (usually boron) and overcoating for environmental passivation.

However, this is the extent of the similarity between this type of device and those of the selenium family.

7.2. Manufacturing Process and Properties

Manufacturing is by glow discharge (plasma decomposition) of silane into disilane in dc/rf/microwave plasma onto a rigid substrate. This process has major disadvantages: slow deposition rate of about 10–15 μm per hour compared with the 60 μm for selenium; need for tight environmental control within the process to ensure low levels of surface defects. This leads to a high process acquisition cost, which explains the relatively low market penetration rate, despite the advantages of the device. These include high panchromaticity, the sensitivity at 800 nm being equivalent to As_2Se_3 at 650 nm. Mechanical hardness (10 times that of selenium), coupled with high resistance to environmental aspects (heat and solvents), often leads to life in excess of one million panels.

7.3. Hydrogenated Amorphous Silicon

1985 saw the introduction of a-Si:H. The chemical structure changes to band gaps, localized states, and dangling bands, are described by Mort, in his paper on amorphous silicon photoreceptors.[3]

Doped a-Si:H has a similar spectral response to that of a-Si (Fig. 7). The multilayer device employs a boron-doped a-Si:H barrier layer, lightly doped (B) bulk layer, often with a-Si:N_x or a-Sic$_x$ overcoat for chemical passivation.

7.4. Market Share

Table 1 discusses the advantages and disadvantages of the a-Si:H photoreceptor. Historically the type has been utilized by only a few manufacturers, again Japanese restricted, and only time will tell whether the efforts to improve the manufacturing process and a move toward the high-volume systems, where the intrinsic long life of the device will be relevant, will lead to larger market share.

8. ORGANIC PHOTORECEPTORS (OPC)

8.1. Manufacturing

Developed in the early 1980s the photoreceptor material set is dependant on the manufacturing process chosen, explaining the wide range of available types in today's marketplace. In addition, the chemical constituent ratio within any given photoreceptor family can be varied to achieve the specific requirements of the xerographic system, more so than with the selenium group of products.

Two types of substrate are employed for this range of device. The first is polyester, the organic layers progressively laid down by continuous web coating, a process similar to that employed in the photographic film industry. The other substrate is the conventional aluminum, although here the substrate is thinner, partially enabling the low unit cost of the type of device required at the low end of

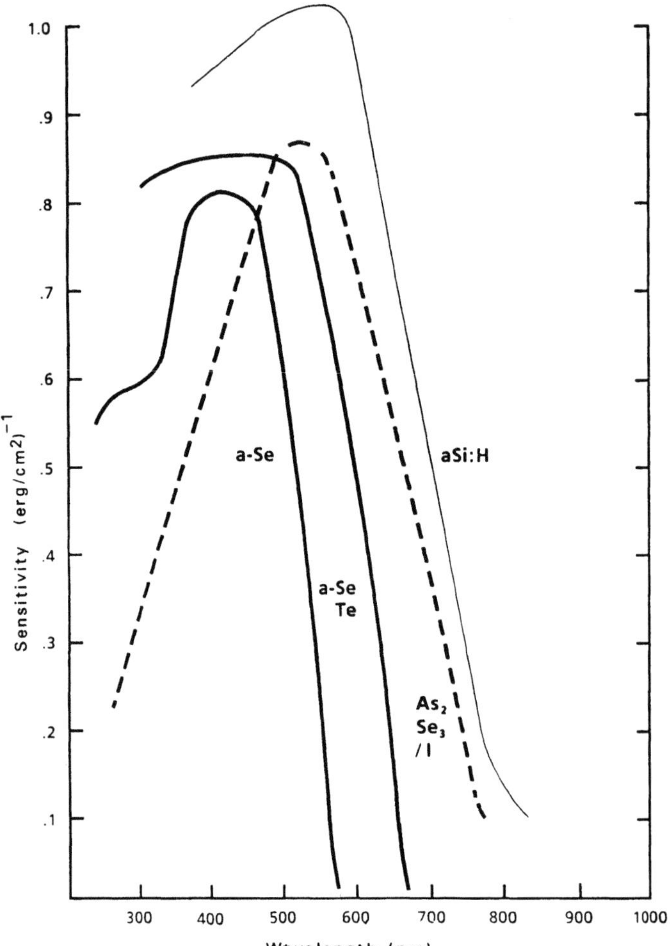

FIGURE 7.   Spectral response for a-Si:H.

TABLE 1.   a-Si:H Characteristics

Advantages	Disadvantages
Resistance to crystallization 500–600°C (Se, 55°C; As₂Se₃, 180°C)	Cost
Hardness	Manufacturability
Cycling stability	Lack of flexibility (rigid substrate only)
Disposability	High surface defect levels (print quality)
High-purity raw material availability	Deletions with corona/humidity
Long life	

the market. Various processes for the coating of aluminum substrates are available, although dip or air spray are usually the preferred routes.

8.2. Xerographic Characteristics

The xerographic process for an organic photoreceptor is essentially the same as that employed by a selenium alloy device, with the exception that here the photoreceptor has negative polarity (selenium is usually positive).

The earliest devices had effective spectral response in the 400–700-nm band, but currently the trend is to add a constituent, typically vanadyl phathalocyanine (VPC), to extend the effective performance to 850 nm, enabling infrared diode laser exposure (Fig. 8).

When compared to a-Si:H and As_2Se_3, organic devices in general have lower life due to process fatigue and poor wear characteristics. Typically, they will achieve

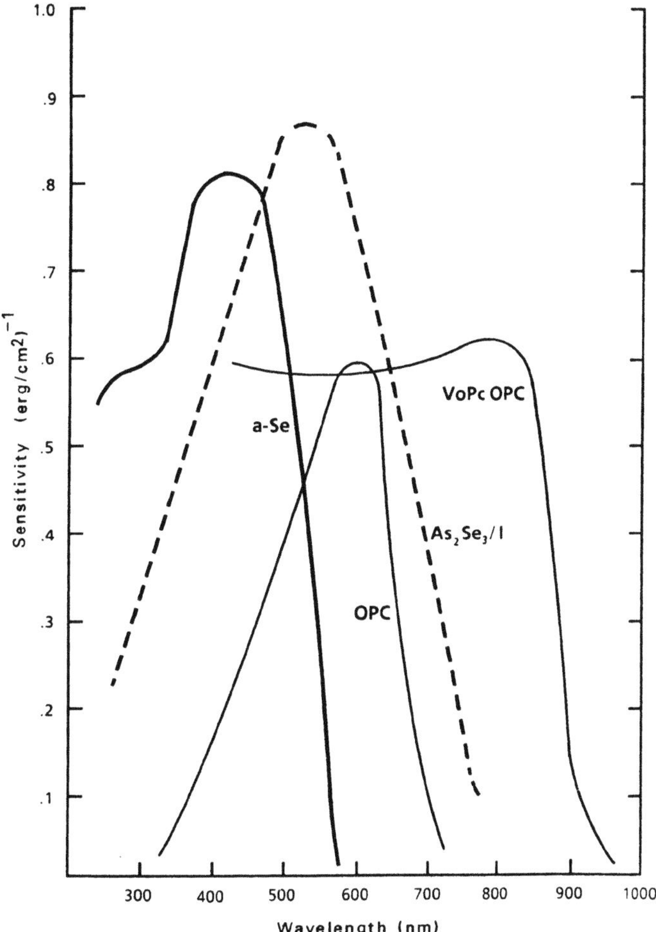

FIGURE 8. Spectral response for organic photoreceptors (OPC, and VPC OPC).

10–25k copies on (low cost) low volume and up to 200–300k on high volume, multilayer products.

8.3. Structure

The generic device is a two-layer structure, binder/charge generator layer and organic charge transport layer. The unit cost is low compared with selenium alloys, making it ideal for an increasing trend in disposable cartridges. Recent indications are for a development of single-layer structures, enabled by the inclusion of a charge generating pigment in the transport layer, this further reducing the basic cost.

The multilayer web photoreceptor typical structure is shown in Fig. 9 (five layers).

The materials that are used in organic photoreceptors are diverse (binder-generator, and transport layers), when considering the number of derivations reported in the literature. Table 2 examplifies these materials.

It should be noted that the organic generating pigments tend to be more difficult to prepare in the high purity usually required for the xerographic process, and indeed pigments of the same chemical structure but differing crystal structure can exhibit very different xerographic properties.

8.4. Market Share

It has been obvious for some time that total organic photoreceptor usage would overtake that of selenium alloys in the 1986–1988 timeframe, and it is thus useful to compare the relative shortfalls of the two types (Table 3).

The most recent major change in direction in the industry has been the introduction of laser reprographics (high-speed printing). The requirements for this process

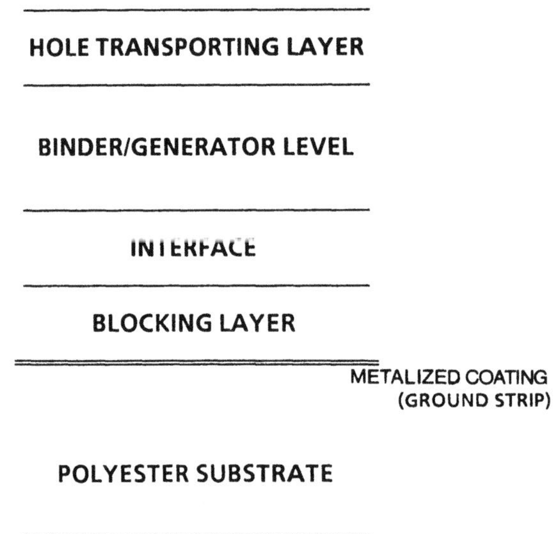

FIGURE 9.   Schematic structure of multilayer web coated OPC.

TABLE 2. Organic Photoreceptor Materials

Substrate	Polyester web	Aluminum drum
Groundstrip	Aluminum	Titanium
Binder/ generator	Trigonal selenium phthalocyanines (i.e., VPC); polycyclic quinones (i.e., dibromo-anthanthrone); squarylium	a-Si Azo compounds (bis, tri, tetra); pyrylium salts
Transport	M-tolyl bipenyl diamine; polysilane; phenyl stilbenes	Polyvinyl ketone; aromatic amines (i.e., hydrazones, pyrazolines, oxadiazoles)

are detailed in the following section, but an illustration of how the photoreceptor manufacturer is modifying structures to speed up xerographic response capability by increasing hole mobility is given in Fig. 10 and the effects on the various parameters in Fig. 11 (see Ohta[4]).

9. LASER REPROGRAPHICS

9.1. Operating Characteristics

Up to the arrival of the laser in the field of xerographics, exposure of the charged photoconductive surface had been by a light-lens system.

Now that laser exposure is available, the spectral response requirements are significantly changed in that the (current) device—diode laser—works at 780 nm,

TABLE 3. Selenium Alloy/Organic Photoreceptors Shortfalls

Selenium alloys	Organics
Spectral response (ir)	Life/mechanical hardness but suitable for cartridge use overcoating for protection
Cost/copy ($0.10–$0.15/copy)	Spectral response (ir) can be achieved by material substitution
Nondisposability	Cost/copy trend towards lower cost devices offset by cartridge (serviceability).
Cartridge compatability	Chemical contamination
Manufacturability	High volume
Crystallization/cold flow	Marginal flexibility (stress-free mode).
Mechanical hardness	High acquisition cost
Light shock (but reversible)	Cycle 1–2 stability
	Life (approaches life of machine only at lower copy volumes) Improving trend
	Mechanical hardness
	Chemical contamination

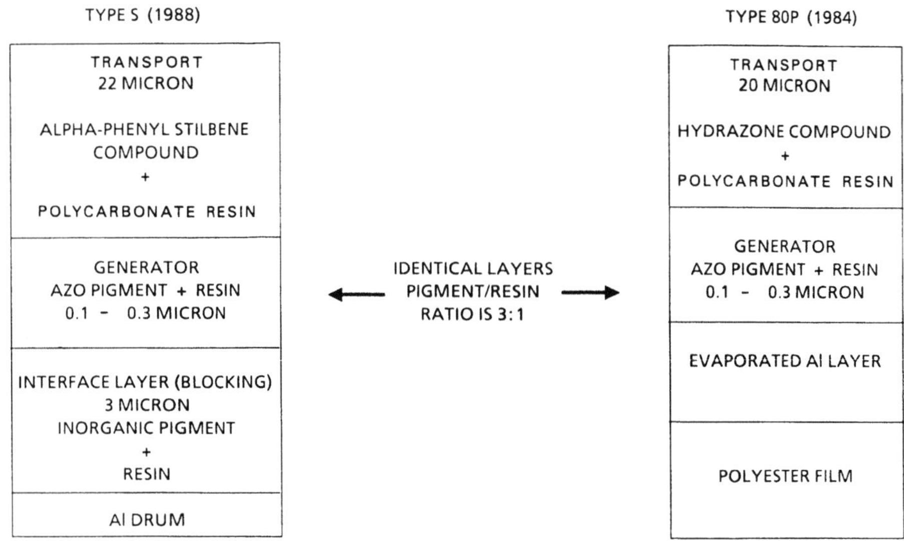

FIGURE 10. Schematic structure for laser reprographics enabled OPC (Reproduced by kind permission of Diamond Research Corporation.)

putting the selenium alloy-cadmium sulfide photoreceptor out of contention for this process.

The diode laser is attractive to the designer in terms of very low cost ($10–20) and long life (20–30 kh) but a recent advance is towards devices operating at 680 nm, which would bring arsenic tri-selenide into reach, while avoiding use of the other currently available lasers, all of which are very high cost/relatively low life: helium–neon (633 nm), argon (488 nm), helium–cadmium (442 nm).

The laser reprographics area is projected to be the fastest-growing segment of the copying market, especially in the 11 plus pages per minute printer range.

The intrinsic properties of the photoreceptor in laser printing systems are very similar to those required in the original xerographic system developed some 30 years previously, with the exception of operability in the infrared region. To restate these, they are high contract potential (high charge acceptance, coupled with low residual—postexposure—voltage), low dark decay (good voltage retention in the dark, pre-exposure interval), high xerographic—photoresponse—speed and high life (low process and environmental fatigue, mechanical strength, wear resilience)

As described in previous sections, no type of photoreceptor is available to provide all these requirements at maximum levels and the art of compromise is still required.

10. INDUSTRY TRENDS

10.1. Volume

1988 Dataquest[2] surveys continue to indicate that organic photoreceptors will increasingly dominate the market at the low and high volume ends as indicated in Table 4.

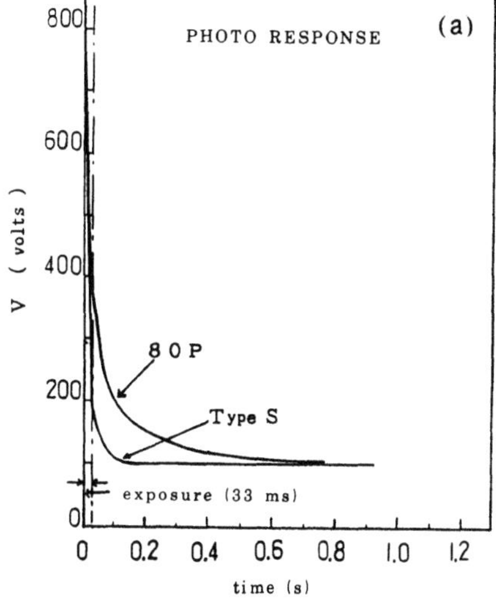

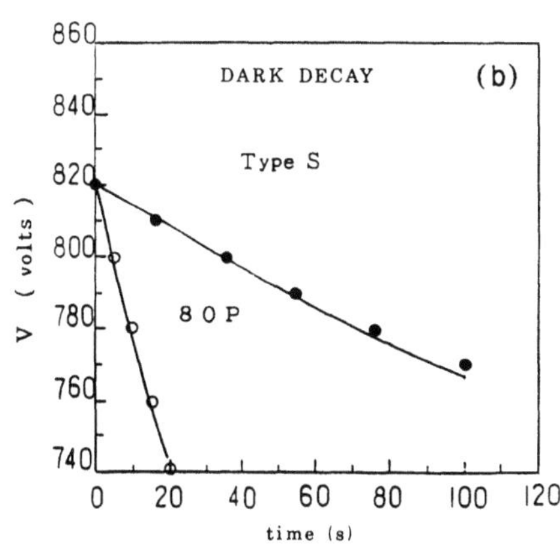

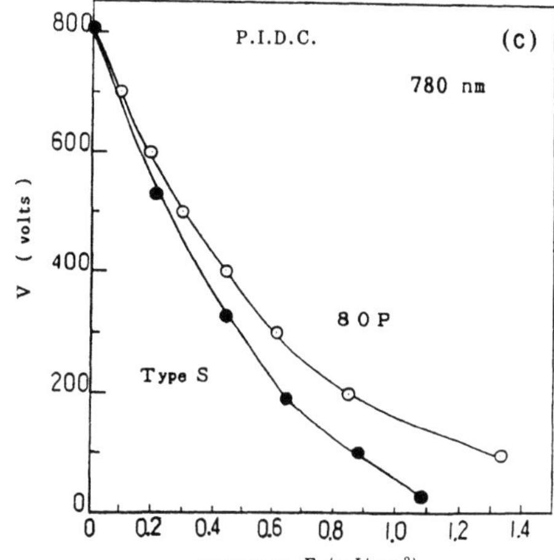

FIGURE 11. Xerographic characteristics comparison between conventional and laser reprographics enabled OPC. (Reproduced by kind permission of Diamond Research Corporation.)

The trend toward organics is accelerating faster than earlier predicted, as in 1984, the 1989 population for organic photoreceptors was expected to be only 6.8 m.

The overall worldwide annual growth rate is expected to slow down as the available market becomes saturated, being 15% in 1982–1987, and anticipated to be 7% in 1987–1992. The growth rate will be significantly higher in printers, with respect to conventional copiers.

TABLE 4. Worldwide Volume Trends

	Million units		
	1984	1987	1992
Organics	1.3	8.8	21.2
Selenium alloys	4.5	3.3	2.5
Cadmium sulfide	0.8		

TABLE 5. Photoreceptor Industry Trends; 1986 Market—A Partial Survey

P/R Type	Number of products	Number of suppliers	Trend
Se–Te	415	53	<<
Se–As	37	13	<<<
As_2Se_3	88	21	<
Cds	86	16	<<
ZnO	41	12	<<
A-Si:H	5	2	?
O.P.C.	85	28	>>>

The use of cartridge (low/mid-volumes) will be increasingly prominent, owing to their customer "friendliness." However, it is expected that selenium alloy devices will remain viable for at least the next five years, especially in the mid-volume product range (30–70 copies/minute).

Correspondingly, photoreceptor revenue, estimated at \$2.3 billion, will have increased to \$4.2 billion in 1992.

10.2 Technology

In terms of technology advance, the work on inorganic photoreceptors will concentrate on the achievement of longer life, infrared capability and for a-Si:H, higher manufacturing speeds (microwave disposition).

For organics, there will be increased emphasis on polycyclic quinones (generator layer), overcoating, and transport/generator layer optimization for xerographic electrical properties.

10.3 Market Share

In conclusion, Table 5 illustrates the 1986 photoreceptor coverage from a survey of some 750 products, including variants. This illustrates the high level of selenium-based systems at that time and subsequently how quickly organic photoreceptors have made their presence felt.

REFERENCES

1. T. P. J. Baker-Self, Xerographic process, Lecture to the Paper, Printing, and Publishing Group, Institute of Physics (April 1987).
2. Dataquest Copying and Duplicating Industry Conference, Sqaw Peak, Phoenix, Arizona, (March, 1988).
3. Dr. J. Mort, From selenium to silicon photoreceptors: A comparative study, Fourth Annual Photoreceptor Industry Conference, Diamond Research Corporation, Santa Barbara, California (April, 1988).
4. K. Ohta, A high-sensitivity OPC for laser diode printers, Fourth Annual Photoreceptor Industry Conference, Diamond Research Corporation, Santa Barbara, California (April, 1988).

19

Piezoelectric and Pyroelectric Materials and Their Applications

R. W. Whatmore

1. INTRODUCTION

The phenomenon of piezoelectricity, the release of electric charge under the application of mechanical stress, occurs in all noncentrosymmetric materials. Pyroelectricity, the release of charge due to a material's change of temperature, occurs in all materials that belong to a polar crystal symmetry class. It should be noted that, as not all noncentrosymmetric classes are polar (222, 4, 422, $\bar{4}$2m, 32, 6, 622, $\bar{6}$m2, 23, and $\bar{4}$3m are the nonpolar, noncentrosymmetric classes), not all piezoelectric crystals are pyroelectric. However, all pyroelectric crystals are piezoelectric. (As a final note, crystals belonging to the noncentrosymmetric cubic class 432 are nonpiezoelectric because all the piezoelectric moduli vanish under the operation of the symmetry elements.) Ferroelectrics form a subset of the set of pyroelectrics because they are polar materials in which the direction of the polar axis can be changed by the application of an electric field. As a consequence they are both pyroelectric and piezoelectric. As many of the largest pyroelectric and piezoelectric effects occur in ferroelectric materials, they have become very important technologically. Furthermore, the fact that the polar axis can be reoriented by the application of a field means that polycrystalline ferroelectrics in which the crystallographic axes of the component crystallites are randomly oriented can be made to show polar properties by applying a sufficiently large electric field (the process of "poling").

There are a very wide range of ferroelectric materials open to the potential user, from oxide single crystals such as lithium niobate ($LiNbO_3$) (which is widely used in, for example, piezoelectric surface acoustic wave (SAW) devices for rf, VHF, and UHF filtering and signal processing) through oxide ceramics such as the lead zirconate titanate family (which is used in an enormous range of piezoelectric devices and pyroelectric sensors) to polymers such as polyvinylidene fluoride—PVDF (which is used in both piezoelectric and pyroelectric devices).

This chapter reviews the basic pyroelectric and piezoelectric effects and considers one or two examples of how they can be used in electronic devices.

R. W. Whatmore • Plessey Research Caswell Limited, Towcester, Northamptonshire NN12 8EQ, U.K.

283

2. PIEZOELECTRIC MATERIALS AND DEVICES

The direct piezoelectric effect can be described as follows:

$$P_i = d_{ijk} \cdot \sigma_{jk} \tag{1}$$

where P_i is the change in the dielctric polarization due to the application of the stress σ_{jk}. (This equation uses the repeated suffix notation such that all repeated suffixes are repeated over the range 1–3.) As P_i is a vector and σ_{jk} is a second-rank tensor, d_{ijk} is a third-rank tensor.

Nye[1] has described the crystal physics of the effect. d_{ijk} has 27 independent coefficients, although in most cases crystal symmetry markedly reduces these. Using the reduced suffix notation, Eq. (1) reduces to

$$P_i = d_{ij} \cdot \sigma_j \qquad (i = 1\text{–}3; j = 1\text{–}6) \tag{2}$$

The converse piezoelectric effect describes the generation of strain (ε_j) in a piezoelectric due to the application of an electric field E_i, viz;

$$\varepsilon_j = d_{ij} \cdot E_i \tag{3}$$

Passive piezoelectric devices making use of the direct effect for sensing, for example, sound or mechanical displacement have been known for many years. Such devices include hydrophones, microphones, phonograph pickups, etc. Active devices making use of the converse effect for generating sound include buzzers (modern watch alarms use a disk of piezoelectric ceramic bonded to the back face of the watch as the sounding element), loudspeakers (particularly at high frequencies), and depth sounders and sonar transducers. The converse effect is also used in bulk and surface wave piezoelectric resonators for frequency stabilization and filtering over a very wide range of frequencies. Both active and passive devices are extensively discussed by Herbert.[2]

Piezoelectric polymers offer an interesting example of ferroelectric materials with unusual electrical and mechanical properties, which have been developed to make possible some novel device configurations. PVDF is a polymerized form of vinylidene fluoride ($CH_2 = CF_2$). As formed, the polymer chain takes the form shown in Fig. 1a. The structure of the polymer consists of spherulitic crystals embedded in an amorphous matrix. Although the carbon–fluorine bond is polar, the arrangement of the molecules in the unit cell of the crystallites, as shown in Fig. 1b, gives a nonpolar structure. This is called alpha-phase PVDF. If the polymer is stretched and electrically poled by applying a field, the molecules are reoriented to give the structure shown in Fig. 2, called beta-phase. This is polar and exhibits piezoelectric properties. The material has been applied to a range of active and passive devices (see Wang et al.[3]).

It has been noted that the need for mechanical stretching in this material is a disadvantage, as it limits the shapes that can be made, and presents considerable problems when thick sections (say, >1 mm) of material are required. However, it is possible to make copolymers of vinylidene fluoride with trifluoroethylene (VDF-TrFE). The inclusion of a few percent TrFE causes steric hindrance on the polymer

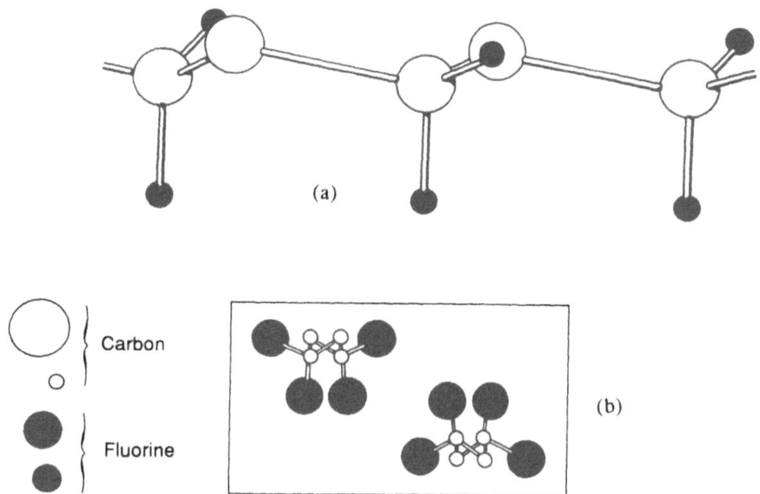

FIGURE 1.   Structure of alpha phase PVDF.

chain, which forces the material to crystallize directly into the beta phase without any requirement for mechanical stretching, although of course electrical poling is still required to make the material piezoelectric and polar. Thus the material offers the possibility for fabricating a wider range of devices because the full range of polymer shaping techniques can be applied to it (e.g., solvent casting, injection moulding, vacuum forming, etc). Table 1 compares the piezoelectric properties of

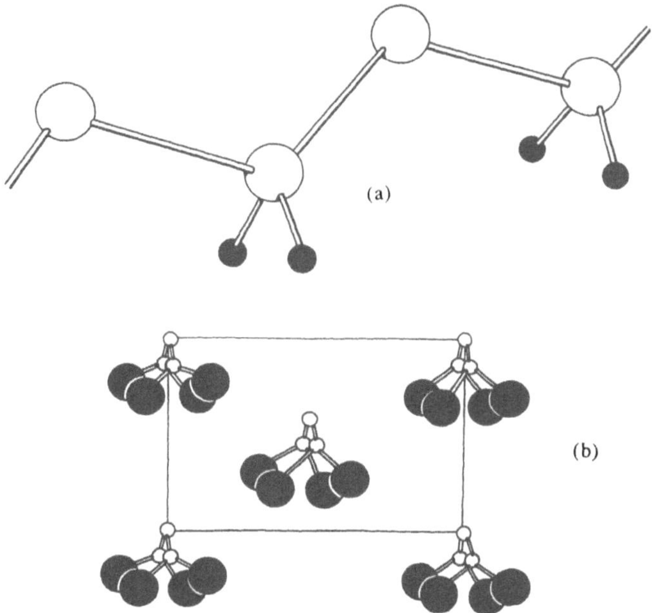

FIGURE 2.   Structure of beta phase PVDF.

TABLE 1. Comparison of Piezoelectric Properties of PVDF and PVDF-TrFE Copolymers

Property	PVDF (uniaxially stretched)	56:44 VDF:TrFE solvent cast	70:30 VDF:TrFE solvent cast	80:20 VDF:TrFE solvent cast	56:44 VDF:TrFE solvent cast
$T_c(°C)$	>Melting temperature	70	99	121	99
$\tan \delta$	0.02	0.021	0.025	0.022	0.025
ε_r	11	11	10.7	9.6	10.7
$d_{31}(pC/N)$	17.9	21.4	17.1	10.0	32.6
$d_{32}(pC/N)$	0.9	21.4	17.1	8.6	20.9
$d_H(pC/N)$	−8.3	−9.2	−6.4	−7.6	−8.8
$d_{33}(pC/N)$	−27.1	−52	−40.6	−26.2	−62.3
Aging in d_{31} at 70°C (percent/time)		65% in 1 h	23% in 24 h	13% in 24 h	23% in 24 h

PVDF and various VDF-TrFE copolymers. It can be seen that the 70:30 copolymer exhibits roughly twice the activity of PVDF. As an example of a type of microphone that is quite difficult to make using PVDF, Fig. 3 shows a dome microphone formed by vacuum forming a copolymer sheet over a hemispherical former. The dome (16 mm in diameter by 0.7 mm thick) is poled electrically, then afixed to a backing plate. External pressure variations acting on the dome cause stresses in the dome wall, which are converted into an electrical voltage output. The performance of this microphone with frequency is shown in Fig. 4. The output is very high and flat with frequency, with some resonances at around 10 kHz. Coating the device with a thin layer of lacquer reduces the output to some degree, but damps out the resonances. It is considered that PVDF and the copolymers have the potential for application to a wide range of piezoelectric sensors.

3. PYROELECTRIC MATERIALS AND DEVICES

The pyroelectric effect is described by

$$P_i = p_i \cdot \Delta T \tag{4}$$

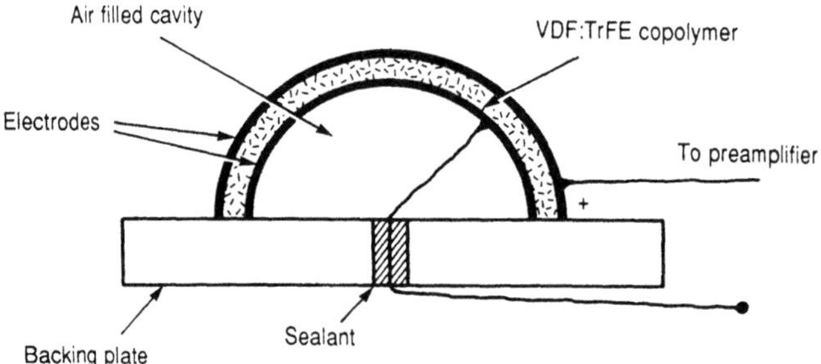

FIGURE 3. Air-backed copolymer dome microphone.

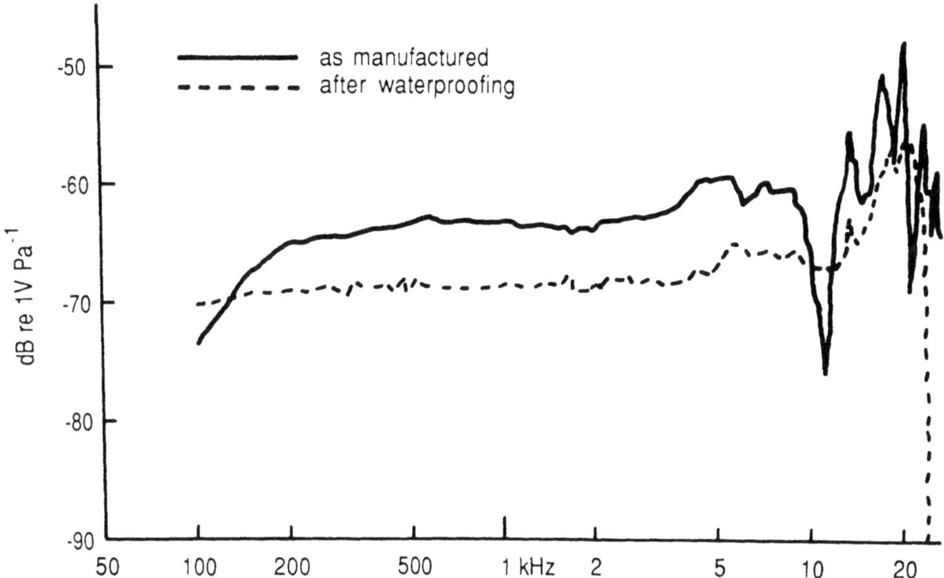

FIGURE 4. Frequency response of dome microphone, as manufactured and after coating with a waterproofing lacquer.

where P_i is the polarization change due to a change in temperature ΔT. p_i is the pyroelectric coefficient, which is a vector.

The effect and its applications have been extensively reviewed by Whatmore.[4] The effect of a temperature change on a pyroelectric material is to cause a current i_p to flow in an external circuit, such that

$$i_p = Ap(dT/dt) \tag{5}$$

where A is the electroded area of the material, p the component of the pyroelectric coefficient normal to the electrodes, and dT/dt the rate of change of temperature with time.

Pyroelectric devices detect changes in temperature in the sensitive material and as such are detectors of supplied energy. It can be seen that the effect is a temporally dependent one and that in order to make a pyroelectric device work, it is necessary to modulate the source of energy. As energy detectors, they are most frequently applied to the detection of incident electromagnetic energy, particularly in the infrared waveband from 8–14 μm, for which there is an atmospheric window. This is also the waveband in which the black-body emission by objects in the region of 300 K peaks. Such devices are used for applications such as intruder detection, fire prevention, energy conservation, pollution monitoring, and thermal imaging. The selection of a pyroelectric material depends strongly upon the application. It is possible to formulate a number of figures of merit which describe the contribution of the physical properties of a material to the performance of a device. For example, the current responsivity is proportional to F_i:

$$F_i = p/c' \tag{6}$$

where c' is the volume specific heat.

The voltage response for a pyroelectric element feeding into a high input impedance, unity gain amplifier (such as a source follower FET) is proportional to F_v:

$$F_v = p/c'\varepsilon\varepsilon_0 \tag{7}$$

where ε is the relative permittivity of the pyroelectric.

The signal-to-noise ratio for a voltage detector of this type will depend upon the nature of the dominant noise sources. However, it is usually found that the noise in the most commonly used frequency band (10 Hz–500 Hz) is dominated by the Johnson noise in the ac resistor represented by the lossy capacitance of the detector element. In this case the detectivity is proportional to F_D:

$$F_D = p/c'(\varepsilon\varepsilon_0 \tan \delta)^{1/2} \tag{8}$$

where $\tan \delta$ is the dielectric loss tangent of the pyroelectric.

The pyroelectric vidicon uses a plate of pyroelectric material for thermal imaging. A thermal image focused onto the surface of the material causes the formation of a pattern of pyroelectric charges which are "read" by means of an electron beam. In this case thermal spreading of the pattern on the target is important and the relevant figure of merit is F_{vid}:

$$F_{\mathrm{vid}} = F_v/K \tag{9}$$

where K is the thermal conductivity of the pyroelectric.

It should be noted that the use of these merit figures must be tempered with a knowledge of the type of detector the material is to be used in. It is necessary, if possible, to match the capacitance of the detector to the input capacitance of the amplifier. Hence, low-permittivity materials are better suited to large-area detectors, and conversely arrays of small-area detectors are better served by materials with a high permittivity.

Table 2 lists the pyroelectric properties of several different materials, single crystals, ceramics, and polymers. It can be seen that triglycine sulfate (TGS) and its deuterated isomorph (DTGS) exhibit the highest value of F_v and are frequently used for high-performance single-element detectors and are the preferred materials for pyroelectric vidicon targets. However, these materials are water soluble, difficult to handle, and show poor long-term stability, both chemically and electrically, because of their low Curie temperatures. Furthermore, their dielectric loss is rather high, so that the F_D figures are not so favorable. Lithium tantalate, on the other hand, is an oxide single-crystal material that possesses a relatively low value of F_v, but a very low loss, so that F_D is favorable. The material is very stable and is now widely used for single element detectors. Its thermal conductivity is quite high so that it is not a good material for the pyroelectric vidicon. The ferroelectric polymers possess relatively low pyroelectric coefficients and low relative permittivities with high losses, so that their figures of merit are also quite low. Their low thermal conductivities make them quite favorable for use in the pyroelectric vidicon, and

TABLE 2. Pyroelectric Properties of Selected Materials

Material (temperature)	p $(10^{-4}\,\mathrm{C\,m^{-2}\,K^{-1}})$	Dielectric properties (1 kHz)		c' $(10^{6}\,\mathrm{J\,m^{-3}\,K^{-1}})$	K $(10^{-7}\,\mathrm{m^{2}\,s^{-1}})$	F_v $(\mathrm{m^{2}\,C^{-1}})$	F_D $(10^{-5}\,\mathrm{Pa^{-1/2}})$	F_{vid} $(10^{6}\,\mathrm{s\,C^{-1}})$
		ε	$\tan\delta$					
TGS (35°C)	5.5	55	0.025	2.6	3.3	0.43	6.1	1.3
DTGS (40°C)	5.5	43	0.020	2.4	3.3	0.60	8.3	1.8
PVDF polymer	0.27	12	0.015	2.43	0.62	0.10	0.88	1.6
LiTaO$_3$ crystal	2.3	47	0.005	3.2	13.0	0.17	4.9	0.13
Modified PZ[a] ceramic	3.8	290	0.003	2.5		0.06	5.8	
Modified PT[b] ceramic	3.8	220	0.011	2.5		0.08	3.3	

[a] PZ, PbZrO$_3$.
[b] PT, PbTiO$_3$.

the fact that they are commercially available in thin section (down to 6 μm) at low cost, removing any requirement for expensive lapping and polishing, makes them attractive for some low-cost detectors. Their low permittivities make them particularly well suited to large-area detectors. The ceramic materials modified lead zirconate and modified lead titanate are interesting in that they possess high pyroelectric coefficients with relatively high permittivities and low losses. Hence, while the F_v values are relatively small, the F_D values are as good as most of the single-crystal materials. They are well suited to small-area detectors, such as those used in thermal imaging arrays. They are also relatively cheap to manufacture, and their electrical properties can be modified over a large range by the inclusion of selected dopants. These not only can control such parameters as p, ε, and tan δ but also can be used to alter the electrical conductivity, which can be extremely useful in the control of electrical impedance in the amplifier circuit. These materials are now finding use in a wide range of the device market, from low-cost intruder alarms to high-value imaging arrays.

ACKNOWLEDGMENT

This work has been supported by the Procurement Executive, Ministry of Defence (Royal Signals and Radar Establishment) and partly sponsored by the Royal Armaments Research and Development Establishment (Fort Halstead).

REFERENCES

1. J. F. Nye, *Physical Properties of Crystals*, Oxford U.P, Oxford (1957).
2. J. M. Herbert, *Ferroelectric Transducers and Sensors*, Electrocomponent Science Monographs, Vol. 3, Gordon and Breach, New York (1982).
3. T. T. Wang, J. M. Herbert, and A. M. Glass (eds.), *The Applications of Ferroelectric Polymers*, Blackie, London (1987).
4. R. W. Whatmore, Pyroelectric devices and materials, *Rep. Prog. Phys.* **49**, 1335–1386 (1986).

Principles of Nonlinear Optical Response

R. W. Munn

1. INTRODUCTION

This chapter describes the nature and origins of nonlinear optical response in materials. It aims to provide a general background to materials behavior for optoelectronics and nonlinear optics, except that the important topic of electromagnetic wave propagation in nonlinear materials is not covered. More specific accounts dealing with particular phenomena, materials, and devices are given in other chapters. The following section gives a general phenomenological description of nonlinear response, considering aspects of frequency dependence and symmetry. Next second-harmonic generation and the Pockels effect are treated briefly as examples in order to illustrate how the general description applies to specific cases each with its own specialized notation. After that, the microscopic origins of nonlinear optical response are considered. By understanding these origins, one can begin to design new materials. This is a particularly attractive possibility in molecular materials, as discussed in the final section.

2. PHENOMENOLOGICAL DESCRIPTION

Nonlinear optics is concerned with the electric polarization **P** (dipole moment per unit volume) induced in a material by an electric field **E**. Usually **P** is proportional to **E** and oscillates at the same frequency, so that the response is linear. However, as the magnitude of the field increases, **P** deviates from proportionality and new nonlinear phenomena occur.

Since the magnitude of the electric field is controllable and nonlinear effects are generally weak, it is appropriate to express **P** as a series of terms in ascending powers of **E**. Thus one can write[1]

$$P_i/\varepsilon_0 = \sum_j \chi_{ij}^{(1)} E_j + \sum_{j,k} \chi_{ijk}^{(2)} E_j E_k + \sum_{j,k,l} \chi_{ijkl}^{(3)} E_j E_k E_l + \cdots \tag{1}$$

R. W. Munn • Department of Chemistry and Centre for Electronic Materials, UMIST, Manchester M60 1QD, U.K.

(where it is conventional to omit the numerical coefficients that would make this a Maclaurin series). The indices i, j, k, and l denote components referred to some set of Cartesian axes; associated with each is also a frequency, as exemplified in more detail later. In each term, the output frequency of the polarization must equal the sum of the input frequencies of the electric fields. (Alternatively, the output frequency may be regarded as negative and the sum of the input and output frequencies must then be zero.)

The quantities $\chi^{(n)}$ are the *electric susceptibilities* of order n: linear for $n = 1$, quadratic for $n = 2$, and cubic for $n = 3$. Terms beyond cubic are not usually required. The susceptibilities are tensor quantities and the quantities $\chi_{ij\ldots}^{(n)}$ are their components in the chosen axis system. A susceptibility of order n has components with $n + 1$ indices and hence has 3^{n+1} components in all. Components with indices not all the same need not be zero, and so the polarization need not be parallel to the field. This applies even in the linear case where $\chi_{ij}^{(1)} = \varepsilon_{ij} - \delta_{ij}$, with ε_{ij} a relative permittivity component and δ_{ij} the Kronecker delta (1 if $i = j$ and 0 otherwise). However, in the principal optical axis system $\chi^{(1)}$ is diagonal with components $\chi_{ii}^{(1)} = n_i^2 - 1$, where n_i is a principal refractive index. Hence in these axes the polarization and the field are parallel for linear response.

As the foregoing indicates, tensor components vary according to the axis system used—indeed, tensors can be defined in terms of the way their components change in different axis systems.[2] If two sets of axes are related by a symmetry operation of the material, then the components must be the same to make the material properties the same in symmetrically equivalent directions. This allows the number of independent nonzero components to be determined and tabulated. In particular, a material with a center of symmetry has all components of $\chi^{(2)}$ zero: it has no quadratic nonlinearity (see Fig. 1). This is clearly of central importance in attempts to design $\chi^{(2)}$ materials.

The symmetry of the susceptibility components also depends on the frequencies concerned. Subscripts i, j, etc. corresponding to the same frequency may be interchanged without changing the component. For the static susceptibilities, with all frequencies zero, the value of a component is then the same for any ordering of a given set of subscripts. Thus $\chi_{123}^{(2)} = \chi_{312}^{(2)} = \chi_{231}^{(2)} = \chi_{321}^{(2)}$ and so on. The theory of the nonlinear susceptibilities shows that the same symmetry will arise when all frequencies are sufficiently far from absorption frequencies of the material. This *Kleinman*

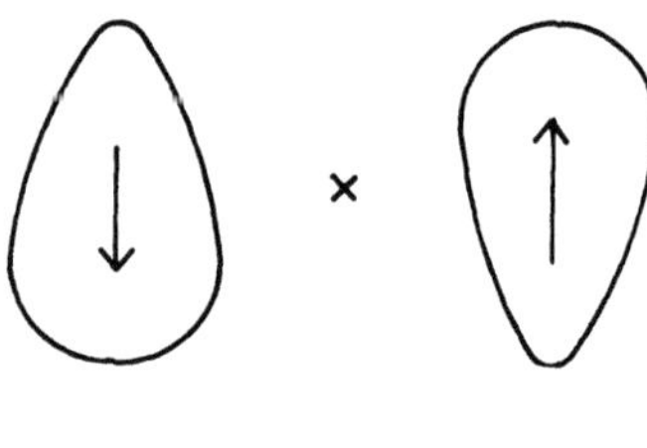

FIGURE 1.   Centrosymmetry precludes quadratic nonlinearity. The two molecules are related by a center of symmetry. Each acquires a quadratic induced dipole moment which depends on the square of the field E and hence not on the direction of E. As a result, the induced moments are equal and opposite, so that the net moment is zero.

symmetry[3] is often assumed in analyzing experiments at frequencies where materials are transparent, although components treated as equal may in fact differ by as much as a factor of 2.

3. SPECIAL CASES

Quadratic nonlinearity gives rise to a variety of different phenomena, and cubic nonlinearity gives rise to even more.[4] Of the quadratic phenomena, second-harmonic generation and the Pockels effect are particularly important.

Second-harmonic generation, usually abbreviated SHG, is the process where two photons of frequency ω yield one of frequency 2ω, or

$$\omega + \omega \to 2\omega \tag{2}$$

Alternatively this can be referred to as frequency doubling. In SHG, Eq. (1) can be specialized to

$$P_\alpha(2\omega)/\varepsilon_0 = \sum_{\beta,\gamma} \chi^{(2)}_{\alpha\beta\gamma}(2\omega; \omega, \omega)E_\beta(\omega)E_\gamma(\omega) \tag{3}$$

Here all frequencies have been written explicitly, and the subscripts α, β, γ describe just Cartesian components.

Because both electric field components in Eq. (2) refer to the same frequency, the SHG susceptibility component $\chi^{(2)}_{\alpha\beta\gamma}(2\omega; \omega, \omega)$ is symmetric under interchange of β and γ. It is therefore conventional to describe SHG by a coefficient $d_{\alpha\lambda}$, where α can take the values 1, 2, 3 and λ can take the values $1, \ldots, 6$ derived from β and γ by the scheme $11 \to 1$, $22 \to 2$, $33 \to 3$, 23 or $32 \to 4$, 31 or $13 \to 5$, 12 or $21 \to 6$. (This is referred to in other contexts as the Voigt contracted notation for $\beta\gamma$.) It is also conventional to relate $d_{\alpha\gamma}$ to the electric field amplitudes rather than the components themselves,[4] which introduces a factor of $\frac{1}{2}$ between $\chi^{(2)}_{\alpha\beta\gamma}$ and $d_{\alpha\lambda}$, so that, for example, $d_{14} = \frac{1}{2}\chi^{(2)}_{\alpha\beta\gamma}$.

The Pockels effect or linear electro-optic (LEO) effect corresponds to the process

$$\omega + 0 \to \omega \tag{4}$$

which is described by Eq. (1) in the specialized form

$$P_\alpha(\omega)/\varepsilon_0 = \sum_{\beta,\gamma} \chi^{(2)}_{\alpha\beta\gamma}(\omega; \omega, 0)E_\beta(\omega)E_\gamma(0) \tag{5}$$

Thus one field is optical and the other is static (or in practice low frequency). As a result, the process is usually viewed as the change produced in the linear optical response by a static electric field.

The Pockels effect as such actually refers to a change in the optical indicatrix linear in the static field.[4] The optical indicatrix tensor is the inverse of the relative permittivity tensor **ε**, with components $1/n_i^2$ in the principal axes. Since it is subscripts α and β that now refer to the same optical frequency, it is conventional to combine these two into a single subscript λ according to the same scheme as for the SHG coefficient. The Pockels coefficient $r_{\lambda\gamma}$ then describes the variation of the λ element of the indicatrix with the γ component of the static field. In the optic axes one then obtains relations between $r_{\lambda\gamma}$ and $\chi^{(2)}_{\alpha\beta\gamma}$ involving the refractive indices, e.g., $r_{63} = \chi^{(2)}_{123}/n_1^2 n_2^2$. One consequence is that Kleinman symmetry of $\chi^{(2)}_{\alpha\beta\gamma}$ does not lead to a corresponding symmetry of $r_{\lambda\gamma}$.

4. MICROSCOPIC ORIGINS

Macroscopic nonlinearity of a material originates in microscopic nonlinearity of the entities that make up the material. In a crystal one may always choose the unit cell as the microscopic entity, although one would usually prefer to choose the separate atoms, ions, or molecules in the unit cell. For such an entity, the electric dipole moment induced by a field has an expansion analogous to (1):

$$p_i = \sum_j \alpha_{ij} F_j + \sum_{j,k} \beta_{ijk} F_j F_k + \sum_{j,k,l} \gamma_{ijkl} F_j F_k F_l + \cdots \tag{6}$$

Here the indices i, j, k, and l again denote Cartesian components and associated frequencies. The coefficients in this expansion are the polarizability α_{ij} and the first and second hyperpolarizabilities β_{ijk} and γ_{ijkl}, all referring to the response in the crystal environment.[5,6] The quantities F_i are the local electric fields, which come from the macroscopic electric fields E_i plus the fields of the dipole moments induced in the surrounding entities. As indicated by Eq. (6), it is the local fields that polarize the entities.

The relation between the macroscopic and microscopic response depends on the fact that the polarization in Eq. (1) is the induced dipole moment per unit volume from Eq. (6). Given the local fields in terms of the macroscopic fields, one can obtain microscopic expressions for the susceptibilities $\boldsymbol{\chi}^{(n)}$ by comparing the coefficients of E^n in the two equations. Relating the local field to the macroscopic field is a standard problem in condensed matter, at least for linear response. The more difficult problem of nonlinear response has also been solved.[5,7] The details are beyond the scope of this chapter, but it is worth noting that $\mathbf{F}$ becomes a nonlinear function of $\mathbf{E}$.

The full expressions for the nonlinear susceptibilities are complicated (see the original papers[5,7]). For present purposes it will be sufficient to give the results in a highly simplified form which shows their basic structure. Thus ignoring all tensor subscripts and other details one finds[1]

$$\chi^{(1)} = (\alpha/\varepsilon_0 v)d \tag{7}$$

$$\chi^{(2)} = (\beta/\varepsilon_0 v)d^3 \tag{8}$$

$$\chi^{(3)} = (\gamma/\varepsilon_0 v)d^4 + (\beta/\varepsilon_0 v)^2 d^5 \tag{9}$$

Here d is the dimensionless local-field tensor which relates $\mathbf{F}$ to $\mathbf{E}$ for linear response. It can be calculated from α and lattice dipole sums known as Lorentz-factor tensors. In the simplest case of a cubic crystal it is given by $1/(1 - \alpha/3\varepsilon_0 v)$ or $(n^2 + 2)/3$, and it is often approximated by $(n_i^2 + 2)/3$ in the principal optical axes of other crystals.

The linear susceptibility depends on the local field tensor and the dimensionless polarizability density $\alpha/\varepsilon_0 v$. The quadratic susceptibility depends similarly on the hyperpolarizability density $\beta/\varepsilon_0 v$ and on the cube of the local-field tensor rather than the square as might have been expected. This feature arises from the nonlinear dependence of $\mathbf{F}$ on $\mathbf{E}$ already mentioned. The same nonlinearity also accounts for the form of $\chi^{(3)}$. It contains not only a "direct" contribution depending on $\gamma/\varepsilon_0 v$, the hyperpolarizability density of the same order, but also a "cascading" contribution quadratic in $\beta/\varepsilon_0 v$. In the absence of special features, d is typically of the order of 1.5, so that the local field factors contribute about one order of magnitude to the nonlinear susceptibilities. However, a large nonlinear response clearly requires a high microscopic nonlinearity, as well as a suitable arrangement of the microscopic entities [one of the factors simplified out of Eqs. (7)–(9)].

At this stage, it is convenient to consider the dimensions and units of nonlinear response coefficients. All the equations written here imply a four-dimensional rationalized system of units such as the SI. Then in Eq. (1), P_i/ε_0 has the dimensions of electric field. Thus $\chi^{(1)}$ is dimensionless, $\chi^{(2)}$ has the dimensions of 1/field, and $\chi^{(3)}$ the dimensions of $1/(\text{field})^2$. In Eq. (6), α has the dimensions of electric dipole/field, β the dimensions of dipole/(field)^2 and γ the dimensions of dipole/(field)^3. The combination $\alpha/\varepsilon_0 v$ in Eq. (7) is dimensionless like $\chi^{(1)}$; $\beta/\varepsilon_0 v$ in Eqs. (8) and (9) has the dimensions of 1/field like $\chi^{(2)}$; and $\gamma/\varepsilon_0 v$ in Eq. (9) has the dimensions of $1/(\text{field})^2$ like $\chi^{(3)}$.

Unfortunately, nonlinear response is also often described in terms of the three-dimensional unrationalized electrostatic system of units or esu. Suitable conversions between esu and SI units are as follows, where a prime denotes the esu quantity:

$$\alpha'/10^{-24} \, \text{esu} = \alpha/1.11265 \times 10^{-40} \, \text{F m}^2$$

$$\beta'/10^{-30} \, \text{esu} = \beta/3.71140 \times 10^{-51} \, \text{F m}^3 \, \text{V}^{-1}$$

$$\gamma'/10^{-36} \, \text{esu} = \gamma/1.23799 \times 10^{-61} \, \text{F m}^4 \, \text{V}^{-1}$$

$$\chi'^{(1)} = \chi^{(1)}$$

$$\chi'^{(2)}/10^{-9} \, \text{esu} = \chi^{(2)}/0.419169 \times 10^{-12} \, \text{m V}^{-1}$$

$$\chi'^{(3)}/10^{-15} \, \text{esu} = \chi^{(3)}/1.11265 \times 10^{-24} \, \text{m}^2 \, \text{V}^{-2}$$

[These conversions assume that the version of Eq. (1) for esu does not contain the factor of 4π that is found in other contexts.]

5. DESIGN OF NONLINEAR MATERIALS

To obtain highly active materials for nonlinear optics, one clearly has to design microscopic entities with large hyperpolarizabilities. It is also necessary to satisfy other criteria such as the absence of centrosymmetry for $\chi^{(2)}$ materials and the absence of absorptions at inconvenient frequencies—clearly, it is not much use designing a material that has very high SHG coefficients but also absorbs the second harmonic frequency very strongly.

Design is easiest and most attractive in molecular materials, where the non-linearity is predominantly electronic and fast rather than nuclear and slow in origin.[8,9] The theory of hyperpolarizability shows that it is large when the electronic excitations of the molecule are intense and give rise to large dipole moment changes.[10] The response is also enhanced for frequencies near absorption or resonance, but as already noted this may not be unequivocally beneficial. To obtain these features, one seeks to make molecules that contain a strong electron donor group, a strong electron acceptor group, and a conjugated pathway between them designed to facilitate electron transfer on excitation. Sometimes these are called "push–pull" molecules.

Many donors are based on the amino $-NH_2$ group (see Fig. 2), which can donate the lone pair electrons from nitrogen to form a double bond and a positively charged nitrogen. Correspondingly, many acceptors are based on the nitro $-NO_2$ group (see Fig. 2), which can accept electrons to form a single bond and a second negatively charged oxygen. Conjugated pathways between donors and acceptors are usually provided by combinations of aromatic rings and polyenes. Extending the conjugation path clearly increases the possible dipole moment change and hence the possible hyperpolarizabilities β and γ. However, since the susceptibilities depend on the hyperpolarizability per unit volume, extending the conjugation path must increase β and γ faster than it increases the molecular volume if it is to increase $\chi^{(2)}$ and $\chi^{(3)}$.

Exploring the effects of different combinations of donors, acceptors, and pathways experimentally is very time-consuming. Information about hyperpolarizabilities can be obtained from electric field induced SHG (EFISH) in solution,[11] but since a range of concentrations must be studied to eliminate the effect of the solvent, such measurements are not rapid. Fortunately, molecular orbital calculations

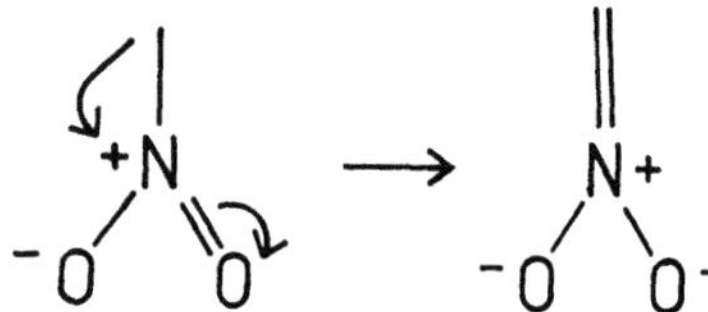

FIGURE 2.   Donor and acceptor groups. The amino $-NH_2$ group can donate an electron from the nitrogen lone pair, while the nitro $-NO_2$ group can accept an electron on the oxygen. (Note that the nitro group is properly depicted as a superposition of two structures, namely, the one illustrated on the left and its equivalent with the oxygens interchanged, so that the oxygens are equivalent.)

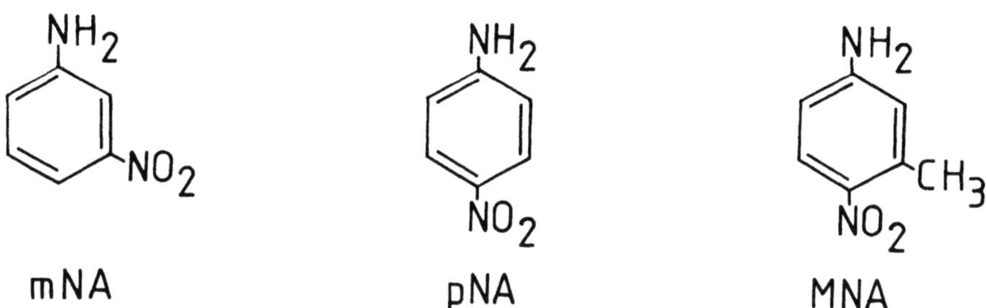

FIGURE 3.   Molecular structures of *meta*-nitroaniline (mNA), *para*-nitroaniline (pNA), and 3-methyl–4–nitroaniline (MNA).

can be used to obtain information on the relative effectiveness of different substitutents.[12] One can thereby screen out the most promising molecules for synthesis and experimental study.

In designing $\chi^{(2)}$ materials, it is necessary to make molecules that not only have a high β but also form noncentrosymmetric crystals. Fig. 3 illustrates how this may be achieved in the nitroanilines. These molecules use the amino and nitro donor and acceptor groups by simply attaching them to a benzene ring. Three distinct arrangements are possible, of which two are illustrated, *meta*-nitroaniline (mNA) and *para*-nitroaniline (pNA). Because the donor and acceptor groups are farther apart in pNA than in mNA, it is found that pNA has a much higher β: 45 as opposed to 19×10^{-51} F m^3 V^{-1}.[12] However, when crystals are grown, it is found that pNA is centrosymmetric and hence $\chi^{(2)}$ inactive whereas mNA is not, with $d_{14} = 34 \times 10^{-12}$ m V^{-1} [13]: although the pNA molecules respond readily to an electric field, their contributions exactly cancel in the crystal. By adding a methyl $-CH_3$ group to pNA, one obtains the third structure in Fig. 3, MNA. This has a β even higher than pNA at 50×10^{-51} F m^3 V^{-1} [12] and a shape more reminiscent of mNA which leads to a noncentrosymmetric crystal with a very high $\chi^{(2)}$, $d_{11} = 250 \times 10^{-12}$ m V^{-1}.[13]

A similar example is provided by the nitropyridine-*N*-oxides shown in Fig. 4. The simple compound (labeled PO) is centrosymmetric and hence $\chi^{(2)}$ inactive; the compound with the added methyl group (labeled POM) is noncentrosymmetric and hence $\chi^{(2)}$ active, though with $d_{11} = 10 \times 10^{-12}$ m V^{-1} [13] it is not as good as MNA.

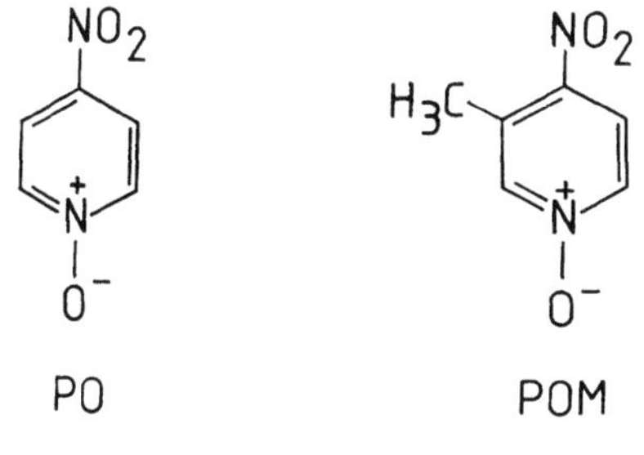

FIGURE 4. Molecular structures of 4-nitropyridine-1-oxide (PO) and 3-methyl-4-nitropyridine-1-oxide (POM).

These examples suggest that an "awkward" molecular shape helps to prevent centrosymmetry. There have been suggestions that the permanent molecular dipole moments tend to favor centrosymmetry. However, energy calculations show that dipole interactions may help to stabilize crystals whether or not the structures are centrosymmetric[14,15]; predicting the occurrence or absence of centrosymmetry is evidently not simple.

The possibility of "molecular engineering" of the sort illustrated here in a very simple form has stimulated extensive work on organic materials for nonlinear optics. Apart from work on crystals, there is considerable interest in polymers. Doping with molecules of high β followed by poling the polymer near its glass transition temperature in a strong electric field to align the dopants in a noncentrosymmetric arrangement is a promising technique for producing $\chi^{(2)}$ materials,[16] while various conjugated polymers have high $\chi^{(3)}$ values.[17] Langmuir–Blodgett films also offer a means of constructing noncentrosymmetric materials composed of alternating layers of molecules designed to have reinforcing high β values and interactions favoring stability of the film.[18] Other attractions of organic materials include the high susceptibilities and figures of merit achievable, the fast predominantly electronic response, and a high optical damage threshold. As a result, nonlinear optics of molecular materials constitutes one of the most promising areas of molecular electronics.

REFERENCES

1. R. W. Munn, Theory of molecular opto-electronics: From the molecule to the crystal, *J. Molec. Electron.* **4**, 31–36 (1988).
2. H. Jeffreys, *Cartesian Tensors*, Cambridge U.P., London (1963).
3. D. A. Kleinman, Nonlinear dielectric polarization in optical media, *Phys. Rev.* **126**, 1977–1979 (1962).
4. Y. R. Shen, *The Principles of Nonlinear Optics*, Wiley, New York (1984).
5. M. Hurst and R. W. Munn, Theory of molecular opto-electronics. I. Macroscopic and microscopic response, *J. Molec. Electron.* **2**, 35–41 (1986).
6. M. Hurst and R. W. Munn, Theory of molecular opto-electronics. II. Environmental effects on molecular response, *J. Molec. Electron.* **2**, 43–47 (1986).
7. J. A. Armstrong, N. Bloembergen, J. Ducuing, and P. S. Pershan, Interactions between light waves in a nonlinear dielectric, *Phys. Rev.* **127**, 1918–1939 (1962).
8. J. Zyss, Nonlinear organic materials for integrated optics, *J. Molec. Electron.* **1**, 25–45 (1985).
9. D. S. Chemla and J. Zyss (eds.), *Nonlinear Optical Properties of Organic Molecules and Crystals*, Academic, Orlando (1987), 2 Vols.
10. J. F. Ward, Calculation of nonlinear optical susceptibilities using diagrammatic perturbation theory, *Rev. Mod. Phys.* **37**, 1–18 (1965).
11. J. F. Nicoud and R. J. Twieg, Organic EFISH hyperpolarizability data, in Ref. 9, Vol. 2, pp. 255–267.
12. D. Pugh and J. O. Morley, Molecular hyperpolarizabilities of organic materials, in Ref. 9, Vol. 1, pp. 193–225.
13. J. F. Nicoud and R. J. Twieg, Design and synthesis of organic compounds for efficient second-harmonic generation, in Ref. 9, Vol. 1, pp. 227–296.
14. M. Hurst and R. W. Munn, Theory of molecular opto-electronics. VI. Comparisons between nitroanilines, in *Organic Materials for Nonlinear Optics* (R. A. Hann and D. Bloor, eds), pp. 3–11, Royal Society of Chemistry, London (1989).
15. M. Hurst and R. W. Munn, Theory of molecular opto-electronics. VIII. Calculations for POM and MAP, in *Molecular Electronics—Science and Technology* (A. Aviram, ed.), pp. 267–273, Engineering Foundation, New York (1989).

16. K. D. Singer, M. G. Kuzyk, and J. E. Sohn, Second-order nonlinear optical processes in orientationally ordered materials: Relationship between molecular and macroscopic properties, *J. Opt. Soc. Am. B* **4**, 968–976 (1987).
17. P. Calvert, Polymers that make light work, *Nature* **337**, 408–409 (1989).
18. I. R. Girling, P. V. Kolinsky, N. A. Cade, J. D. Earls, and I. R. Peterson, Second harmonic generation from alternating Langmuir–Blodgett films, *Optics Commun.* **55**, 289–292 (1985).

21

Electro-optic Materials and Applications

Simon Allen

1. INTRODUCTION

The electro-optic, or "Pockels," effect is the modification of the refractive index of a material by the application of a dc or low-frequency (i.e., much lower than optical frequency) electric field. As in the case of piezoelectricity, described in the previous chapter, electro-optic effects are observed only in materials that are noncentrosymmetric. The effect has been characterized in a wide range of materials, including inorganic crystals (mostly ferroelectric compounds), organic and molecular crystals, and crystalline and noncrystalline polymers.

The magnitude δn of the change in refractive index produced is related to the applied electric field E by the equation

$$\delta n = -\tfrac{1}{2}n^3 rE \tag{1}$$

where r is the *electro-optic coefficient.* For background reading on the electro-optic effect and electro-optic devices the reader is referred, for example, to Ref. 1.

2. PROPERTIES OF ANISOTROPIC MEDIA

2.1. Optical Properties

By definition, electro-optic materials are not isotropic. A description of the physics of the process therefore requires an understanding of the optics of anisotropic media. Equation (1) above is in fact an oversimplification. In an anisotropic medium parameters such as refractive index n, dielectric constant ε, and electro-optic coefficient r will depend on

  i. The direction of propagation of light through the medium, as indicated by the wave vector $\mathbf{k}$;
 ii. The polarization of the optical wave $\mathbf{E}$;
iii. The direction of the applied field $\mathbf{E}_0$.

Simon Allen • ICI Wilton Materials Research Centre, Wilton, Middlesbrough, Cleveland TS6 8JE, U.K.

A full description must use tensor notation. For example, consider the dielectric permittivity ε. In an isotropic medium, the displacement field $\mathbf{D}$ is parallel to the applied electric field $\mathbf{E}$:

$$\mathbf{D} = \varepsilon_0 \varepsilon \mathbf{E} \tag{2}$$

but in an anisotropic material this is no longer the case, and we must write

$$D_i = \varepsilon_0 \varepsilon_{ij} E_k \tag{3}$$

for each component ($i = 1, 2, 3$) of $\mathbf{D}$. ε is now a second-rank tensor:

$$\varepsilon = \begin{pmatrix} \varepsilon_{11} & \varepsilon_{12} & \varepsilon_{13} \\ \varepsilon_{21} & \varepsilon_{22} & \varepsilon_{23} \\ \varepsilon_{31} & \varepsilon_{32} & \varepsilon_{33} \end{pmatrix} \tag{4}$$

which has in principle nine independent coefficients. In fact, the tensor is symmetric, having $\varepsilon_{ij} = \varepsilon_{ji}$, and the number of independent coefficients may in general be reduced even further by symmetry constraints. It is also always possible to choose a set of (orthogonal) axes, known as the *principal axes*, within the material such that the tensor has the diagonalized form

$$\varepsilon = \begin{pmatrix} \varepsilon_1 & 0 & 0 \\ 0 & \varepsilon_2 & 0 \\ 0 & 0 & \varepsilon_3 \end{pmatrix} \tag{5}$$

with ε_1, ε_2, ε_3 being the principal values of the permittivity tensor.

A useful method of visualizing properties such as this, described by a second-rank tensor, utilizes the *representation quadric* as described in some detail by Nye.[2] Consider any symmetric second-rank tensor S_{ij}. The equation

$$S_{ij}x_i x_j = 1 \tag{6}$$

is that of a generalized quadric (i.e., ellipsoid or hyperboloid) referred to the axes (x_1, x_2, x_3). It can be shown that the coefficients of the quadric (6) transform in the same way as those of the tensor S_{ij}. Under an appropriate transformation the quadric can be referred to the *principal axes* described above, and it then has the simplified form

$$S_1 x_1^2 + S_2 x_2^2 + S_3 x_3^2 = 1 \tag{7}$$

By comparing this with the standard equation for an ellipsoid:

$$(x^2/a^2) + (y^2/b^2) + (x^2/c^2) = 1 \tag{8}$$

it can be seen that the semiaxes of the representation quadric are of lengths $1/\sqrt{S_1}$, $1/\sqrt{S_2}$, $1/\sqrt{S_3}$.

The refractive indices of an anisotropic material are not in fact the components of a tensor, but are related to the principal components of the *impermeability tensor* by

$$(1/n_i^2) = \kappa_{ii} \tag{9}$$

Their values, for any given direction of propagation of light through a material, can be found with the help of the *optical indicatrix*, which is another name for the representation quadric of the impermeability tensor. Rewriting Eq. (6) in full we have

$$\kappa_{11}x^2 + \kappa_{22}y^2 + \kappa_{33}z^2 + 2\kappa_{23}yz + 2\kappa_{12}xy + 2\kappa_{13}xz = 1 \tag{10}$$

or, when referred to the principal axes

$$\kappa_{11}x^2 + \kappa_{22}y^2\kappa_{33}z^2 = 1 \tag{11}$$

or

$$(x^2/n_1^2) + (y^2/n_2^2) + (z^2/n_3^2) = 1 \tag{12}$$

and the lengths of the semiaxes are just n_1, n_2, and n_3. The indicatrix has the important property illustrated in Fig. 1. For light propagating in an arbitrary direction (OP) through the material, there are two allowed wave fronts. The central section of the indicatrix perpendicular to OP is an ellipse, and the refractive indices of the allowed waves are equal to the lengths of the semiaxes (X and Y) of this ellipse. The wave with refractive index X is polarized parallel to OX and that with refractive index Y is polarized parallel to OY.

The shape of the indicatrix is determined to a certain extent by the symmetry of the medium, which can belong to one of three broad categories:

Isotropic: $n_1 = n_2 = n_3$. n is independent of $\mathbf{k}$, so no birefringence is observed.

Uniaxial: $n_1 = n_2 \neq n_3$. $n_1(=n_2)$ is known as the ordinary refractive index n_0 and n_3 the extraordinary refractive index n_e. For $\mathbf{k}$ along z (the "optic axis") there is no birefringence. For $\mathbf{k}$ in the x–y plane there are two refractive indices: n_0 for light polarized in the x–y plane and n_e for light polarized along z. For other directions there are also two refractive indices. Again light polarized in the x–y plane has refractive index n_0, but light polarized perpendicular to this (in the plane containing $\mathbf{k}$ and the optic axis) has refractive index $n_e(\Phi)$, where Φ is the angle between $\mathbf{k}$ and the optic axis and

$$1/n_e(\Phi)^2 = (\sin^2 \Phi)/n_0^2 + (\cos^2 \Phi)/n_e^2 \tag{13}$$

The birefringence is $\Delta n = n_e(\Phi) - n_0 \ (< n_e - n_0)$.

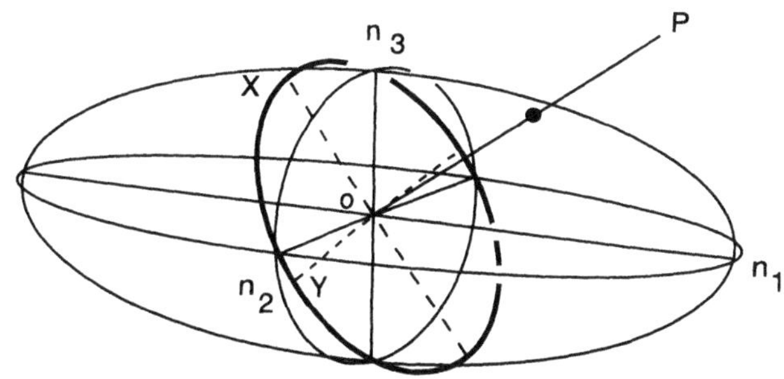

FIGURE 1. The optical indicatrix for a biaxial crystal. For propagation of light along OP two waves, with refractive indices X and Y, and polarized along OX and OY, are allowed.

Biaxial: $n_1 \neq n_2 \neq n_3$. This is the most complicated case and is described fully by Hartshorn and Stuart.[3]

2.2. Electro-optic Coefficients

The electro-optic coefficient r is a third-rank tensor, and is defined in terms of the change in the impermeability tensor κ induced by an electric field E:

$$\delta\kappa_{ij} = r_{ijk}E_k \tag{14}$$

In general there are 27 independent components of r, but in practice for any given material many are equal to each other or zero. Reduced, or "Voigt," notation[1] is often used to simplify the representation, wherein the three subscripts are reduced to two by the contraction

$$
\begin{array}{ccccccc}
i,j & 11 & 22 & 33 & 23 & 13 & 12 \\
\downarrow & & & & & & \\
l & 1 & 2 & 3 & 4 & 5 & 6
\end{array}
$$

so that Eq. (14) becomes

$$\delta\kappa_l = r_{lk}E_k \qquad (l = 1\text{-}6,\ k = 1\text{-}3) \tag{15}$$

Since $(1/n_l^2) = \kappa_l (l = 1, 2, 3)$ we get $\delta n_l = -\frac{1}{2}n^3\delta\kappa_l$, and hence the origin of Eq. (1) is clear

$$\delta n_l = -\tfrac{1}{2}n_l^3 r_{lk}E_k \tag{1'}$$

2.3. Propagation of Light Through an Anisotropic Medium

Consider a slab of birefringent material as shown in Fig. 2, having refractive indices n_x and n_y. Light propagates a distance l through the material, along the z axis. The electric field of the light wave, at position z, can be written

$$E(z) = E(0)\,e^{i(\omega t - kz)} = E^0\,e^{-i\phi} \tag{16}$$

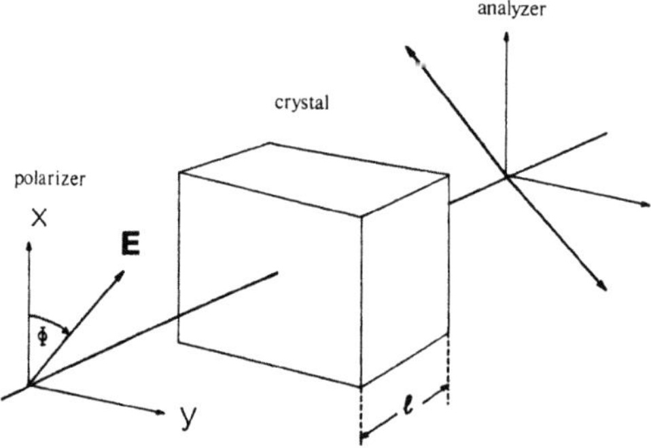

FIGURE 2. Propagation of light through an anisotropic slab.

where ϕ is the phase of the wave with respect to that at $z = 0$, given by

$$\phi = kz = 2\pi(nz)/\lambda \tag{17}$$

For light plane-polarized along x, the refractive index is n_x, and the wave emerges from the slab still polarized along x having traveled an optical path length $n_x l$, and with a phase $\phi_x = 2\pi n_x l/\lambda$. Likewise light plane-polarized along y passes through with its polarization state unchanged and emerges with a phase $\phi_y = 2\pi n_y l/\lambda$. Light polarized at some angle Φ to the x axis has components along both x and y. These travel at different speeds, owing to the birefringence $n_x - n_y$ ($= \Delta n$), so that a phase difference (known as the retardation Γ) occurs between the two components:

$$\Gamma = \phi_x - \phi_y = 2\pi \Delta n l/\lambda \tag{18}$$

Because of this phase difference between the two components, the output wave is in general elliptically polarized. The "phase-modification" described above can be converted into "amplitude-modification" by the addition of a second polarizing element (the analyzer) at the output from the material. Thus if the analyzer is oriented at right angles to the initial polarizer (i.e., at $90 - \Phi$ degrees to the x axis) the transmission ($T = I_{\text{out}}/I_{\text{in}}$) of the system has the form

$$T = \sin^2(\Gamma/2) \tag{19}$$

These equations show the basis for many electro-optic applications. If the electro-optic effect is used to modify one or both of the refractive indices n_x and n_y, then through Eq. (18) it can be seen that the retardation Γ will also be modified, and hence the transmitted intensity [Eq. (19)] will also change.

3. ELECTRO-OPTICAL MATERIAL PARAMETERS

A number of different parameters are commonly used to compare the relative merits of electro-optic materials. The r-coefficients have already been described, and relate the change in the impermeability tensor to the applied electric field. The figure of merit (f.o.m.) gives a direct measure of the change in refractive index for a given applied electric field. From Eq. (1) this has the form

$$\text{f.o.m.} = \tfrac{1}{2}n^3 r$$

Finally, the half-wave voltage (V_π) is the parameter used in conjunction with amplitude-modulation applications. It is defined as the voltage required to induce a phase change of π radians in the retardation Γ. From Eq. (19) it can be seen that this will also be the voltage required to switch the transmitted intensity from minimum to maximum. The exact definition of V_π in terms of electro-optic coefficients depends on the geometry of the interaction between the light wave and the medium. For the geometry already considered in Fig. 2, with a voltage applied along the x-axis between electrodes separated by a distance d (so $E_x = V/d$) then from Eq. (18).

$$\delta\Gamma = (2\pi l/\lambda)(\delta n_x - \delta n_y)$$

$$= (2\pi l/\lambda)(-\tfrac{1}{2}n_x^3 r_{11} + \tfrac{1}{2}n_y^3 r_{21})E_x \tag{20}$$

and so for a change of π in Γ

$$\pi = (\pi l/\lambda)(n_y^3 r_{21} - n_x^3 r_{11})(V_\pi/d)$$

$$V_\pi = (\lambda/n_y^3 r_{\text{eff}})(d/l) \tag{21}$$

where

$$r_{\text{eff}} = r_{21} - (n_x/n_y)^3 r_{11} \tag{22}$$

Thus in this case the half-wave voltage depends explicitly on the dimensions of the device, through the ratio (d/l). Comparison of different materials is often made using a *reduced half-wave voltage* (V_π^*), which is just the half-wave voltage for a sample of unit dimensions $(d = l)$.

It can be seen from the above equations that high figures of merit and low half-wave voltages are obtained by a combination of high r coefficients and high refractive indices. It should, however, be noted that the large values of the dielectric constant ε associated with high refractive indices result in limitations to the switching speeds that can be obtained from an electro-optic device.

4. APPLICATIONS

4.1. Bulk Crystals

Bulk crystals of inorganic compounds such as KDP and lithium niobate have been used for many years in the applications listed below.

i. Amplitude Modulators. Here the main applications are in the field of optical communications, where a signal is impressed onto a light beam via the modulation of its amplitude. Although, using diode lasers, the modulation is often carried out at source (by varying the bias current), there are a number of applications where the *external* modulation of the beam is preferable.[4] From Eq. (18)

$$T = \sin^2(\Gamma/2) = \sin^2(\pi V/2 V_\pi)$$

and so the modulation is obtained by the application of a suitable voltage waveform (V) across the crystal.

ii. Electro-optic Shutter (Q-Switch). Q-switches are used within the cavity of high-power pulsed lasers to open or shut the optical feedback path, and control the build-up of optical power. Performance is essentially as in the case of the amplitude modulator, except that the voltage applied is exactly V_π, to switch between full and zero transmission.

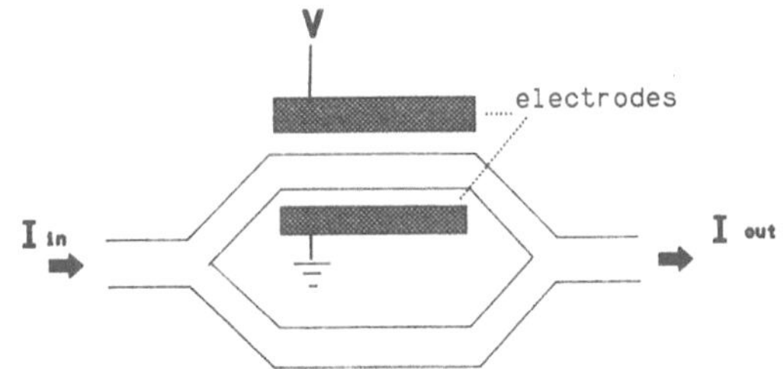

FIGURE 3.   A waveguide Mach–Zender interferrometer.

iii. Phase Modulator. In this case there is no output polarizer incorporated in the device. The phase of the output wave is modified by the applied field. This also has applications in the field of optical communications.

4.2. Waveguides

There has recently been a great deal of interest in the development of electro-optic waveguide devices for use in integrated optics. Most applications involve either the modulation of the amplitude of a guided light beam, or the routing of light between different waveguides. A number of lithium niobate waveguide devices have recently become commercially available. Examples of electro-optic waveguide devices are given below.

i. Mach–Zender Interferometer.[5] This is an amplitude modulation device, having two arms of equal length as shown in Fig. 3. Light input into the device enters both arms via the y junction. When no voltage is applied, the light emerges from the two (identical) arms in phase, and is recombined constructively at the second junction to give maximum output. When a voltage V is applied across one of the arms, it changes the refractive index in that arm and a phase difference is introduced between the light from the two arms. Destructive interference now occurs on recombination at the second junction, and the intensity of the transmitted light is reduced.

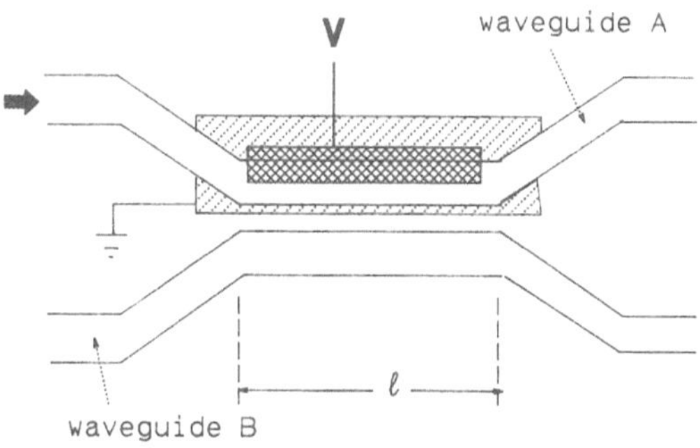

FIGURE 4.   An electro-optic waveguide coupler.

ii. Waveguide Coupler (Fig. 4). Two channel waveguides run parallel (a few micrometers apart) over a distance *l*. If light is input into guide A, then because a portion of the energy is actually outside the defined channel (the *evanescent wave*), light is coupled from channel A to channel B within the interaction region *l*. If the length *l* is chosen correctly, all the light may be output from B. If a voltage is now applied across one of the channels, the coupling condition is altered and the intensities output from A and B change. Such a device can thus be used to switch light from channel B to channel A.

These modulators and switches can be combined on a single substrate to produce complicated switching networks for optical communications.[6]

5. INORGANIC MATERIALS—KDP AND LITHIUM NIOBATE

Many inorganic materials have been shown to have high electro-optic coefficients (see, for example, Table 1), but the most widely used to date have been potassium dihydrogen phosphate (KDP) and its isomorphs, and lithium niobate ($LiNbO_3$), with KDP primarily being used in the bulk form and lithium niobate in waveguides. Both materials are *ferroelectric*. The high electro-optic coefficients arise partly because the positions of the ferroelectric atoms are readily affected by the application of electric fields, giving changes in the electron distribution within the material.

5.1. KDP

Crystals of KDP have symmetry $42m$, and the r-tensor has the particularly simple form

TABLE 1. Properties of Some Electro-optic Materials at 633 nm

Material	Crystal symmetry	r coefficients (pm/V)	n	ε
Inorganic				
KDP	$42m$	$r_{41} = 8$	$n_0 = 1.5074$	$\varepsilon_1 = 44$
(KH_2PO_4)		$r_{63} = 11$	$n_e = 1.4669$	$\varepsilon_3 = 21$
KD*P	$42m$	$r_{63} = 24$		$\varepsilon_3 = 48$
(KD_2PO_4)			$n_e = 1.462$	
$LiNbO_3$	$3m$	$r_{13} = 9.6$	$n_0 = 2.286$	$\varepsilon_1 = 78$
		$r_{41} = 30.9$	$n_e = 2.200$	$\varepsilon_3 = 32$
BSN	$4mm$	$r_{13} = 67$	$n_0 = 2.312$	
		$r_{33} = 1340$	$n_e = 2.299$	
GaAs	$43m$	$r_{41} = 0.97$	$n = 3.32$	$\varepsilon \approx 10$
Organic				
MNA	m	$r_{11} = 65$		
DCNP	m	$r_{33} = 82$	$n_x = 1.9$	
			$n_z = 2.8$	

$$\mathbf{r} = \begin{pmatrix} 0 & 0 & 0 \\ 0 & 0 & 0 \\ 0 & 0 & 0 \\ r_{41} & 0 & 0 \\ 0 & r_{41} & 0 \\ 0 & 0 & r_{63} \end{pmatrix} \qquad (23)$$

with $r_{41} = 8$ pm/V and $r_{63} = 11$ pm/V. The crystals are uniaxial, with the optic axis parallel to z. To utilize the largest coefficient, $r_{63}\ (= r_{123})$, light propagates along the z axis, and is polarized at $45°$ to the x and y axes. The electric field is also applied along the z axis (usually using ring electrodes at either end of the sample). This is an example of the *longitudinal* electro-optic effect, with E and k being parallel. In the absence of an applied field the indicatrix has the standard uniaxial form

$$(x^2/n_0^2) + (y^2/n_0^2) + (z^2/n_e^2) = 1 \qquad (24)$$

When a voltage V is applied to produce a field $E_z = V/l$ this becomes

$$(x^2/n_0^2) + (y^2/n_0^2) + (z^2/n_e^2) + 2r_{63}E_z xy = 1 \qquad (25)$$

The cross term in x and y shows that the application of the field has brought about a rotation of the principal dielectric axes by an angle φ about z. The symmetry in x and y shows that φ must equal $45°$, and so to return the indicatrix to the form of Eq. (24) we must make a change of coordinates from (x, y, z) to (x', y', z):

$$x = x' \cos 45 - y' \sin 45$$

$$y = x' \sin 45 + y' \cos 45$$

giving

$$(x'^2/n_{x'}^2) + (y'^2/n_{y'}^2) + (z^2/n_e^2) = 1 \qquad (26)$$

where

$$(1/n_{x'}^2) = (1/n_0^2) + r_{63}E_x \qquad (27a)$$

$$(1/n_{y'}^2) = (1/n_0^2) - r_{63}E_x \qquad (27b)$$

Since $r_{63}E_z \ll 1/n_0^2$, $\delta n \approx -\tfrac{1}{2}n^3\delta(1/n^2)$ and so

$$n_{x'} = n_0 - \tfrac{1}{2}n_0^3 r_{63} V/l \qquad (28a)$$

$$n_{y'} = n_0 + \tfrac{1}{2}n_0^3 r_{63} V/l \qquad (28b)$$

and the induced birefringence Δn is

$$\Delta n = n_0^3 r_{63} V / l \tag{29}$$

The half-wave voltage will in this case be

$$V_\pi = \lambda / (2n_0^3 r_{63}) \tag{30}$$

Thus in the case of the longitudinal electro-optic effect the half-wave voltage is independent of the crystal dimensions. For KDP $V_\pi = 8.4\,\text{kV}$ at 633 nm.

5.2. Lithium Niobate

Crystals of lithium niobate have symmetry $3m$, and are also uniaxial with the optic axis along z. The r tensor is a little more complicated than that of KDP:

$$r = \begin{pmatrix} 0 & -r_{22} & r_{13} \\ 0 & r_{22} & r_{13} \\ 0 & 0 & r_{33} \\ 0 & r_{51} & 0 \\ r_{51} & 0 & 0 \\ -r_{22} & 0 & 0 \end{pmatrix} \tag{31}$$

with $r_{13} = 9\,\text{pm/V}$, $r_{33} = 31\,\text{pm/V}$, and $r_{51} = 28\,\text{pm/V}$. The indicatrix, in the absence of an applied field has the form given by Eq. (24). We will consider the propagation along the crystal's x axis of a light beam polarized at 45° to the z and y axes, with the electric field being applied along the z axis. This is an example of the *transverse* electro-optic effect, as $\mathbf{k}$ and $\mathbf{E}$ are perpendicular to each other. In the absence of an applied field the birefringence is

$$\Delta n^0 = n_e - n_0 \tag{32}$$

and the retardation is

$$\Gamma^0 = 2\pi \Delta n^0 l / \lambda \tag{33}$$

The propagation distance is l, and a voltage V is applied between electrodes a distance d apart. The indicatrix then has the form

$$x^2(1/n_0^2 + r_{13}E) + y^2(1/n_0^2 + r_{13}E) + z^2(1/n_e^2 + r_{33}E) = 1 \tag{34}$$

Because there are no cross terms, there has been no rotation of the principal dielectric axes, and in fact the crystal remains uniaxial, with

$$n_x = n_y = n_0 - \tfrac{1}{2}n_0^3 r_{13} V / d \tag{35a}$$

$$n_z = n_e - \tfrac{1}{2}n_e^3 r_{33} V/d \tag{35b}$$

The birefringence is now

$$\Delta n = n_z - n_y$$

$$= \Delta n^0 - \tfrac{1}{2}(n_e^3 r_{33} - n_0^3 r_{13}) V/d \tag{36}$$

and the retardation is

$$\Gamma = \Gamma^0 - (n_e^3 r_{33} - n_0^3 r_{13})\pi Vl/\lambda d \tag{37}$$

so that the half-wave voltage is

$$V\pi = (\lambda/[n_e^3 r_{33} - n_0^3 r_{13}])(d/l) \tag{38}$$

6. ORGANIC CRYSTALLINE ELECTRO–OPTIC MATERIALS

There has recently been a great deal of interest in organic compounds for electro-optical and nonlinear optical applications.[7] Certain types of molecules that are asymmetric and conjugated (i.e., have structures containing alternating single and double carbon–carbon bonds) have been shown to have very high *molecular* nonlinearities β. If such molecules are arranged within a crystal in such a way that their molecular nonlinearities are additive, very high electro-optic coefficients may occur. One such material (DCNP) has recently been described.[8] In this organic crystal, all the molecules lie parallel to each other. The crystal symmetry of DCNP is m, so that the crystal is biaxial. In general, in the biaxial point groups, the r tensor is complex, with many independent coefficients. However, in this case, because all the molecules lie parallel to each other and to the optical z axis, the r_{33} coefficient for this compound will be much greater than any of the other components. Measurements were made on samples oriented for propagation along the y axis, of light polarized at 45° to the x and z axes. The electric field was applied along the z axis, in the transverse configuration described in Section 5.2. In this geometry the half-wave voltage is given by

$$V_\pi = (\lambda/n_z^3 r_{33})(d/l) \tag{39}$$

and was measured as 145 V for a crystal having $l/d = 2.55$. This corresponds to a reduced half-wave voltage of 370 V at 633 nm, compared with a value of about 3 kV for lithium niobate. This large reduction in half-wave voltage demonstrates the potential of organic materials. In fact, measurements of the electro-optic parameters as a function of wavelength show that at longer wavelengths the advantage of DCNP over lithium niobate is reduced.[9]

7. POLYMERIC ELECTRO-OPTIC MATERIALS

Although organic crystals show great promise for bulk devices, there are a number of problems associated with their development:

i. Many molecules known to have high molecular nonlinearities β crystallize in centrosymmetric or pseudocentrosymmetric structures, which result in low or zero r coefficients.
ii. Growth of crystals is time consuming.
iii. Single-crystal growth is not readily adapted to the production of thin films for waveguides.
iv. Mechanical properties tend to be poor.

Over the last few years work has focused on methods for the production of polymeric materials in which the highly nonlinear molecules are dissolved in or attached to an inert polymeric matrix.[10] The nonlinear molecules are aligned by the application of a strong dc electric field at high temperatures (where the polymer matrix is rubbery), and then this alignment is "frozen" in by cooling the material down into its glassy state with the field still applied. This technique has a number of advantages:

i. Standard techniques such as spin coating or dip coating can be readily used for the production of thin films. Waveguides can be produced by lithographic techniques, or formed by the birefringence induced during poling.[11]
ii. Orientation can in principle be controlled: the degree of alignment is governed by the molecular dipole moment, field strength, and temperature, and can be predicted.
iii. Polymer films are mechanically and thermally robust, and cheap to produce. They may also have very good optical properties, in terms of low scattering and absorption losses.

Waveguide devices have recently been demonstrated in these materials, and r coefficients comparable to those of lithium niobate are claimed.[12]

REFERENCES

1. A. Yariv and P. Yeh, *Optical Waves in Crystals*, Chap. 7 and 8, Wiley, New York (1984).
2. J. F. Nye, *Physical Properties of Crystals*, Oxford U.P., London (1957).
3. N. H. Hartshorne and A. Stuart, *Practical Optical Crystallography*, Edward Arnold, London (1964).
4. D. W. Smith, Techniques for multigigabit coherent optical transmission, *J. Lightwave Tech.* **LT-5**, 1466–1478 (1987).
5. M. Papuchon, Integrated optical modulation and switching, in *New directions in guided wave and coherent optics*, Vol. II, (D. B. Ostrowsky and E. Spitz, eds.), pp. 371–404, Martinus Nijhoff, The Hague (1984).
6. P. Granestrand, B. Stoltz, L. Thylen, K. Bergvall, W. Dodlissen, H. Heinrich, and D. Hoffman, *Electron. Lett.* **22**, 816–817 (1986).
7. D. S. Chemla and J. Zyss (eds.) *Nonlinear Optical Properties of Organic Molecules and Crystals*, Vol. 1 and 2, Academic, Orlando (1987).
8. S. Allen, T. D. McLean, P. F. Gordon, B. D. Bothwell, M. B. Hursthouse, and S. A. Karaulov, A novel organic electro-optic crystal: 3-(1,1-dicyanoethenyl)-1-phenyl-4,5-dihydro-1H-pyrazole, *J. Appl. Phys.* **64**, 2583–2590 (1988).

9. S. Allen, Electro-optic properties of a new organic pyrazole crystal, in *Organic Materials for Nonlinear Optics* (R. A. Hann and D. Bloor, eds.), R.S.C., London (1989).
10. R. N. DeMartino, E. W. Choe, G. Khanarian, D. Hass, T. Leslie, G. Nelson, J. Stamatoff, D. Stuetz, C. C. Teng, and H. Yoon, Development of polymeric nonlinear optical materials, in *Nonlinear Optical and Electroactive Polymers* (P. N. Prasad and D. R. Ulrich, eds.), pp. 169–188, Plenum Press, New York (1988).
11. J. I. Thackara, G. F. Lipscombe, M. A. Stiller, A. J. Ticknor, and R. Lytel, Poled electro-optic waveguide formation in thin-film organic media, *Appl. Phys. Lett.* **52**, 1031–1033 (1988).
12. R. Lytel, G. F. Lipscombe, M. Stiller, J. I. Thackara, and A. J. Ticknor, Organic electro-optic waveguide modulators and switches, in *Nonlinear Optical Properties of Organic Materials* (G. Khanarian, ed.), Proc. SPIE 971, pp. 218–229 (1988).

22

Nonlinear Waveguides

M. J. Goodwin

1. INTRODUCTION

The use of monomode optical fibers in optical communications systems is widespread, and their transmission losses at the communications wavelengths of 1.3 and 1.55 μm are now well below 1 dB/km, facilitating the use of very long, high-capacity data links. Optoelectronic components for producing optical signals and converting them back to electronic signals are well established, though technological advancement continues. The existence of optical data links provides strong motivation for developing integrated optical devices capable of manipulating optical signals directly without the need to return to electronics. These devices utilize nonlinear optical effects in guided wave geometries compatible with fiber optic systems, and are capable of performing increasingly complex high-speed signal-processing functions.

High-speed optical waveguide modulators and large, high-capacity optical switch arrays will find immediate applications in telecommunications systems, providing data multiplexing and signal-routing functions. Their use will allow higher data rates to be used, offering the capability to provide additional services such as high-definition cable television, video phones, and computer links. However, their use is not limited to telecommunications systems. Integrated optics offers devices capable of reducing the wavelength of laser diode sources to the blue region of the optical spectrum at which higher data storage capacity can be achieved in optical memories such as compact disks. A wide variety of sensor devices are being developed, ranging from highly sensitive position sensors for use in optical gyroscopes to disposable biosensors for blood testing.

Longer-term applications of integrated optics will utilize the next generation of nonlinear waveguide devices based on "all-optical" switching. These will be ultrafast optical switches operating with femtosecond ($<10^{-12}$ seconds) response times, in which optical beams are used to directly control other optical beams, providing the optical logic gates and routing components required for the development of optical computing and signal-processing systems.

M. J. Goodwin • Plessey Research Caswell Limited, Allen Clark Research Centre, Caswell, Towcester, Northamptonshire NN12 8EQ, U.K.

315

1.1. Nonlinear Optics

Nonlinear optical interactions occur when the electric fields associated with optical beams propagating through a medium are large enough to produce polarization fields proportional to two or more of the incident fields. These nonlinear polarization fields are responsible for a wide range of effects. Formally, the polarization $\mathbf{P}$ is given by

$$\mathbf{P} = \chi(1)\mathbf{E} + \chi(2)\mathbf{E} \cdot \mathbf{E} + \chi(3)\mathbf{E} \cdot \mathbf{E} \cdot \mathbf{E} + \cdots \tag{1}$$

where $\mathbf{E}$ is the electric field vector and $\chi(1)$, $\chi(2)$, and $\chi(3)$, are the first-, second-, and third-order macroscopic susceptibilities of the medium. On a molecular level, this can be written as

$$\mathbf{p} = \alpha\mathbf{E} + \beta\mathbf{E} \cdot \mathbf{E} + \gamma\mathbf{E} \cdot \mathbf{E} \cdot \mathbf{E} + \cdots \tag{2}$$

where in this case α, β, and γ are the first-, second-, and third-order microscopic susceptibilities and $\mathbf{p}$ the molecular dipole moment. One important constraint is that only materials possessing an ordered, noncentrosymmetric structure will display a nonzero $\chi(2)$ value, whereas all materials will display nonzero $\chi(1)$ and $\chi(3)$ values. This has significant consequences on the choice of nonlinear material and the device fabrication techniques that can be employed.

Nonlinear effects resulting from $\chi(2)$ are termed second-order nonlinearities, while those due to $\chi(3)$ are third-order nonlinearities. Table 1 details a number of the more important operations that can be performed using nonlinear optical effects. It is clear that the utilization of nonlinear effects enables a range of significant device applications to be accessed. Second and third harmonic generation and frequency mixing are widely employed to produce additional operating wavelengths for lasers, and for this the efficiency of the nonlinear interaction is of great importance. The linear electro-optic effect has been employed for many years to produce optical shutters and modulators. More recently, the field of integrated optics has taken advantage of the large electro-optic coefficients of lithium niobate in the

TABLE 1. Nonlinear Optical Effects

Second-order nonlinearity	Second-harmonic generation	$\omega + \omega \rightarrow 2\omega$
	Frequency mixing	$\omega_1 + \omega_2 \rightarrow \omega_3$
	Pockels effect (linear electro-optic effect)	$\omega + 0 \rightarrow \omega$
	Parametric amplification	$\omega_3 \rightarrow \omega_1 + \omega_2$
Third-order nonlinearity	Third-harmonic generation	$\omega + \omega + \omega \rightarrow 3\omega$
	Four-wave mixing	$\omega_1 + \omega_2 + \omega_3 \rightarrow \omega_4$
	Kerr effect (intensity-dependent refractive index)	$n = n_0 + n_2 I$
	Optical bistability	
	Quadratic electro-optic Effect	$\omega + 0 + 0 \rightarrow \omega$

production of efficient high-speed waveguide switches and modulators, in which the output of the device is controlled by an externally applied electric field. It has also been realized that many integrated optic devices can be operated in an "all-optical" mode by using materials displaying an intensity-dependent refractive index. In this case, the device output is controlled by the intensity of the incident optical fields.

2. WAVEGUIDE STRUCTURES

The efficiency of nonlinear interactions is normally proportional to the local optical intensity and to the distance over which the interaction takes place. In bulk optical devices, there is invariably a trade-off between the local optical intensity produced by focusing the beam, and the distance over which is the focused beam can be maintained. By contrast, very small beam sizes and hence correspondingly large optical intensities can be produced in waveguide geometries,[1] and additionally such geometries maintain the small beam size over distances limited only by material loss (typically between a few centimeters and many kilometers). For example, the second harmonic generation (SHG) efficiency in a bulk device can be shown[2] to vary as

$$\eta_{\text{bulk}} \sim \frac{L P_i(\omega)}{\lambda} \tag{3}$$

where L is the interaction length, $P(\omega)$ is the fundamental beam power, and λ its wavelength. The corresponding expression for a planar guided wave configuration of thickness t is

$$\eta_{\text{wg}} \sim \frac{L^2 P_i(\omega)}{t^2} \tag{4}$$

Thus, the bulk SHG efficiency is dependent on the total power, the wavelength, and the interaction length, whereas the guided wave SHG efficiency varies with the power density $[P_i(\omega)/t^2]$ and is quadratic in length. Further, assuming a waveguide thickness of order λ, one obtains

$$\frac{\eta_{\text{wg}}}{\eta_{\text{bulk}}} \sim \frac{L}{\lambda} \tag{5}$$

so that for $\lambda = 1\mu$m and a 1-cm interaction length, the waveguide configuration would theoretically be some 10^4 times more efficient than the corresponding bulk device configuration.

In general, a waveguide consists of a region of high refractive index material, bounded in one or two dimensions by material of a lower refractive index, such that light propagating in the high-index region will be confined or trapped in one or two dimensions by material of a lower refractive index, such that light propagating in the high-index region will be confined or trapped in one or two dimensions, and thus guided within that region.

Figure 1 shows a number of commonly used waveguide geometries. The first is the planar or slab waveguide (Fig. 1a), formed by depositing or forming a thin film of high refractive index on a flat substrate. A propagating beam will be trapped in the film, but can spread laterally within the film, so that light is confined in one dimension only. A channel or stripe waveguide (Fig. 1b) provides confinement in two dimensions, thus increasing the potential optical intensities, and providing the means for forming more complex waveguide device patterns. The third structure (Fig. 1c) is the optical fiber, in which light is trapped in the central high index core, again providing two-dimensional confinement. Optical fibers represent the state-of-the-art in linear waveguides, for which propagating losses are measured in dB/km rather than the dB/cm typical for planar or channel waveguides.

If one or more of the materials used to form the waveguide is optically nonlinear, the resulting structure can be used to perform nonlinear functions. The choice of waveguide structure will be determined to a large extent by the processibility and absorption of the nonlinear material being used. Clearly, optical fibers formed from fused silica enable optical fields to be confined to cross-sectional areas of a few tens of square micrometers, and propagated for hundreds of meters with negligible attenuation. However, fused silica has a zero $\chi(2)$ coefficient and a very small $\chi(3)$ coefficient. By contrast, lithium niobate has large second-order nonlinear coefficients, but waveguide lengths are limited by attenuation to about 10 cm. This illustrates the two key elements in current research into nonlinear waveguides: the synthesis and characterization of optically nonlinear materials, and the fabrication of waveguide structures of the necessary quality.

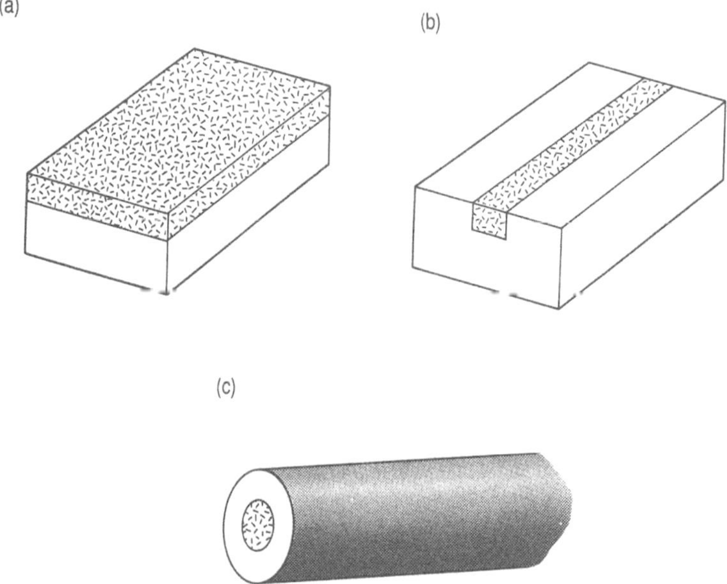

FIGURE 1.   Waveguide structures. (a) Planar waveguide; (b) channel waveguide; (c) optical fiber.

3. MATERIALS FOR NONLINEAR WAVEGUIDES

The demonstration of nonlinear waveguide devices requires that waveguiding structures can be fabricated from nonlinear materials and that the properties of the waveguides be suitable. The material requirements depend on the type of nonlinear device being fabricated, but generally it is necessary to be able to define high-quality refractive index structures on the micrometer scale, that the waveguide loss be sufficiently low to·allow an interaction length of the required size, and that the waveguide can withstand the high optical power densities required. The fabrication of waveguides for second-order nonlinear devices is now well established and uses two principal material systems, though other materials are being investigated. The first of these is titanium indiffused lithium niobate,[3] in which the selective indiffusion of titanium metal at high temperatures ($\sim 1000°C$) into the surface of polished crystaline lithium niobate wafers locally increases the refractive index, forming a waveguide structure. Complex, efficient low-loss waveguide devices have been demonstrated in this material, examples of which will be given in the next section. The second material system that is the subject of much research work utilizes epitaxially grown III–V semiconductors such as GaAs and InP, and their alloys.[4] This material system offers several attractive features. Firstly, the growth techniques used, such as metallo-organic chemical vapor deposition (MOCVD) and molecular beam epitaxy (MBE), enable very thin layers of well-controlled composition to be deposited, with which very accurate refractive index structures can be formed. Secondly, semiconductor lasers, detectors, and high-speed electronic devices can all be fabricated with these materials, implying that all of these devices could eventually be integrated onto a common substrate, and thus bring advantages in speed, cost, reliability, and stability.

For third-order nonlinear waveguide devices, a number of additional requirements need to be satisfied, such as the saturation value of the nonlinear index change, the response time of the nonlinearity, and the stability of the waveguide under high optical intensity. A wide range of material systems have been investigated using an equally wide range of waveguide formation techniques. Table 2 presents a selection of materials and techniques, together with references from which further details can be obtained. However, to illustrate some of the relevant issues one technique will be described in more detail: the formation of doped spin-coated

TABLE 2. Third-Order Nonlinear Waveguide Fabrication Techniques

Material	Waveguide fabrication	Reference
Lead doped glass	Slab and fiber guides	5
Semiconductor doped glass	Potassium ion exchange	6
GaAs/GaAlAs multiple quantum wells	VPE, MOVPE	7
Zinc oxide, ZnO	RF sputtering	8
Polydiacetylene, nitroanilines	Thin-film crystal growth	9, 10
Nitroanilines	Crystal cored fibers	11
Preformed polymers	Langmuir–Blodgett films	12
Nitroanilines	Solvent-assisted indiffusion	13

polymer film waveguides.[14] This technique involves forming a solution containing a nonlinear organic dopant material and a polymer which will form a suitable matrix. Thin films of doped polymer are then formed by spin-coating this solution onto an inert glass substrate and baking it to remove the solvent. The main attraction of this technique is that it enables high-quality thin films to be formed with the nonlinear organic material worked into an inert, robust polymer. The technique is simple, cheap, and appropriate for mass production, involving simple processes standard in semiconductor circuit fabrication. Of importance for device design is the flexibility and control of waveguide refractive index, and this is achieved by choice of dopant and polymer refractive index and film thickness. Waveguide propagation loss is minimized by using high-purity compounds, optically polished glass substrates, and the optimal choice of solvent. Monomode waveguides with high doping concentrations have been produced by this technique with less than 1 dB/cm loss. A variety of waveguide patterning techniques can be used with this material system including ion-beam milling, photopolymerization, photolocking (of the dopant), and additional dopant indiffusion. This fabrication technique results in amorphous waveguides with no orientational structure, and consequently can only be used for third-order nonlinear devices. However, additional processing can be used to cause an alignment of the nonlinear dopant molecules within the polymer host,[15] after which second-order nonlinear effects can be accessed. This poling technique involves heating the doped-polymer film above its glass transition temperature to increase the mobility of the dopant molecules, and applying an intense electric field across the film. The dopant molecules align themselves to the electric field, and the alignment is then "frozen in" by cooling the sample down again.

4. NONLINEAR WAVEGUIDE DEVICES

4.1. Second-Order Waveguide Devices

The second-order waveguide device in which the most significant progress has been made is the electro-optic switch or modulator, whereby an applied electric field is used to control the optical output from the waveguide. A variety of materials have been used to fabricate these devices, including inorganic dielectrics such as lithium niobate, lithium tantalate, and potassium niobate, semiconductors such as gallium arsenide and indium phosphide, and organic materials such as nitroaniline derivatives and nonlinear side-chain polymers. Of these, lithium niobate waveguides formed by titanium indiffusion are perhaps the most advanced, with devices now becoming available commercially. Figure 2 illustrates a state-of-the-art lithium niobate electro-optic device, an 8×8 waveguide switch array.[16] This device takes eight separate inputs and reroutes them to eight outputs, using 28 separate switching elements. The switching voltages are 26 V, and crosstalk between channels is better than 20 dB (limited by the measurement technique). Overall the device loss was 5.5 dB, of which 3 dB was propagation loss (0.46 dB/cm), 1.5 dB was reflection loss and 1 dB was bend loss. Such switching arrays will find widespread application for telecommunications and local area networks. Lithium niobate waveguide devices have also been used to demonstrate high-speed modulation,[17] multiplexing–demultiplexing,[18] second-harmonic generation,[19] and parametric oscillation.[20]

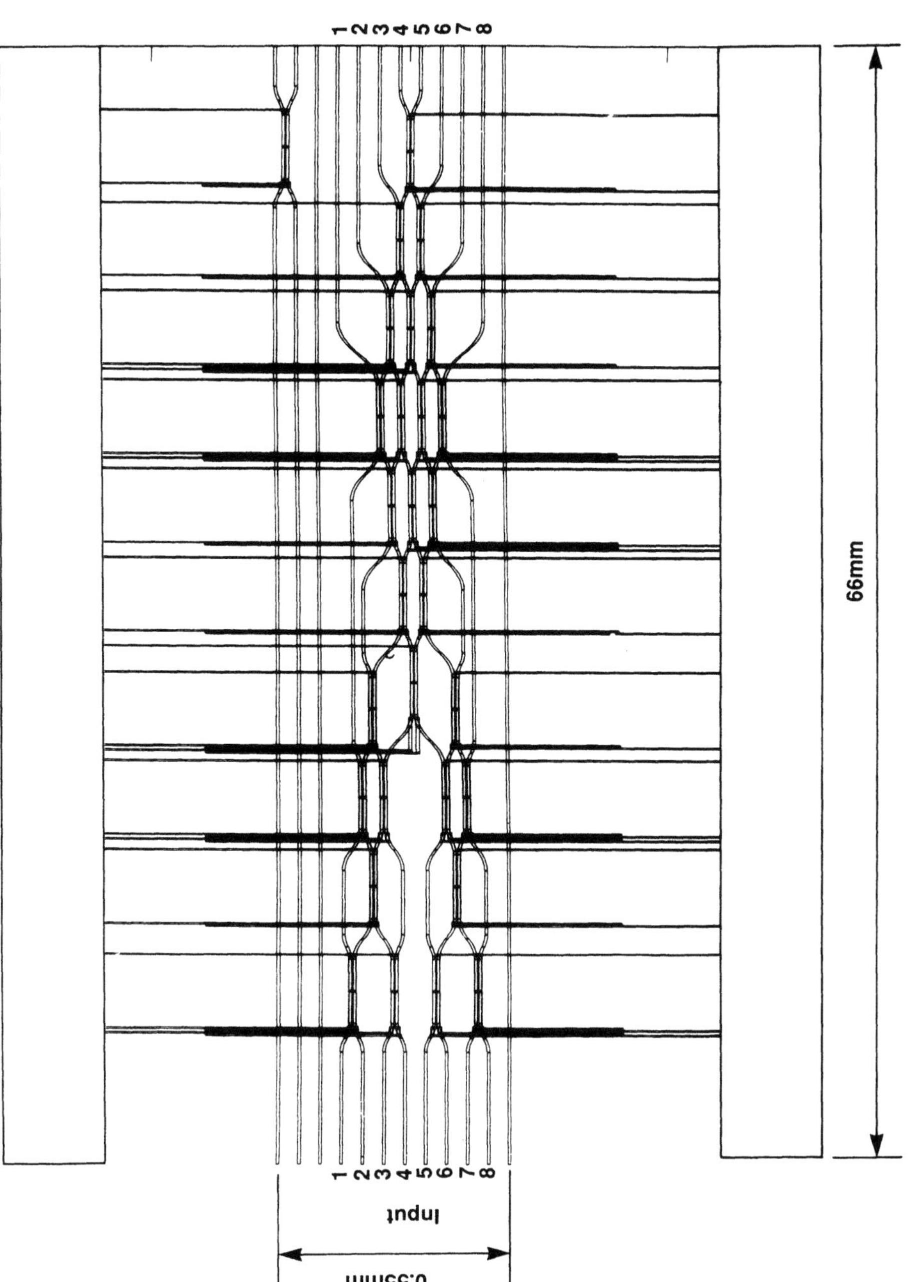

FIGURE 2. 8 × 8 Lithium niobate switch.

Loss–loss III–V semiconductor waveguide devices have been demonstrated including high-speed Mach–Zehnder modulators,[21] multiple quantum well (MQW) modulators,[22] and integrated distributed feedback (DFB) laser/modulator structures.[23]

4.2. Third-Order Waveguide Devices

The majority of third-order nonlinear waveguide devices are based on materials exhibiting an intensity-dependent refractive index[24] (though examples of waveguide degenerate four-wave mixing have been reported[25]). In these materials, the local refractive index, n, varies as

$$n = n_0 + n_2 I \tag{6}$$

where n_0 is the low-power refractive index, n_2 is the nonlinear refractive index, and I is the local intensity. The velocity of light guided in a waveguide is determined by the effective index β of the guide, and in an intensity-dependent material this will also be a function of the local optical intensity, where

$$\beta = \beta_0 + \Delta\beta_0 P_i \tag{7}$$

Here P_i is the guided wave power, and $\Delta\beta_0$ is determined by n_2 and the distribution of light within the waveguide. There are two principal ways in which Eq. (7) affects the operation of a waveguide device. A cumulative nonlinear phase shift $\Delta\phi$, where

$$\Delta\phi = \frac{2\pi\Delta\beta_0 LP}{\lambda} \tag{8}$$

occurs over a propagation distance L, which has the consequence of making interference (i.e., phase-dependent) effects power dependent. Secondly, any function that requires matching the velocity of optical waves, such as distributed waveguide coupling, will also become power dependent.

The nonlinear response of a variety of intensity-dependent waveguide devices is shown in Fig. 3. In each case, the device output is seen to vary with the incident power, and "all-optical" switching is observed. In practice the response of such devices is strongly influenced by material properties such as loss and saturation of Δn.

To illustrate the operation of third-order nonlinear waveguide devices, two examples will be described. The first of these utilizes the large nonlinearity offered by certain organic materials.[26] In this case the nonlinear response is a result of loosely bound, delocalized electron clouds in the molecule, which can move rapidly in response to the applied optical field. Extremely fast response times have been measured with these materials,[27] making them potentially attractive for sub-picosecond ($<10^{-12}$) optical switching. Nonlinear prism coupling[28] was demon-

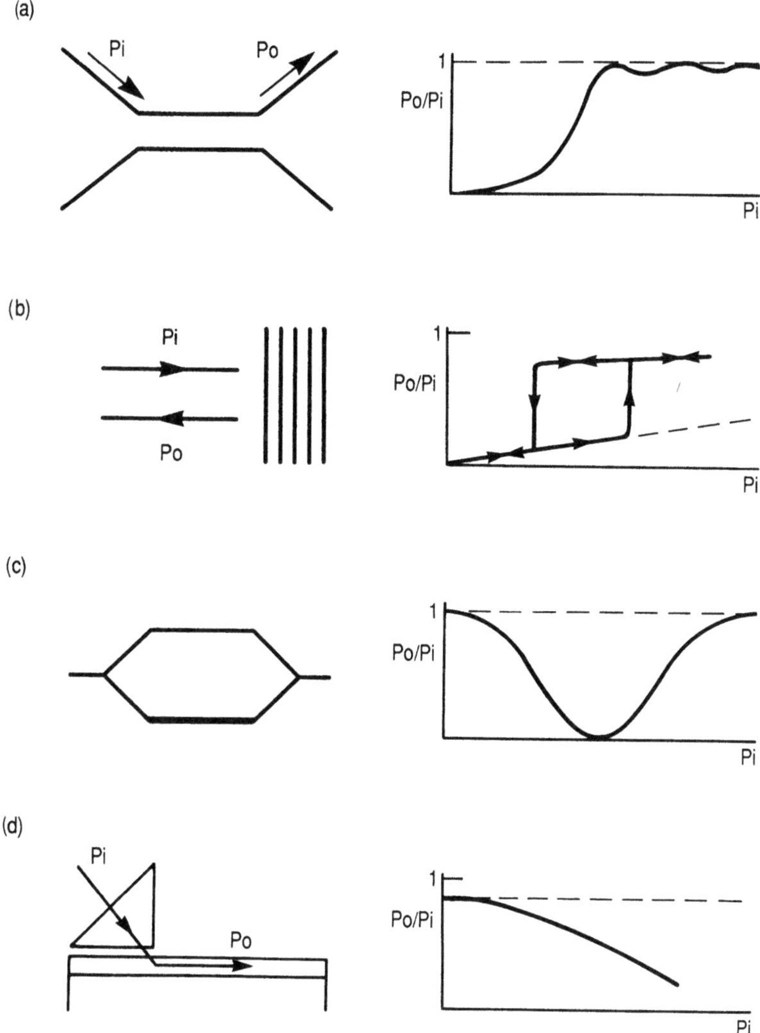

FIGURE 3. Linear and nonlinear response of waveguide devices. (a) Directional coupler; (b) distributed feedback grating; (c) Mach–Zehnder interferometer; (d) prism coupler. (After Ref. 24.)

strated using a planar waveguide formed by spin-coating a very thin polymer film, highly doped with the nonlinear organic material methyl-nitroaniline (MNA). High-intensity optical pulses from a Nd:YAG laser operating at 1.06 μm wavelength were launched into the waveguide using a prism coupler. The coupling efficiency of a prism coupler is a function of the effective index β of the guide and the angle at which light is incident on the prism, as illustrated in Fig. 4. Maximum efficiency occurs when the following condition is satisfied:

$$\beta = \frac{2\pi n_p}{\lambda} \sin\left[Y + \sin^{-1}\left(\frac{n_a}{n_p} \sin \alpha\right)\right] \tag{9}$$

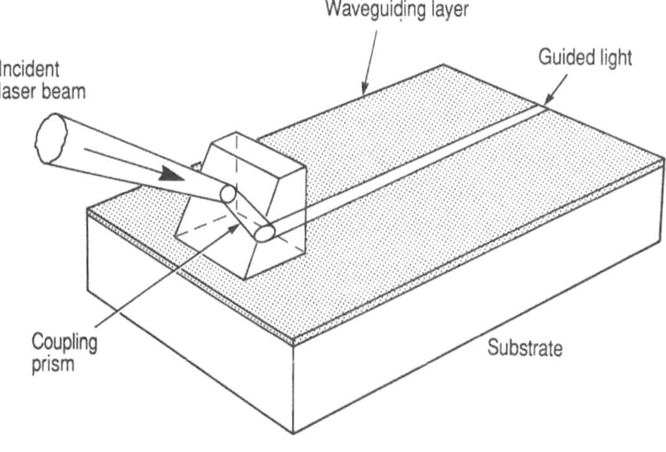

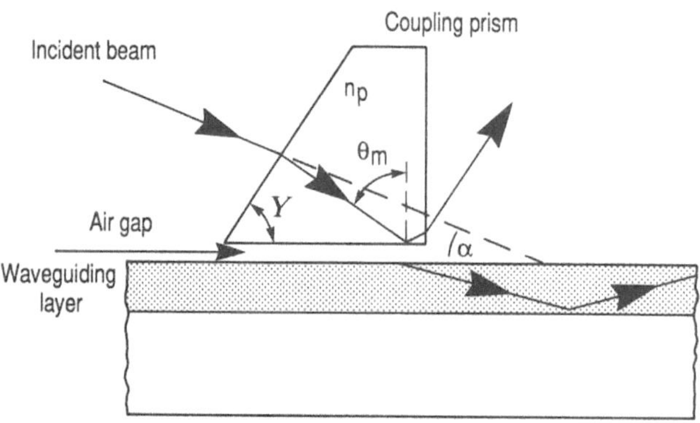

FIGURE 4. Configuration for prism coupling.

where n_p and n_a are the prism and air refractive indices, respectively, Y is the prism angle, and α is the angle of incidence. In conjunction with Eq. (7), it can be seen that as the input power is increased from an initial level for which Eq. (9) is satisfied, the effective index changes and the angle α, for which maximum coupling efficiency is obtained, changes. Consequently, if α is held constant, the efficiency of coupling decreases with increasing intensity. At high input intensities, the coupling efficiency can be largely restored by adjusting α to satisfy Eq. (9) for the new value of β.

A typical response is shown in Fig. 5, in which a marked reduction in coupler efficiency was observed as the input light intensity was increased.[14] Significant recovery in coupler efficiency was obtained by realigning the input beam.

The second example illustrates nonlinear optical switching in a waveguide directional coupler.[29] The coupler, shown schematically in Fig. 6, was formed using a short length of dual-core silica fiber, the two cores being approximately $8\ \mu$m

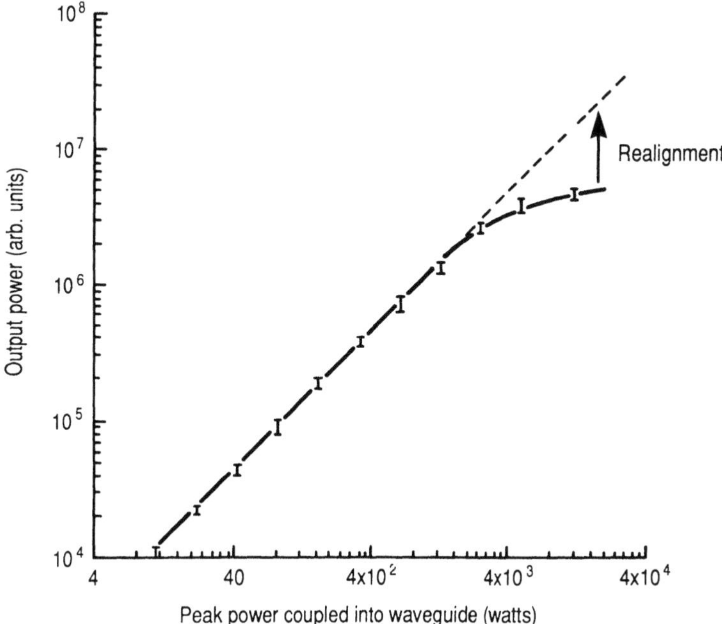

FIGURE 5. Nonlinear response for a prism-coupled MNA-polymer waveguide.

apart. Low-intensity light coupled into waveguide 1 couples across the gap to waveguide 2, the transfer being complete after a distance L. If the input intensity is increased, the intensity-dependent refractive index of silica causes the phase of the guided light to change, interfering with the coupling process and preventing transfer to waveguide 2. Thus, below a critical power, the light emerges from waveguide 2; above the critical power, it emerges from waveguide 1. The performance of this device is shown in Fig. 7. The device was excited using extremely short pulses (100 fs) from a colliding-pulse–mode-locked (CPM) dye laser, and, as predicted,

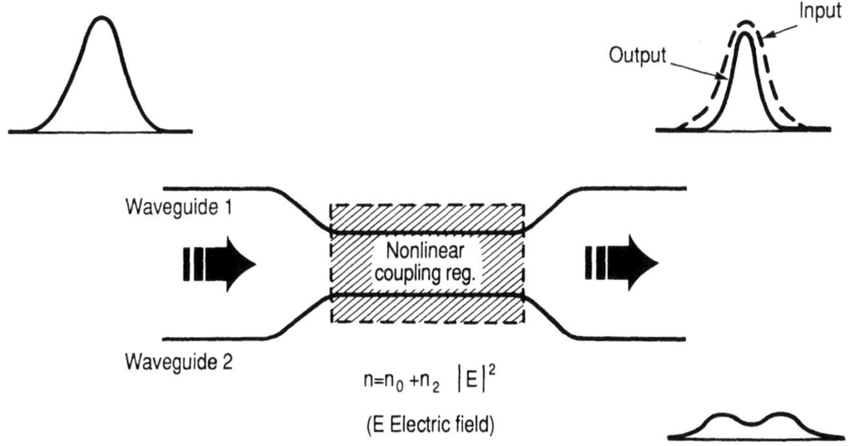

FIGURE 6. Schematic diagram of a nonlinear directional coupler.

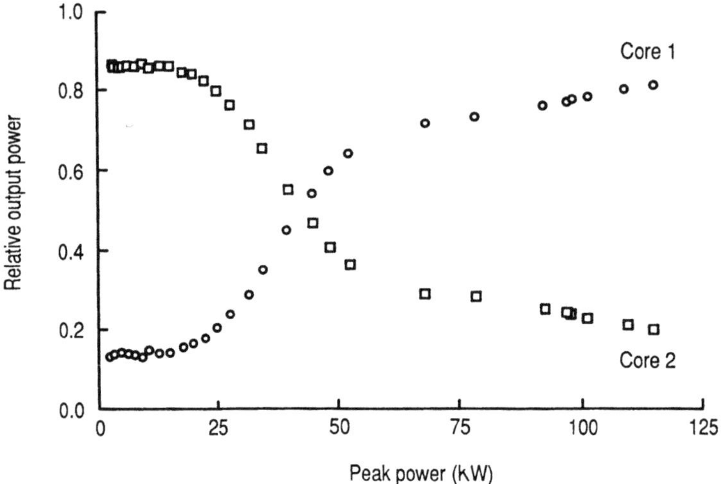

FIGURE 7.   Measured response of the nonlinear fiber directional coupler.

optical switching occurs with a critical power of approximately 35 kW. These results show that the response time of the nonlinearity in silica is faster than 100 fs, and it represents the fastest all-optical switch reported to date.

A variety of other third-order nonlinear waveguide devices have been reported, including nonlinear grating couplers,[30] nonlinear Mach–Zehnder modulators,[31] nonlinear waveguide birefringence,[14] and nonlinear guided modes.[32]

5. SUMMARY

The use of optical nonlinearities in waveguiding structures has found widespread applications in optical communications and signal processing. Integrated optical waveguide modulators and switches are now reaching the stage of commercial exploitation, and efficient waveguide frequency conversion devices have been demonstrated in research laboratories. The investigation of third-order nonlinearities for all-optical switching devices is still at an early stage, but promises ultrafast switching and optical logic functions for the next generation of integrated optical waveguide device.

This chapter has attempted to provide a brief introduction to this exciting research area, with illustrative examples of recent experimental results. Further information can be found in the references provided, but for a general view of the state of current research, readers are directed to the recent *Journal of Lightwave Technology* Special Issue on Integrated Optics (June 1988), edited by J. T. Boyd, and the *Journal of the Optical Society of America B* Special Issue on Nonlinear Guided-Wave Phenomena (February 1988), edited by G. I. Stegeman and R. H. Stolen.

REFERENCES

1. G. I. Stegeman and C. T. Seaton, Nonlinear integrated optics, *J. Appl. Phys.* **58**, R57 (1985).
2. J. Zyss, Nonlinear organic material for integrated optics: A review, *J. Molec. Electron.* **1**, 25 (1985).
3. R. Schmidt and I. Kaminow, Metal diffused optical waveguides in $LiNbO_3$, *Appl. Phys. Lett.* **25**, 458 (1974).
4. S. Wang and S. Lin, High speed III–V electro-optic waveguide modulators at $\lambda = 1.3 \mu$m, *IEEE J. Lightwave Tech.* **6**, 758 (1988).
5. S. R. Friberg and P. W. Smith, Nonlinear optical glasses for ultrafast optical switches, *IEEE J. Quantum Electron* **QE-23**, 2089 (1987).
6. T. J. Cullen, C. N. Ironside, C. T. Seaton, and G. I. Stegeman, Semiconductor-doped glass ion-exchanged waveguides, *Appl. Phys. Lett.* **49**, 1403 (1986).
7. M. Erman, P. Jarry, R. Gamonal, P. Autier, J. P. Chane, and P. Frijlink, Mach–Zehnder modulators and optical switches on III–V semiconductors, *IEEE J. Lightwave Tech.* **6**, 837 (1988).
8. R. M. Fortenberry, G. Assanto, R. Moshrefzadeh, C. T. Seaton, and G. I. Stegeman, Pulsed excitation of nonlinear distributed coupling into zinc oxide optical guides, *IEEE J. Lightwave Tech.* **5**, 425 (1988).
9. For examples, M. Thakur, and S. Meyler, Growth of large-scale thin-film single crystals of poly (diacetylenes), *Macromolecules* **18**, 2341, (1985); H. Itoh, K. Hotta, H. Takara, and K. Sasaki, The growth of 2-methyl-4-nitroaniline single crystalline thin films for phase-matched frequency-doubling, *Opt. Commun.* **59**, 299 (1986).
10. S. Tomaru, M. Kawachi, and M. Kobayashi, Organic crystals growth of optical channel waveguides, *Opt. Commun.* **50**, 154 (1984).
11. B. K. Nayer, Nonlinear optical interactions in organic crystal cored fibres, in *Nonlinear Properties of Organic and Polymeric Materials*, ACS Symposium Services 233, American Chemical Society, Washington, D.C., p. 153 (1983) and references cited therein.
12. R. H. Tredgold, M. C. J. Young, P. Hodge, and E. Khoshdel, Lightguiding in Langmuir–Blodgett films of performed polymers, *Thin Solid Films* **151**, 441 (1987).
13. M. J. Goodwin, R. Glenn, and I. Bennion, Organic nonlinear optical waveguides formed by solvent assisted indiffusion, *Electron. Lett.* **22**, 789 (1986).
14. M. J. Goodwin, C. Edge, C. Trundle, and I. Bennion, Intensity dependent birefringence in nonlinear organic polymer waveguides, *J. Opt. Soc. Am. B* **5**, 419 (1988).
15. K. D. Singer, J. E. Sohn, and S. J. Lalama, Second harmonic generation in poled polymer films, *Appl. Phys. Lett.* **49**, 248 (1986).
16. P. J. Duthie and M. J. Wale, Rearrangeably nonblocking 8×8 guided wave optical switch, *Electron. Lett.* **24**, 594 (1988).
17. S. Korotky, G. Eisenstein, R. Tucker, J. Veselka, and G. Raybon, Optical intensity modulation to 40 GHz using a waveguide electro-optic switch, *Appl. Phys. Lett.* **50**, 1631 (1987).
18. S. K. Korotky, A. Gnauk, B. Kasper, J. Campbell, J. Veselka, J. Talman, and A. McCormick, 8-Gbit/s transmission experiment over 68 km of optical fibre using a Ti : $LiNbO_3$ external modulator, *IEEE J. Lightwave. Tech.* **5**, 1180 (1985).
19. R. Rogener and W. Sohler, Efficient second harmonic generation in Ti : $LiNbO_3$ channel waveguide resonators, *J. Opt. Soc. Am. B* **5**, 267 (1988).
20. W. Sohler and H. Suche, in *Integrated Optics III*, (L. D. Hutcheson and D. G. Hall, eds.) Proc. SPIE **408**, 163 (1983).
21. S. Wang, S. Lin, and Y. Houng, GaAs traveling-wave polarization electro-optic waveguide modulator with bandwidth in excess of 20 GHz at 1.3 μm, *Appl. Phys. Lett.* **51**, 83 (1987).
22. T. Wood, E. Carr, C. Burrus, R. Tucker, T. Chiu, and W. Tsang, High-speed waveguide optical modulator made from GaSb/AlGaSb multiple quantum wells (MQWs), *Electron. Lett.* **23**, 540 (1987).
23. Y. Kawamura, K. Wakita, Y. Yoshikuru, Y. Itaya, and H. Asahi, Monolithic integration of InGaAsP/InP DFB lasers and InGaAs/InAlAs MQW optical modulators, *Electron. Lett.* **22**, 242 (1986).
24. G. I. Stegeman, E. M. Wright, N. Finlayson, R. Zanoni, and C. T. Seaton, Third order nonlinear integrated optics, *IEEE J. Lightwave Tech.* **6**, 953 (1988).
25. A. Gabel, V. W. Delong, C. T. Seaton, and G. I. Stegeman, Efficient degenerate four-wave mixing in an ion exchanged semiconductor-doped glass waveguide, *Appl. Phys. Lett.* **51**, 1682 (1987).

26. D. J. Williams, ed., *Nonlinear Properties of Organic and Polymeric Materials*, ACS Symposium Series 233, American Chemical Society, Washington, D.C. (1983).
27. G. M. Carter, J. V. Hryniewicz, M. K. Thakur, Y. J. Chen, and S. E. Meyler, Nonlinear optical processes in polydiacetylene measured with femtosecond duration pulses, *Appl. Phys. Lett.* **49**, 988 (1986).
28. C. Liao, G. I. Stegeman, C. T. Seaton, R. L. Shoemaker, and J. D. Valera, Nonlinear distributed waveguide couplers, *J. Opt. Soc. Am. A* **2**, 590 (1985).
29. S. R. Friberg, A. M. Weiner, Y. Silberberg, B. G. Sfez, and P. W. Smith, Femtosecond switching in a dual-core-fiber nonlinear coupler, *Opt. Lett.* **13**, 904 (1988).
30. M. J. Goodwin, C. J. Rowe, I. Bennion, and C. Trundle, Periodic grating structures in nonlinear optical waveguides, Proceedings of Fourth European Conference on Integrated Optics (Glasgow, Scotland), p. 182 (1987).
31. P. Li Kam Wa, P. N. Robson, J. P. R. David, G. Hill, P. Mistry, M. A. Pate, and J. S. Roberts, All-optical switching effects in a passive GaAs/GaAlAs multiple quantum well waveguide resonator, *Electron. Lett.* **22**, 1129 (1986).
32. H. Vach, C. T. Seaton, G. I. Stegeman, and I. C. Khoo, Observation of intensity-dependent guided waves, *Opt. Lett.* **9**, 238 (1984).

23

Second-Harmonic Generation

Malcolm H. Dunn

1. INTRODUCTION

In optical second-harmonic generation coherent optical radiation of frequency 2ω is generated, through the use of a medium with the appropriate nonlinear susceptibility, from incident coherent optical radiation of frequency ω. For the process to take place efficiently, the phase matching condition (colinear geometry)

$$k(2\omega) = 2k(\omega) \tag{1}$$

must be fulfilled, where $k(\omega)$ and $k(2\omega)$ are the wave numbers, in the nonlinear medium, for the coherent radiations at frequencies ω and 2ω, respectively.

Over the last few years significant developments have enhanced the potential of this technique in terms of both new wavelengths generated and improved conversion efficiencies.

These developments can be classified into (i) the availability of new nonlinear materials [such as beta barium borate (BBO), lithium triborate (LBO), urea, organics such as dimethylamino nitroacetanilide (DAN), and polymerics] which exhibit markedly improved nonlinear coefficients and optical damage properties, and (ii) improved primary sources that provide incident radiation with improved coherence properties (mode quality, frequency stability) allowing resonant enhancement techniques to be used to greatly improve conversion efficiencies in the nonlinear process. Current developments relating to both these aspects will be reviewed later in this chapter. We now develop the basic physics underlying second-harmonic generation, and then apply this to the practical design of second-harmonic generation devices.

2. FORMULATION

In the case of linear optics, the interaction of optical radiation with a medium is usually described through the induced linear polarization in the medium given by

$$P = \varepsilon_0 \chi^{(1)} E, \tag{2}$$

Malcolm H. Dunn ● Department of Physics and Astronomy, University of St. Andrews, North Haugh, St. Andrews, Fife KY16 9SS, Scotland, U.K.

where $\chi^{(1)}$ is the linear susceptibility of the medium and E is the electric field of the optical radiation. Strictly the equation as written is applicable only to the case where E describes a harmonic field of some given frequency, $\chi^{(1)}$ being a function of frequency. In the case where the applied electric field strength is large (approaching the intra atomic electric field, for example), the above linear relation no longer suffices, and (2) is replaced by a power series in the field, namely,

$$P = \varepsilon_0 \chi^{(1)} E + \varepsilon_0 \chi^{(2)} E^2 + \varepsilon_0 \chi^{(3)} E^3 + \cdots \tag{3}$$

Again, the equation, as written, is only strictly applicable to the case where E describes a wave that is a collection of harmonic waves, the values of the χ's being dependent on those particular harmonic components relevant to the particular nonlinear process under consideration (again, this aspect will become clearer when it is discussed more fully later).

In the current situation we are concerned only with the first nonlinear term $\chi^{(2)}$. This is referred to as the second-order nonlinear susceptibility. Since it is only finite in anisotropic media, it is necessary to write Eq. (3) in terms of vector rather than scalar electric fields and polarizations, and generalizing in this way we hence obtain

$$\mathbf{P} = \varepsilon_0 \mathbf{X}^{(1)} \cdot \mathbf{E} + \varepsilon_0 \mathbf{X}^{(2)} : \mathbf{EE} \tag{4}$$

where the $\mathbf{X}$'s are tensors. The meaning of this equation will become clearer when we discuss its component form later.

We write the harmonic waves in complex notation adopting the usual convention of linear optics, namely,

$$\mathbf{E}(\mathbf{k}, \omega) = \tfrac{1}{2}(\mathscr{E}\, e^{i\mathbf{k}\cdot\mathbf{r}-i\omega t} + \mathscr{E}^*\, e^{-i\mathbf{k}\cdot\mathbf{r}+i\omega t})$$

$$= \mathrm{Re}(\mathscr{E}\, e^{i\mathbf{k}\cdot\mathbf{r}-i\omega t}) \tag{5}$$

where Re indicates the real part of the expression following. The amplitude of the wave is $|\mathscr{E}|$. By allowing $\mathscr{E}$ to be complex, the initial phase of the harmonic wave can be incorporated since

$$\mathrm{Re}(\mathscr{E}\, e^{i\mathbf{k}\cdot\mathbf{r}-i\omega t}) = |\mathscr{E}|\cos(\mathbf{k}\cdot\mathbf{r} - \omega t - \phi) \tag{6}$$

where

$$\mathscr{E} = |\mathscr{E}|\, e^{-i\phi} \tag{7}$$

[The reader is warned that the convention adopted in Eq. (5) is by no means universal. The factor of $\tfrac{1}{2}$ is often omitted from (5), particularly in the early papers in the field.]

In the more general case where we are concerned with a collection of N harmonic fields, the real field is written as

$$\mathbf{E} = \tfrac{1}{2}\sum_{1=-N}^{N} \{\mathscr{E}_1\, e^{i(\mathbf{k_1}\cdot\mathbf{r}-\omega_1 t)}\} \tag{8}$$

where we adopt the convention that

$$\omega_{-1} = -\omega_1, \qquad \mathbf{k}_{-1} = -\mathbf{k}_1$$

$$\mathscr{E}_{-1} = \mathscr{E}_1^*　\tag{9}$$

As before, the real field amplitude of the harmonic component of frequency ω_1 is $|\mathscr{E}_1|$.

3. ELECTROMAGNETIC WAVE EQUATIONS IN NONLINEAR MEDIA

We develop here a general description for second-order nonlinear processes, which we will then specialize to the particular case of second-harmonic generation.

We start by writing down Maxwell's wave equation for a medium containing no free charge and with zero conductivity, namely,

$$\nabla^2 \mathbf{E} = \mu_0 \varepsilon_0 \frac{\partial^2 \mathbf{E}}{\partial t^2} + \mu_0 \frac{\partial^2 \mathbf{P}}{\partial t^2} \tag{10}$$

where we have adopted MKS units. The polarization in the above equation now contains a nonlinear component as described previously by Eq. (4).

Equations (10) and (4) describe the behavior of the real electric field (we will emphasize this by use of the subscript R) in terms of a real polarization. We wish to develop a description in terms of complex functions. The link between the two is somewhat simplified in that we are concerned here only with harmonic fields or, more correctly, with the superposition of a finite number of harmonic fields of different angular frequency.

Consider first of all the case of the linear polarization. Each of the harmonic waves gives rise to an induced linear polarization at the angular frequency of the wave. Consider first of all the case of a single component at frequency ω, say. In general the polarization might not be in phase with the inducing field; such a case occurs, for example, when there is loss in the medium. Se we write the real linear polarization in terms of the real field as

$$\mathbf{P}_R^{(1)} = \varepsilon_0 \mathbf{\chi}_R^{(1)} \cdot |\mathscr{E}| \cos(\mathbf{k} \cdot \mathbf{r} - \omega t - \phi - \theta) \tag{11}$$

This is the value of the linear polarization to be substituted into the Maxwell wave equation, and the above equation may be regarded as a definition for $\mathbf{\chi}_R^{(1)}$. Any phase difference between field and polarization is taken care of by the angle θ in the argument of the cosine. The above relation may be writen in terms of complex functions describing the field as

$$\mathbf{P}_R^{(1)} = \tfrac{1}{2}\varepsilon_0 \mathbf{\chi}_R^{(1)} \cdot [\mathscr{E}\, e^{i(\mathbf{k}\cdot\mathbf{r} - \omega t - \theta)} + \text{c.c.}] \tag{12}$$

where, as previously, the complex phase factor arising from ϕ is taken care of by the complex form of $\mathscr{E}$. If we now replace $\mathbf{\chi}_R^{(1)}$ by complex components defined by

$$\mathbf{\chi}^{(1)} = \mathbf{\chi}_R^{(1)} e^{-i\theta}$$

$$\mathbf{\chi}^{(1)*} = \mathbf{\chi}_R^{(1)} e^{+i\theta} \tag{13}$$

then the equation may be rewritten in the form

$$\mathbf{P}_R^{(1)} = \tfrac{1}{2}\varepsilon_0[\mathbf{x}^{(1)} \cdot \boldsymbol{\mathscr{E}}\, e^{i(\mathbf{k}\cdot\mathbf{r}-\omega t)} + \text{c.c.}] \tag{14}$$

where, of course,

$$|\mathbf{x}^{(1)}| = |\mathbf{x}_R^{(1)}| \tag{15}$$

We can define a complex linear polarization $\mathbf{P}^{(1)}$ such that

$$\mathbf{P}_R^{(1)} = \text{Re}\{\mathbf{P}^{(1)}\} = \tfrac{1}{2}\{\mathbf{P}^{(1)} + \mathbf{P}^{(1)*}\} \tag{16}$$

and hence

$$\mathbf{P}^{(1)} = \mathbf{x}^{(1)} \cdot \boldsymbol{\mathscr{E}}\, e^{i(\mathbf{k}\cdot\mathbf{r}-\omega t)} \tag{17}$$

There is now no difficulty in generalizing to the case of a collection of harmonic fields, namely,

$$\mathbf{P}_R^{(1)} = \tfrac{1}{2}\omega_0 \sum_{1=-N}^{1=+N} \mathbf{x}^{(1)}(\omega_1) \cdot \boldsymbol{\mathscr{E}}_1\, e^{i(\mathbf{k}_1\cdot\mathbf{r}-\omega_1 t)} \tag{18}$$

We have considered the case of complex notation in relation to the linear polarization in some detail, because of the increased difficulty encountered in applying it to the case of the nonlinear polarization, which we now tackle.

First of all, to simplify the situation, suppose we are concerned with only three fields of frequencies ω_1, ω_2 and ω_3, such that

$$\omega_3 = (\omega_1 + \omega_2) \tag{19}$$

Further, for the time being, we will omit writing in explicitly the spatial dependences of the waves.

If we consider the nonlinear polarization produced by the interaction of the two waves at frequencies ω_1 and ω_2, then we expect this to be of the form

$$\mathbf{P}_R^{(2)} = 2\varepsilon_0\mathbf{x}_R^{(2)}:|\boldsymbol{\mathscr{E}}_1|\cos[\omega_1 t + \phi_1 + \theta_1]|\boldsymbol{\mathscr{E}}_2|\cos[\omega_2 t + \phi_2 + \theta_2] \tag{20}$$

The reason for what appears to be an arbitrary act of introducing the factor of 2 on the right-hand side will become apparent shortly. However, it is worth noting that we are perfectly free to do this, since the above may be taken as a definition of $\mathbf{x}_R^{(2)}$. We can examine the frequency components present in the polarization by rewriting the product of the two fields as follows:

$$|\boldsymbol{\mathscr{E}}_1|\cos(\omega_1 t + \phi_1 + \theta_1) \cdot |\boldsymbol{\mathscr{E}}_2|\cos(\omega_2 t + \phi_2 + \theta_2)$$

$$= \tfrac{1}{2}|\boldsymbol{\mathscr{E}}_1||\boldsymbol{\mathscr{E}}_2|\cos[(\omega_1 + \omega_2)t + \phi_1 + \phi_2 + \theta_1 + \theta_2]$$

$$+ \tfrac{1}{2}|\boldsymbol{\mathscr{E}}_1||\boldsymbol{\mathscr{E}}_2|\cos[(\omega_1 - \omega_2)t + \phi_1 - \phi_2 + \theta_1 - \theta_2] \tag{21}$$

From this it is apparent that there will be a sum frequency component at frequency $(\omega_1 + \omega_2)$, and a difference frequency component at a frequency $(\omega_1 - \omega_2)$. The response of the medium, as described through the nonlinear susceptibility, is unlikely to be the same at what may be two largely different frequencies, so we must rewrite our equation in the form

$$\mathbf{P}_R^{(2)} = \varepsilon_0 \mathbf{\chi}_R^{(2)}(\omega_1 + \omega_2) : |\mathscr{E}_1||\mathscr{E}_2| \cos[(\omega_1 + \omega_2)t + \phi_1 + \phi_2 + \theta_1 + \theta_2]$$

$$+ \mathbf{\chi}_R^{(2)}(\omega_1 - \omega_2) : |\mathscr{E}_1||\mathscr{E}_2| \cos[(\omega_1 - \omega_2)t + \phi_1 - \phi_2 + \theta_1 - \theta_2] \quad (22)$$

where the reason for inclusion of the factor of 2 previously is now apparent. We will limit our considerations to the polarization at the sum frequency $(\omega_1 + \omega_2)$, which is given by

$$\mathbf{P}_R^{(2)} = \mathbf{P}_R^{(2)}(\omega_1 + \omega_2)$$

$$= \varepsilon_0 \mathbf{\chi}_R^{(2)}(\omega = \omega_1 + \omega_2) : |\mathscr{E}_1||\mathscr{E}_2| \cos[(\omega_1 + \omega_2)t + \phi_1 + \phi_2 + \theta_1 + \theta_2] \quad (23)$$

This may now be written terms of complex fields as

$$\mathbf{P}_R^{(2)}(\omega_1 + \omega_2) = \varepsilon_0 \mathbf{\chi}_R^{(2)}(\omega = \omega_1 + \omega_2) : \tfrac{1}{2}[\mathscr{E}_1\mathscr{E}_2 \, e^{-i(\omega_1+\omega_2)t} \, e^{-i(\theta_1+\theta_2)} + \text{c.c.}] \quad (24)$$

This equation may now be expressed in terms of a complex susceptibility defined by

$$\mathbf{\chi}^{(2)} = \mathbf{\chi}_R^{(2)} \, e^{-i(\theta_1+\theta_2)} \quad (25)$$

as

$$\mathbf{P}_R^{(2)}(\omega_1 + \omega_2) = \frac{\varepsilon_0}{2} [\mathbf{\chi}^{(2)} : \mathscr{E}_1\mathscr{E}_2 \, e^{-i(\omega_1+\omega_2)t} + \text{c.c.}] \quad (26)$$

Further if we define a complex nonlinear polarization $\mathbf{P}^{(2)}(\omega_1 + \omega_2)$ such that

$$\mathbf{P}_R^{(2)}(\omega_1 + \omega_2) = \tfrac{1}{2}[\mathbf{P}^{(2)}(\omega_1 + \omega_2) + \mathbf{P}^{(2)*}(\omega_1 + \omega_2)] \quad (27)$$

then we see that

$$\mathbf{P}^{(2)} = \varepsilon_0 \mathbf{\chi}^{(2)} : \mathscr{E}_1\mathscr{E}_2 \, e^{-i(\omega_1+\omega_2)t} \quad (28)$$

It is apparent from the above that

$$|\mathbf{\chi}_R^{(2)}| = |\mathbf{\chi}^{(2)}| \quad (29)$$

We can define a general real polarization in the presence of the superposition of a number (N) of harmonic fields in the following form:

$$\mathbf{P}_R^{(2)} = \frac{\varepsilon_0}{2} \sum_{m=-N}^{m=N} \sum_{l=-N}^{l=N} \mathbf{\chi}^{(2)}(\omega = \omega_1 + \omega_m) : \mathscr{E}_l \mathscr{E}_m \, e^{-i(\omega_l + \omega_m)t} \tag{30}$$

If we now restore the spatial dependence to the description, then the above equation becomes

$$\mathbf{P}_R^{(2)}(\mathbf{r}, t)$$

$$= \frac{\varepsilon_0}{2} \sum_{m=-N}^{m=N} \sum_{l=-N}^{l=N} \mathbf{\chi}^{(2)}(\mathbf{k} = \mathbf{k}_l + \mathbf{k}_m ; \omega = \omega_1 + \omega_m) : \mathscr{E}_l \, e^{i(\mathbf{k}_l \cdot \mathbf{r} - \omega_l t)} \mathscr{E}_m \, e^{i(\mathbf{k}_m \cdot \mathbf{r} - \omega_m t)} \tag{31}$$

Having carefully defined all our real functions (electric field, linear and nonlinear polarizations) in terms of associated complex functions, we can proceed to derive the wave equation in terms of these complex functions.

4. WAVE EQUATIONS FOR THREE-WAVE MIXING

In the case of second-order nonlinear processes we are usually concerned with the interaction of three harmonic waves, of angular frequencies ω_1, ω_2, and ω_3, where

$$\omega_3 = (\omega_1 + \omega_2)$$

The real electric field is hence of the form

$$\mathbf{E}_R = \tfrac{1}{2}[\mathscr{E}_1 \, e^{i(\mathbf{k}_1 \cdot \mathbf{r} - \omega_1 t)} + \mathscr{E}_2 \, e^{i(\mathbf{k}_2 \cdot \mathbf{r} - \omega_2 t)} + \mathscr{E}_3 \, e^{i(\mathbf{k}_3 \cdot \mathbf{r} - \omega_3 t)} + \text{c.c.}] \tag{32}$$

and there are associated with it real linear and nonlinear polarizations, given by Eq. (18) and (31). We proceed by substituting these into Maxwell's wave equation (10), and then separating out the components in the resulting equation at the different angular frequencies, which separately must satisfy the equation. Thus we obtain, for the case of the complex component at angular frequency ω_n.

$$-[\nabla^2 + \mu_0 \omega_n^2 \varepsilon_0 + \mu_0 \varepsilon_0 \omega_n^2 \mathbf{\chi}^{(1)}(\omega_n).]$$

$$\mathscr{E}_n \, e^{-\mathbf{k}_n \cdot \mathbf{r}} = -\mu_0 \varepsilon_0 \omega_n^2 \mathbf{\chi}^{(2)}(\omega_n = \omega_l + \omega_m) : \mathscr{E}_l \mathscr{E}_m \, e^{i(\mathbf{k}_l + \mathbf{k}_m) \cdot \mathbf{r}} \tag{33}$$

when n, l, m can take on values $\pm 1, 2, 3$, referring to the different fields present, but $n = l + m$. In deriving the above we have assumed that the complex amplitudes, $\mathscr{E}_l$, are all time independent, the only time dependence being in the complex exponential term (i.e., $e^{-i\omega t}$).

5. SLOWLY VARYING AMPLITUDE APPROXIMATION

We restrict our attention to the case where the waves are infinite plane waves propagating along the z axis, so that the spatial factor in the complex exponential, $\mathbf{k} \cdot \mathbf{r}$, is replaced by kz. If the waves at the different frequencies are interchanging energy through the nonlinear polarization terms, then the wave amplitudes will also change with positions, so that the $\mathscr{E}$s are functions of z. We hence write the fields in the form

$$\mathscr{E}_l(z)\, e^{i(k_l z - \omega_l t)}, \qquad \text{etc.}$$

We assume that the field amplitudes vary little over distances comparable to a wavelength (the adiabatic approximation). This leads to a considerable simplification in the analysis, as will become apparent shortly.

Since we are considering plane waves propagating parallel to the z axis, then ∇^2 in the wave equation can be replaced by $(\partial^2/\partial z^2)$. Consider now this operator acting on the field component $\mathscr{E}_m \exp[i(k_m z - \omega_m t)]$; we have

$$\frac{\partial^2}{\partial z^2}[\mathscr{E}_m(z)\, e^{ik_m z}] = \frac{\partial}{\partial z}\left(\frac{\partial \mathscr{E}_m}{\partial z} e^{ik_m z} + ik_m \mathscr{E}_m e^{ik_m z}\right)$$

$$= \left(\frac{\partial^2 \mathscr{E}_m}{\partial z^2} + 2ik_m \frac{\partial \mathscr{E}_m}{\partial z} - k_m^2 \mathscr{E}_m\right) e^{ik_m z} \qquad (34)$$

If the field changes only slowly over a wavelength then we have

$$\left|\frac{\partial^2 \mathscr{E}_m}{\partial z^2}\right| \ll \left|k_m \frac{\partial \mathscr{E}_m}{\partial z}\right| \qquad (35)$$

and the first term in the above may be omitted (adiabatic approximation). Further, for a freely propagating wave in the medium we have that

$$k_m^2 = \mu_0 \omega_m^2 \varepsilon_m$$

$$\varepsilon = \varepsilon_0[1 + \chi^{(1)}(\omega_m)] \qquad (36)$$

so that the wave equation (33) reduces to

$$\frac{\partial \mathscr{E}_n}{\partial z} = -\frac{i\mu_0 \varepsilon_0 \omega_n^2}{2k_n} \mathbf{\chi}^{(2)}(\omega_n = \omega_l + \omega_m) : \mathscr{E}_l \mathscr{E}_m\, e^{i(k_l + k_m - k_n)z} \qquad (37)$$

We can readily write down a similar equation for each of the field components $\mathscr{E}_m$. The resulting set of equations is extremely important, and forms the "launching point" for the analysis of a wide range of optical situations involving second-order nonlinear effects.

For convenience we write down explicitly the three coupled equations for the case of three wave mixing, where

$$\omega_3 = (\omega_1 + \omega_2)$$

namely,

$$\frac{\partial \mathscr{E}_1}{\partial z} = -\frac{i\varepsilon_0}{2}\left(\frac{\mu_0}{\varepsilon_1}\right)^{1/2}\omega_1 \mathbf{\chi}^{(2)}((\omega_1 = \omega_3 - \omega_2):\mathscr{E}_3\mathscr{E}_2^* \, e^{i(k_3-k_2-k_1)z}$$

$$\frac{\partial \mathscr{E}_2}{\partial z} = -\frac{i\varepsilon_0}{2}\left(\frac{\mu_0}{\varepsilon_2}\right)^{1/2}\omega_2 \mathbf{\chi}^{(2)}(\omega_2 = \omega_3 - \omega_1):\mathscr{E}_3\mathscr{E}_1^* \, e^{i(k_3-k_2-k_1)z}$$

$$\frac{\partial \mathscr{E}_3}{\partial z} = -\frac{i\varepsilon_0}{2}\left(\frac{\mu_0}{\varepsilon_3}\right)^{1/2}\omega_3 \mathbf{\chi}^{(2)}(\omega_3 = \omega_1 + \omega_2):\mathscr{E}_1\mathscr{E}_2 \, e^{-i(k_1+k_2-k_3)z} \tag{38}$$

6. NONLINEAR EQUATIONS IN COMPONENT FORM

It is often convenient when embarking on calculations related to particular geometries to write the equations in terms of field components rather than in terms of vector fields. If we consider the case of the complex susceptibility defined by Eq. (28), this may be written in component form as

$$P_i^{(2)}(\omega = \omega_1 + \omega_2) = \varepsilon_0 \chi_{ijk}^{(2)}(\omega = \omega_1 + \omega_2)\mathscr{E}_{1j}\mathscr{E}_{2k} \, e^{-i(\omega_1+\omega_2)t} \tag{39}$$

where $P_i^{(2)}$ is the component of the induced second-order nonlinear polarization along the direction i ($i = x, y, z$) due to electric field components $\mathscr{E}_{1j}$ and $\mathscr{E}_{2k}$, and summation over repeated indices is implied.

7. SYMMETRY CONDITIONS APPLIED TO $\chi^{(2)}$

Certain symmetry relations apply to the second-order nonlinear susceptibility when it is expressed in component form, and these can prove extremely useful in practical situations.

Using arguments based on energy conservation, it can be shown that for the case of a lossless medium

$$\chi_{ijk}^{(2)}(\omega_3 = \omega_1 + \omega_2) = \chi_{jik}^{(2)}(\omega_1 = \omega_3 - \omega_2) = \chi_{kij}^{(2)}(\omega_2 = \omega_3 - \omega_1) \tag{40}$$

Put in words, this is equivalent to the statement that the nonlinear susceptibility is unchanged on interchanging the tensor indices (i, j, k) providing their corresponding frequencies are interchanged with them while maintaining the correct signs as defined through Eq. (19).

If as well as being lossless, the medium is also nondispersive in the vicinities of the three frequencies ω_1, ω_2, ω_3 then the susceptibility is unchanged on interchanging the tensor indices while keeping the frequencies fixed. This is called Kleinman symmetry, an example of which is

$$\chi_{ijk}^{(2)}(\omega_3 = \omega_1 + \omega_2) = \chi_{jik}^{(2)}(\omega_3 = \omega_1 + \omega_2) \tag{41}$$

When we restrict our attention to the case of second-harmonic generation, certain other symmetry properties occur due to the degeneracy between the two fields at ω_1 and ω_2, which are now, of course, components of the same field. In this case it is convenient to introduce a reduced notation—the d-notation, in a way analogous to the case of piezoelectricity, which the effect resembles with regard to its symmetry properties. In the case of degenerate fields ($\omega_1 = \omega_2$), the tensor describing $\chi^{(2)}$ is invariant under interchange of the last two indices; in other words we have

$$\chi_{ijk}^{(2)}(\omega_3 = \omega_1 + \omega_1) = \chi_{ikj}^{(2)}(\omega_3 = \omega_1 + \omega_1) \tag{42}$$

Hence we may use the reduced notation

$$\chi_{ijk}^{(2)}(\omega_3 = \omega_1 + \omega_1) \equiv d_{ip} \tag{43}$$

where the relations between the values of j and k and p are given by

jk	11	22	33	23, 32	13, 31	12, 21
p	1	2	3	4	5	6

Obviously only 18 components, at most, of the nonlinear susceptibility tensor (i.e., 18 values of $\chi_{ijk}^{(2)}$) are required to describe the case of second-harmonic generation, as opposed to 27 components in the general case when ω_1 and ω_2 are not equal.

Conditions imposed by crystal symmetry further reduce the number of nonzero components from the full 18 for the degenerate case of second-harmonic generation. We do not intend to discuss this aspect at length here, but certain important relations will be considered.

If a medium possesses inversion symmetry, there can be no second-harmonic generation; in other words all the d's are zero in centrosymmetric crystals. In noncentrosymmetric crystals, the d's that are then nonzero depend on the crystal class. For example, KDP (KH_2PO_4), ADA ($NH_4H_2AsO_4$), and urea belong to the crystal point group $\bar{4}2m$ when

$$d_{14} = d_{25} \neq 0$$

$$d_{36} \neq 0$$

and the rest of the d's are equal to zero.

The general symmetry properties discussed earlier interact with those imposed through crystal symmetry, and this results in additional elements becoming zero. For example, quartz, which belongs to the point group 32, has

$$d_{11} = -d_{12} = -d_{26} \neq 0,$$

$$d_{14} = -d_{25} \neq 0$$

with the rest equal to zero, owing to crystal symmetry. If the quartz is used at visible wavelengths, where the dispersion is very small, Kleinman symmetry also applies and hence

$$d_{14} \equiv d_{123} = d_{213} \equiv d_{25}$$

If both requirements are to be fulfilled simultaneously, it follows that both d_{14} and d_{25} must be zero in this case.

8. SECOND-HARMONIC GENERATION

In this case the waves at ω_1 and ω_2 have the same frequency, equal to ω, say, and the sum wave at ω_3 has a frequency 2ω. If we further assume that the conversion efficiency to the second harmonic is so low that the fundamental beam amplitude does not change significantly, then Eq. (38) can be rewritten as

$$\frac{\partial \mathscr{E}(2\omega)}{\partial z} = -\frac{i\varepsilon_0}{2}\left(\frac{\mu_0}{\varepsilon_{2\omega}}\right)^{1/2} 2\omega d_{\text{eff}} |\mathscr{E}(\omega)|^2 \, e^{i\Delta\mathbf{k}\cdot\mathbf{z}} \tag{44}$$

where

$$\Delta k = 2k(\omega) - k(2\omega) \tag{45}$$

and we have, for the moment, neglected the vector properties of the fields, and replaced $\chi^{(2)}$ by an effective d coefficient. Since $\mathscr{E}(\omega)$ is independent of position, we can integrate Eq. (44) and, assuming that $\mathscr{E}(2\omega)$ is initially zero at the point $z = 0$, obtain the value of $\mathscr{E}(2\omega)$ as

$$\mathscr{E}(2\omega) = (1 - e^{i\Delta\mathbf{k}\cdot L})\omega\left(\frac{\mu_0}{\varepsilon_{2\omega}}\right)^{1/2} \varepsilon_0 d_{\text{eff}} |\mathscr{E}(\omega)|^2 \tag{46}$$

Remembering that the Poynting vector is related to the fields through an expression of the form

$$S = \tfrac{1}{2}\varepsilon_0 n \mathscr{E}\mathscr{E}^* \ (W\,m^{-2}) \tag{47}$$

and that

$$n = (\varepsilon/\varepsilon_0)^{1/2} \tag{48}$$

where n is the appropriate index of refraction, we obtain

$$S(2\omega) = 2\left(\frac{\mu_0}{\varepsilon_0}\right)^{3/2} \frac{\omega^2 d_{\text{eff}}^2 \varepsilon_0^2 L^2}{n^2(\omega)n(2\omega)}\left[\frac{\sin^2\left(\dfrac{\Delta k \cdot L}{2}\right)}{(\Delta k \cdot L/2)^2}\right]S^2(\omega) \tag{49}$$

where $S(\omega)$ or $S(2\omega)$ is measured in $\mathrm{W\,m^{-2}}$, and all other quantities have been defined. The power in a beam is related to the flux by

$$P = AS \tag{50}$$

where A is the cross-sectional area of the beam, so that the conversion efficiency for second-harmonic generation can be written as

$$\eta_{\mathrm{SHG}} = P(2\omega)/(P(\omega))$$

$$= 2\left(\frac{\mu_0}{\varepsilon_0}\right)^{3/2} \frac{\omega^2 d_{\mathrm{eff}}^2 L^2 \varepsilon_0^2}{n^2(\omega)n(2\omega)} \left[\frac{\sin^2\left(\dfrac{\Delta k\cdot L}{2}\right)}{(\Delta k\cdot L/2)^2}\right]\frac{P(\omega)}{A} \tag{51}$$

The sinc function in the above (i.e., the term in square brackets) describes the effect of phase matching—a topic we will discuss fully in the next section. Assuming this term attains its maximum value (i.e., unity), we can calculate the conversion efficiency in second-harmonic generation under typical conditions. Consider the case of an unfocused beam with a cross section of $4\,\mathrm{mm}^2$ ($4 \times 10^{-6}\,\mathrm{m}^2$) and of power 10 MW from a Nd : YAG laser passing through a crystal of length 1 cm with a nonlinear coefficient of the order of 0.6 pm/V; then from (51) above we have

$$\eta_{\mathrm{eff}} \sim 3.5 \times 10^{-8}\, P\,(\text{watts})$$

so for a 10-MW beam, the conversion efficiency to the second harmonic is of the order of 35%. (At this level of conversion the assumption of a constant flux at the fundamental wavelength would not strictly apply.)

9. DETERMINATION OF d_{eff}

We now discuss more fully how d_{eff} is calculated once the crystal class, the nonzero nonlinear coefficients, and the geometry are known.

As our example we calculate the nonlinear susceptibility of a $\bar{4}2m$ material when the fundamental beam propagates as an ordinary ray at an angle θ to the optic axes (3-axis) and where the plane containing the ray and the optic axis makes an angle ψ with the 1-axis (see Fig. 1).

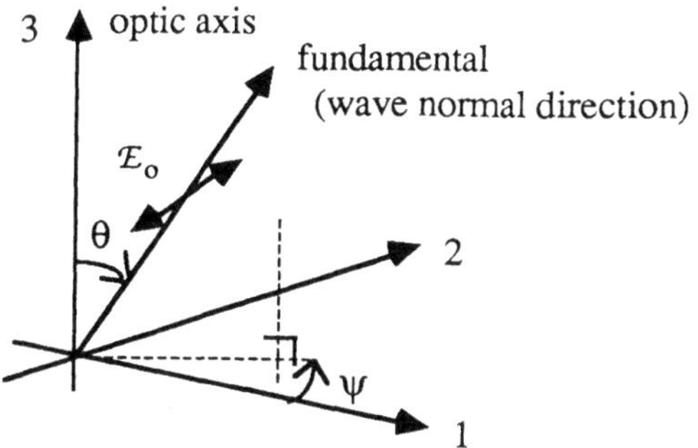

FIGURE 1. Calculation of the effective non-linear susceptibility d_{eff} for the case where the fundamental beam propagates as an ordinary ray at an angle θ to the optic axis (3-axis) and where the plane containing this ray and the optic axis makes an angle ψ with the 1-axis.

The fundamental wave has components $\mathscr{E}_1 = \mathscr{E}_0 \sin \psi$, $\mathscr{E}_2 = \mathscr{E}_0 \cos \psi$, $\mathscr{E}_3 = 0$. Hence only the element d_{36} need concern us here, and so the induced nonlinear polarization is of the form

$$P_3^{(2)} = d_{36} 2 \mathscr{E}_1 \mathscr{E}_2 = 2 d_{36} \mathscr{E}_0^2 \sin \psi \cos \psi \tag{52}$$

It is the component of $P_3^{(2)}$ perpendicular to the propagation direction that is responsible for the generation of the SH wave and this has magnitude $P_3^{(2)} \sin \theta$. From the figure it can be seen that this component generates an extraordinary wave. Such a wave, being orthogonally polarized to the fundamental, is suitable for phase matching through the use of crystal birefringence. This aspect will be discussed in a later section. The final driving polarization is therefore

$$2 d_{36} \sin \psi \cos \psi \sin \theta \, \mathscr{E}_0^2$$

and hence the nonlinear susceptibility for substitution into (51) is

$$d_{\text{eff}} = 2 d_{36} \sin \psi \cos \psi \sin \theta \tag{53}$$

It is apparent that to optimize d_{eff}, ψ should be made 45°. The angle θ is fixed by phase match considerations—this we discuss in the next section.

The distinction between **D** and **E** fields in anisotropic media needs to be commented on here. In such a medium these two vector quantities are no longer in the same direction as is the case in an isotropic medium. The **D** field vector is the one that is orthogonal to the wave normal direction (**k**-vector direction), and is the one used to define the polarization state of the light. It is also the vector described by construction based on the ellipsoid of wave normals. On the other hand, the **E** fields are orthogonal to the ray direction (direction of energy flux), which in an anisotropic medium generally differs from the wave normal direction, the angle between the two directions being called the walk-off angle [see Section 10, Eq. (62)]. In equations such as (38) describing the nonlinear coupling between waves, we are concerned with their **E** fields, and hence the fields involved in the evaluation of d_{eff} are also **E** fields. However, in Fig. 1, where we work in terms of the wave normal direction (**k**-vector direction), the orthogonal fields are the **D** fields. Our evaluation of d_{eff} is therefore an approximation to the extent that **D** and **E** field directions differ. Usually the walk-off angle is so small that the two field directions are close to parallel and the approximation is reasonable. We discuss the evaluation of the walk-off angle in Section 10.

10. PHASE MATCHING

We now consider more fully the term in square brackets in Eqs. (49) and (51). Since the term is a "sinc"-type function, it reaches its maximum value when $(\Delta k \cdot L/2)$ is equal to zero. Since L cannot be zero for finite generation efficiency, we therefore require that Δk be equal to zero, or, in other words

$$\Delta k = 2k(\omega) - k(2\omega) = 2k_1 - k_2 = 0$$

a condition known as phase matching. Remembering that

$$\omega/k = c/n$$

this condition is equivalent to requiring that

$$n_1 = n_2$$

namely, that the phase velocities of the second harmonic and the fundamental waves be equal. In this case the phase of the induced nonlinear polarization in the medium, determined by the phase of the fundamental wave, and the phase of the reradiated second harmonic wave from this polarization keep in step. The contributions of the individual radiators then add together in phase in the forward scattering direction, and a large field builds up. It is apparent on this picture that if there is a phase lag ($\Delta k \neq 0$), the contributions of individual radiators are not then in phase and destructive interference can occur. At certain critical lengths this can lead to the net generation being zero. Clearly this occurs in the case when $(\Delta k \cdot L) = 2\pi$ [see Eq. (49)]. Half the above length, i.e., $L_c = \pi/\Delta k$, is referred to as the coherence length, since as Eq. (49) also shows, it is the length of material for maximum second harmonic conversion in the absence of a complete phase match (i.e., $\Delta k \neq 0$).

Because of dispersion Δk is not usually zero, unless special precautions are taken. One technique is to offset the dispersion with the birefringence of the anisotropic medium. In type I phase matching in negative uniaxial crystals the fundamental ray is propagated as an ordinary ray, and the second harmonic ray as an extraordinary ray. (We have shown above that for this arrangement there is a suitable induced polarization depending on d_{36} for the class $\bar{4}2m$.)

The refractive index of the extraordinary ray varies with the angle of its wave normal to the optic axis θ according to

$$\frac{1}{n_e^2(\theta)} = \frac{\cos^2\theta}{n_0^2} + \frac{\sin^2\theta}{n_e^2} \tag{54}$$

where n_e is the extraordinary refractive index and n_o is the ordinary refractive index. Phase matching is obtained in a wave normal direction such that

$$n_{e2}(\theta_{pm}) = n_{o1} \tag{55}$$

where $n_{e2}(\theta_{pm})$ is the refractive index of the extraordinary ray at the second harmonic frequency (2ω) and at angle θ_{pm}, and n_{01} is the refractive index of the ordinary ray at fundamental frequency (ω). From (54) it can be seen that this angle is given by

$$\sin^2\theta_{pm} = \frac{(n_{o2}^2 - n_{o1}^2)n_{e2}^2}{(n_{o2}^2 - n_{e2}^2)n_{o1}^2} \tag{56}$$

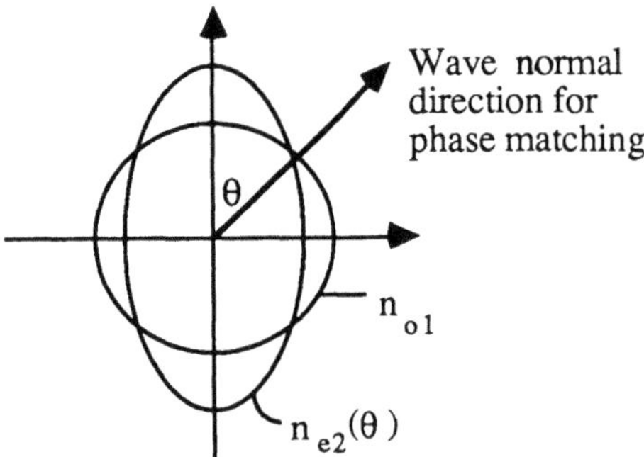

FIGURE 2.   Phase matching using birefringence to offset dispersion. For the wave normal direction shown, the refractive index of an extraordinary second harmonic wave equals the refractive index of an ordinary fundamental wave.

The phase match condition is illustrated in Fig. 2.

The above situation applies to a negative uniaxial crystal ($n_e < n_o$). For type I phase matching in a positive uniaxial crystal, the fundamental would be propagated as the extraordinary ray and the second harmonic as the ordinary ray.

In type II phase matching the fundamental propagates as a mixture of ordinary and extraordinary waves, and the second harmonic wave is either an ordinary or extraordinary wave.

Figure 3 shows a typical case of type II phase matching in a $\bar{4}2m$ crystal.

The fundamental propagates in the (3, 1) plane at an angle θ to the optic axis (3), and is assumed to be an equal mixture of ordinary and extraordinary waves. It readily follows that the three components of the field are

$$\mathscr{E}_1 = -(\mathscr{E}_0/\sqrt{2}) \cos\theta, \quad \mathscr{E}_2 = \mathscr{E}_0/\sqrt{2}, \quad \mathscr{E}_3 = +(\mathscr{E}_0/\sqrt{2}) \sin\theta$$

For the case where the SH wave is the extraordinary wave, the driving polarization is given by

$$P_3^{(2)} \sin\theta - P_1^{(2)} \cos\theta$$

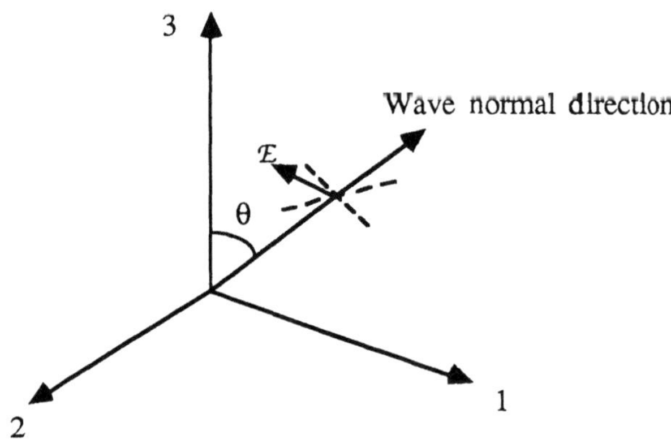

FIGURE 3.   Type-II phase matching geometry.

Evaluating $P_1^{(2)}$ and $P_3^{(2)}$ in terms of $\mathscr{E}_1$, $\mathscr{E}_2$, and $\mathscr{E}_3$ we obtain

$$d_{\text{eff}} = -\tfrac{1}{2}(d_{14} + d_{36})\sin\theta\cos\theta \tag{57}$$

For the case where the crystal exhibits Kleinmann symmetry, $d_{14} = d_{36}$, and so we have

$$d_{\text{eff}} = -d_{36}\sin 2\theta \tag{58}$$

The phase match condition is now

$$k_{o1} + k_{e1} = k_{e2} \tag{59}$$

where k_{o1} and k_{e1} are the wave numbers of the ordinary and extraordinary waves, respectively, at fundamental frequency, and k_{e2} is the wave number of the extraordinary wave at second harmonic frequency.

The choice between type I and type II phase matching depends on the wavelength range to be covered with the nonlinear crystal. Because refractive indices vary with wavelength, if phase matching is to be achieved over an extended range, then either the angle θ must be varied as wavelength is varied (angle tuning) or the temperature of the crystal must be varied (temperature tuning).

Since phase matching depends on angle θ, it is obvious that the acceptance angle of the crystal is limited. We can estimate this angle by calculating the change in angle required for the generated power to drop to zero. This occurs when $(\Delta k \cdot L) = 2\pi$.

Using the relation

$$\Delta k = \frac{2\pi}{\lambda_1}[n_{01} - n_{e2}(\theta)]$$

for the case of type I phase matching and using (54) to determine the angular dependence of $n_{e2}(\theta)$, then the variation in Δk with change in angle $\Delta\theta$ about the phase match angle θ_{pm} is

$$\Delta k \approx \frac{4\pi}{\lambda_1}(n_{o2} - n_{e2})\sin(2\theta_{\text{pm}})\,\Delta\theta$$

hence the *full* angular width to the zero power points is

$$\Delta\theta \approx \frac{\lambda_1}{(n_{o2} - n_{e2})L\sin(2\theta_{\text{pm}})} \tag{60}$$

For the special case when $\theta = \pi/2$ (referred to as the condition for noncritical phase matching), $\sin(2\theta_{\text{pm}})\Delta\theta$ can be approximated by $2(\Delta\theta)^2$ to give

$$\Delta\theta(\theta_{\text{pm}} = \pi/2) \approx \left[\frac{\lambda_1}{2L(n_{o2} - n_{e2})}\right]^{1/2} \tag{61}$$

Substitution of typical values into the above [$(n_{o2} - n_{e2}) \approx 10^{-2}$, $\lambda_1 \approx 600$ nm, $L \approx$ 1 cm] gives

$$\Delta\theta \approx 6/\sin(2\theta_{pm}) \text{ mrad}$$

in general case, and

$$\Delta\theta \approx 50 \text{ mrad}$$

for the case of noncritical phase matching. The angular aperture is thus an order of magnitude larger if noncritical phase matching is employed. This is extremely important if focusing is used to increase field intensity and so improve conversion efficiency. For the case of noncritical phase matching, tuning must be by changing the crystal temperature, whereas for critical phase matching either temperature or angle tuning is possible.

A similar analysis for type II phase matching shows that with comparable parameters the angular aperture is a factor of 2 greater.

So far in our treatment of phase matching we have considered only the wave normal direction of the waves in the crystal; the fundamental and SH waves being collinear in the sense of having the same wave normal directions. In an anisotropic medium, however, the wave normal direction is not in general the same as the ray direction, which is the direction of energy propagation. In fact the ray direction can readily be derived from Fig. 2, since it can be shown that it is parallel to the normal to the surface at the point of intersection. The construction for finding the ray direction is shown in Fig. 4. By simple geometry, the angle between wave normal and ray direction, ρ, can be shown to be given by

$$\tan(\rho + \theta) = \left(\frac{n_o}{n_e}\right)^2 \tan\theta \tag{62}$$

{In Fig. 4 the length OA is proportional to $n_e(\theta)$, and from (54) it readily follows that the curve it defines is an ellipse. Since $|\mathbf{k}| = n_e(\theta) \cdot (\omega/c)$, it also follows that

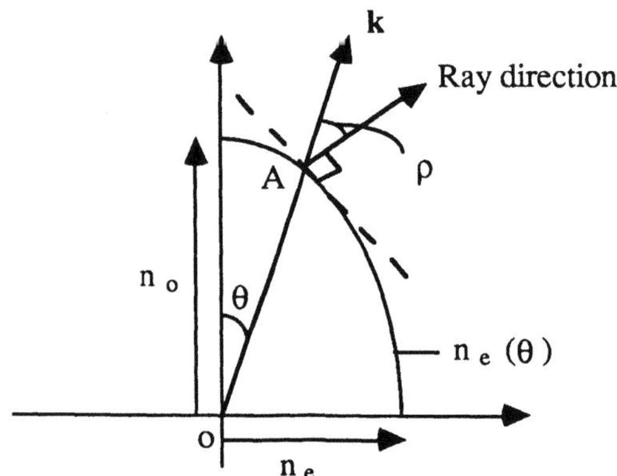

FIGURE 4. Construction for determining the ray direction from the wave normal direction for an extraordinary ray.

the ellipse is the locus of the wave vector of the extraordinary ray. The surface of which the above ellipse is a cross section is sometimes referred to as the normal (index) surface. Some caution should be exercised here, however, since this term is also applied to the surface which is the locus of the phase velocity vector $[|\mathbf{v}_p| = c/n_e(\theta)].\}$

For a general angle θ, the generated SH wave "walks-off" from the generating wave over a length in the crystal given by

$$l_a \sim \pi^{1/2} w/\rho \tag{63}$$

where w is the radius of the beam. The length l_a is usually referred to as the aperture length, and the effect is referred to as Poynting vector walk-off. It is particularly significant with focused beams, since in this case w is small, and can lead to a significant reductions in SH generation efficiency. For the case of noncritical phase matching ($\theta = \pi/2$), the aperture length becomes infinite—i.e., the wave normal and ray directions are the same. This is another significant advantage of noncritical phase matching.

11. GENERATION WITH FOCUSED BEAMS

The most comprehensive treatment of SH generation involving a focused fundamental beam is due to Boyd and Kleinman,[1] who also review earlier work. The fundamental beam is assumed to be a Gaussian beam characterized by a beam waist w_0 in the crystal. Two parameters are then defined to characterize the interaction: (i) the focusing parameter ξ, defined by

$$\xi = L/b = L/w_0^2 k_1, \tag{64}$$

where L is the crystal length and b is the confocal parameter of the fundamental beam in the crystal ($b = w_0^2 k_1$, where k_1 is the wave number of the fundamental beam in the crystal); and (ii) the double refraction parameter B, defined by

$$B = \rho(Lk_1)^{1/2}/2 \tag{65}$$

where ρ is the walk-off angle given by (62).

The generated second-harmonic power is deduced using an exact integral expression. Assuming negligible absorption in the crystal at both fundamental and SH wavelengths and assuming that the beam waist is located at the center of the crystal, then at optimum phase match, the generated second harmonic power is found to be given by

$$P_2 = KP_1^2 Lk_1 h_m(B, \xi) \tag{66}$$

where (in MKS units)

$$K = \frac{\omega_1^2 d_{\text{eff}}^2}{2\pi n_1^2 n_2 \varepsilon_0 c^3} \tag{67}$$

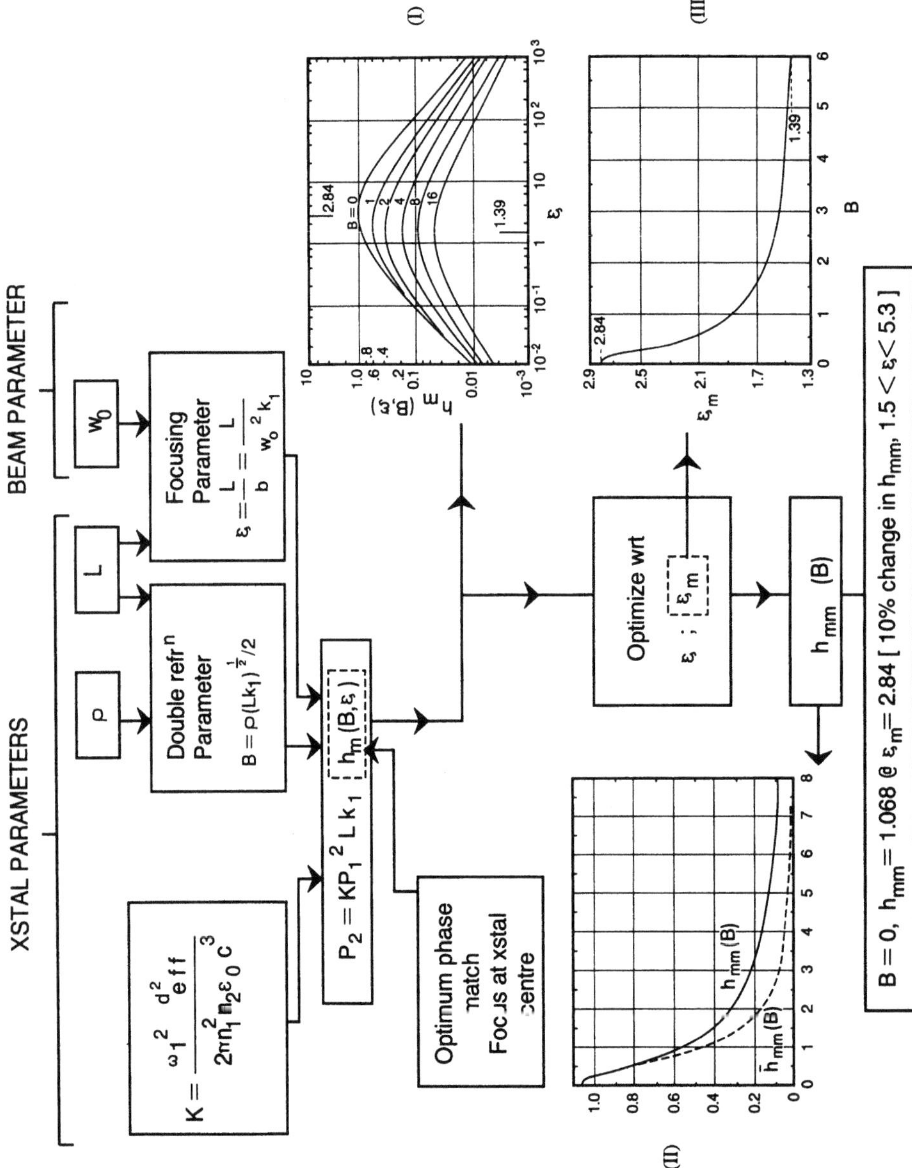

FIGURE 5. Second harmonic generation with focused beams. Based on Boyd & Kleinman.[1] (Reproduced with permission of the American Institute of Physics.)

and the function $h_m(B, \xi)$ is determined numerically. In Fig. 5(I), $h_m(B, \xi)$ is plotted as a function of ξ with B as parameter. It has a maximum value of 1.068 when $\xi = \xi_m = 2.84$ and $B = 0$. Around this maximum, $h_m(B, \xi)$ is only a weak function of ξ; for example, in the case when $B = 0$, it only changes by 10% for the range $1.5 < \xi < 5.3$. The optimum value of $h_m(B, \xi)$ with respect to ξ, $h_{mm}(B)$, is plotted as a function of B in Fig. 5(II). For $B = 1$, corresponding to a walk-off angle of 8 mrad in a 1-cm crystal, $h_{mm}(B)$ drops to about a half its maximum value at $B = 0$. At the same time, as may be seen from Fig. 5(III), ξ_m is reduced from 2.84 at $B = 0$ to 1.9.

One interesting feature of SH generation in the presence of focused beams is that optimum phase match now no longer occurs at $\Delta k = 0$, the effect becoming more pronounced as ξ becomes larger (stronger focusing).

In the above treatment of SH generation using focused beams, thermal effects due to absorption of the fundamental radiation by the crystal were neglected. In the case of intracavity generation, where high circulating fields are involved, thermal effects can seriously limit conversion efficiency, generally through thermal phase mismatching. In the presence of absorption the focused fundamental beams set up a nonuniform temperature distribution in the crystal, so that optimum phase matching is no longer maintained across the profile. The problem has been analyzed by Okada and Ieiri[2] and Mikhina et al.[3] The importance of the effect is characterized by a parameter (power), P_T, given by

$$P_T = \frac{\lambda_1 K_T}{\delta_1 L} \left(\frac{dn_2}{dT} - \frac{dn_1}{dT} \right)^{-1} \tag{68}$$

where K_T is the thermal conductivity of the crystal and δ_1 is the absorption per unit length at fundamental wavelength λ_1. For the case $P_1 \ll P_T$ thermal effects are negligible and (66) can be used to describe generation. For $P_1 \gg P_T$, the coherence length of the beam in the crystal is significantly altered, and generated second-harmonic power is then given by

$$P_2 = (\pi/2) K h_m L k_1 P_T P_1 \tag{69}$$

Note that in this case the generated power becomes linearly dependent on the fundamental power as opposed to quadratically dependent in the absence of thermal effects. The intermediate region is described by including in (66) a parameter h_T, thus we have

$$P_2 = K h_m k_1 L h_T P_1^2 \tag{70}$$

which has been numerically determined as a function of (P_1/P_T).

Substituting into (68) values appropriate to a typical $42m$ crystal ($K_T \sim 2 \times 10^{-2}$ W cm^{-1} K^{-1}, $\delta \sim 0.005$ cm^{-1}, $[dn_2/dT - dn_1/dT] \sim 4 \times 10^{-5}$ K^{-1}) gives a value for P_T of the order of 6 W for a 1-cm crystal.

12. INTRACAVITY SECOND-HARMONIC
GENERATION AND RESONANT ENHANCEMENT

Because SH generation is quadratically dependent on fundamental power, an improvement in conversion efficiency is to be expected if the crystal is placed inside the laser cavity where the circulating field is high. However, the inclusion of an additional element inside the cavity invariably introduces loss, both parasitic and due to conversion to SH, and this can have a significant effect on intracavity power, particularly if the device is operating near threshold. Also thermal effects, discussed in the previous section, limit conversion efficiency in the presence of high fields. Design optimization is therefore required.

The problem of intracavity SH generation in solid state lasers has been extensively treated. The case of a laser with a homogeneously broadened transition and SH generation quadratic in intracavity power has been considered by R. G. Smith.[4] The conclusion of this simple theory, which does not include absorption in the nonlinear element, is that for a given laser with a given linear loss there is an optimum coupling of second-harmonic radiation out of the cavity which is independent of the pump power. Furthermore, the second-harmonic power at optimum coupling is equal to the power output of the same laser without the nonlinear element but with an output mirror of optimally chosen transmission. When absorption of the cavity radiation by the nonlinear element is included in the analysis this not only increases the linear loss of the cavity but also introduces a thermal effect in the nonlinear element, with consequent reduction in conversion efficiency. This situation has been considered in the context of the frequency doubling of cw dye lasers by Ferguson and Dunn.[5]

To illustrate the requirement for optimization in the case of intracavity doubling, we present a simplified treatment for the case of a four-level, homogeneously broadened laser (e.g., dye laser) and consider the limiting case where SH conversion is not restricted by thermal effects, but where crystal absorption is present.

For the case of the four-level laser illustrated in Fig. 6, the population of the upper laser level, $N_1(\text{vol}^{-1})$, and hence the inversion, is determined by the simplified

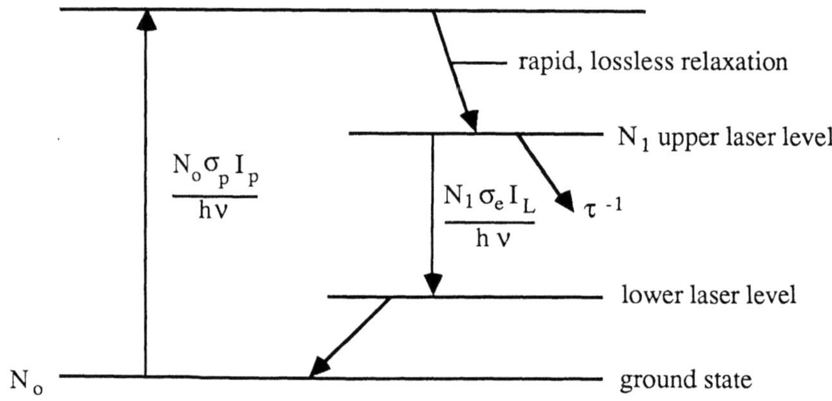

FIGURE 6.   Excitation and relaxation processes in a four-level laser.

rate equation

$$\frac{dN_1}{dt} = \frac{N_0\sigma_p I_p}{h\nu} - \frac{N_1}{\tau} - \frac{N_1\sigma_e I_L}{h\nu} \tag{71}$$

where N_0 is the population of the ground state, assumed constant, I_p is the intensity of the pump radiation exciting the upper level, σ_p is the molecular cross section for absorption of the pump radiation, τ is the lifetime of the upper level to all processes other than stimulated emission, σ_e is the cross section for stimulated emission, and I_L is the intensity of the laser radiation within the active medium.

The intensity, I_L, within the active medium is related to the fundamental power within the cavity through

$$I_L = P_1/\pi w_0^2 \tag{72}$$

where w_0 is the beam waist within the medium.

The energy balance (per round trip) for the radiation within the cavity is

$$\delta P_1 = N_1\sigma_e P_1 t - \delta_1 L P_1 - \gamma P_1 - \Phi \tag{73}$$

where t is the thickness of the active medium, δ_1 is the absorption per unit length of the nonlinear crystal, γ is the linear loss due to mirrors and other optical components, and Φ is the loss due to useful conversion to second harmonic. This latter is of the form

$$\Phi = k_1 K h_m L P_1^2 = \eta L P_1^2 \tag{74}$$

in the absence of thermal effects.

If we solve (71) and (72) for the steady state case, we obtain the following expression for the fundamental power within the cavity:

$$\frac{\alpha P_1}{1 + \beta P_1} - \delta_1 L P_1 - \gamma P_1 - \Phi = 0 \tag{75}$$

where

$$\alpha = \frac{\tau N_0\sigma_p P_p\sigma_e t}{\pi w_0^2 h\nu} \tag{76}$$

$$\beta = \frac{\tau\sigma_e}{\pi w_0^2 h\nu} \tag{77}$$

If we neglect crystal absorption ($\delta_1 = 0$), then from (75) and (76) we obtain the following expression for generated SH power:

$$\beta P_2 + \frac{(\gamma\beta + \eta L)}{(\eta L)^{1/2}} P_2^{1/2} + (\gamma - \alpha) = 0 \tag{78}$$

We now optimize the generated SH power by differentiating the above expression with respect to L and putting $\partial P_2/\partial L = 0$. This determines the optimum crystal length given by

$$L_{\text{opt}} = \gamma\beta/\eta = \gamma\tau\sigma_e/(\pi w_0^2 h\nu k_1 K h_m) \tag{79}$$

Note that the optimum length is independent of pump power, so that once the laser parameters are known an optimum length suitable for all pump powers can be specified. If (79) is substituted back into (78) we determine the maximum generated SH power as

$$P_{2,\text{max}} = (\sqrt{\alpha} - \sqrt{\gamma})^2 / \beta \tag{80}$$

It is interesting to not that this is the same as the maximum fundamental power that can be coupled out of the laser cavity. This is readily shown by replacing γ in (73) by $(\gamma + \gamma^1)$, where γ is the parasitic linear loss from the cavity and γ^1 is the transmission of the output coupler. The useful extracted power at fundamental is then $P_{1,\text{out}} = \gamma^1 P_1$.

Maximum output is found to occur when

$$\gamma^1_{\text{opt}} = (\alpha\gamma)^{1/2} - \gamma \tag{81}$$

[Note that this coupling *is* a function of pump power, and if this value is substituted back into (75) the expression for maximum extracted fundamental power is found to be the same as (80) above.]

If absorption in the nonlinear element is included (δ_1, length^{-1}), then by a similar procedure to the above, we find that optimum crystal length is now given by

$$L_{\text{opt}} = \frac{\gamma\beta}{(\eta + \delta_1\beta)} \tag{82}$$

and is still independent of the pump power. The maximum generated SH power using a crystal of the above length is

$$P_2 = \frac{\left\{ \left[\alpha + \delta_1\left(\frac{1}{\eta} - \frac{\gamma\beta}{(\eta + \delta_1\beta)}\right) \right]^{1/2} - \left[\frac{\gamma(\eta + \delta_1\beta)}{\eta} \right]^{1/2} \right\}^2}{\beta} \tag{83}$$

In this case the maximum extractable power at SH is now less than the maximum extractable power at fundamental wavelength, and is a function of the pump power (which comes into α).

Intracavity second-harmonic generation has become of widespread interest recently in the context of diode-laser pumped solid-state lasers (all-solid-state or "holosteric" lasers). The use of the technique to generate green light from diode-laser pumped 1064-nm Nd:YAG and related lasers is now well established and has already led to commercial products (see, for example, Baer,[6] who used potassium titanium phosphate (KTP) and Fan *et al.*,[7] who used MgO:LiNbP$_3$ as the intracavity, nonlinear medium to frequency double the Nd:YLF laser). The technique has been extended to diode-laser pumped Nd:YAG lasers operating at more unusual wavelengths, for example, at 946 nm when blue light at 473 nm is generated. Owing to the lower gain cross section of this transition compared to that at 1064 nm and limitations due to reabsorption of the 946-nm light in the gain medium (the lower laser level for the 946-nm transition is on the ground state manifold), operation of this laser is more difficult compared to the 1064-nm laser. However, the reduction in thermal effects through the use of diode laser pumping schemes considerably alleviates such problems, and 946-nm lasers are now practical devices. Intracavity frequency doubling of such a laser has resulted in the generation of 473-nm blue light at continuous-wave power levels of the order of 10 mW,[8] and such performance is likely to improve significantly in the near future as more powerful diode lasers become available for pumping the 946 nm laser. Potassium niobate is a suitable material for second-harmonic generation at this wavelength and two phase match schemes are possible: noncritical phase matching when propagation is along the *a* axis of the material and the crystal must be heated to approximately 180°C; or alternatively critical phase matching with angle tuning and operation at room temperature. The latter scheme does, however, have the disadvantage of beam walk-off, reduced effective nonlinear coefficient and reduced acceptance angle compared to the noncritical phase matched case. As an alternative to intracavity techniques for enhancing second-harmonic conversion efficiency in the 946-nm laser, Hong *et al.*[9] have recently demonstrated the use of a Q-switched 946-nm laser with single pass doubling using an external crystal of potassium niobate to generate blue light at 473 nm. The considerable increase in peak power associated with the Q-switched pulse (in this case to the 50–100-W level) improves the single-pass conversion efficiency so as to be comparable to that in the intracavity case [see Eq. (51), for example].

An elegant alternative to intracavity frequency doubling that still exploits resonator enhancement effects to build up large field intensities, thus maintaining high nonlinear conversion efficiencies, is the external cavity resonant enhancement technique first proposed by Ashkin *et al.*[10] Here the nonlinear crystal is placed outside the laser cavity in an external cavity of its own that can be resonant at either the fundamental or second-harmonic frequency. Such an approach offers one means of avoiding the amplitude instabilities reported in the case of internal, intracavity frequency doubling which result from the nonlinear conversion processes in the crystal being interactive with the laser gain medium since the gain and nonlinear media are now effectively decoupled (see, for example, Baer,[6] but also Oka and Kubota,[11] who show that careful design can minimize such effects while still retaining the internal geometry). However, the need for laser sources of high beam quality (TEM$_{00}$ and single longitudinal mode) and high stability has meant the exploitation of the full potential of this technique has had to await the development of refined laser sources such as diode-laser pumped solid-state lasers. The seminal device developed by Kozlovsky, Nabors, and Byer[12] is illustrated in Fig. 7. The external resonator associated with the nonlinear crystal, in this case MgO:LiNbO$_3$, is both a ring resonator and monolithic. The crystal coatings are such as to increase the circulating fundamental power by a factor of the order of 60 over that of the incident power. When some 52 mW of single frequency laser power, from a laser-diode pumped Nd:YAG single-mode ring laser, is incident on the device, the intracavity circulating field established on resonance at 1064 nm is of the order of 3 W, when some 30 mW of second harmonic power is generated in the green. This corresponds to an effective conversion efficiency of 56%, and efficiencies exceeding 80% are anticipated in the near future.

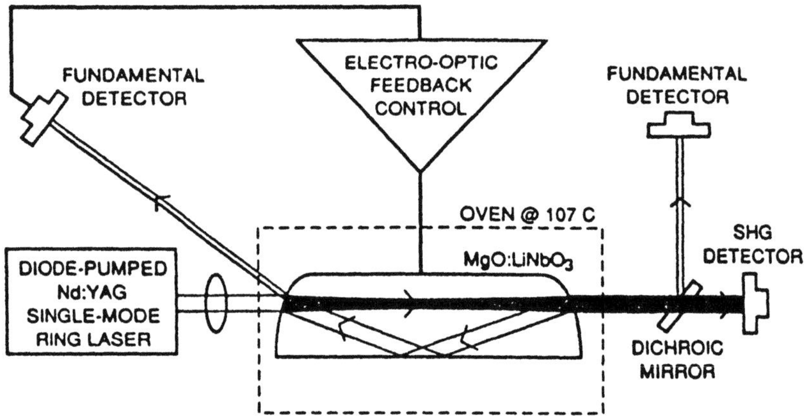

FIGURE 7. Monolithic, external cavity, resonantly enhanced frequency doubling device, due to Kozlovsky, Nabor and Byer.[12] (Reproduced with permission of Laser Focus World.)

External resonant cavity enhancement techniques have now been extended to the direct frequency doubling of diode lasers with good spatial and spectral (single-frequency) mode properties. Potassium niobate has been the most favored nonlinear material because it can be noncritically phase matched at room temperature for wavelengths around 860 nm (propagation along the *a* axis). For example, Goldberg and Chun[13] have reported the generation of 24 mW of blue light at 420 nm from 167 mW of radiation at 840 nm from a diode laser. No attempt was made to lock the frequencies of the diode laser and resonant cavity together, so generation was only observed in sweep mode. More recently this has been done by Kozlovsky and Lenth[14] using electronic stabilization when 40 mW of true continuous-wave light at 430 nm was generated from a diode laser emitting 140 mW, and by Dixon *et al.*[15] using optical feedback techniques when 0.2 mW was generated at 430 nm from 12 mW of diode laser light.

13. NEW MATERIALS

From the previous discussion, important properties of a nonlinear optical material in relation to its use in second harmonic generation include:

i. High effective nonlinear coefficient. This depends on both the usable nonlinear coefficients of the material as well as the cut required to fulfill the phase matching requirement (see Sections 8 and 9).

ii. High optical damage threshold (see Section 8).

iii. Ability to phase match (either angle or temperature tuning) over a wide wavelength range coupled with good transparency at both fundamental and second harmonic wavelengths (see Section 10, but the latter is also particularly important in relation to intracavity generation, see Section 12).

iv. Potential for noncritical matching leading to large angular aperture and to avoidance of walk-off effects (see Section 10).

v. Stability (chemical/mechanical/environmental), and ease of growth to large size (both aperture and length) with good optical quality.

Recent years have seen a host of new materials offering significant improvements in many of these key properties when compared to the "classical" materials of nonlinear optics (for example, KDP/ADP). Here we will concentrate on new nonlinear materials with applications in the near ultra-violet, visible and near infrared (0.2 to 5 μm). Lin and Chen[19] have recently carried out a comprehensive survey of the newer nonlinear materials and Table 1 is taken from their paper. The table summarises transparency range, phase match range, damage threshold and relative nonlinear figure of merit for various nonlinear crystals. The relative figure of merit is based on Eq. (51) being given by

$$\frac{d_{\mathrm{eff}}^2 L^2 \omega^2}{n^2(\omega)n(2\omega)}$$

but normalised to the case of type-I phase matching in KDP at 1.064 μm. The table is a convenient means of making the initial selection of a nonlinear material for a particular application. However, in so far as equation (51) applies to the case of plane waves only, optimum focusing conditions (see Section 11),[11] the effects of walk-off on generation efficiency (see Section 10) and the limiting effects of thermal phase mismatching (Section 11) and ultimately optical damage, are not taken into account. Further, the spatial quality of the fundamental beam may be such that the acceptance angle of the crystal needs to be considered explicitly (see Section 10), and in extreme cases the linewidth of the fundamental laser may upset the phase

TABLE 1. Nonlinear Crystals for Second Harmonic Generation. Based on a Table due to Lin and Chen.[19] (Reproduced with permission of Lasers and Optronics).

Material	Transparency range (nm)	Phase-matching range (Type)	Damage threshold (GW/cm^2)	Relative figure of merit
KDP	200–1500	517–1500 (I)	0.20	1.0
KDP	200–1500	732–1500 (II)	0.20	1.8
D-CDA	270–1600	1034–1600 (I)	0.50	1.7
Urea	210–1400	472–1400 (I)	1.50	6.1
Urea	210–1400	600–1400 (II)	1.50	10.6
BBO	198–3300	400–3300 (I)	5.00	26.0
BBO	198–3300	526–3300 (II)	5.00	15.0
LAP	220–1950	440–1950 (I)	10.00	40.0
LiIO$_3$	300–5500	570–5500 (I)	0.50	50.0
m-NA	500–2000	1000–2000 (I)	0.20	60.0
MgO : LiNbO$_3$	400–5000	800–5000 (I)	0.05	105.0
KTP	350–4500	1000–2500 (II)	1.00	215.0
POM	414–2000	830–2000 (I)	2.00	350.0
MAP	472–2000	900–2000 (I)	3.00	1600.
KNbO$_3$	410–5000	840–1065 (II)	0.35	1460/1755
COANP	480–2000	960–2000		4690.0
DAN	430–2000	860–2000		5090.0
Structure-modulated LiNbO$_3$	400–5000	800–5000	0.05	2460 N^2 ^{a}

^{a} N is the number of periods.

match. Potentially suitable materials therefore require a more detailed assessment in relation to the exact conditions and constraints of usage than is provided by Table 1.

14. NEW AREAS AND FUTURE PROSPECTS

The use of optical waveguides for second-harmonic generation is currently an area of intense activity and rapid progress. Waveguides offer the twin benefits of increased interaction length at high optical intensity by beating the diffraction limit to which free space, focused beams are subject, as well as new phase matching opportunities. With regard to the latter, phase matching may be extended beyond the limits imposed by birefringence in bulk materials by

i. Exploitation of waveguide dispersion effects;
ii. Radiation-mode phase-matching schemes (Cerenkov radiation[16]);
iii. Quasi-phase-matching schemes in which periodic variations in material composition may be used to overcome phase mismatching through

$$\Delta k_1 l_1 + \Delta k_2 l_2 = 2\pi N$$

where Δk_1 is the phase mismatch in region 1 of length l_1, Δk_2 is the phase mismatch in region 2 of length l_2, and N is an integer.[17]

So far most activity in this area has been based on waveguides formed in the nonlinear material lithium niobate through titanium in-diffusion or proton exchange,[16,17] but recently significant advances have been reported with regard to waveguide formation and quasi-phase-matching in KTP using rubidium exchange techniques.[18] Phase matching for second-harmonic generation in bulk KTP is normally limited to wavelengths in excess of 994 nm through lack of sufficient birefringence to overcome dispersion. However, using periodic variations in material composition (5 μm period) to effect quasi-phase-matching in channel waveguides, the generation of 2 mW of radiation in the blue at 426 nm from 80 mW of fundamental radiation at 852 nm has recently been reported for a nonresonant geometry.[18]

REFERENCES

1. G. D. Boyd and D. A. Kleinman, Parametric interaction of focused Gaussian light beams, *J. Appl. Phys.* **39**, 3597–3639 (1968).
2. M. Okada, and S. Ieiri, *IEEE J. Quantum Electron* **QE7**, 469 (1971).
3. T. V. Mikihina, A. P. Sukhorukov, and I. A. Tomov, *Zh. Prikl, Spectrosk.* **15**, 1001 (1971).
4. R. G. Smith, *IEEE J. Quantum Electron.* **QE6**, 215 (1970).
5. A. I. Ferguson and M. H. Dunn, Intracavity second harmonic generation in continuous-wave dye lasers, *IEEE J. Quantum Electron.* **QE13**, 751–756 (1977).
6. T. Baer, *J. Opt. Soc. Am.* **B3**, 1175 (1986).
7. T. Y. Fan, G. J. Dixon, and R. L. Byer, *Opt. Lett.* **11**, 788 (1986).
8. W. P. Risk, *Appl. Phys. Lett.* **54**, 1625–1627 (1989).
9. J. Hong, B. D. Sinclair, W. Sibbett, and M. H. Dunn, Proc. CLEO '90, Anaheim, California, Paper CWC7 (1990).

10. A. Ashkin, G. D. Boyd, and J. M. Dziedzic, *J. Quantum Electron.* **QE2**, 109 (1966).
11. M. Oka and S. Kubota, *Opt. Lett.* **13**, 805 (1988).
12. W. J. Kozlovsky, C. D. Nabors, and R. L. Byer, *IEEE J. Quantum Electron.* **QE24**, 913 (1988).
13. L. Goldberg and M. K. Chun, *Appl. Phys. Lett.* **55**, 218 (1989).
14. W. J. Kozlovsky and W. Lenth, Proc. OSA Annual Meeting, Orlando, Florida, Postdeadline Paper PD24 (1989).
15. G. J. Dixon, C. E. Tanner, and C. E. Wieman, *Opt. Lett.* **14**, 731 (1989).
16. T. Tanuichi and K. Yamamoto, *Oyo Buturi* **52**, 1637 (1987).
17. E. J. Lim, M. M. Fejer, R. L. Byer, and W. J. Kozlovsky, *Electron. Lett.* **25**, 132 (1989).
18. J. D. Bierlein, Proc. CLEO '90, Anaheim, California, Paper CWC1 (1990).
19. J. T. Lin and C. Chen, Choosing a nonlinear crystal, *Lasers Opt.* **6**(11), 58 (1987).

24

Materials for Nonlinear Optical Signal Processing

A. M. Scott

1. INTRODUCTION

In the previous chapter nonlinear optical waveguides were discussed and in many aspects the aim of this work is to try to emulate digital electronic circuits—switching, serial processing, and routing. In this chapter we are considering materials for a different aspect of optical signal processing. In analog optical signal processing we take an analog image or two-dimensional array of information and pass it once through a device, to produce an output image that has been processed in some way. The analog image will have a great deal of information embedded in it, and even if our optical processor is relatively slow it may be able to carry out a great number of parallel operations that would otherwise require fast sophisticated electronics.

Before we can discuss materials for optical signal processing and phase conjugation we have to understand what is the physical process we are using and how we might use it in signal processing. The key to nonlinear optical signal processing and phase conjugation is that we use a transparent material with a refractive index which varies according to the intensity of light passing through it. If the material is illuminated by a "signal" beam so that the intensity varies across the medium then the refractive index will also vary, and when a second beam is passed through it, this beam will be modified, and in effect information is transferred from the first to the second beam.

The most practical technique for optical signal processing and phase conjugation is degenerate four-wave mixing, and we shall review this in detail.[1,2] When we restrict ourselves to phase conjugation an alternative technique becomes important: stimulated Brillouin scattering (SBS). We shall also briefly touch on this subject.

2. PHASE CONJUGATION AND OPTICAL SIGNAL PROCESSING

The technique of degenerate four-wave mixing can be used to carry out various

---

A. M. Scott • Royal Signals and Radar Establishment, St. Andrews Road, Malvern, Worcestershire WR14 3PS, U.K.

analog optical operations on a beam bearing an image. It can also produce one of the most dramatic effects, optical phase conjugation.

In essence four-wave mixing is a form of real time holography (Fig. 1). A reference "pump" beam (1) and signal beam (3) overlap in a medium to produce interference fringes. If the medium has an intensity-dependent refractive index then the intensity variations associated with the interference fringes will produce refractive index variations in the medium, in the form of a refractive index grating. Pump beam 1 is sometimes called the "write" beam. If a second pump beam (beam 2, sometimes called the "read" beam) is now shone at the medium in the opposite direction to beam 1, then it will scatter off the refractive index grating to form a fourth beam, which travels in the opposite direction to the signal beam 3.

We can show this mathematically. We have two "pump" beams 1 and 2 traveling in opposite directions (Fig. 1, beams 1 and 2) and some arbitrary signal beam 3. The electric fields of these three beams can be written as

$$\mathscr{E}_j(\mathbf{r}, t) = \tfrac{1}{2} E_j \exp[i(\omega t - \mathbf{k}_j \cdot \mathbf{r})] + \text{c.c.} \qquad \text{for } j = 1, 2, 3 \tag{1}$$

where the counterpropagating pump beams means $\mathbf{k}_2 = -\mathbf{k}_1$. The wavevector of the signal beam $\mathbf{k}_3$ is at some small angle 2θ to beam 1. The overall light intensity is

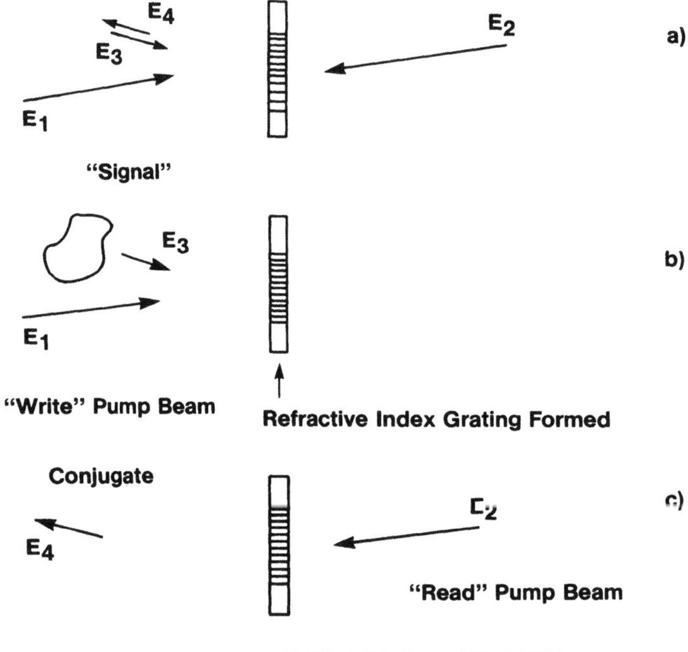

FIGURE 1.   Geometry for degenerate four-wave mixing. The process (Fig. 1a) can be considered as being in two parts, (b) and (c). (a) High-power pump beams 1 and 2 and signal beam 3 are simultaneously incident on nonlinear medium. Nonlinear interaction produces beam 4. (b) Interference between "write" beam 1 and "signal" 3 produces intensity fringes in medium. Intensity-dependent refractive index leads to refractive index grating in medium. (c) "Read" beam 2 scatters off grating to form conjugate beam 4.

given by $|\mathscr{E}|^2$, and if we consider the interference pattern formed by signal beam 3 overlapping beams 1 and 2, we obtain

$$|\mathscr{E}|^2 = |E_1 + E_2 + E_3 + E_4|^2$$

$$= \tfrac{1}{2}E_1 E_3^* e[i(\mathbf{k}_3 - \mathbf{k}_1) \cdot \mathbf{r}] + \tfrac{1}{2}E_2 E_3^* \exp[i(\mathbf{k}_3 - \mathbf{k}_2) \cdot \mathbf{r}] + \text{c.c.} + \text{other terms} \quad (2)$$

These two terms represent two sets of interference fringes; one due to beam 3 overlapping with pump beam 1 and one due to it overlapping with pump beam 2. If beam 3 intersects beam 1 at a small arbitrary angle $[2\theta \sim 10\ \mathrm{mrad}\ (\text{Fig. 2})]$ then the fringes formed by beams 3 and 1 have a long wavelength ($\mathbf{k}_L = \mathbf{k}_3 - \mathbf{k}_1$; grating wavelength $\Lambda_L = \lambda/2 \sin\theta$; for $2\theta = 10\ \mathrm{mrad}$, $\Gamma_L = 100\ \lambda$), while the fringes formed by beams 2 and 3 have a short wavelength ($\mathbf{k}_s = \mathbf{k}_3 - \mathbf{k}_2$, grating wavelength $\Lambda_s = \lambda/2 \cos\theta$; for $2\theta = 10\ \mathrm{mrad}\ \Lambda_s = \lambda/2$).

If we have a material with an intensity-dependent refractive index, i.e.,

$$n = n_0 + n_2|\mathscr{E}|^2 \quad (3)$$

then the density variations in the interference pattern will produce variations in the refractive index which are given by

$$\Delta n = \tfrac{1}{2}n_{2L} E_1 E_3^* \exp[i(\mathbf{k}_3 - \mathbf{k}_1) \cdot \mathbf{r}] + \tfrac{1}{2}n_{2S} E_2 E_3^* \exp[i(\mathbf{k}_3 - \mathbf{k}_2) \cdot \mathbf{r}] + \text{c.c.} \quad (4)$$

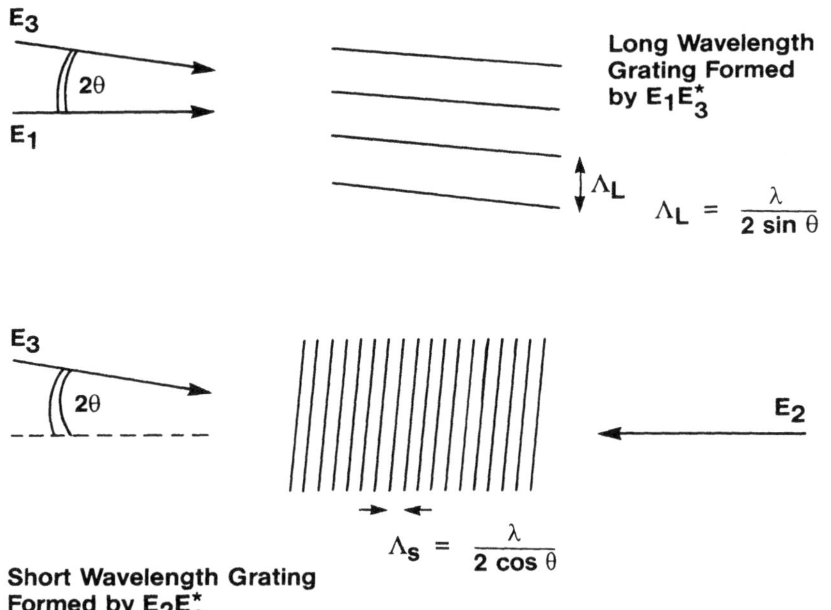

FIGURE 2. Top: Long-wavelength fringes. Bottom: Short-wavelength fringes. Both sets of fringes are formed simultaneously when beams 1, 2, and 3 are present; but one of the gratings may dominate in the four-wave mixing process.

We have written subscripts L and S to indicate the refractive index gratings formed by the long- and the short-wavelength fringes.

The refractive index gratings will scatter the pump beams to form a new fourth beam E_4 (hence the name degenerate four-wave mixing), which is given by

$$\partial E_4 = \beta \mathscr{E} \Delta n \partial r \tag{5}$$

where β is a constant of proportionality. (This can be derived more rigorously.[1])

$$E_4 = \tfrac{1}{2}\beta L(n_{2L} + n_{2S})E_1 E_2 E_3^* \exp[i\omega t = (\mathbf{k}_1 - \mathbf{k}_3 + \mathbf{k}_2) \cdot \mathbf{r}]$$

$$I_4 \propto \tfrac{1}{4}\beta^2 L^2 [n_{2L} + n_{2S}]^2 I_1 I_2 I_3 \tag{6}$$

Since $\mathbf{k}_1 = -\mathbf{k}_2$, this means that whatever the precise direction of the input signal $\mathbf{k}_3$, a beam E_4 will be produced that is traveling in precisely the opposite direction $\mathbf{k}_4 = -\mathbf{k}_3$. This beam will retrace the path of the signal back through any aberrations to its source. This phenomenon is called phase conjugation because the spatial part of the E_4 is proportional to the complex conjugate of the input field, i.e.,

$$E_4\, e^{-i\mathbf{k}_4 \cdot \mathbf{r}} \propto \{E_3\, e^{-i\mathbf{k}_3 \cdot \mathbf{r}}\}^* \tag{7}$$

The device consisting of the cell and the counterpropagating pump beams is called a phase conjugate mirror and the conjugate beam 4 is then regarded as a reflection of beam 3. However, unlike a real mirror, where the reflected light appears to come from a virtual source (Fig. 3a), the conjugate mirror always returns light back to its source (Fig. 3b), rather like a corner cube retroreflector. As a result we can take a highly collimated laser beam, pass it through a piece of frosted glass to distort it, and then direct it onto a phase conjugate mirror. The conjugate beam will retrace the path back through the frosted glass and emerge as a highly collimated beam (Fig. 4). The property makes phase conjugation useful in improving the performance of laser systems (Fig. 5). Images can also be passed through an abberation and back, and appear undistorted (Fig. 6).

A second feature of this type of phase conjugate mirror is that the conjugate beam can be stronger than the signal beam (since its power comes from the pump beam). This means that its "reflectivity" can be greater than 100%. Reflection coefficients of more than 10^6 have been achieved using a variation of this technique.[13]

Returning to signal-processing applications, we note that the conjugate beam is proportional to the product of three beams, i.e.

$$E_4 \propto E_1 E_2 E_3^* \tag{8}$$

This means that images (or two-dimensional arrays of information) can be multiplied together. Furthermore, we can use the property of a simple lens, which is that the intensity distribution at one focal plane is the Fourier transform of the intensity distribution at the other focal plane (Fig. 7). When lenses are combined with four-wave mixing we can multiply Fourier transforms and then apply an inverse

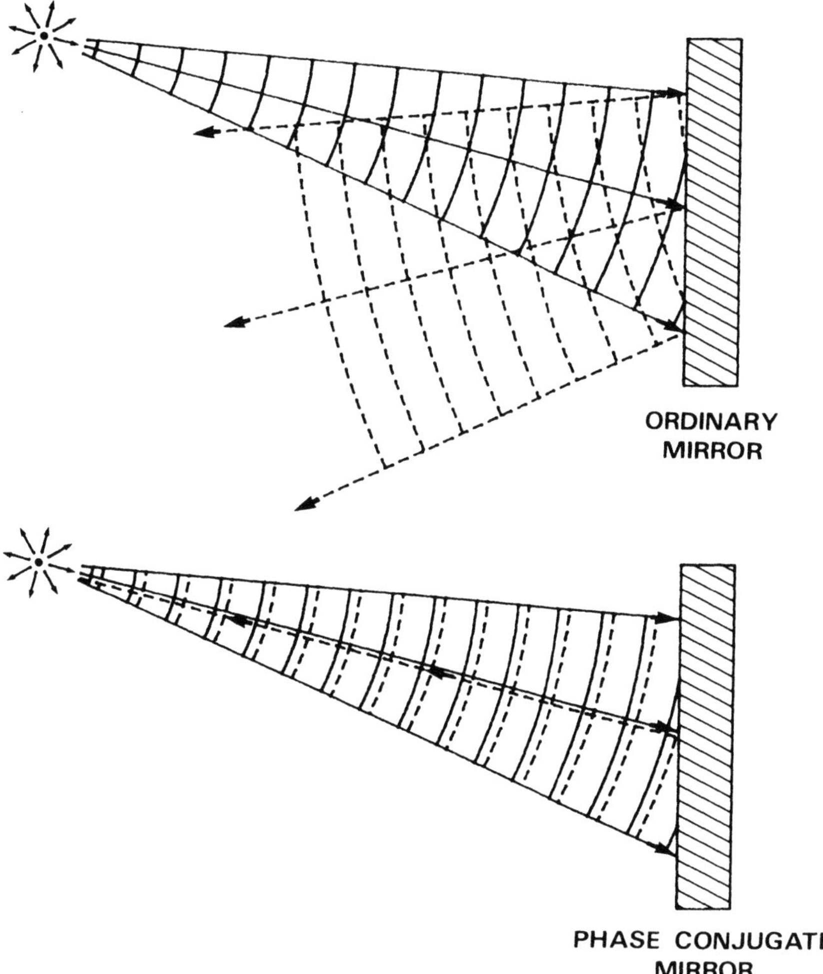

FIGURE 3.   Light reflected by an ordinary mirror continues to diverge. Light reflected by a conjugate mirror returns to its source.

transform (Fig. 8). This leads to further signal-processing operations being possible, such as forming convolutions and correlations between images.[1,2,8,9]

3. PHYSICAL MECHANISMS FOR PRODUCING FOUR-WAVE MIXING

In all the above applications the basic concept is that three overlapping beams produce two sets of interference patterns: the long- and short-wavelength fringes of Fig. 2. The intensity variations associated with these patterns produce variations in the materials refractive index. Any mechanism that can cause an intensity-dependent refractive index can be used for this purpose. Although this can formally

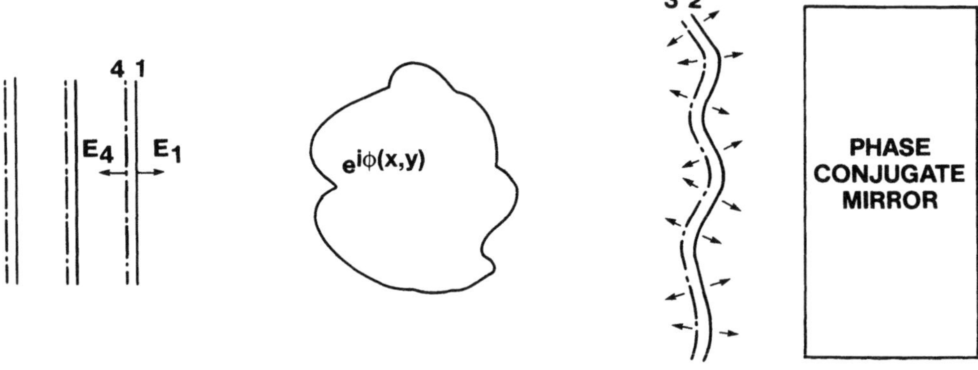

$$E_1 \exp\{i(\omega t - kz)\} \quad \rightarrow \quad E_2 \exp\{i[\omega t - kz + \phi(x,y)]\}$$

$$E_4 \exp\{i(\omega t + kz)\} \quad \leftarrow \quad E_3 \exp\{i[\omega t + kz - \phi(x,y)]\}$$

FIGURE 4. A plane wave front (1) becomes distorted (2) after passing through a phase aberration. The conjugate wave (3) has the same wave front (3) but propagates in the opposite direction and emerges after the aberration as an undistorted wave front (4).

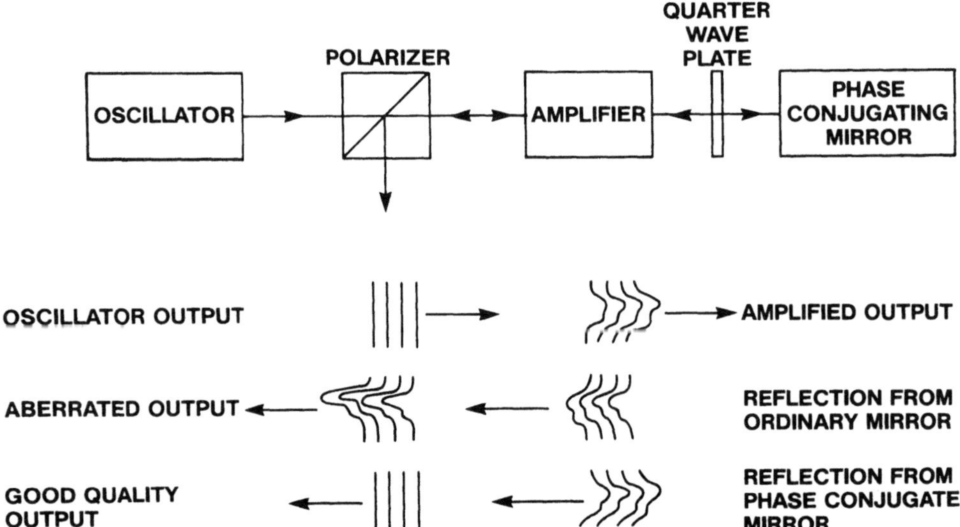

FIGURE 5. Application of phase conjugation to a laser system. The high-gain amplifier will distort the beam while amplifying it, and a phase conjugate mirror (based on stimulated Brillouin scattering) makes the beam pass back through the amplifier giving further gain and correcting for the distortions produced in the first path.

FIGURE 6.   If an image is passed through an aberrator and is reflected by a conventional mirror (a), it emerges as a blur. If it is reflected by a phase conjugate mirror (b), the distortions are removed (photograph courtesy Dr. C. West; Crown Copyright).

be treated as a $\chi^{(3)}$ process,[1] it is not a particularly useful approach to take. We shall consider three possible mechanisms for producing refractive index variations: thermal expansion due to absorption, molecular reorientation, and the photorefractive effect.

Before discussing these mechanisms we shall briefly review the temporal response of these refractive index gratings.

3.1. Temporal Response

The refractive index induced by an interference fringe at time t decays exponentialy with some characteristic time constant, τ_g. More precisely, the refractive index

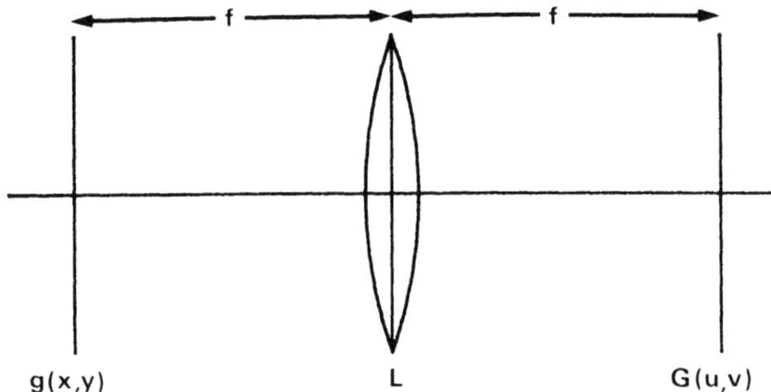

FIGURE 7. A simple lens converts a coherent field at one focal plane into its Fourier transform at the second focal plane.

change produced by a long-wavelength interference pattern (say) is given by

$$\Delta n_L(t) \simeq \frac{n_{2L}}{\tau_L} \int_{-\infty}^{t} E_3^*(\tau) E_2(\tau) \, e^{-(t-\tau)/\tau_L} \, d\tau \tag{9}$$

where the grating time constant for the long-wavelength grating is given by τ_L. A similar expression gives the refractive index change produced by a short-wavelength interference pattern, where Δn_L, n_{2L}, and τ_L are replaced by Δn, n_{2S}, and τ_S.

If we use cw laser beams, Eq. (9) integrates to give

$$\Delta n_L = n_{2L} E_2 E_3^* \tag{10}$$

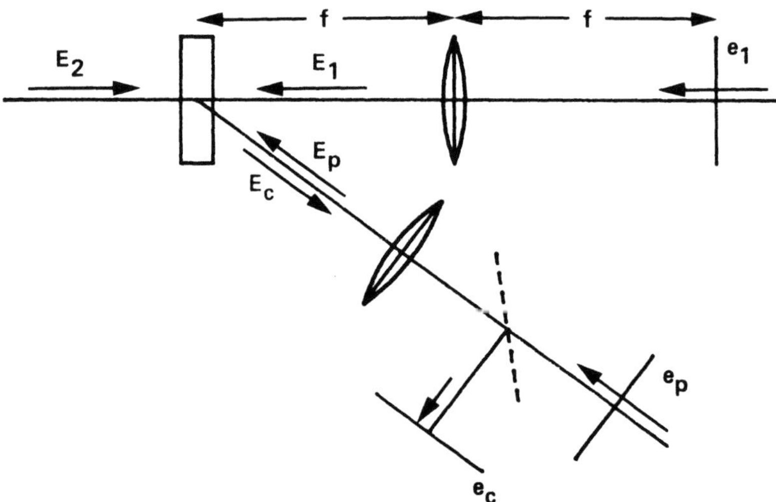

FIGURE 8. The combination of using lenses to produce Fourier transforms and four-wave mixing to multiply fields together allows a variety of processes such as convolution and correlation to be carried out. A signal e_p is transformed and multiplied by the transform of image e_1 on pump beam 1. The inverse transform of the conjugate beam produces the correlation function of e_p and e_1 in the plane e_c.

and the refractive index change depends only on n_{2L} and the intensity of the interference pattern. If we use pulsed laser beams, where the laser pulse is much shorter than τ_L, then

$$\Delta n_L(t) = \frac{n_{2L}}{\tau_L} \int_{-\infty}^{\tau} E_3^*(\tau) E_2(\tau)\, d\tau \tag{11}$$

The refractive index change in this case depends on the ratio of the nonlinear refractive index to the grating time constant, and the integrated energy density of the interference pattern "writing" the grating.

The significance of this is that n_2/τ (where n_2/τ can mean n_{2S}/τ_S or n_{2L}/τ_L) is a measure of the refractive index change produced by a given energy density of radiation and can be regarded as a measure of the sensitivity of a material. The grating decay time allows the medium to integrate the refractive index change over a period τ. We will find that materials with a large nonlinear response to cw beams are usually characterized by a long grating lifetime, τ, rather than a large value of n_2/τ. In extreme cases the grating lifetime τ can be seconds or even days.

The ratio n_2/τ is the key measure of the nonlinearity rather than n_2; and n_2 and τ will often vary greatly (say with temperature or for similar materials) while n_2/τ stays relatively constant.

The time constant may depend on the grating wavelength, and when n_2/τ is the characteristic property of a material then Δn_L and Δn_S may be significantly different.

4. THERMALLY INDUCED GRATINGS

If we have a weakly absorbing medium, then the medium will absorb more power and become hotter in regions of high intensity than it will in regions of low intensity. This will cause the material to expand and the refractive index to change and form a refractive index grating.[3] The difference in the refractive index between the peak and the trough of the interference pattern will depend on the temperature difference, and this in turn will depend on how thermal conductivity damps out the temperature variations in the medium. The thermal variations will decay exponentially with a time constant given by

$$\tau_g = \frac{\rho_0 C_p}{K k_g^2} \quad \begin{cases} \propto \dfrac{\lambda^2}{\sin^2 \theta} & \text{(long-wavelength grating)} \\[2ex] \propto \lambda^2 & \text{(short-wavelength grating)} \end{cases} \tag{12}$$

where ρ_0 is the medium density, C_p is the specific heat, K is the thermal conductivity and k_g is the grating wavevector, and λ is the optical wavelength.

From the above it is clear that for, say, $2\theta = 10\,\text{mrad}$, τ_R is 10^4 times longer for the long-wavelength grating than for the short-wavelength one.

The intensity of the conjugate beam (in its simplest form) in the steady state case is

$$I_4 = Q^2 \tau_R^2 I_1 I_2 I_3 \, e^{-\alpha L}(1 - e^{-\alpha L})^2 \tag{13}$$

where Q is a material figure of merit, which in fact is proportional to n_2/τ_R, τ_R is the decay time, α the absorption coefficient, and L the path length.

From the above we find that for the above example the intensity scattered by the long-wavelength grating (at $2\theta = 10$ mrad) may be four orders of magnitude more than that scattered by the short-wavelength one, and the scattering efficiency varies at $1/\sin^{-4}\theta$. This makes the efficiency very angle dependent and limits the usefulness of thermal gratings in optical signal processing. However, thermal gratings are often present even when we wish they were not, and it is important to take them into account when considering four-wave mixing.

The figure of merit Q is given by

$$Q = \frac{2\pi(dn/T)_p}{\lambda \rho_0 C_p} \tag{14}$$

Thus thermal gratings can be seen in any moderately absorbing material ($\alpha L \sim 1$), including gases, liquids, and solids. If there is a need for a simple four-wave mixing mirror, the simplest approach is to take a solvent with a large value of Q and add an absorbing dye to obtain the desired absorption coefficient.[3]

5. MOLECULAR REORIENTATION

The second mechanism we consider is that of molecular reorientation. In this case an intense oscillating electric field will induce an oscillating dipole in an asymmetric molecule such as CS_2. The field will then couple with the induced dipole and apply a torque, which will cause the molecule to align itself geometrically with the applied field. In the case of a material such as CS_2 the molecules move independently and the alignment is damped by the random collisions between molecules. The time constant is therefore about $\tau_R = 2 \times 10^{-12}$ s and the nonlinear refractive index is $n_2 = 1.2 \times 10^{-11}$ esu, or $n_2 = 3.1 \times 10^{-14}$ cm^2/W. This gives $n_2/\tau_R = 6$ esu/s or 5×10^{-3} cm^2/J. Other simple molecules such as nitrobenzine, toluene, benzene, etc. have n_2 values as shown in Table 1. Thus molecular reorientation requires high intensities for these simple molecules[4,5] and powers of

TABLE 1. Simple Molecules with a Large Kerr Coefficient

Material	n_2 $(10^{-14}\,\mathrm{cm}^2/\mathrm{W})$	τ $(10^{-12}\,\mathrm{s})$	n_2/τ $(10^{-3}\,\mathrm{cm}^2/\mathrm{J})$
CS_2	28	2	140
Benzene	3.2	38	84
Nitrobenzene	2.1	48	44

$\sim$500 MW/cm^2 are typically required for, say, 1% reflection coefficient in a 2-cm sample length.[4]

5.1. Liquid Crystals

If we consider molecules that have a nematic liquid crystal phase (such as 5 cyanobiphenyl) we find that they are large and elongated. As a result, even in the *isotropic* phase, the molecules do not act completely independently but will tend to weakly align themselves with their nearest neighbors.

The time constant for the molecular reorientation and the nonlinear refractive index are both strongly temperature dependant and behave as

$$\tau_R = \frac{C}{(T - T^*)} \exp(-E/kT), \qquad n_2 = \frac{B}{(T - T^*)} \tag{15}$$

where B and C are constants, T is the temperature, and T^* is the temperature of the isotropic–nematic phase transition. The reorientation time can vary from 2 to 100 ns and the nonlinear refractive index n_2 can become very large (say, $n_2 = 3 \times 10^{-12}$ cm^2/W when $T^* = T = 5°$C).[15] However, the ratio n_2/τ does not vary strongly with temperature and is typically 2×10^{-5} cm^2/J at room temperature (Table 2).

In the nematic phase the interaction between molecules extends over very long ranges, of the order of 100 μm or more. However, the cell thickness cannot be more than about 200 μm; otherwise the very slight variations in molecular alignment accumulate and the resulting fluid looks milky because of scattering by refractive index variations in the sample.

The molecular reorientation time in this case is determined by the long-range elastic forces, and the reorientation time can be very long, of the order of 100 ms or more. It is in fact affected by the characteristics of the cell, since this affects how the elastic forces affect the molecular alignment.

Although the time constant may be a factor 10^6 longer than before,[7] the ratio n_2/τ is not greatly affected in going from the isotropic to the nematic phase; so the nonlinear refractive index becomes perhaps 10^6 larger than in the isotropic phase, and nonlinear effects are seen with very modest powers (10 W/cm^2).

TABLE 2. Nematogens with a Large Kerr Coefficient

Material	T^* (°C)	B (10^{-2} cm^2 W^{-1})	n_2/τ (cm^2/J)	C (ns K^{-1})
10CB	42	3.8	7.6×10^{-6}	150
5CB	35	9.5	2×10^{-5}	150
3CB	25	13.1	3.4×10^{-5}	125
ME75	29	2.9	3.8×10^{-6}	250
Pyr7	50	9.0	1.4×10^{-5}	195
CS$_2$	—	$n_2 = 3.1 \times 10^{-14}$ cm^2/W	1.5×10^{-8}	$\tau = 2 \times 10^{-3}$ ns

However, the requirement for very thin samples reduces the usefulness of nematic liquid crystals in nonlinear optics. Firstly the intensity I_4 is proportional to L^2 (see earlier) so the thin sample requirement leads to a reduced coupling efficiency. Secondly the orientational nonlinearity is often weaker or slower than competing thermal nonlinearities. Thirdly the very thin cell can lead to the production of anomalous beams. This is because the thin sample behaves like a thin conventional refractive index grating, and scatters light to form first-order, second-order, and other high diffraction orders. In a thick medium (i.e., for typical thicknesses of 1 mm or more), constructive interference throughout the medium ensures that only the first-order scattered beam is produced, while destructive interference reduces the intensity of the higher-order beams to negligible levels. Thus a reasonably thick nonlinear medium is needed to ensure that higher-order beams do not produce anomalies in optical signal processing.

6. PHOTOREFRACTIVE MEDIA

Finally we consider photorefractive media. We consider an insulating material such as $BaTiO_3$ or $Bi_{12}SiO_{20}$ which contains crystal defect or some impurity doping such as Fe^{2+}. When light is shone into the crystal, donors will be ionized by the incoming beam and the electrons will move around the crystal by diffusion before being trapped by another site. Here they may be photoexcited and again move to another site. If there are intensity variations within the crystal then electrons will tend to be photoexcited in regions of high intensity and be trapped in regions of low intensity. This produces an internal electric field between regions of high and low electron density and hence between regions of different light intensity. If the crystal exhibits the Pockels effect (i.e., an applied voltage changes the refractive index), then this internal electric field will produce refractive index variations in the crystal. The refractive index change will in fact depend on $\Delta \cdot I$ rather than I and the electric field will be largest halfway between an intensity maxima and an intensity minimum (Fig. 9).

A comprehensive treatment of the photorefractive effect is not possible in the space available, and a list of references are provided.[8-11] However, we can comment on a number of aspects.

The refractive index change depends on $\Delta n \propto n_0^3 r_{eff} E$, where n_0 is the normal refractive index, r_{eff} is the effective value of the electro-optic coefficient, and E is the strength of the internal field. This emphasizes that the effect will depend on the orientation of the crystal to the interfering beams. The electric field itself will depend on a number of material properties. The time constant of the internal field depends on the electron density in the crystal, and since this depends on the overall intensity, we find

$$\tau_g \propto \frac{1}{I_0} \tag{16}$$

where I_0 is the average light intensity.

The two most widely used materials are $BaTiO_3$ and $Bi_{12}SiO_{20}$ (BSO). The first one has an extremely large slow nonlinearity where beams of a few milliwatts

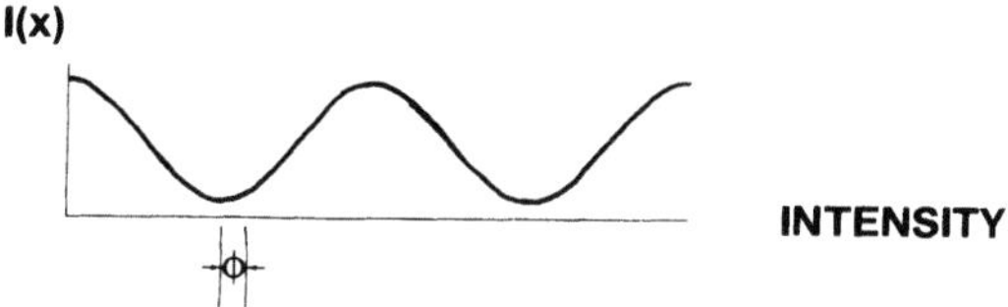

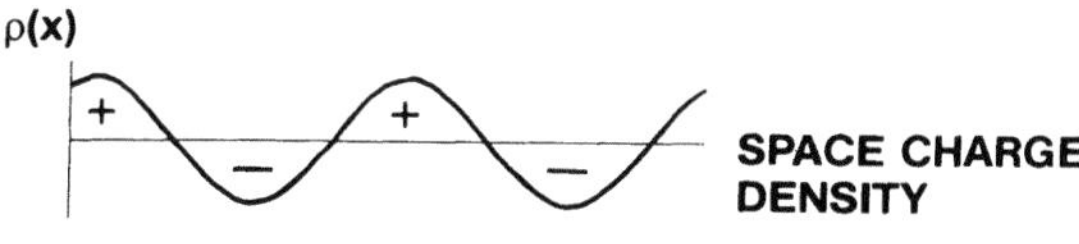

Figure 9. Photorefractive effect. (a) Signal and "write" pump beam produce interference fringes in the crystal. (b) Electrons migrate from regions of high intensity and are trapped in regions of low intensity. (c) Electric field build up between regions of high and low space charge. (d) The Pockels effect produces a refractive index change proportional to the electric field.

TABLE 3. Characteristics of Various Photorefractive Materials

	Material			
	BSO	InP	GaAs	BaTiO$_3$
Writing—response time (ms)	2	<0.05	<0.05	300
Write energy—sensitivity (mJ/cm^2)	0.3	<0.3	3.2	6
Wavelength (nm)	514–632	800–1800	800–1800	450–500
Scattering efficiency	3%	0.5%	0.1%	200% 30% self pumped
Index change Δn	5×10^{-6}	10^{-6}	3×10^{-6}	5×10^{-5}
Absorption (cm^{-1})	1 at 514	2.7	1.5	—
Storage time τ_d (s)	10^{-2}	10^{-4}	10^{-4}	3×10^4
Resistivity (Ω^{-1} cm^{-1})	1.6×10^{-5} 2×10^{-14}	10^{-8}	1.6×10^{-8}	— 1×10^{-12}
Pockels coefficient (r_{ij})	4.5 at 450	−1.32	−1.33	1640

(W/cm^2) can combine to produce a remarkable array of anomalous effects. However, the time constant is very slow and can typically be several seconds at low pumping intensities, and if a crystal is left in darkness, gratings can be present days after they were first formed. They can then be erased by illuminating the crystal with a uniform beam of light.

In Table 3 we list some common photorefractive crystals. The response time assumes that the write beam is about 100 mW/cm^2, and the sensitivity is a measure of the integrated energy density required to get 1% scattering efficiency. Also listed are the changes in refractive index associated with the grating, the absorption coefficient of the medium, and the lifetime of the grating when the beams are all switched off. The resistivity and Pockels coefficients are also shown.

The figures emphasize that although BaTiO$_3$ is very attractive for producing unusual effects, the very slow response time makes it unsuitable in optical signal processing.

BSO is somewhat less exotic and has a response time of typically a few milliseconds and provides weaker coupling between beams. However, its more rapid response and high sensitivity makes it ideal to use in optical signal processing applications. Beam coupling can be greatly enhanced by applying an external electric field and getting a beam to provide moving fringes.

The materials InP and GaAs have been investigated to a lesser extent but may be useful for four-wave mixing in the infrared.

7. STIMULATED BRILLOUIN SCATTERING (SBS)

If an intense pulsed laser beam is focused into a cell containing, for example, high-pressure xenon or liquid carbon disulfide, the intense fields can generate an intense acoustic wave, which will act as a moving mirror and reflect the incoming beam.[12] Surprisingly, it turns out that the reflected light is often phase conjugate to the incoming beam and will retrace the path of the incoming beam back through any aberrations.

Mathematically it is very difficult to show exactly why SBS is effective in producing a phase conjugate reflection. In essence, interference effects cause the wave front of the acoustic wave to align themselves with the wave fronts of the optical beam. When this happens each optical ray sees a mirror at normal incidence and is exactly retroreflected.

Physically SBS is a much simpler method of achieving phase conjugation than four-wave mixing—one simply has to take a cell filled with a suitable medium and focus a coherent intense laser beam into it. As a result, SBS is probably much closer to being regarded as a technology appropriate for laser systems[12] than four-wave mixing is regarded as a technology for optical signal processing. It is also possible to use a combination of four-wave mixing and SBS in a technique called "Brillouin enhanced four-wave mixing" and hence build phase conjugate mirrors with "reflection coefficients" of greater than 10^6.[13]

We will briefly discuss stimulated Brillouin scattering.[14]

In any medium, density variations arise naturally as a result of thermal fluctuations, and these density variations have associated refractive index variations that can scatter light. The density variations propagate at the speed of sound in the medium and act as weak acoustic gratings. Light of frequency ν scattered by the grating is doppler shifted or Brillouin shifted by an amount $\Delta\nu/\nu = 2n\nu/c \approx 10^{-5}$,[15-17] where n is the refractive index, ν is the acoustic velocity, and c is the speed of light. If the incoming beam is of low intensity the scattering process does not change the amplitude of the acoustic wave and the intensity of the scattered light will be low (typically $I_{\text{scattered}} \approx e^{-25} I_{\text{input}}$).

If we consider the interference pattern formed by the incoming beam $\mathscr{E}_1$ and a second counterpropagating beam $\mathscr{E}_2$ which is frequency shifted, the interference term is of the form

$$E_1 E_2^* \exp i(\Delta\omega t - 2k_1 r)$$

(where we have assumed $\mathscr{E}_1 = E_1 \exp[i(\omega_1 t - \mathbf{k}_1 \cdot \mathbf{r})] + \text{c.c.}$ and $\mathscr{E}_2 = E_2 \exp\{i[(\omega_1 - \Delta\omega)t + \mathbf{k}_2 \cdot \mathbf{r}]\} + \text{c.c.}$, and $|k_1| \approx |k_2|$ because $\Delta\omega/\omega$ is very small, typically 10^{-5}).

The interference pattern is moving at a speed $\Delta\omega/2k_1$, and if this equals the acoustic velocity ν it will tend to drive an acoustic wave. It is straightforward to show that the frequency difference $\Delta\omega$ is given by $\Delta\omega/\omega = 2n\nu/c$, where n is the refractive index, and c is the speed of light. In the steady state regime the acoustic wave will be proportional to the strength of the moving interference pattern, i.e.,

$$u \propto E_1 E_2^*$$

This will scatter light as

$$\partial E_2 \propto E_1 u^* \partial r$$

which we can write as

$$\frac{\partial E_2}{\partial r} = \tfrac{1}{2} g_1 |E_1|^2 E_2$$

where g_1 is taken as a constant of proportionality.

This has a solution

$$E_2 = E_{20} \exp(\tfrac{1}{2} g_1 |E_1|^2 L); \qquad I_2 = I_{20} \exp(g_B I_1 L)$$

This emphasizes that an intense field tends to amplify the acoustic wave, which in turn scatters light from beam 1 into beam 2, leading to an effective amplification of beam 2. The factor g_1 in the above equations is the Brillouin gain coefficient, which depends on the medium.[15-17]

In practice one laser beam is usually focused into a cell and no external beam 2 is applied. Spontaneous Brillouin scattering occurs at the far end when light is scattered off random thermal acoustic fluctuations. The intensity of this scattered light is low ($I_{\text{scatt}} \approx e^{-25} I_{\text{input}}$), and it is amplified by stimulated Brillouin scattering.

Thus we have

$$I_2 = I_{\text{noise}} \times \exp(g_B I_1 L)$$

$$= e^{-25} I_1 \exp(g_B I_1 L)$$

As a result, stimulated backscattering has a threshold response, with very low scattered intensities observed until $g_B I_1 L \approx 25$, and above this input intensity the backscattered light will rapidly become comparable in intensity to the incoming beam.

The Brillouin gain coefficient is the key parameter, and it depends on a number of the material properties.

The parameter g_B is given by[16]

$$g_B = \frac{2\pi^2 Y^2 \tau}{\lambda^2 cn\rho v}$$

where Y is the electrostrictive coefficient, τ the acoustic decay time, λ the wavelength, c the speed of light, n the refractive index, ρ the density, and v the velocity of sound.

The electrostrictive coefficient is given by[16]

$$Y = \frac{1}{3}\left(\frac{n^2 - 1}{n^2 + 2}\right)$$

and it can be seen that materials with a high refractive index have a large value of Y.

The acoustic decay time τ is given by

$$\tau = \frac{\rho}{\pi\eta}$$

where ρ is the material density and η the viscosity. The effective viscosity is frequency dependent and in fact $\tau \propto \lambda^2$. As a result the Brillouin gain coefficient is relatively insensitive to wavelength but the acoustic decay time becomes much longer for long-wavelength scattering. If the incoming beam is monochromatic the scattered light will be broadened by an amount $\delta\nu = 1/(2\pi\tau)$.

In Table 4 we list a number of materials and the values of g_B, τ, and β_B for 1.06-μm laser radiation. The table shows that gases have a smaller shift than liquids

TABLE 4.   Brillouin Parameters of various Media

Medium	g (cm/GW)	τ_p (ns)	ν_B (GHz)
CH_4 100 atm	65	17	0.15
N_2 135 atm	30	22	
SF_6 22 atm	35	24	0.3
Xe 39 atm	44	30	
CS_2	68	4.6	3.8
Acetone	16	2.7	3.0
CCl_4	4	0.6	2.8
$TiCl_4$	14	1.5	3.1
Glass	~1		15

and a longer acoustic decay time. They also emphasize that high gas pressures are required in order to obtain usable gain coefficients.

Solids can also be used in principle, but in practice they are not convenient. The threshold for Brillouin scattering is often close to the damage threshold in all media. This is not a major problem in gases or liquids as the damage is "self-healing," but in solids any damage is permanent and stops the medium being used further.

8. CONCLUSION

We have shown that nonlinear image processing and optical phase conjugation can be achieved by using the technique of degenerate four-wave mixing. Any phenomenon that leads to an intensity-dependent refractive index can produce refractive index gratings for four-wave mixing. However, the most significant mechanisms are thermal expansion, molecular reorientation, and the photorefractive effect. The photorefractive effect is probably the most significant of the three, but even with photorefractive materials, the techniques of image processing are still very much in the research phase.

In the area of high-power pulsed lasers it transpires that stimulated Brillouin scattering appears to show promise for a technological solution to a number of problems. In this case the materials used are relatively simple highly transparent liquids and high-pressure gases.

REFERENCES

1. A. Yariv, Phase conjugate optics and real time holography, *J. Quantum Electron.* **14**, 650 (1978).
2. Reviews: see *Optical Phase Conjugation*, R. A. Fisher (ed.), Academic, New York (1983); D. M. Pepper, *Opt. Eng.* **21**, 156–183 (1983) (and other articles in that book).
3. H. J. Hoffman, Thermally induced phase conjugation by transient real-time holography: A review, *J. Opt. Soc. Am. B* **3**, 253 (1986).
4. B. R. Sudyam and R. A. Fisher, Transparent response of Kerr-like phase conjugators: A review, *Opt. Eng.* **21**, 184 (1982).
5. T. Y. Chang, Fast self-induced refractive index changes in optical media: A survey, *Opt. Eng.* **20**, 220 (1981).
6. P. A. Madden, F. C. Saunders, and A. M. Scott, *J. Quantum Electron.* **22**, 1287 (1986).
7. I. C. Khoo and Y. R. Shen, Liquid crystals: Nonlinear optical properties and processes, *Opt. Eng.* **24**, 579 (1985).
8. P. Gunther, Holography, coherent light amplification and optical phase conjugation with photorefractive materials, *Phys. Lett. C, Phys. Rep.* **93**, 199–299 (1982).
9. P. D. Foote, T. J. Hall, N. B. Aldridge, and A. G. Levenston, Photorefractive materials and their applications in optical image processing, *IEE Proc.* **133**, P + J, 83 (1986).
10. T. J. Hall, R. Jaura, L. M. Connors, and P. D. Foote, The photorefractive effect—A review, *Prog. Quantum Electron.* **10**, 77–146 (1985).
11. J. O. White and A. Yariv, Photorefractive crystals as optical devices elements and processors, *Proc. SPIE* **464**, 7 (1984).
12. D. A. Rockwell, A review of phase conjugate solid state lasers, *J. Quantum Electron.* **24**, 1124 (1988).
13. A. M. Scott and K. D. Ridley, A review of Brillouin enhanced four wave mixing, *J. Quantum Electron.* **25**, 438–459, (1989).
14. B. Ya. Zeldovich, N. F. Pilipetsky and V. V. Shkunov, *Principles of Phase Conjugation*, Springer-Verlag, Berlin (1984).

15. W. Kaiser and M. Maier, Stimulated Rayleigh Brillouin and Raman spectroscopy, p. 1077 in *Laser Handbook* (F. T. Arecchi and E. O. Schulz Dubois, eds.), North-Holland, Amsterdam (1972).

16. A. I. Erokhin, V. I. Kovalev, and F. S. Farzullov, Determination of the parameters of liquids in an acoustic resonance region by the method of non-degenerate four wave mixing, *Sov. J. Quantum Electron.* **16**, 877 (1986).

17. M. J. Damzen, M. H. R. Hutchinson, and W. A. Schroeder, Direct measurement of the acoustic decay times of hypersonic waves generated by SBS, *J. Quantum Electron.* **23**, 328 (1987).

25

The Chemistry of Liquid Crystals

David Coates

1. INTRODUCTION

The liquid crystal phase was first recognized in 1888 by Reinitzer,[1] who studied cholesteryl benzoate [structure (1)], and Lehmann,[2] who studied 4-azoxyphenetole [structure (2)]. Both compounds exhibited a turbid melt, which at a higher temperature became clear. Reinitzer and Lehmann soon identified more compounds that showed this "in between" state. Some sceptics were doubtful that this "in between" or liquid crystal phase represented a new state of matter; their arguments were based on emulsification and colloidal effects. These arguments were completely suppressed in 1922 when Friedel,[3] as the result of careful optical microscopic studies, was able to distinguish and describe the molecular organization of three distinct types of liquid crystal phase, which he termed nematic, smectic, and cholesteric.

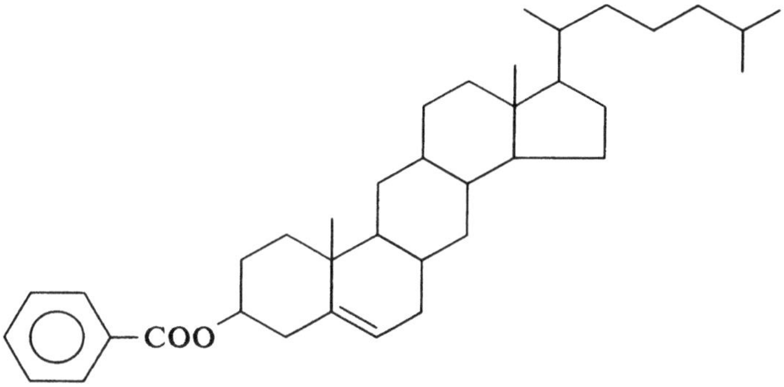

1

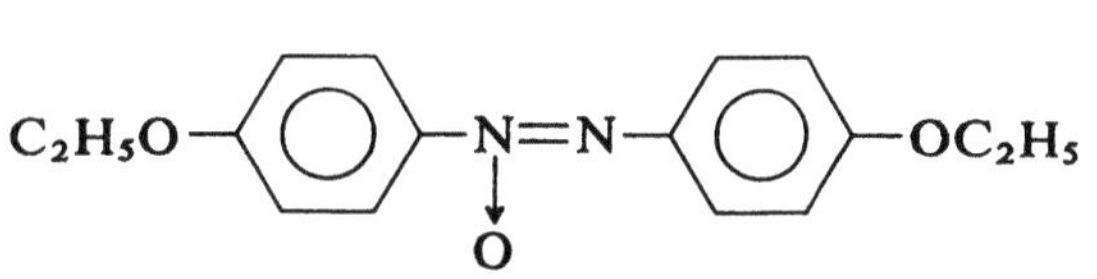

$$C_2H_5O-\langle\bigcirc\rangle-N=N-\langle\bigcirc\rangle-OC_2H_5$$

2

David Coates • BDH Limited, Broom Road, Poole, Dorset BH12 4NN, U.K.

Many thousands of compounds have now been made that exhibit one or more liquid crystal phases. However, it is only in the last seventeen years, and due to the discovery of several stable compounds that exhibit a suitable liquid crystal phase (nematic) at room temperature, that the liquid crystal phase has become of commercial importance. Over 380 million liquid crystal displays are now made each year.

2. CLASSIFICATION AND FORMATION OF LIQUID CRYSTALS

There are two general types of liquid crystal:

(*i*) *Thermotropic Liquid Crystals.* These are regarded as electronic materials (and are the subject of this chapter), they are formed by the action of heat on certain solids. Also included in this class are the recently discovered[4] discotic liquid crystal phases formed by disklike molecules. The "disks" pack into columns and give rise to a range of discotic liquid crystal phases, analogous to the phases that are discussed here. However, as these phases cannot, as yet, be considered as electronic materials they will not be discussed further.

A second type of liquid crystal, which has probably been recognized longer (before 1888), is the following:

(*ii*) *Lyotropic Liquid Crystals.* These are formed by the action of a solvent, usually water, on a solid. The solid is usually a soaplike material capable of forming micelles; the arrangement of these micelles leads to liquid crystal phases. Such phases are important in domestic cleaning products and in biological systems, e.g., lipid layers.

Nearly all thermotropic liquid crystal materials consist of molecules having a rod-shaped molecular structure. The assymmetric shape of the molecules leads to assymmetric attractive forces between the ends and the sides of the molecules, which are responsible for the unusual melting behavior of these solids. Figure 1 schematically depicts the events that can occur when a compound composed of elongated rods is heated. It should be borne in mind that this is a schematic drawing; in reality the molecules are undergoing thermal motion.

When the solid, depicted in Fig. 1, is heated, and if the attractive forces are such that the terminal attractive forces are weaker than the lateral ones, then at some temperature T_1 "layers of molecules" will be created which are relatively free to slide over one another; this is the smectic phase. When heated further, and at some temperature T_2, the lateral attractive forces may loosen and allow the layers

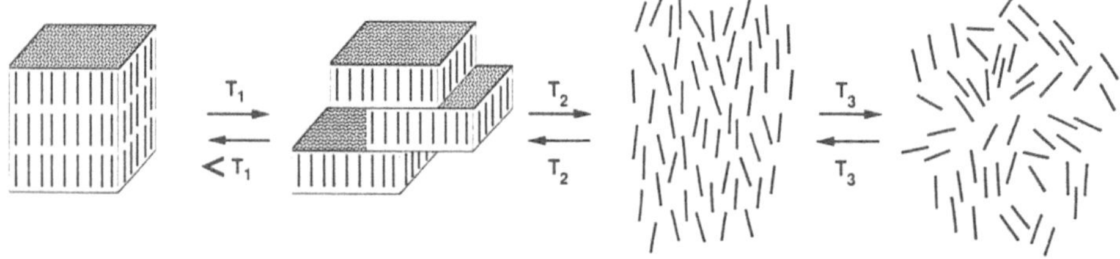

FIGURE 1.   Schematic drawing showing the molecular ordering of the liquid crystal phases formed on heating a solid to the isotropic liquid.

of molecules to interpenetrate, thus forming the nematic phase. When the molecules contain a chiral centre, a special case of the nematic phase occurs; this is called the cholesteric phase (see Section 6.1). When heated further the residual attractive forces loosen at a temperature T_3 and the isotropic liquid forms. On cooling, and at the same temperatures, the reverse sequence of events occurs, but crystallization to the solid often involves some supercooling. This phase reversibility is a characteristic of liquid crystal phases.

If the lateral forces are relatively strong, then only a smectic phase may be exhibited; alternatively if the terminal forces are strong then only a nematic phase may be exhibited.

The temperatures T_1, T_2, and T_3 are generally related to the shape and size of the molecule: a length to breadth ratio of at least 5:1 is usually required for a liquid phase to exist. Longer molecules have higher T_2 and T_3 values, but the melting point T_1 is not readily predicted. The phase type is often determined by the detailed molecular structure, i.e., what groups are where on the molecule.

When a compound melts it takes in heat (heat of fusion); likewise at a liquid crystal phase transition heat is taken in on heating or given out on cooling. This heat exchange can be detected by differential thermal or scanning calorimetry (DTA or DSC) and thus a determination of heat input versus temperature is a means of detecting these transitions; such a trace for 4-cyano-4'-n-octyloxy-biphenyl is shown in Fig. 2.

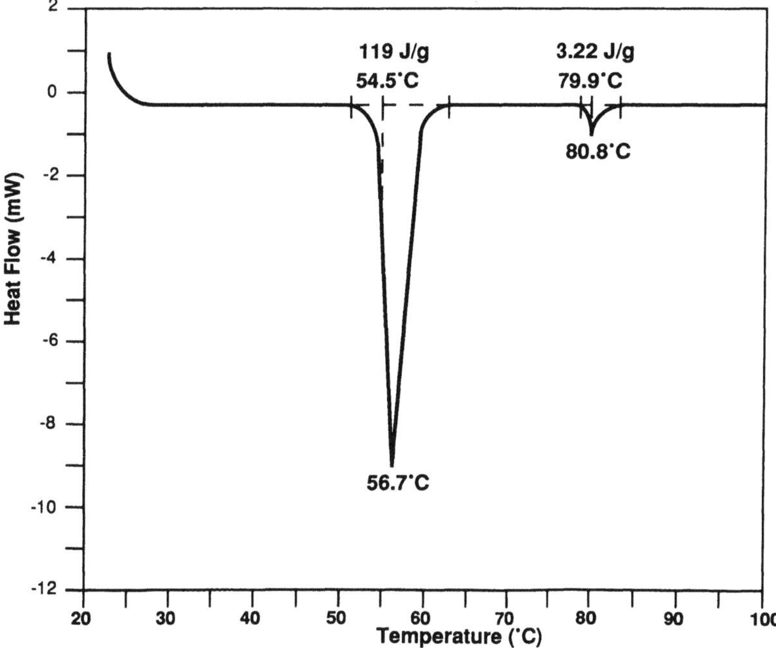

FIGURE 2.   Differential scanning calorimetric trace of the heating cycle for 4-cyano-4'-octyloxy-biphenyl.

3. PHASE RECOGNITION

Several types of liquid crystal phase can exist; therefore it is important to be able to identify them. This is usually done using DSC followed by optical microscopic examination with the sample in an accurately controlled hot stage.

3.1. The Nematic Phase

This is the commonest, and currently the most commercially important, liquid crystal phase; it is the least ordered and most fluid liquid crystal phase. Microscopically, between crossed polarizers (100 × magnification), it appears to flow and "flash" when a thin sample, sandwiched between two microscope slides, is gently sheared. A typical texture (microscopic appearance) is shown in Fig. 3. In this "homogeneous" texture the long molecular axes of the molecules are lying essentially parallel but random to the glass slides. Molecules with a strongly polar terminal group such as Cyano tend to lie with their long molecular axis perpendicular to the glass. This orientation produces a texture that appears black (homeotropic texture) until the sample is sheared; it then appears to "flash."

3.2. The Smectic Phase

Although only one nematic phase is known, the smectic mesophase is far more complex. Friedel recognized only one smectic phase; by about 1970 four were recognized, and today many more have been recognized. They are designated as smectic A (S_A), smectic B (S_B), etc., the subscripts having no significance other than

FIGURE 3. Homogeneous nematic texture.

that they are roughly in the order of recognition, which coincidently is roughly in order of least to most ordered. The evolution of phase classification has been a complex one; some reclassification caused confusion during the 1970s, but this has recently[5] been reviewed. It is now considered that five true smectics (S_A, S_B, S_C, S_I, and S_F), one cubic phase (S_D), and six soft crystals formerly called smectics (S_B crystal, S_E, S_G, S_H, S_J, and S_K) can be recognized.

Although the subject of smectic polymorphism is too complex for this chapter, some basic principles can be understood from Fig. 4. When the molecules are arranged randomly in the layers they can be orthogonal to the layer plane (the S_A phase) or tilted (the S_C phase). These are the two most useful smectic phases. Alternatively, the molecules could be hexagonally ordered within the layers, in which case the S_B or S_F and S_I phases are produced. Further correlation of order within the layers and between the layers leads to further even more viscous phases.

An enthalpy change occurs between the phases, and the various phases are immiscible with each other—indeed, this is the basis of the smectic classification system. No single compound has been found that contains all the smectic phases; usually up to four smectic phases may occur in a compound. The ability to microscopically recognize these phases requires considerable experience. Figure 5 illustrates the natural "focal-conic" texture of a smectic A phase, while Fig. 6 shows a smectic C phase in a schlieren texture. Generally the differences are less readily seen; for example, Fig. 7 shows the S_B phase that occurred on cooling the S_A phase shown in Plate 2! Careful examination shows that the fans in Fig. 5 are "rougher" than those of Fig. 7. The absolute molecular orientation can be determined by x-ray scattering studies, but even this technique has difficulties differentiating some phases, e.g., S_I and S_F.

4. MOLECULAR STRUCTURE AND LIQUID CRYSTAL PROPERTIES

The proviso that for a compound to exhibit a liquid crystal phase it must be fairly rigid and elongated may seem somewhat inhibiting from a chemist's viewpoint, but this has not prevented many thousands of such compounds from being synthesized. Initially chemists were concerned with the questions: Why do these phases form, and what molecular features contribute to their formation? Later, during the late 1960s and early 1970s, the major effort was to synthesize compounds that exhibit

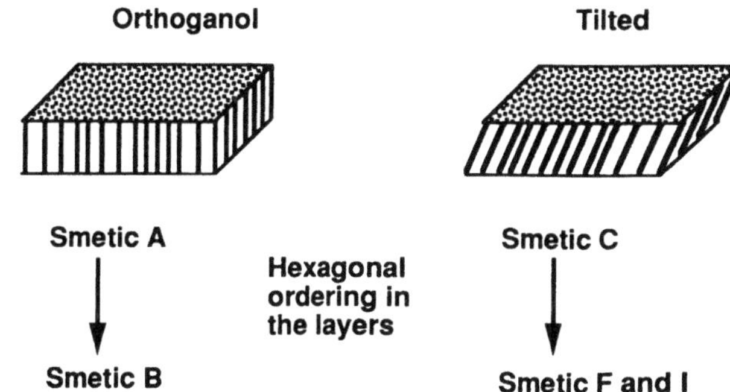

FIGURE 4. Molecular arrangement in the smectic layers can be orthoganol or tilted and disordered or ordered.

FIGURE 5.   Focal-conic or "fan" texture of a smectic A phase.

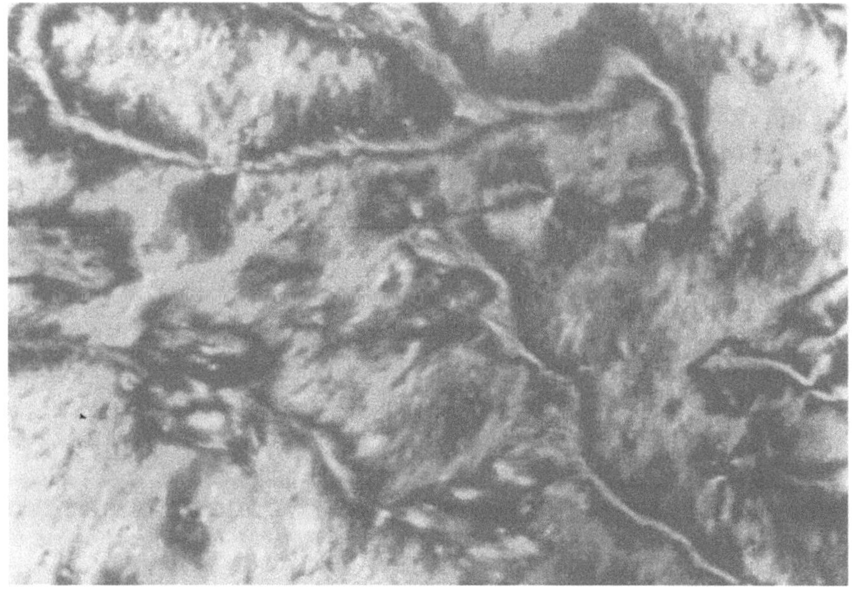

FIGURE 6.   Schlieren texture of a smectic C phase.

FIGURE 7.   Focal conic texture of a smectic B phase.

a nematic phase at room temperature; the widespread interest in displays then began. In recent years the many other physical properties of the nematic, and very recently the smectic A and C phases, have been related to molecular structure, thus leading to the optimization of the newer modes of display. Generally a liquid crystal molecule can be represented by the structure (3), where X and Y are terminal groups, L, are optional lateral groups, Z, is a linking group (on some occasions more than one linking is interspaced between the ring systems), A, are ring systems and can be aromatic, heterocyclic or alicyclic, and, n and m are usually 1, 2, or 3.

$$X \left[\begin{array}{c} L \\ | \\ \text{(A)} \end{array} \right]_n Z \left[\begin{array}{c} L \\ | \\ \text{(A)} \end{array} \right]_m Y$$

3

Several notable compounds do not fit into this general structure, for example, cholesteryl esters [structure (1)] and alkane 2,4-dienic acids [structure (4)].

$$R-CH{=}CH-CH{=}CH-C \overset{\displaystyle O}{\underset{\displaystyle O-H}{\big\langle}} \quad \overset{\displaystyle H-O}{\underset{\displaystyle O}{\big\rangle}} C-CH{=}CH-CH{=}CH-R$$

4

In both these systems the essential core rigidity is produced by either fused alicyclic rings or extended conjugation and hydrogen bonding in the carboxylic acid. It is usually found that carboxylic acids form dimers and thus exhibit an anomalously high liquid crystal stability.

Each of the molecular structural units in structure (3) will be briefly considered.

4.1. The Effect of Terminal Groups

Very many groups have been studied and compared, simple easily modified systems were usually studied, e.g., 4-*p*-substituted benzylideneamino 4-*n*-alkoxy biphenyls [structure (5)].

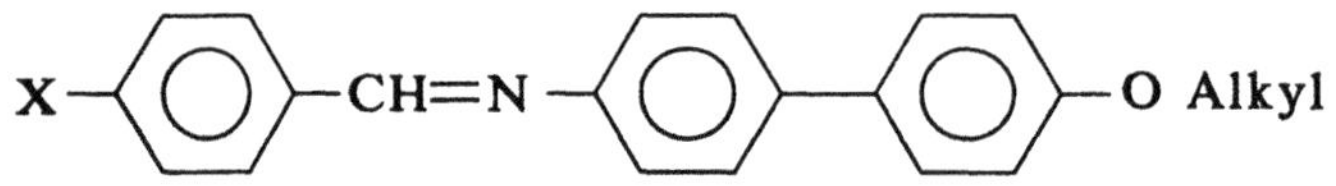

$$X \!-\!\!\langle \bigcirc \rangle\!-\! CH\!=\!N \!-\!\!\langle \bigcirc \rangle\!-\!\langle \bigcirc \rangle\!-\! O\ Alkyl$$

5

For the nematic phase, strong terminal attractions and/or extended length of the molecule are desirable[6]; thus phenyl and cyano groups are better than halogens or hydrogen.

Commonly one terminal group is alkyl, and in some cases both terminal groups can be alkyl. Short alkyl chains favor nematic phases while longer alkyl chains enhance smectic phases.

Of the smectic phases only two have been considered; the smectic A phase is the commonest, and older references nearly always refer to this phase. In this case the terminal group efficiency order[6] can be variable, but generally small polar groups such as halogens and polarizable groups such as phenyl are the most efficient. In recent years smectic C phases have become very important. In this context terminal alkoxy groups and branched alkyl chains[7] are advantageous.

4.2. Lateral Groups

This is a more complex situation as the position of the lateral group on the molecule can also be very important. Lateral groups tend to hold the molecules apart, thus reducing the effectiveness of lateral attractions. This lowers the thermal stability of most liquid crystal phases and particularly smectic phases; however, in some cases the S_C phase is not destabilized as much as the other phases.[8]

The effect of a lateral fluorine on nematic phase thermal stability and the effect of its position on smectic A thermal stability is shown for structure (6) in Table 1.

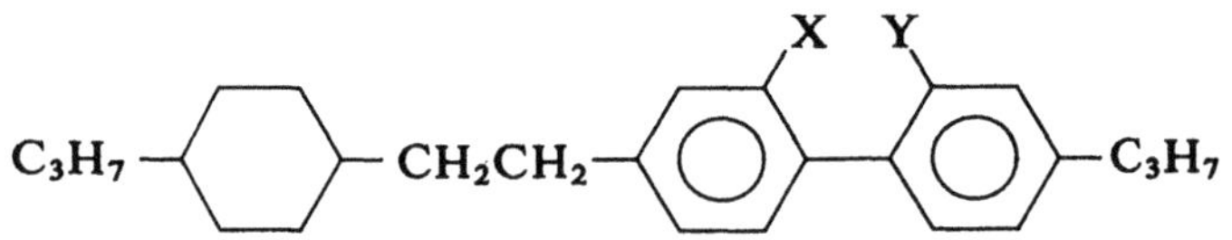

$$C_3H_7 \!-\!\!\langle \bigcirc \rangle\!-\! CH_2CH_2 \!-\!\!\langle \bigcirc \rangle\!-\!\langle \bigcirc \rangle\!-\! C_3H_7$$

6

TABLE 1.   The Effect of Fluorine Substitution on Nematic and Smectic A Thermal Stability

Compound	X	Y	C-S_A (°C)	S_A-N (°C)	N-I (°C)
6a	H	H	67	119	144
6b	F	H	59	[34]	108
6c	H	F	40	—	108

4.3. Linking Groups

A wide variety of groups that preserve the linear nature of the molecule have been investigated.[6] The commonest are $-COO-$, $CH=CH$, $-C\vdots C-$, $-CH\vdots N$, $-CH_2CH_2-$, $-N\vdots N-$, and a carbon–carbon single bond.

4.4. Ring Systems

The effect of several different ring systems on the nematic phase thermal stability is shown in Table 2.

The nature of the ring system has a significant effect upon other physical properties such as viscosity and birefringence of the nematic phase as well as on the thermal stability, and the best ring system for one property is not necessarily the best for another. This situation is also the case for terminal, lateral, and linking groups, therefore many factors have to be considered in the design of new molecules.

TABLE 2.   Physical Properties for Several Related Typical Mesogens. Nematic–Isotropic (N–I), Melting Point (C–N), Birefringence (Δn), Viscosity (η), and Dielectric Anisotropy ($\Delta\varepsilon$)

	C-N (°C)	N-I (°C)	η (cSt)	Δn	$\Delta\varepsilon$
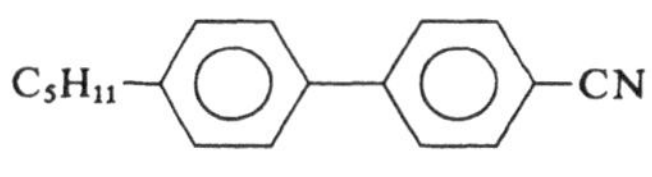 C_5H_{11}—(benzene)—(benzene)—CN	23	35	25	0.21	+12
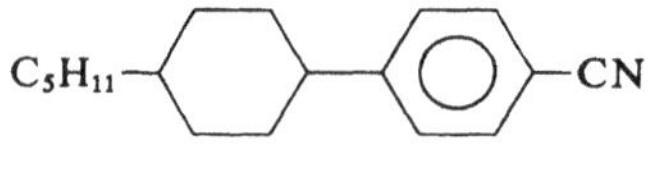 C_5H_{11}—(cyclohexane)—(benzene)—CN	30	55	23	0.13	+10
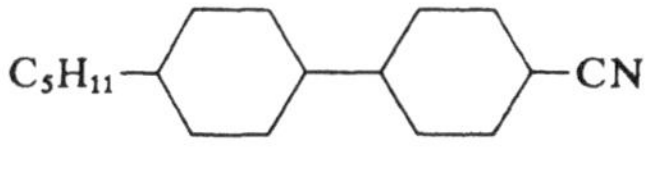 C_5H_{11}—(cyclohexane)—(cyclohexane)—CN	62	85	65	0.06	+4.5
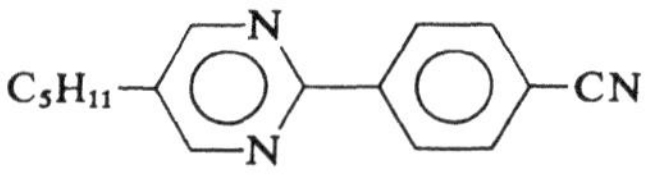 C_5H_{11}—(pyrimidine)—(benzene)—CN	71	(53)	55	0.22	+20
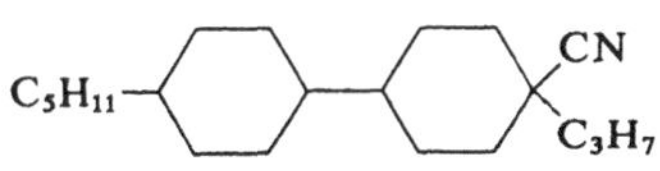 C_5H_{11}—(cyclohexane)—(cyclohexane)(CN)(C_3H_7)	32	(42)	71	0.03	−7.7

5. RELATIONSHIP OF MOLECULAR STRUCTURE TO DISPLAY PARAMETERS

Many different types of electro-optic display based on the nematic phase are known. The optimization of each type of display usually requires different physical properties from the liquid crystal. Some of the more important physical parameters are as follows:

Viscosity (η). This property, like most others involving liquid crystals, is anisotropic. The viscosity of the nematic phase has a major influence on the response times, especially the OFF response time, and is very temperature dependent, varying by a factor of 3–5 over a 20°C temperature range. Compounds containing polar groups, and especially polar lateral groups, generally have higher viscosities. (See Table 2.)

Birefringence (Δn). This property determines the optical properties of the system. The nematic phase behaves as a positive uniaxial crystal; thus there are two refractive indices n_e and n_o. These can be measured using an Abbé refractometer, the difference between their values being the birefringence (Δn). It is influenced by the extent of aromatisation and π bonding in the molecule. (See Table 2.)

Dielectric Anisotropy ($\Delta\varepsilon$). This is the difference between the dielectric permittivities along ($\varepsilon_\parallel$) and across ($\varepsilon_\perp$) the molecule. The major contribution to $\Delta\varepsilon$ is from permanent dipoles. Materials with a large positive $\Delta\varepsilon$ and thus large $\varepsilon_\parallel$ contain polar terminal groups, such as cyano; these compounds exhibit a low threshold voltage in displays. For example, a $\Delta\varepsilon$ of about +16 gives a threshold voltage of about 1.2 V. Table 2 illustrates how this property varies wih molecular structure.

Elastic Constants. It is only in recent years, with the advent of cheap, powerful computers, that these parameters have been extensively calculated from experimental data. Three elastic constants—splay (k_{11}), bend (k_{33}), and twist (k_{22})—describe the forces governing the deformation of the nematic liquid crystal; the ratios of these, rather than their absolute values, are important in displays. For example, a low k_{33}/k_{11} ratio is required to produce a sharp electro-optic response. At present only empirical rules are known; for instance, long alkyl chains and the use of aromatic rings with no lateral substituents lower k_{33}/k_{11}.

Nematic Phase Range and Mixtures. Even the simplest display requires the nematic phase to exist from −10 to +60°C; it must also have the requisite birefringence, dielectric anisotropy, viscosity, and elastic constants. No single compound has been found that has all the desired properties. However, by mixing together a range of compounds the melting point can be lowered and the N-I raised, thus broadening the nematic phase range (Fig. 8). This is optimum at the eutectic composition (E).

Most commercial mixtures contain at least four and up to perhaps 20 components. At its most basic, some components provide a low melting point, while other components serve to raise the N-I, others may enhance specific other properties, e.g., giving high $\Delta\varepsilon$ to lower the threshold voltage. It should be realized, however, that all the components must be mutually soluble at −10°C (or less). In multicomponent mixtures the theoretical eutectic composition can be calculated[9] knowing the melting point and enthalpy of fusion of each component. In practice this is only a guide, as nonideal behavior nearly always occurs.

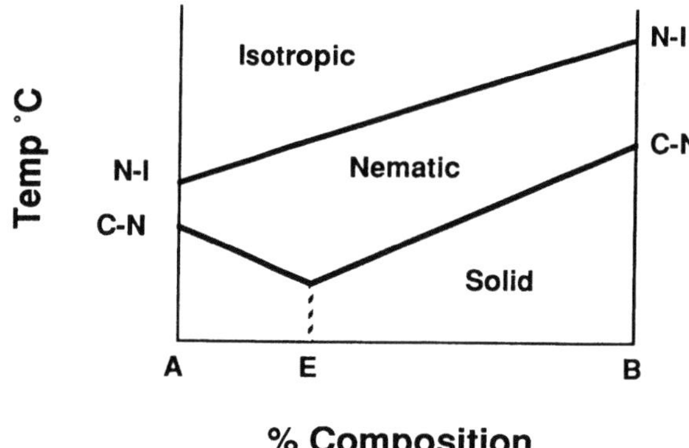

FIGURE 8. Phase diagram of compounds A and B, showing the eutectic composition (E) which exhibits the lowest melting point for the system and a wide nematic phase temperature range.

6. CHIRAL LIQUID CRYSTAL PHASES

6.1. The Cholesteric Phase

The term "cholesteric" originates from the first compounds, esters of cholesterol, which were found to exhibit this phase. Friedel[3] showed that it is a special case of the nematic phase and is produced by compounds that contain one or more chiral centers. If a compound contains a carbon atom that has four different groups attached to it, they can be arranged in two distinct ways in space, these two isomers being mirror images of each other. They cannot be overlaid and have different "force fields" associated with them, thus in the bulk the molecules exist in an asymmetric environment. If nematic-type ordering also exists then a structure as shown in Fig. 9 is formed. Sheets of molecules, in which the molecules are aligned as in the nematic phase, are imagined, the "director" of each sheet being at a slight angle to the previous sheet, so that a helical arrangement of the molecules or sheets is formed. The molecular structural properties giving rise to the cholesteric phase are of course identical to those discussed for the nematic phase. However, a further parameter now exists—the helical pitch. Apart from cholesteryl esters and a few other steryl esters, most cholesteric phases are produced by the use of a terminal chiral alkyl

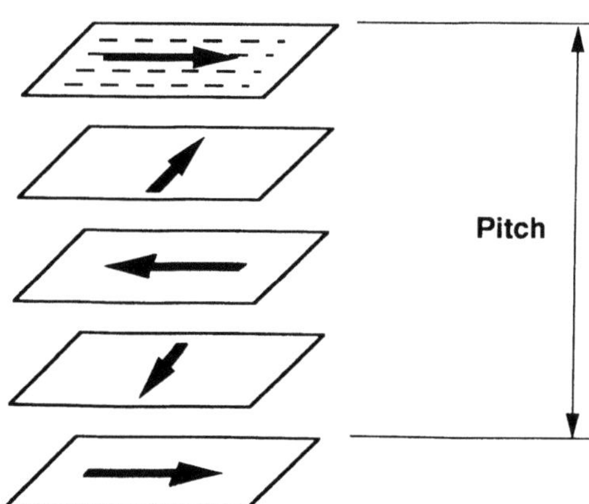

FIGURE 9. Schematic representation of the twisted nematiclike layers of the cholesteric phase producing a helix of pitch P.

chain, e.g., 2-methylbutyl such as in compound (7). The closer the chiral center is to the aromatic core the shorter is the pitch length P. An empirical relationship[10] between the sense of helix (left or right handed), the position of the asymmetric center, and the absolute configuration of the chiral center allows prediction of the helix sense.

$$C_6H_{13}O-\text{〈O〉}-COO-\text{〈O〉}-CH_2\overset{\overset{\displaystyle CH_3}{|}}{\underset{\bullet}{C}}HC_2H_5$$

7

When light impinges on the helical arrangement of molecules a particular wavelength (λ), which is related to the helical pitch (P) by $\lambda = nP$ (n = refractive index) is selectively reflected, the other wavelengths being transmitted.

The pitch length of the helix can also be varied by a change in temperature; this is particularly pronounced close to a transition into a smectic phase when the helix must "unwind" to accommodate the smectic structure. When λ is in the visible wavelength region (0.4–0.7 μm) the cholesteric phase can be used as a visible temperature sensor and is ideally suited for many thermal mapping applications.

Another application of cholesterics arises when small amounts of a pure cholesteric material are dissolved in a nematic liquid crystal. This dilution lengthens the helical pitch; two general cases can be identified: (i) at pitch lengths of a few microns the "chiral nematic phase" is the basis for an electro-optic effect called the dyed phase change display[11]; (ii) at longer pitch lengths the phase is used in most "twisted nematic" displays, e.g., watches, calculators, etc.

6.2. Chiral Smectic Phases

In general only tilted smectic phases are affected if their constituent molecules possess a chiral center. The chiral smectic C phase is the most important. It was predicted and shown to be ferroelectric[12] in 1975.

In 1979 it was shown that in very thin display cells (2 μm cell spacing) this phase could be suitably aligned and would respond to an electric field about 1000 times faster than a nematic liquid crystal, the two switched states also being bistable. These properties promise complex displays operating at video frame rates. This has led to intense research activity and many new smectic C host mesogens and chiral dopants, which make the S_c phase ferroelectric, have been made. Some typical smectic C hosts are structures (8) and (9).

$$C_9H_{19}-\text{〈O〉}_N^N-\text{〈O〉}-OC_9H_{19}$$

8

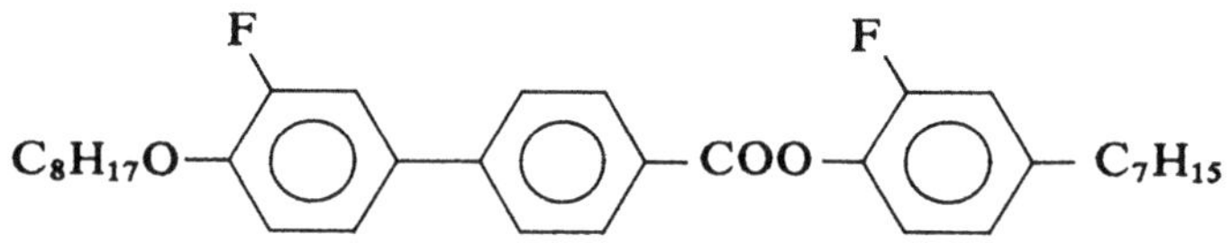

$$C_8H_{17}O\text{—}\langle\langle\bigcirc\rangle\rangle\text{—}\langle\bigcirc\rangle\text{—COO—}\langle\bigcirc\rangle\text{—}C_7H_{15}$$

9

Some typical chiral dopants which produce a large ferroelectric effect (high spontaneous polarisation) in smectic C mixtures are structures (10) and (11).

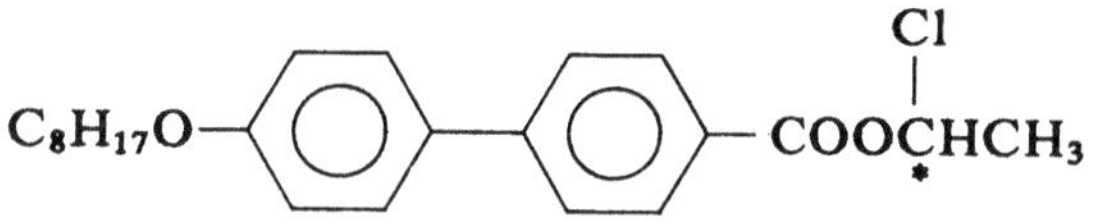

$$C_8H_{17}O\text{—}\langle\bigcirc\rangle\text{—}\langle\bigcirc\rangle\text{—COOCHCH}_3$$

10

$$C_8H_{17}O\text{—}\langle\bigcirc\rangle\text{—}\langle\bigcirc\rangle\text{—COOCHCH}_3$$

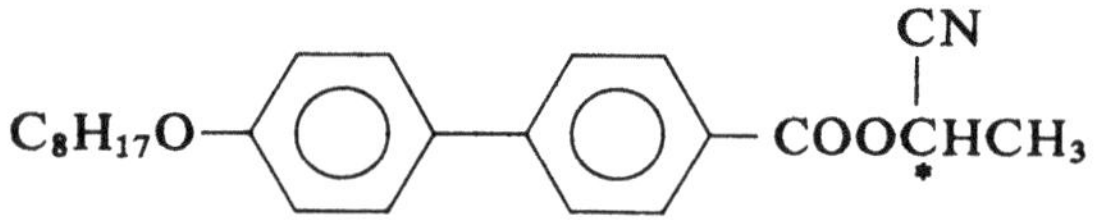

11

7. DICHROIC DYES

Most liquid crystal displays use crossed polarizers to allow the change in molecular orientation, caused by the application of an electric field, to be visible. However, polarizers are expensive and an alternative cheaper method has been sought for many years. White and Taylor[11] discovered that a dichroic dye, dissolved in a chiral nematic liquid crystal of a few micrometers pitch length, appears colored, but on the application of a field the liquid crystal molecules align with the field (if they are of the required positive dielectric anisotropy). The dyes then align in the liquid crystal and appear colorless. This is the dyed phase change display. Many dichroic dyes have been synthesized, dyes 12 and 13 being representative. The perfect dye has to dissolve well, have an high extinction coefficient (so that only a small amount of it is needed), have an high order parameter (S) in the nematic host so that the change from colored to colorless is optimum, and be stable. Such dyes are scarce, and owing to other display problems this type of display has found only limited use.

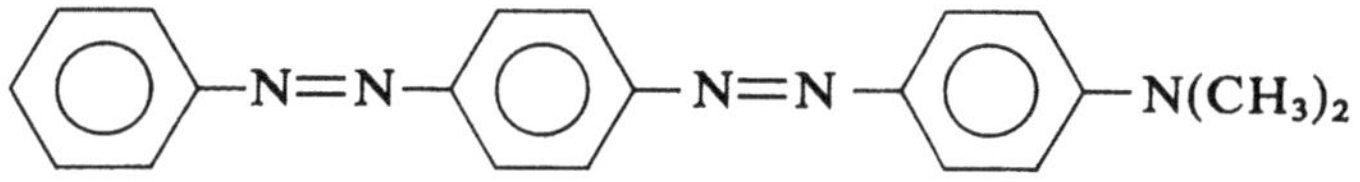

$$\langle\bigcirc\rangle\text{—N=N—}\langle\bigcirc\rangle\text{—N=N—}\langle\bigcirc\rangle\text{—N(CH}_3)_2$$

12

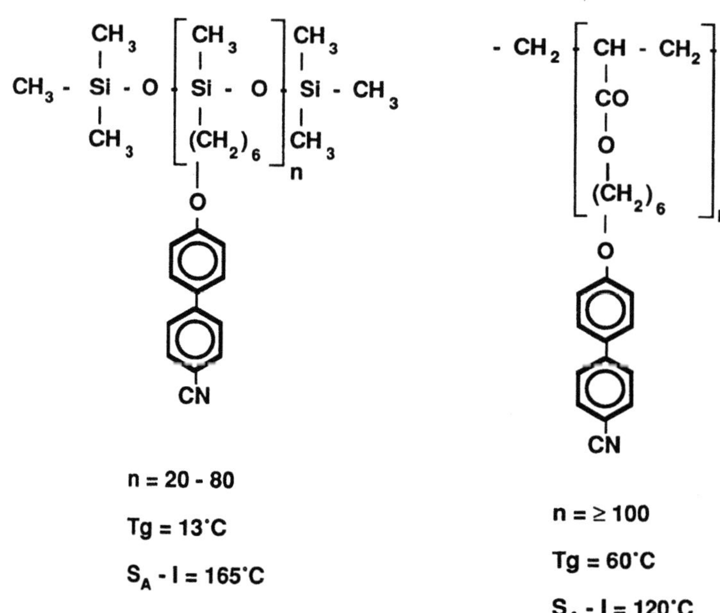

13

8. LIQUID CRYSTAL POLYMERS

Two general types of liquid crystal polymer can be identified: (i) main chain and (ii) side chain.

Main chain polymers are composed of rigid groups linked by flexible units; on heating to high temperatures they melt to produce a liquid crystal phase. These polymers are used for high-strength fibers (e.g., Kevlar) and in precision and specialist injection mouldings.

Side chain liquid crystal polymers consist of a flexible backbone, usually of relatively short chain length (<100 units), onto which are attached typical liquid crystal monomers. Figure 10 shows the same monomer on two of the commonest backbones, polysiloxane and polyacrylate.

FIGURE 10.   A polysiloxane and a polyacrylate polymer both with the same cyanobiphenyl side chain or pendant group.

The glass transition temperature (T_g) of the polyacrylate is usually higher but the liquid crystal phase is less stable than in the polysiloxane. The linking group between the monomer pendant group and the polymer backbone has a great influence on T_g and melting point. For a liquid crystal phase to form, the linking alkyl chain length is usually between 2 and 12 carbons long; at short chain lengths the polymer tends to crystallize and at long chain lengths the ordering is usually not sufficient to produce liquid crystal ordering. The T_g also decreases as the linking group is lengthened.

8.1. Applications

Liquid crystal polymers respond to an electric field in the same way as monomer liquid crystals, but very slowly. Their application as an optical recording medium using a laser beam to "write" on thin a film of the polymer has been demonstrated.[13] To absorb the laser energy a dye is incorporated into the polymer.

Another potential application is as a host in which suitable rod-shaped nonlinear optical materials can be aligned by poling the polymer with an electric field; on cooling below T_g the induced ordering is frozen in.

9. CONCLUSION

Liquid crystals are a multidisciplinary subject. The development of liquid crystals has been due to the active interaction of chemists, theoreticians, physicists, device engineers, and electronics engineers. The next two chapters demonstrate how effective this interaction has been and how each discipline has had to express its desires and achievements in a form that could be understood by the other disciplines.

REFERENCES

1. F. Reinitzer, *Monatsh. Wiener Chem. Gesell.* **9**, 421 (1988).
2. O. Lehmann, *Z. Phys. Chem. (Leipzig)* **4**, 462 (1889).
3. G. Friedel, *Ann Phys. (Paris)* **18**, 273 (1922).
4. S. Chandrasekhar, *Phil. Trans. R. Soc. London A* **309**, 93 (1983); C. Destrade, H. Gasporoux, P. Foucher, N. H. Tinh, J. Malthete, and J. Jacques, *J. Chim. Phys.* **80**, 137 (1983).
5. G. W. Gray and J. W. Goodby, *Smectic Liquid Crystals*, Leonard Hill, Glasgow and London (1984).
6. G. W. Gray and P. A. Winsor, *Liquid Crystals and Plastic Crystals*, Vol. 1, Ellis Horwood, Chichester (1974).
7. D. Coates, *Liquid Crystals* **2**(1), 63 (1987).
8. D. Coates, *Liquid Crystals* **2**(4), 423 (1987).
9. D. S. Hulme and E. P. Raynes *J. Chem. Soc., Chem. Commun.* 98 (1974).
10. G. W. Gray and D. G. McDonnell *Mol. Cryst. Liquid Cryst.* **34**, 211 (1977).
11. D. L. White and G. N. Taylor, *J. Appl. Phys.* **45**, 4718 (1974).
12. R. B. Meyer, L. Liebert, L. Strzelecki, and P. Keller, *J. Phys. (Paris) Lett.* **36**, 69 (1975).
13. H. J. Coles and R. Simon, *Recent Advances in Liquid Crystalline Polymers* (L. L. Chapoy, ed.), Elsevier, New York (1985), p. 323.

26

Electro-optic Effects in Liquid Crystals

E. P. Raynes

1. INTRODUCTION

Liquid crystal displays (LCDs) have become very well known and widely used in the past few years. They depend for their operation on a range of electro-optic effects found in oriented thin layers of liquid crystals (LCs). The previous chapter has described the chemistry and structures involved in the formation of liquid crystal phases, and the next chapter discusses the applications of liquid crystals. In this chapter the major electro-optic effects found in liquid crystals will be described and explained. All these electro-optic effects arise from the anisotropic physical properties induced by the orientational ordering of the rodlike molecules that constitute liquid crystals.

2. THE LIQUID CRYSTAL PHASE

The thermotropic liquid crystal phases used in fabricating electro-optic devices are found over a range of temperature in simple, low-molecular-weight molecules. The chemical structures involved in these phases were discussed in detail in the previous chapter. There are three main subdivisions of the thermotropic liquid crystal phase; these are shown schematically in Fig. 1.

Nematic. This is the most widely used liquid crystal phase, possessing only orientational order of the molecular axes.

Cholesteric. The cholesteric phase is similar to the nematic phase, but with a helical ordering imposed with a pitch length, P, which can be as short as 0.1 μm.

Smectic. These are several smectic liquid crystal phases, which can show various degrees of positional order; however, the common feature is a tendency to order the centers of gravity of the molecules into layers.

E. P. Raynes • Royal Signals and Radar Establishment, St. Andrews Road, Malvern, Worcestershire, WR14 3PS, U.K.

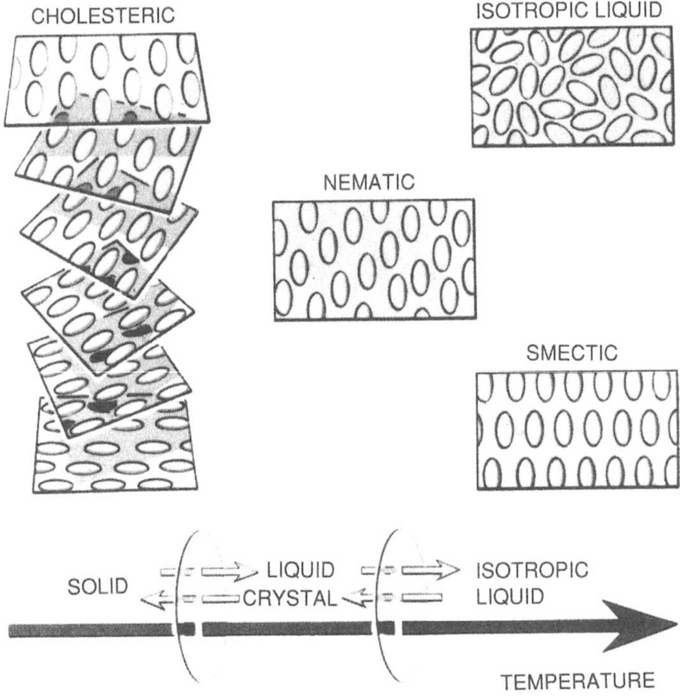

FIGURE 1.   Phase structure of thermotropic liquid crystals.

3. THE ANISOTROPIC PHYSICAL PROPERTIES OF LIQUID CRYSTALS

The anisotropic properties of liquid crystals are dealt with in detail in a number of texts,[1-5] and we will here deal with only the major features of nematic liquid crystals which are relevant to their electro-optic effects.

3.1. Nematic Order Parameter

The molecules in the nematic phase are not perfectly aligned and lie at an angle θ to the average direction usually described by a unit vector **n**, called the director. The total distribution function is usefully described by the order parameter S, which is given by the equation

$$S = \tfrac{1}{2}\langle(3\cos^2\theta - 1)\rangle \tag{1}$$

Typically $0.6 < S < 0.8$ for a nematic liquid crystal. It is useful to note that

Total disorder (isotropic liquid) corresponds to $S = 0$

and

total order (solid crystal) corresponds to $S = 1$

For the measurement and results of the order parameter of nematic liquid crystals, see Tough and Bradshaw.[6]

3.2. Electric Permittivities (ε) and Refractive Indices (n)

There are two components of the relative permittivity, $\varepsilon_\parallel$ (measured along the director) and $\varepsilon_\perp$ (measured perpendicular to the director), which are related to the molecular dipole moments μ_l and μ_t, where μ_l is the molecular dipole along the molecular axis and μ_t is the molecular dipole transverse to the molecular axis, and the anisotropic $(\Delta\alpha)$ and mean (α) molecular polarizability by (7)

$$\Delta\varepsilon = A\left[\Delta\alpha + \frac{B}{kT}\left(\mu_l^2 - \frac{\mu_t^2}{2}\right)\right]S \tag{2}$$

$$\bar\varepsilon = 1 + A\left[\bar\alpha + \frac{B}{3kT}(\mu_l^2 + \mu_t^2)\right] \tag{3}$$

where

$$\Delta\varepsilon = \varepsilon_\parallel - \varepsilon_\perp$$

and

$$\bar\varepsilon = \tfrac{1}{3}(\varepsilon_\parallel + 2\varepsilon_\perp)$$

Similar equations relate $\Delta\alpha$ and α to $\alpha_\parallel$ and $\varepsilon_\perp$, and the factors A and B are combinations of cavity and reaction field factors.[3]

The range found for present-day liquid crystals is given by $-10 < \Delta\varepsilon < 50$, and Fig. 2 shows the permittivities of a selection of typical nematic compounds. Similar

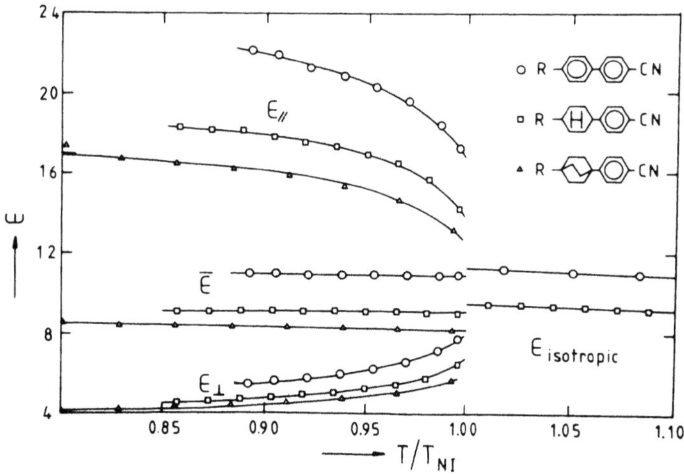

FIGURE 2.   Electric permittivities of some nematic liquid crystals.

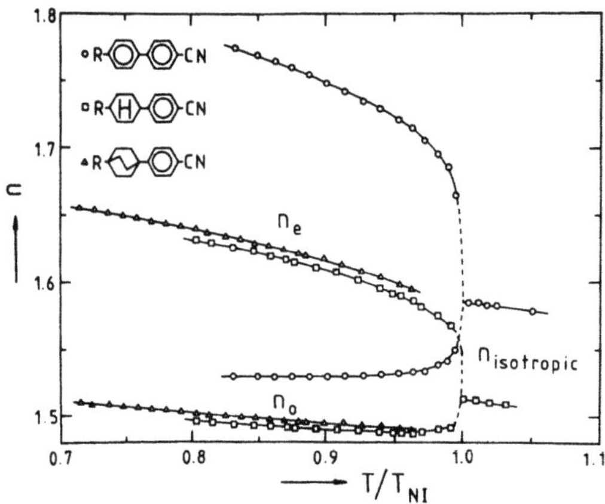

FIGURE 3.   Refractive indices of some nematic liquid crystals.

expressions, but without the effect of molecular dipoles, apply to the refractive indices, with

$$n_o^2 = \varepsilon_\perp$$

and

$$n_e^2 = \varepsilon_\parallel$$

The range found for present-day liquid crystals is given by $0.03 < \Delta n < 0.30$, with some typical results shown in Fig. 3.

3.3. Nematic Continuum Theory

The average direction of nematic and long pitch cholesteric liquid crystal molecules is described by the unit vector **n**, called the director. When the state of uniform alignment is distorted, the free energy density is given by[8]

$$F_K = \frac{1}{2}\left\{ K_{11}(\text{div } \mathbf{n})^2 + K_{22}\left(\mathbf{n} \cdot \text{curl } \mathbf{n} + \frac{2\pi}{P} \right)^2 \right.$$

$$\left. + K_{33}(\mathbf{n} \times \text{curl } \mathbf{n})^2 - \varepsilon_o \Delta\varepsilon(\mathbf{n} \cdot \mathbf{E})^2 \right\} \tag{4}$$

where P is the cholesteric pitch. The three elastic constants are identified with the distortions shown in Fig. 4 and present-day nematic liquid crystals show the following range:

$$5 < K_{11} < 20(pN)$$

$$K_{22}/K_{11} \approx 0.5$$

$$0.5 < K_{33}/K_{11} < 3.0$$

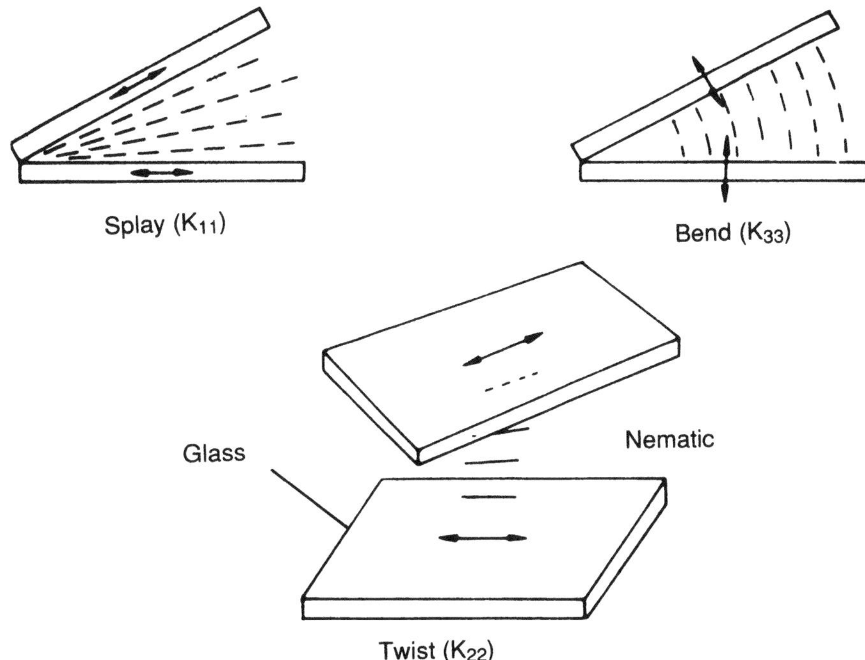

FIGURE 4. Three basic types of elastic deformation in nematic and cholesteric liquid crystals.

The total free energy given by (4) can be minimized using either analytical[9] or numerical[10] techniques, and the results allow the prediction of the properties of aligned nematic and long pitch cholesteric layers with applied electric fields to quite extraordinary accuracy.

4. LIQUID CRYSTAL ALIGNMENT PROPERTIES

The director can be aligned on glass surfaces either by surfactant effects or by the need to minimize the continuum energy (4). The two principal alignments are shown in Fig. 5. The planar alignment is particularly useful, and is usually achieved by simply rubbing or buffing a coating of a tough polymer such as polyimide. It is believed that grooves in the polymer are produced by the buffing process, and that these align the liquid crystal. This is because the director lies along the grooves in order to minimize the bend and splay distortion energy which would result if the director lay across the grooves. An electric field aligns the director through $\Delta\varepsilon$, by minimizing the electrostatic term in the continuum energy (4); the two possible field-induced alignments are shown in Fig. 6, and depend on the sign of $\Delta\varepsilon$. As a result of the form of the electrostatic energy term in (4), the director is sensitive to E^2. Furthermore, the inherent slow response, of the order of 100 ms, means that the director responds to the rms value of any rapidly changing electric field.

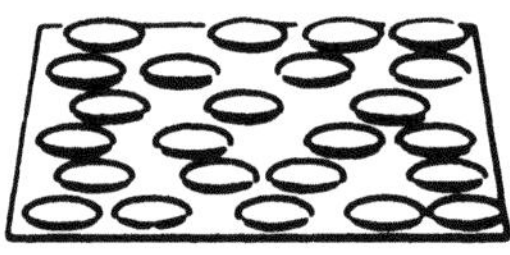

FIGURE 5.   Liquid crystal alignment on glass surfaces.

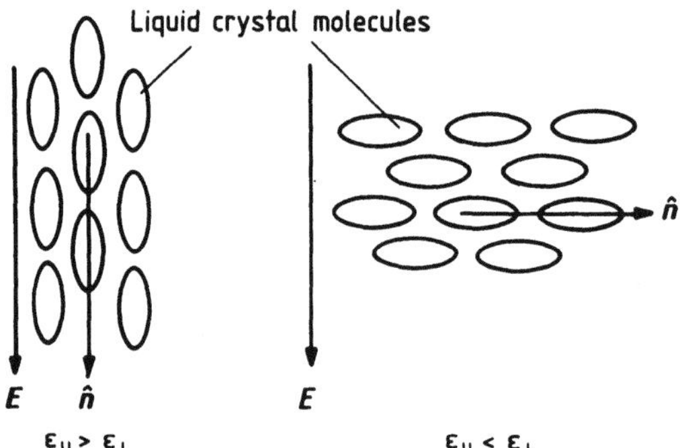

FIGURE 6.   Alignment induced by an electric field.

5. FREEDERICKSZ TRANSITIONS IN NEMATIC LIQUID CRYSTALS

The orientation of a nematic liquid crystal layer aligned by two surfaces can be altered by the application of an electric field.[11] Subsequent removal of the field allows the director to relax back to the original configuration determined by the alignment on the two surfaces. This reorientation is called the Freedericksz transition, after its discoverer, and Fig. 7 shows an example illustrated by the permittivity change of an aligned planar sample with $\Delta\varepsilon > 0$ subject to an increasing voltage applied across the layer.

Solution of Eq. (4) gives the details of Fig. 7, with a threshold voltage V_c that is given by

$$\varepsilon_o \Delta\varepsilon V_c^2 = K_{11}\pi^2 \tag{5}$$

and an initial slope that is inversely proportional to K_{33}/K_{11}.

6. THE TWISTED NEMATIC ELECTRO-OPTIC EFFECT

The twisted nematic (TN) electro-optic effect is the best known example of a Freedericksz transition; it was invented in 1972[12] and is now widely used in display

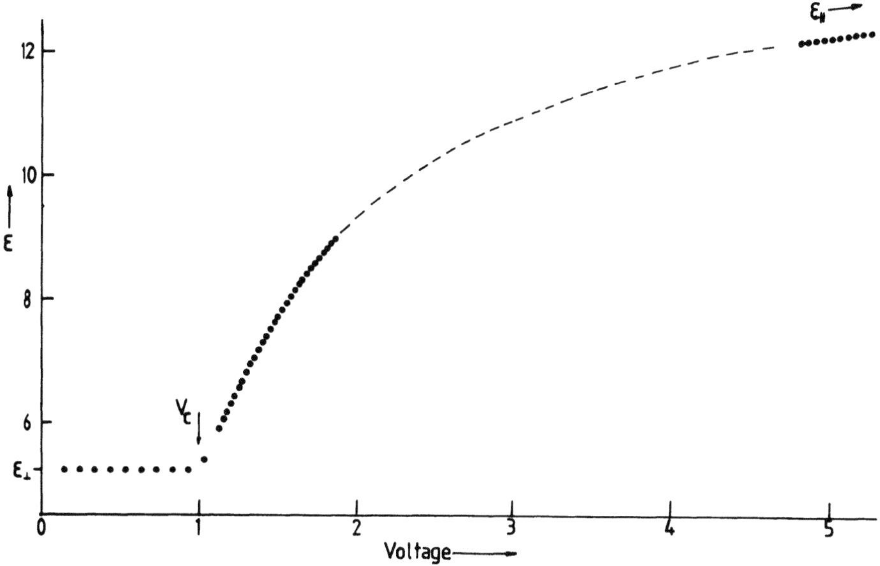

FIGURE 7. Voltage dependence of the permittivity of a planar aligned sample of nematic liquid crystal.

devices. A 90° twist is induced by the surface alignment, as shown in Fig. 8. For perpendicular polarizers, the transmission T is given by[13]

$$T = 1 - \frac{\sin^2(\pi/2)[1 + (2\Delta n \cdot d/\lambda)^2]^{1/2}}{[1 + (2\Delta n \cdot d/\lambda)^2]} \tag{6}$$

and $T = 1$, corresponding to complete transmission by the twisted layer, when

$$(\Delta n \cdot d/\lambda) \gg 1$$

or

$$(\Delta n \cdot d/\lambda) = (m^2 - 1/4)^{1/2}$$

where m is an integer. (In simple terms, this may be explained by noting that the 90° twist of the liquid crystal alignment leads to a 90° rotation of the plane of polarization of the light.) Early TN devices operated with $(\Delta n \cdot d/\lambda) \gg 1$, but in the last few years it has become common to use the better performance of devices operating in the second case with $m = 1$ or 2. These devices are said to be operating in the first, or second, minima, respectively. Above a threshold voltage calculated from Eq. (4) and given by

$$\varepsilon_o \Delta \varepsilon V_c^2 = \pi^2 [K_{11} + \tfrac{1}{4}(K_{33} - 2K_{22}) + 2K_{22} d/P] \tag{7}$$

the layer reorientates, so the 90° rotation of the polarization no longer occurs; this

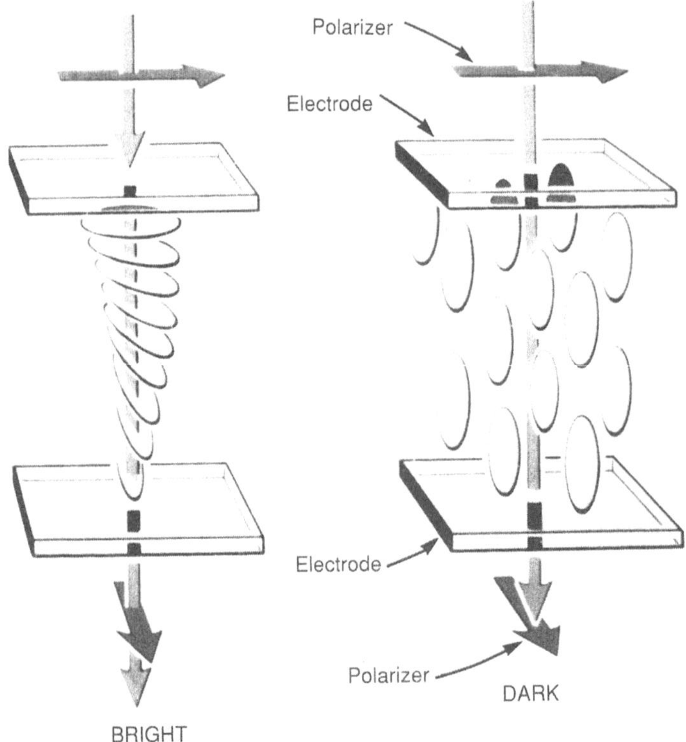

FIGURE 8.   Operation of a conventional twisted nematic device.

results in the optical switching shown in Fig. 9. Removal of the field allows the layer to relax back to its original twisted configuration, so that the plane of polarization of the light is again rotated, giving transmission. TN displays are exceedingly successful as simple display devices, with of the order of 10^9 devices manufactured

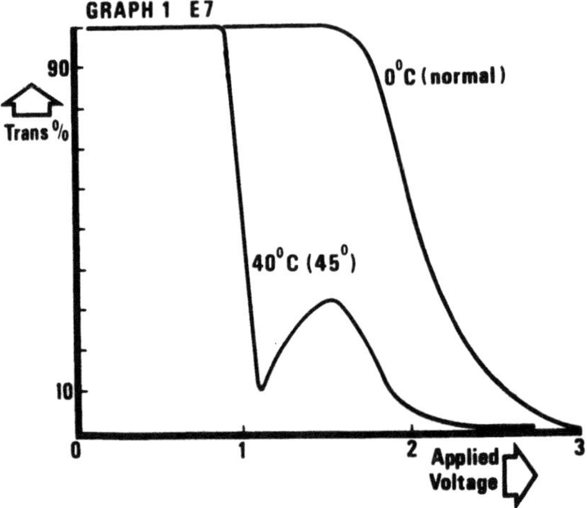

FIGURE 9.   Transmission–voltage curve of a typical TN liquid crystal display.

each year. The major disadvantage of TN LCDs is their poor performance in complex, multiplexed displays. This is a direct consequence of the gentle slope of the transmission voltage curve found in TN LCDs and illustrated in Fig. 9. Variations of viewing angle and temperature are seen to make the problem even worse. One solution to this problem is the integration of nonlinear electronic devices at each picture element in the display. Thin film transistors (TFTs) fabricated using amorphous silicon are the best known and most widely used nonlinear device[14] for this purpose. The applications of the twisted nematic effect are dealt with in detail in the following chapter. There are also two alternative LCD configurations which give improved performance in complex displays and are based on the supertwist and ferroelectric structures.

7. THE SUPERTWISTED NEMATIC ELECTRO-OPTIC EFFECT

Recently "supertwist" layers with twist angles $(\varphi) > 90°$ have been found to possess remarkably steep electro-optic properties[15]; this is illustrated by the device with $\varphi = 270°$ in Fig. 10.

The origins of this steeper response can be found by studying the voltage dependence of θ_{max}, the tilt angle of the director in the middle of the layer. Computer modeling (Fig. 11) clearly shows the effect of increasing φ. The initial slope of the curve just above threshold becomes progressively steeper, becoming infinite and eventually bistable for $\varphi = 270°$; the hysteresis bounded by the dotted lines is observed, rather than the unstable reentrant region of the curve for $\varphi = 270°$.

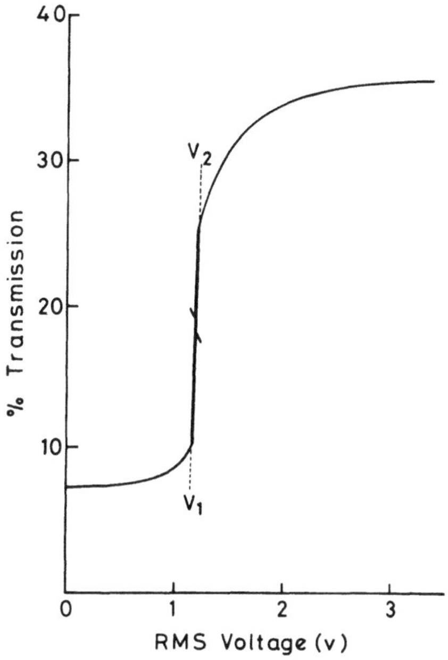

FIGURE 10. Transmission–voltage curve of a supertiwst liquid crystal display.

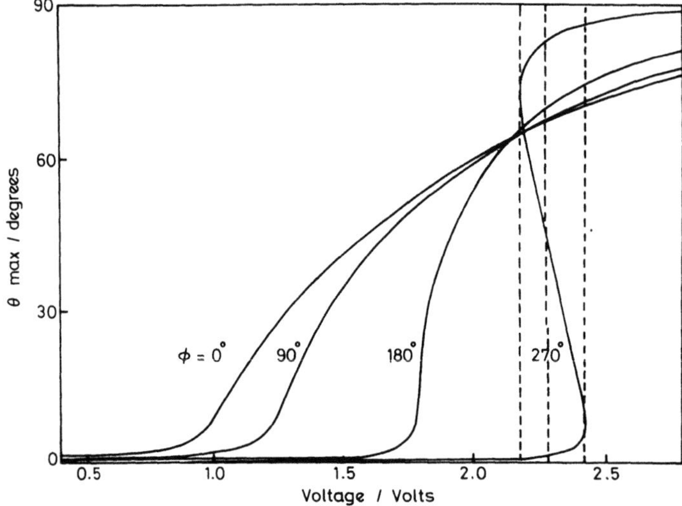

FIGURE 11. Computer calculated voltage dependence of the midplane director tilt angle (θ_{max}) in layers of various twist angle (φ).

Analytical modeling[16] shows that just above threshold

$$\theta_{max}^2 = \frac{4[(V - V_c)/V_c]}{F(K) + \Delta\varepsilon/\varepsilon_\perp} \tag{8}$$

where

$$F(K) = \frac{K_{33} - C\varphi^2}{K_{11} + D\varphi^2}$$

Both C and D are combinations of the elastic constants, and are both positive, so that for increasing φ

$$F(K) \rightarrow 0$$

and eventually

$$F(K) < 0$$

such that

$$F(K) + \Delta\varepsilon/\varepsilon_\perp = 0$$

and the denominator in Eq. (8) is zero, producing an infinite gradient with a steep electro-optic response. The exact value of φ for infinite gradient depends on a number of device and material parameters.

The optical properties are also more complicated than in a TN display, and when the polarizer and analyzer are at 45° to the director, the transmission is given[17] by

$$T = (\cos^2 \pi)\{[(\varphi/\pi)^2 + (\Delta n \cdot d/\lambda)^2]\}^{1/2} \tag{9}$$

The maximum transmission occurs when

$$(\Delta n \cdot d/\lambda)^2 = p^2 - (\varphi/\pi)^2 \tag{10}$$

where p is an integer.

Equations (9) and (10) mean that the value of $(\Delta n \cdot d)$ must be carefully chosen to match the middle of the visible spectrum, and there is also a noticeable color to the display, although several techniques described in the next chapter have recently been suggested to remove this color. Supertwist displays make very successful complex displays, and already many PCs using them are available commercially.

8. FERROELECTRIC LIQUID CRYSTALS

The molecules in smectic liquid crystals prefer to be in layers (Fig. 1). In the smectic A phase the molecules are orthogonal to the layers, and in the smectic C phase they are tilted by about 25° to the layer normal. If the smectic C phase is composed of chiral molecules, it can show ferroelectric properties with a very pronounced spontaneous polarization P_s.[18] The molecular origin of this spontaneous polarization in chiral smectic (S_c^*) layers is shown in Fig. 12. The top diagram shows all possible orientations of S_c^* molecules that have a dipole moment (perpendicular to the page) residing on the chiral center. Clearly two of the possible orientations fit better into the smectic layers, and these are shown in the lower diagram. Both orientations in the lower figure show a positive dipole, resulting in a net positive dipole moment for the layer; tilt to the left, instead of the right as shown, would result in a negative dipole moment. The application of a dc electric field therefore forces the molecules to tilt either to the right or to the left, depending on the polarity of the applied electric field.

Alignment is a critical process in S_c^* device fabrication, and is achieved (Fig. 13) by a combination of cooling through the phase sequence

$$N^* \to S_A \to S_c^*$$

and an elongated N^* pitch[19]:

$$P >_{\cdot} 4d$$

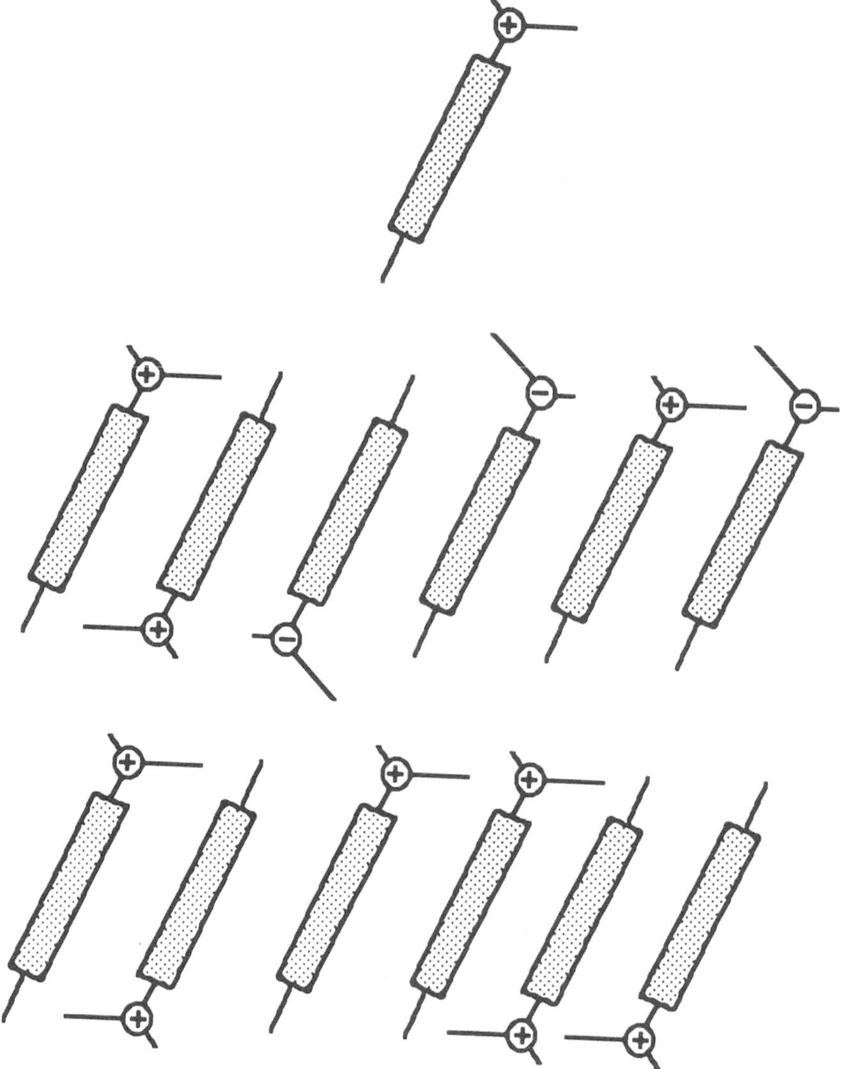

FIGURE 12. The hindrance of molecular rotation and the production of a spontaneous polarization (P_s) in S_C^* layers.

This alignment process, produces two smectic C states, and the application of a dc electric field produces polarity-dependent switching between the two states. There is currently much interest in the switching of ferroelectric S_C^* layers. They show a combination of very fast switching (a few microseconds at room temperature) and bistability which makes them extremely attractive for fast electro-optical devices and for complex displays. The performance and driving of displays based on ferroelectric S_C^* layers will be described in more detail in the following chapter.

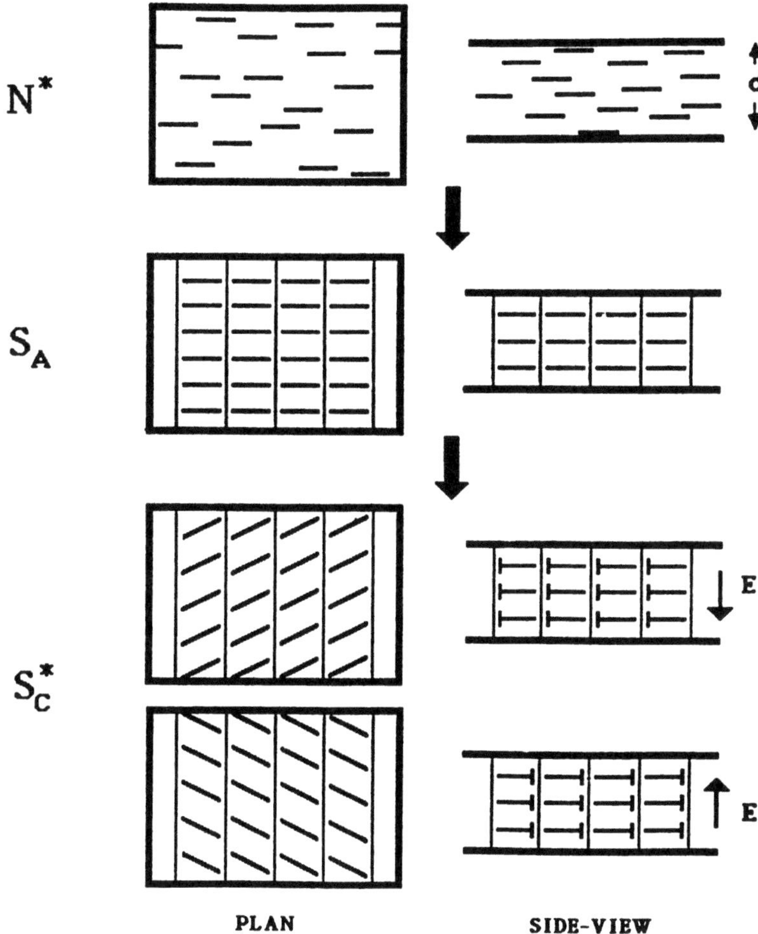

FIGURE 13.   Alignment of S_c^* layers by using a combination of a particular phase sequence and elongated N^* pitch.

REFERENCES

1. S. Chandrasekhar *Liquid Crystals*, Cambridge U. P., Cambridge (1977).
2. P. G. de Gennes, *The Physics of Liquid Crystals*, Oxford U. P., Oxford (1974).
3. W. H. de Jeu, *Physical Properties of Liquid Crystal Materials* Gordon and Breach, London, (1980).
4. *Liquid Crystals: Their Physics, Chemistry and Applications* (C. Hilsum and E. P. Raynes, eds.), The Royal Society of London (1983).
5. E. P. Raynes, In *Electro-optic and Photorefractive Materials* (P. Gunter, ed.), Springer-Verlag, Berlin (1986).
6. R. J. A. Tough and M. J. Bradshaw, *J. Phys. (Paris)* **44**, 447 (1983).
7. W. Maier and G. Meier, *Z. Naturforsch.* **16a**, 262 (1961).
8. F. C. Frank, *Discuss. Faraday Soc.* **25**, 19 (1958).
9. H. J. Deuling, *Mol. Cryst. Liq. Cryst.* **27**, 81 (1974).
10. D. W. Berreman, *Phil. Trans. R. Soc. Lond.* **A309**, 203 (1983).

11. V. Freedericksz and V. Zolina, *Trans. Faraday Soc.* **29**, 919 (1933).
12. M. Schadt and W. Helfrich, *Appl. Phys. Lett.* **18**, 127 (1971).
13. C. H. Gooch and H. A. Tarry, *J. Phys. D* **8**, 1575 (1975).
14. A. J. Snell, K. D. Mackenzie, W. E. Spear, P. G. LeComber, and A. J. Hughes, *Appl. Phys.* **24**, 357 (1981).
15. C. M. Waters, V. Brimmell, and E. P. Raynes, In *Proc 3rd Int. Display Res. Conf.* Kobe, Japan, 396 (1983).
16. E. P. Raynes, *Mol. Cryst. Liq. Cryst. Lett.* **4**, 1 (1986).
17. E. P. Raynes, *Mol. Cryst. Liq. Cryst. Lett.* **4**, 69 (1987).
18. R. B. Meyer, L. Liebert, L. Strzelecki, and P. Keller, *J. Phys.* **36**, L-69 (1975).
19. M. J. Bradshaw, V. Brimmell, and E. P. Raynes, *Liquid Crystals* **2**, 107 (1987).

27

Liquid Crystal Applications

Sally E. Day

1. INTRODUCTION

The widespread use of information technology has led to a greater need for displays to exhibit the information generated. The cathode ray tube (CRT) is a very well established technology, but it is bulky, consumes relatively large amounts of power, and cannot be conveniently scaled up for large-area displays. Until recently, little has been available to replace the CRT. The requirements for different types of display are many and various, including size, complexity or resolution, color, and viewing conditions, both in terms of the ambient light levels and the angle of view. A number of different solutions to these requirements have been found using liquid crystals. Some of the devices are well proven and familiar to everyone; there are also many new devices that are just coming onto the market, and others that are still being developed.

2. TWISTED NEWMATIC (TN) LIQUID CRYSTAL DISPLAYS

The twisted nematic display was the first type of liquid crystal display to be widely used. It is still used in many applications, including watches and calculators.

The operation of the TN device was described in the chapter by Raynes. The TN device is a true field effect device, needing in principle no charge flow, so the display requires very low power.

2.1. The Construction

The construction of a typical cell is shown in Fig. 1. The drive voltage is applied to the electrodes, which are made of the transparent conductor, indium tin oxide (ITO). To maintain the twisted structure of the liquid crystal a thin polymer film is rubbed or buffed unidirectionally on each plate of the cell. In addition the nematic liquid crystal is doped with a small amount of cholesteric liquid crystal to impose

---

Sally E. Day • Thorn EMI Central Research Laboratories, Dawley Road, Hayes, Middlesex UB3 1HH, U.K. *Present address*: Royal Signals and Radar Establishment, St. Andrews Road, Malvern, Worcestershire WR14 3PS, U.K.

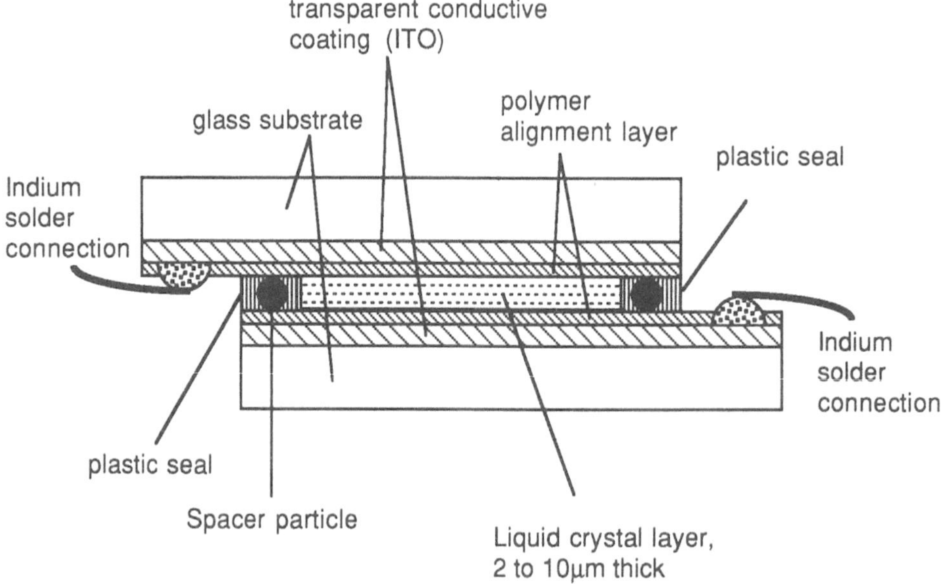

FIGURE 1.   The construction of a liquid crystal cell.

a defined sense of twist on the structure; correctly used, this prevents the formation of unsightly areas of reverse tilt and twist in the display. The thickness of the display is controlled using spacer particles, which may be, for example, polythene balls or Mylar strips. The thickness is usually about 10 μm; thickness uniformity is not as critical as for some of the other displays discussed in later sections.

Polarizing film is laminated onto both surfaces of the display. In many applications the display is used in reflective mode for which the cell is placed in front of a diffuse, polarization-conserving reflector, such as brushed aluminum. The use of the polarizers, which only transmit ~40% of incident unpolarized light, together with the reflector means that these displays have relatively low brightness.

2.2. Driving a Twisted Nematic Display—Multiplexing

The transmission–voltage characteristic for a typical twisted nematic cell is shown in Fig. 9 of Raynes. Nematic liquid crystals respond to the rms value of the voltage. Two values of the voltage are important with regard to driving the display; one is the rms voltage required to just switch the display on, V_s (the switched voltage), the other is the maximum rms voltage that may be applied without switching the display, V_{ns} (the nonswitched voltage). The voltages, V_s and V_{ns}, must be chosen to give good contrast upon switching for reasonable viewing angles and over the required temperature range.

When a high density of picture elements (pixels) is required in a display it is impractical to make a separate connection to each; instead the electrodes may be arranged in a matrix, with row and column strip electrodes, respectively, on the top and bottom surfaces. A single row of pixels may be addressed by applying suitable voltages along both the row and the columns defining those pixels. Since this must be repeated for all the other rows in sequence, each of the N rows is addressed for $1/N$ of the time per frame and receives a constant background voltage for $(N-1)/N$ of the time so that V_s and V_{ns} are rather similar. There is a fundamental limit on the maximum number of rows, N, which may be addressed for a given ratio of V_s/V_{ns}, the so-called Alt and Pleshko limit,[1] where

$$\frac{V_s}{V_{ns}} = \left(\frac{\sqrt{N}+1}{\sqrt{N}-1}\right)^{1/2} \tag{1}$$

A 100-row display, for example, requires a ratio of V_s/V_{ns} of about 1.1. The data to be displayed are applied as a string of voltages of value $\pm D$ to the column electrodes. The row on which the data are to appear is defined by applying a voltage S to that row. An array of pixels is shown schematically in Fig. 2,[2] together with the voltages applied sequentially to select the rows and the data waveforms applied to the columns. The rms voltages are also shown for both a switched and nonswitched pixel. Unwanted electrolytic effects in the liquid crystal are avoided by either changing the polarity of the applied voltages on alternate frames or by using bipolar pulses to address the display.

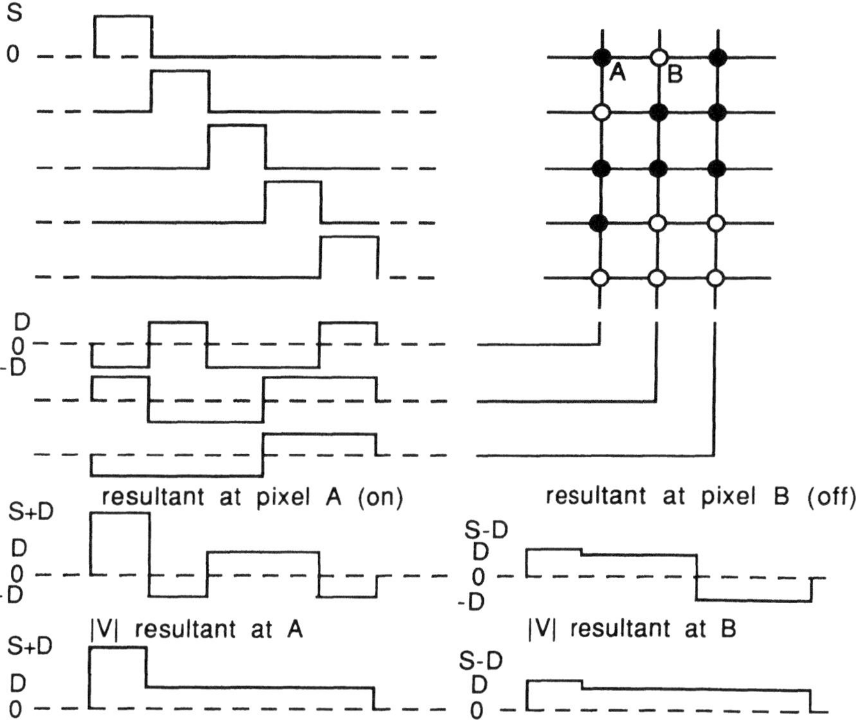

FIGURE 2. Simple implementation of an rms multiplexing scheme.

2.3. Limitations of Twisted Nematic Displays

It is difficult to multiplex TN displays. The values of the switched and non-switched voltages, V_s and V_{ns}, must necessarily be close to obtain multiplexing and as can be seen from Fig. 5 of Raynes poor angle of view results, particularly when the display is required to operate over a reasonable temperature range. For applications requiring high levels of information density, the level of multiplexing must be increased by making the transmission–voltage characteristic steeper. This is achieved in the supertwisted nematic (STN) display described in the following section.

Nevertheless, the TN display is used with nonmultiplexed, direct addressing and with low levels of multiplexing. Color filters are incorporated into individual cells and built up into very large area displays for advertising boards and also for giant video displays in stadia and other outside venues. The displays are generally backlit for these applications and the backlights require large amounts of power because of the high ambient light conditions.

3. THE SUPERTWISTED NEMATIC DISPLAY

The principle behind this type of display has been discussed in the previous chapter. Its importance may now be more fully appreciated in the light of the requirements of multiplexing. More complex displays (i.e., higher levels of multiplexing) mean that the value of V_s/V_{ns} must be closer to 1. The necessary steep transmission–voltage curve is obtained by increasing the amount of twist to between $180°$ and $270°$, as shown in Fig. 11 of Raynes. This increase has a number of implications for the manufacture and performance of STN displays, particularly in their optical properties and switching speed as well as the type of alignment layers that are used.

3.1. Optical Properties

The STN or supertwisted birefringence effect (SBE)[3] display is constructed as before with polarizers on each surface; however, an increase in the cholesteric dopant gives a tightly twisted structure in the liquid crystal, which results in elliptically polarized light being transmitted. The best contrast between the switched and unswitched states is obtained with the polarizers oriented at $\sim 30°$ and $\sim 60°$ to the top and bottom alignment directions, respectively. The ellipticity of the transmitted light depends on the wavelength of light and this orientation of polarizers results in colored states. The display can be operated in two modes, with the polarizers as above, leading to yellow and black states or with one polarizer rotated by a further $90°$ to give blue and white states. This obviously removes the opportunity to incorporate full color. However, an additional advantage of the STN display over the TN display is an improvement in the viewing angle over which good contrast is maintained.

A recent development in STN displays has been the use of optical compensation to give black and white states.[4,5] This is achieved by making a double layer STN, in which the second passive layer has the opposite sense of twist to the active layer, thereby compensating the optical effect of the twist. The thickness of the liquid crystal layer in STN displays must be carefully controlled to maintain a uniform color due to the birefringence. The thickness tolerance of the compensating layer must be the same, which makes the double layer display more difficult to manufacture. However, color may now be incorporated into the pixels to give a limited color display.

A direct consequence of the steep electro-optic characteristic of the STN display is the lack of grey scale. Careful choice of the amount of twist and of the material allows some grey levels by lowering the steepness of the electro-optic characteristic, but at the cost of the number of pixels that can be matrix addressed. This means that the STN display is not suitable for monochrome or color television displays. Another, more important, limit of the STN is the time taken to switch the cell. The response time is relatively long, ≈ 250 ms, compared with around 30 ms for TN devices. Improvements in materials have given response times of less than 150 ms[7] and further reductions to approximately 60 ms[8] have been reported by reducing the thickness, d, of the liquid crystal layer to 4 μm.

3.2. Manufacture of STN Displays

In order to maintain a high degree of twist in the STN cells it has been indicated that a modest amount of cholesteric liquid crystal is added as a dopant to the nematic liquid crystal. However, this leads to a tendency to the formation of a scattering texture.[9] The scattering texture arises when the twist forms along the cell instead of perpendicularly to it. This can be prevented by careful control of the amount of cholesteric dopant, together with the use of a special high tilt alignment.

The high tilt alignment can be achieved in a number of ways. Certain polymers have been found to give a surface tilt to the liquid crystal structure of $\sim 7°$; this is sufficient to maintain twists of around 180–240°. Oblique evaporation of SiO onto the surface of the glass and ITO induces higher amounts of tilt, around 25°, which allows the formation of 270° of twist uniformly across the display. Unfortunately evaporation of SiO is not particularly suitable for mass production methods and so many of the STN displays are made with polymer alignment and 180–240° of twist.

3.3. Applications

The STN device is suitable for more complex displays than the TN device because of the high level of multiplexing that is possible. However, the slow response times and lack of grey scale restrict the applications to devices such as computer terminals where a fast refresh rate is not usually important. The recent developments of double layers to compensate the colored state may mean that displays with limited color capability will soon be on the market. In addition, a reduction in the thickness of the liquid crystal layer may reduce the switching times, although it is unlikely that the STN display will ever be able to show pictures at video rates.

4. THIN-FILM TRANSISTOR (TFT) LIQUID CRYSTAL DISPLAYS

Multiplexed LCDs require an adequate nonlinearity (i.e., steep electro-optic characteristic) in the pixels of the display so that they can be addressed independently from each other. Active matrix displays remove the requirement for nonlinearity from the liquid crystal. Instead, the electrode is made active by incorporating, for example, a thin-film transistor at each pixel.

Many types of nonlinear device have been considered for insertion at each pixel,[11] including metal–insulator–metal (MIM) devices, as well as thin-film transistors. Various materials have been considered for their manufacture[12] including polycrystalline silicon and cadmium selenide. The most widely used device in commercially available LCD televisions is the thin-film transistor, fabricated using a layer of hydrogenated amorphous silicon[13] (α-Si).

4.1. Construction of Amorphous Silicon Thin-Film Transistors

The active area required to make an LCD of this type is a small proportion of the relatively large total area, so it is not worth using expensive crystalline silicon as the substrate. In addition, crystalline silicon cannot be grown to give very large areas. Instead amorphous silicon can be used. The techniques involved in the manufacture of amorphous silicon devices were developed originally for use in solar cells. It was found that the low mobility, caused by dangling bonds, could be increased by partial hydrogenation of the silicon when it is deposited using chemical vapor deposition (CVD).

The construction of the FET for a typical display is shown in Fig. 3. The gate electrode is deposited on the glass substrate and typically is made from Cr. It is connected to one of the select lines, used in driving the display. Silicon nitride (SiN_x) or silicon oxide (SiO_x) is used for the gate insulation layer. The semiconducting layer is then deposited and metal contacts made for the source and drain electrodes. The source is connected to the liquid crystal cell and the drain is connected to the data line used to address the pixel. The amorphous silicon in the channel region is kept thin to give a low leakage current. In the source and drain regions

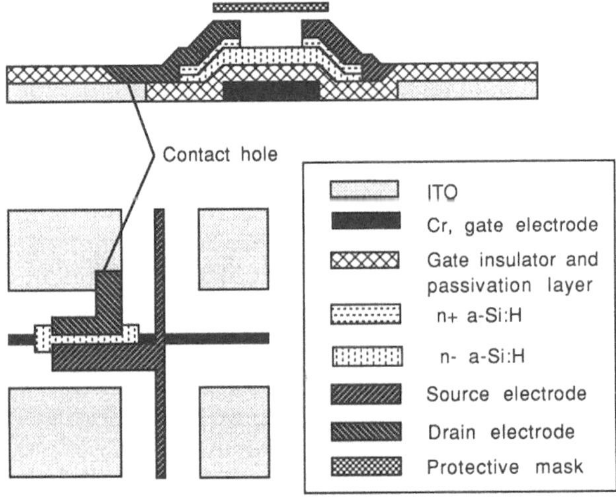

FIGURE 3. Construction of an amorphous silicon thin-film transistor.

additional doped amorphous silicon is deposited to lower the contact resistance. Chemical contamination is prevented by deposition of a passivation layer, such as SiN_x. Finally a protective black mask, of CrO_x, is put over the FET to reduce its photosensitivity and to prevent the generation of a photocurrent.

The drain electrode is connected to the ITO electrode of the pixel through a contact hole etched in the passivation layer. The display is then constructed as normal with alignment layers on the substrates, which are separated using spacers. The polarizers and alignment are the same as used in the simple TN displays; therefore the optical properties are the same and similar optimization may be carried out to improve the contrast and viewing angle.

4.2. Driving the TFT Display

A voltage is applied to a line of gate electrodes to define the line of pixels to be addressed; the data are then applied to the drain electrodes. When the pixels on that line have been switched into the state defined by the data, the next source line is defined for the data waveforms to be applied and so on across the display.

If the data voltage on a particular pixel selects an on state then the transistor is switched so as to charge the liquid crystal cell capacitance. Figure 4 shows the circuit of a single pixel. Storage capacitance may be included to maintain the voltage across the liquid crystal, thereby keeping the switched state of the liquid crystal through the frame scan time.

One of the disadvantages of amorphous silicon is its low carrier mobility. The conductance of the transistor must be sufficient to charge, within the line address period, not only the capacitance of the liquid crystal but also the storage capacitance of the pixel and any parasitic capacitance of the transistor. The conductance depends on the carrier mobility and on the size of the transistor—in both cases the larger

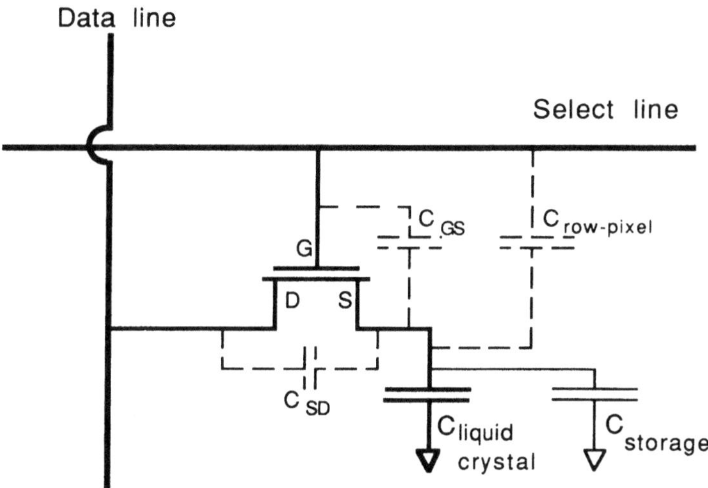

FIGURE 4.   Circuit diagram of a TFT pixel, including the parasitic capacitance.

the better. However, increasing the size of the transistor increases the leakage current and the parasitic capacitance.

4.3. Polycrystalline Silicon TFT

Higher carrier mobility can be obtained without increasing the size of the transistor by using polycrystalline silicon.[12] However, the temperature required ($\sim 600°C$) for deposition of polycrystalline silicon prevents the use of normal glass substrates; quartz or borosilicate glass may be used as the substrate. Alternatively amorphous silicon can be recrystallized using localized laser heating. At the present time this cannot be easily adapted for use in the production of large numbers of displays.

Another important advantage of the higher carrier mobility in polysilicon is that some of the drive electronics can be incorporated into the display, thereby reducing the number of connections to the display. This simplifies production.

4.4. Incorporating Color

The red, green, and blue colors are incorporated in a dyed polymeric, or gelatin film, superimposed over the liquid crystal light shutters. The number of pixels must be increased by a factor of 3 or 4 in order to maintain the resolution of a color display.

A number of parameters can be optimized to give good contrast for each color as well as color purity. One parameter is the backlight, which, if used, can be chosen to match the transmission characteristics of the filters. Another improvement uses a different thickness of pixel for each color. The reasons are as follows: (1) A display constructed with uniform thickness has a viewing cone that varies with the wavelength of light being transmitted. (2) The best contrast is obtained by operating the display in the first minimum (Section 6 of Raynes); however, the required thickness varies with wavelength.

The multigap construction is obtained in some displays by varying the thickness of the polymer film, although obviously the use of such multigaps increases the complexity of production. The improvement in the viewing angle, over which uniform grey scale is maintained, is considerable.

4.5. Production of TFT Displays

TV displays using thin-film transistor technology are in production. The products are hand-held, battery-operated color TVs with screen sizes up to about a 5-in. diagonal. A typical example is shown in Fig. 5.

TFT displays are rather complex to manufacture. The control required in the manufacture is comparable to that of conventional silicon chip processing. The area of the display is, however, considerably larger than required in silicon chip manufacture and the eye is very sensitive to defects and blemishes in displays. It is possible to incorporate some redundancy by fabricating additional elements near the pixel or to reduce the likelihood of faults by using more complex electrode structure. In addition, some repair of the elements is possible by laser welding after assembly, but neither method will correct all of the defects that may arise. Furthermore, laser welding repair is not suitable for large-scale, mass production.

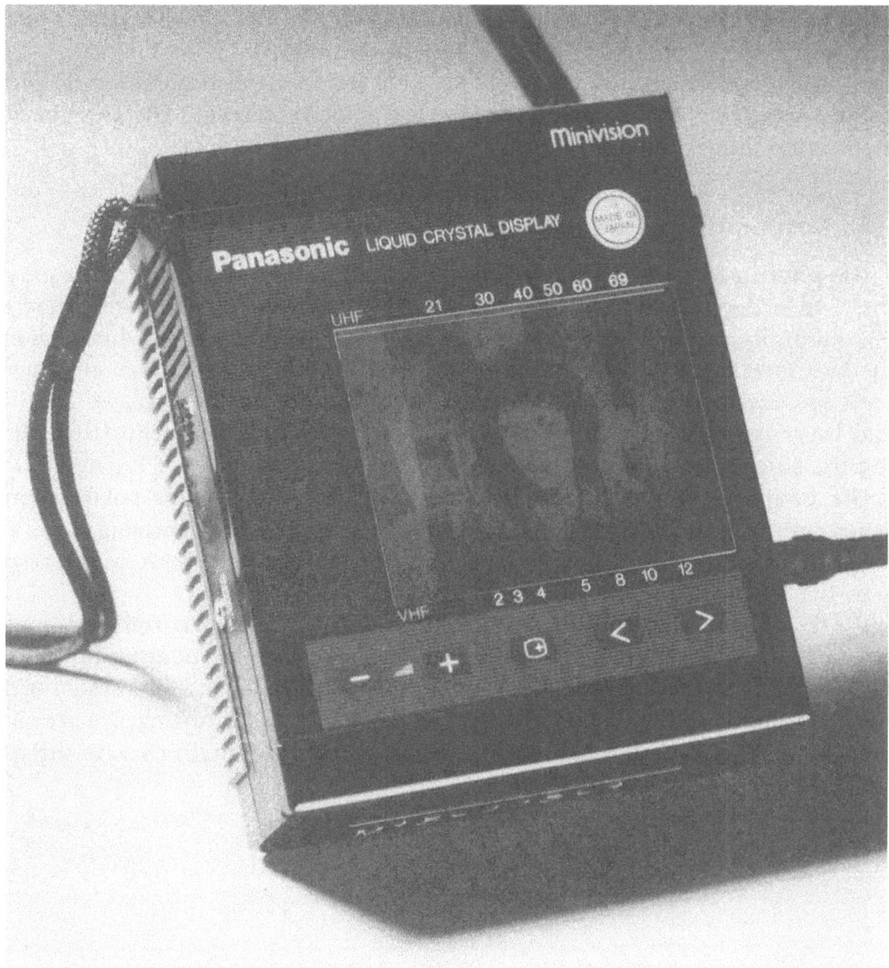

FIGURE 5.   A 3-in. diagonal pocket TV using α-Si TFT, with multigap, color pixellation.

In spite of the difficulties, yields in the manufacture of 3–5-in. displays have improved and a number of companies have this type of pocket television on the markets. It is likely that 14-in. color televisions will shortly be available commercially.

Projection systems using TFT displays seem the most likely option in the near future for high definition television (HDTV) using liquid crystals. Three displays may be used, one for each of the primary colors red, green, and blue, with dichroic beam splitters used to separate the white light for projection through the displays and then to recombine the images.

5. FERROELECTRIC DISPLAYS

In the TFT display the transmission state of the pixel is stored by the unswitched transistor after it has been addressed. In the ferroelectric display[14] the storage is effected by the liquid crystal itself. The configuration of the liquid crystal is bistable and, with the use of polarizers, gives a dark and a light state. This means that once a pixel has been addressed it remains in that state so the limits on multiplexing described in Section 2.2 do not apply to the ferroelectric display. Therefore an active element is no longer required at each pixel.

5.1. Surface Stabilized Ferroelectric LCD

The physics of the ferroelectric liquid crystal has been described in the previous chapter. In order to use the effect in a display it is necessary to remove the twist that is naturally present in the chiral liquid crystal so that the net dipole moment aligns and interacts with an applied filed. This is done using surface alignment in the cell as described in Section 2.1. for nematic liquid crystal displays. The liquid crystal layer is made very thin (1–2 μm), thus unwinding the helical structure[15] giving the so-called "bookshelf geometry" shown in Figure 13 of Raynes.

The coupling of the electric field $\mathbf{E}$, with the spontaneous polarization P_s, depends on the polarity of $\mathbf{E}$, i.e., the response is to the dc or instantaneous value of the applied field. This is in contrast to the nematic liquid crystals, which respond only to the rms value of the applied field.

There are two stable states as shown in Fig. 6. The polarizer and analyzer orientations are shown, giving the light and dark states. In the nontransmitting state the optic axis is aligned parallel with the input polarizer. In the light state the optic axis is at 45° between the two polarizers. The thickness and birefringence are chosen so that the display is then a half-wave retarder for a wavelength close to the center

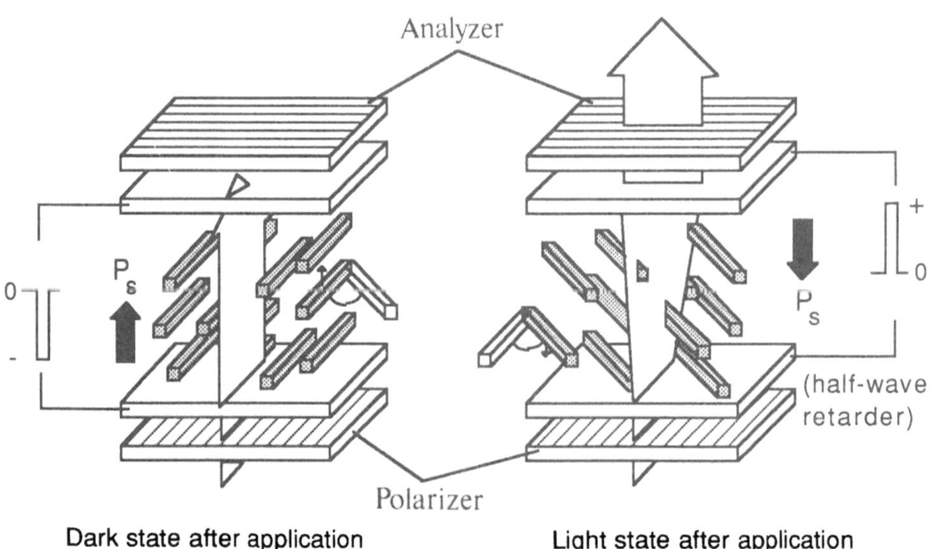

FIGURE 6.   Bistable states of surface stabilized liquid crystal for use in displays.

of the visible spectrum (~ 560 nm). The plane of polarization is rotated by 90° and the light is transmitted by the analyzer.

5.2. Multiplexing SSFLCD

Switching of the ferroelectric liquid crystal occurs when the product of the voltage pulse amplitude and pulse width exceeds a threshold value. The liquid crystal then stays in that state until a pulse of opposite polarity is applied. In practice the bistable states are maintained in a display by superimposing a high-frequency ac stabilization field.

Of great importance for display applications is the very fast switching time of ferroelectric liquid crystals. The shortest times reported are of the order of a microsecond.[16] The bistability means that the display can be written a line at a time and remains in that state; the limit to the number of lines that can be addressed depends on the refresh time per line and the time available to address the entire display. This permits a higher limit to the number of addressable lines than that given by Eq. (1) for TN and STN displays.

5.3. Grey Scale in SSFLCD

The bistability means that grey scale is difficult to obtain. However, there are two methods by which it can be achieved. The first uses the speed available in the switching process. Each pixel is rapidly switched during binary weighted time slots. The intensity seen by the eye is integrated to give 2^m levels of brightness, where m time slots are used. The second method uses subpixellation, which increases the complexity of the display. Each pixel is divided into binary weighted areas which are addressed independently to give the different levels of brightness. Again the number of grey levels is 2^m, where in this case, m is the number of subpixels. It is also possible to use a combination of these methods of "time dither" and "spatial dither" to obtain the required number of grey levels. Color may be incorporated in the same way as for TN displays, in the form of colored filters at each pixel.

5.4. Production of Large Area SSFLCD

Although the ferroelectric liquid crystal display has tremendous potential for large-area TV displays, there are still a number of difficulties to be overcome.

Some of the requirements are hard to achieve in manufacturing, in particular, the production of a uniform 2-μm gap between the glass plates over a large area. Another difficulty is in the operating temperature range; further developments in the materials and drive schemes may, however, overcome this problem. Our understanding of materials will be improved upon the development of a continuum theory for smectic liquid crystals. This will also be useful in the understanding and prevention of alignment defects.

Provided these problems can be overcome, the potential improvements in yield and reduction in cost over the TFT displays make this technology attractive as a means of manufacturing large-area flat televisions.

6. THE FUTURE FOR APPLICATIONS OF LIQUID CRYSTALS

The improvements, such as in the resolution and speed, of liquid crystal displays in the last few years are considerable. From the manufacture of very cheap TN displays, whose use is widespread in such devices as watch and calculator displays, developments have led to light-weight, low-power, portable computer displays and to pocket televisions using amorphous silicon technology. Future research is likely to be directed towards larger-area television for use in HDTV, as well as continuing work on the refinement of materials and manufacturing techniques for existing display technology.

There are many other applications of liquid crystals that have not been mentioned here, such as optical shutters, as well as other effects such as those in thermochromic liquid crystals, which are used to indicate temperature changes. Liquid crystal polymers have diverse applications. For example, some polymers, such as Kevlar, have a high tensile strength; thermo-optical effects are used in laser addressed displays as well as other types of polymer which have applications exploiting nonlinear optical effects. Research into the use of liquid crystals is branching out into areas such as optical interconnects, spatial light modulators, and even optical computing, although orders of magnitude improvement in the response time are required for this.

The exploitation of liquid crystals has been made possible by the combined contributions of chemists, physicists, and engineers. With continued collaboration, it is certain that the new products, such as pocket televisions and full color displays in portable computers, will be followed in the future by many other devices incorporating liquid crystals.

REFERENCES

1. P. M. Alt and P. Pleshko, *IEEE Trans.* **ED-21**, 146–155 (1974).
2. J. R. Hughes, IEE Proc. **133**, 145–151 (1986).
3. T. J. Scheffer and J. Nehring, Appl. Phys. Lett. **45**, 1021 (1984).
4. K. Katoh, Y. Endo, M. Akatsuka, M. Ohhawara, and K. Sawada. Jpn. J. Appl. Phys. **26**, L1784 (1987).
5. N. Kimura, T. Shinomiya, K. Yamamoto, Y. Ichimura, K. Nakahawa, Y. Ishi, and M. Matsunra. SID 19th International Symposium Digest of Technical Papers, p. 49 (1988).
6. E. Jakeman and E. P. Raynes, *Phys. Lett.* **39A**, 69 (1972).
7. K. Asano, K. Arai, and S. Nishi, Proceedings of the 6th International Display Research Conference, p. 392 (1986).
8. H. Watanabe, O. Okumura, H. Wada, A. Ito, M. Yazaki, M. Nagata, H. Takeshita, and S. Morozumi, SID 19th International Symposium Digest of Technical Papers, p. 416 (1988).
9. W. Helfrich, *Appl. Phys. Lett.* **17**, 531 (1970).
10. G. N. Taylor and D. L. White, J. Appl. Phys. **45**, 4718 (1974).
11. E. Kaneko, *Liquid Crystal TV Displays*, KTK, Tokyo (1987).
12. P. Migliorato, Eurodisplay, Conference Proceedings, p. 44 (1987).
13. A. J. Snell, K. D. MacKenzie, W. E. Spear, P. G. Le Comber, and A. J. Hughes, *Appl. Phys.* **24**, 357 (1981).
14. W. A. Crossland, M. Bone, and P. W. Ross, Eurodisplay, Conference Proceedings, p. 44 (1987).
15. N. A. Clark and S. T. Lagerwall, in *Recent Developments in Condensed Matter Physcis*, Vol. 4, p. 309 (1987).
16. J. Constant, E. P. Raynes, and A. K. Samra, to be published at the British Liquid Crystals Conference in Sheffield, April (1989).

28

High-Temperature Superconducting Materials

D. A. Cardwell

1. INTRODUCTION

Since the discovery of superconductivity in 1911 by the Dutch physicist Kamerlingh Onnes, who first observed this phenomenon in mercury at 4.2 K ($-269°C$),[1] as shown in Fig. 1, there has been a succession of materials that superconduct at increasingly higher temperatures. Tens of metallic elements and thousands of compounds are now known to exhibit this propety. Until the mid-1980s, the highest recorded superconducting transition temperature was about 23 K ($-250°C$), in niobium germanate (Nb_3Ge). In 1987, however, a new class of materials that superconduct at temperatures above the boiling point of liquid nitrogen (77 K) were discovered. These so-called HiTc materials, which are usually ceramic in nature (i.e., complex polycrystalline metal oxides), may prove to be of considerable technological importance since they offer great potential for a wide range of both novel and existing applications. Although the superconducting mechanism in HiTc compounds is not yet fully understood, their emergence has been heralded as potentially the most significant scientific event since the advent of the transistor.

The principal feature of a superconducting material is that it offers no Ohmic resistance to the flow of dc electrical current below some critical temperature, T_c. In consequence superconductors can conduct electricity without the accompanying dissipation of energy usually associated with metallic materials (i.e., Joule heating). Thus if a supercurrent is induced in a specimen of toroidal geometry, for example, it will flow indefinitely.

A second manifestation of this phenomenon is the ability of a bulk superconductor to behave as a perfect diamagnet when exposed to a weak magnetic field. This is achieved by the induction of Eddy currents in the sample surface which maintain a state of zero magnetic field in its interior (this effective flux expulsion is known as the Meissner effect and its detection is formally the "acid test" of superconductivity).

D. A. Cardwell ● Plessey Research Caswell Limited, Caswell, Towcester, Northamptonshire NN12 8EQ, U.K.

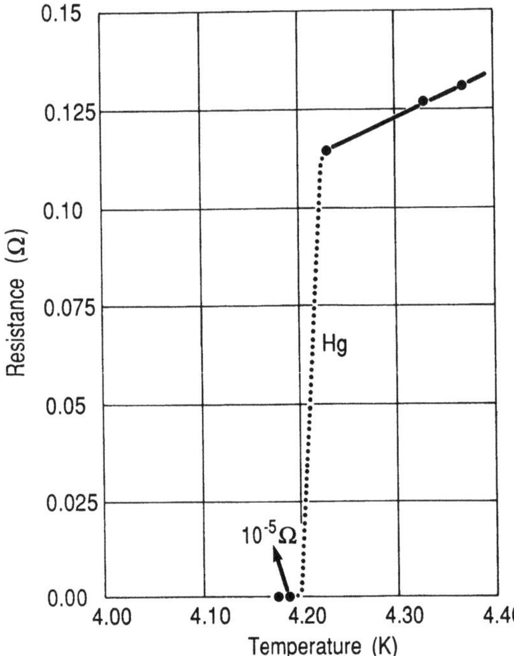

FIGURE 1. The discovery of superconductivity in 1911. Kamerlingh Onnes observed that the electrical resistance of mercury became vanishingly small at 4.2 K.[1]

The final major signature of a superconductor is that it behaves as if its normal (metallic) and superconducting states are separated from each other in terms of energy. In effect this means that an energy gap, which is typically of the order of a few meV and is centered on the Fermi energy, opens up in the one electron density of states at temperatures below T_c. From BCS theory (see Section 2.1), the width of this temperature-dependent energy gap at absolute zero is given approximately by[2]

$$2\Delta \approx 3.5 k_B T_c \qquad (1)$$

where 2Δ represents the superconducting/normal state energy gap and k_B is Boltzman's constant. Evidence for the existence of the energy gap is provided by the presence of an anomaly in the heat capacity of all known superconductors at their transition temperature.[3] The subsequent onset of superconductivity is then achieved by the formation of charged carrier pairs (so-called *Cooper pairs*),[4] with each carrier having an equal and opposite wave vector and opposite spins with respect to that of its partner. These pairs then collectively form a boson gas which condenses from states above the Fermi energy to those below it at the superconducting transition temperature. (N.B.: all pairs may occupy the same condensed energy state since they necessarily have a net integral spin and hence obey Bose–Einstein statistics.) By this mechanism the electronic system may lower its total energy by transformation to the superconducting state.

A full account of the basic properties of superconductors is given in Ref. 3. A more detailed account of this phenomenon is presented in Ref. 4.

All superconducting materials exhibit one of two types of behavior when exposed to a magnetic field of increasing strength, as shown in Fig. 2. Type I

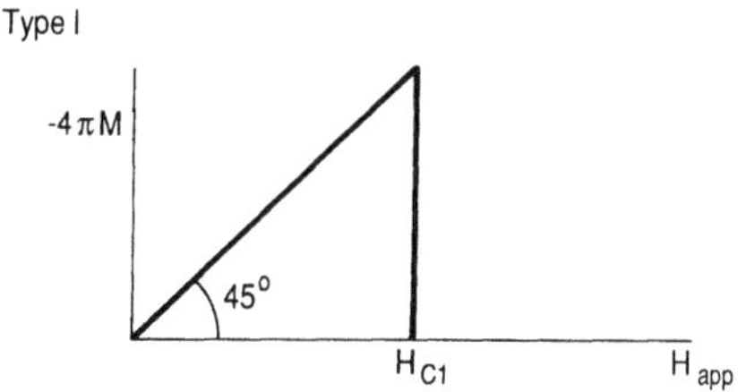

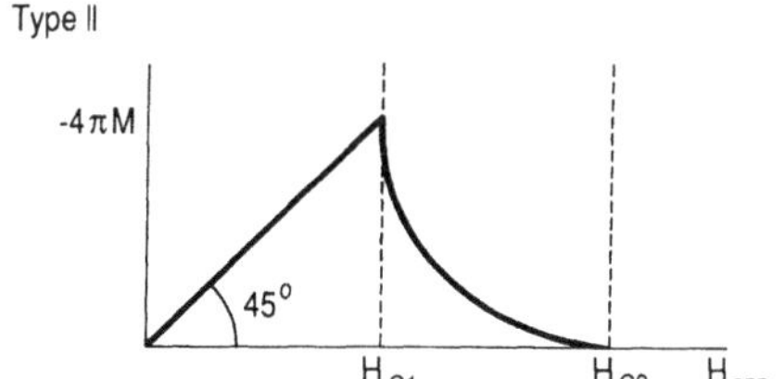

FIGURE 2. The diamagnetic properties of type I and type II superconductors. Type II materials retain their superconducting properties in the presence of a high magnetic field and hence are of considerable technological significance.

superconductors show perfect diamagnetic behavior up to some critical field, H_{c1}, at which point superconductivity ceases and the material recovers its metallic properties. At this field the sum of the magnetic and free energies associated with the superconducting state exceed the free energy of the normal state and hence the latter becomes more favorable energetically. The degree to which the superconducting free energy lies below the normal free energy of a specific material, therefore, determines the magnitude of H_{c1}. Although type II superconductors show similar behavior up to H_{c1}, their degradation from the superconducting state at this critical field is not so abrupt. Instead their diamagnetic susceptibility falls continuously to zero at a second critical field, H_{c2}. In the so-called "mixed-state" (i.e., $H_{c1} < H < H_{c2}$), magnetic flux partially penetrates the sample in the form of thin filaments. Consequently type II materials behave as if they consist of a mixture of superconducting and normal phases for fields in this range. Since H_{c2} in type II superconductors is much greater, typically, than H_{c1} in type I materials, the former are of considerably greater technological significance since the associated current-carrying capabilities and magnetic field stability are increased proportionally. The superconducting properties of both type I and type II materials are also enhanced as their operating temperature is decreased below T_c. Hence the superconducting energy gap, critical current, and associated upper critical field (corresponding to that produced by the critical current) all assume increased values at low temperature (these parameters are identically zero at T_c). This is summarized by Fig. 3, which shows the typical variation of these properties with temperature. From this figure it can be seen that near optimum superconducting performance may be achieved at temperatures below $\approx 2T_c/3$. This places a significant constraint on the operating temperature of superconducting devices.

2. THEORETICAL CONSIDERATIONS

The basic assumption of all theories of superconductivity is that this phenomenon is a cooperative electronic effect (i.e., superconducting behavior cannot

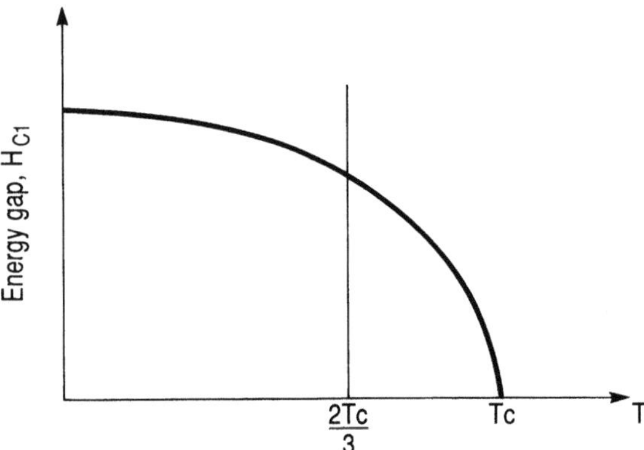

FIGURE 3. Typical variation of the energy gap, critical current, and associated upper critical field with temperature for a superconducting material. An operating temperature of approximately $2T_c/3$ yields close to optimum superconducting performance for most device applications.

be predicted if the electrons responsible for its occurrence are treated as though they exist independently of one another). Consequently a net *attractive* interaction between pairs of electrons is necessary to facilitate the superconducting mechanism.[4] This assertion is verified by solving the Schrödinger equation for a variety of attractive and repulsive coupling coefficients between charge carriers in a nearly free electron gas.[2] Such a study reveals that the total ground state energy of such a configuration may be minimized by the assumption of an attractive interaction between charge carriers, irrespective of how weak it may be. This suggests that the superconducting state is thermodynamically favorable over that of the normal metallic state and hence it will prevail given the appropriate environment. A number of models of the superconducting mechanism have been proposed. Three such theories are outlined in the following sections.

2.1. BCS Theory

Figure 4 illustrates schematically the BCS (Bardeen, Cooper, and Schrieffer[5]) theory of superconductivity which has been applied successfully to a large number of (low-temperature) superconducting materials. As the temperature of the lattice falls, the amplitude of the ionic vibrations about localized sites becomes reduced proportionally. If an electron is then introduced to an interstitial region, a significant distortion of the lattice (i.e., a phonon) can occur, which may be large compared with the intrinsic thermal motion of lattice ions at low temperatures. Hence a second

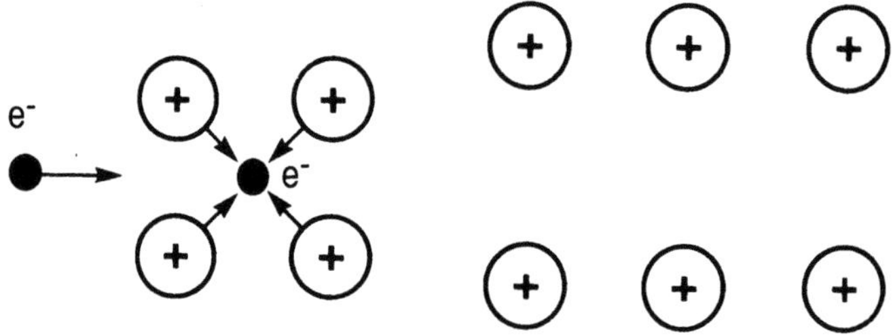

FIGURE 4. Schematic illustration of the electron–lattice coupling which forms the basis of the BCS formulation of superconductivity.

electron in the vicinity of the first will be drawn towards the phonon to form a Cooper pair (i.e., the lattice distortion effectively screens out any Coulomb repulsion between the electrons). Thus the second electron will appear to be attracted to the first (albeit indirectly) via the surrounding lattice. If both electrons are of the same energy but necessarily of opposite spin, in accordance with the Pauli exclusion principle, then the bound Cooper pair may progress unhindered through the lattice and a supercurrent will be generated. Hence the presence of the lattice effectively *facilitates* superconductivity in the BCS formulation. This is in contrast to ordinary metallic conductivity where electron-ion scattering is a major source of electrical resistance. Significant features of this model are (i) that a relatively free electron gas is required to promote the formation of a bound boson state (i.e., a total paired electron state of integral spin—in this case zero) and (ii) the spatial separation, or coherence length, between two carriers that constitute a Cooper pair is typically of the order of 0.1 μm. This reflects the cooperative electronic nature of a superconducting system. In addition, a *low* conduction electron density is an essential precursor of the superconducting state; otherwise Coulomb repulsive forces would screen out any electron–lattice–electron coupling. This is consistent with the observation that all known superconductors behave as poor metals in their normal state, which is a characteristic of a low-density conduction electron gas with strong electron–phonon coupling.

From the full quantum mechanical development of BCS theory, which is based on the simplified overview presented here, the critical temperature for the onset of superconductivity may be expressed in terms of three basic properties of a given superconducting system,[3] i.e.,

$$T_c = 1.14\Theta_D \exp\left[\frac{-1}{U\mathscr{D}(E_F)}\right] \tag{2}$$

where Θ_D is the Debye temperature, $\hbar\omega/k_B$, and is a characteristic of the ionic vibrational spectrum of the lattice; U is a parameter describing the attractive electron-lattice interaction; and $\mathscr{D}(E_F)$ is the density of electron energy states at the Fermi level. From Eq. (2), therefore, a high superconducting transition temperature may, at least in principle, be obtained by maximizing Θ_D, $\mathscr{D}(E_F)$ and U. It should be noted, however, that a major limitation of the BCS model is that it predicts a maximum achievable transition temperature of about 30 K, due principally to the practical inability to obtain a strong electron–lattice coupling via this mechanism at higher temperatures. It should be pointed out, however, that recent observations would suggest that the BCS framework may yet be extended to encompass HiTc phenomena.

In view of the apparent limitations of the BCS model, coupled with the discovery of HiTc materials, more complex, rival theories have emerged. In general, these alternative models assume a strong coupling between charge carriers in order to retain stability of the paired state at temperatures in excess of 100 K (unmodified BCS theory addresses only the weak coupling limit). The HiTc mechanism in such models, therefore, is determined predominantly by the properties of the lattice, rather than the free electron distribution in the case of BCS. In particular, magnetic and polar fluctuation coupling mechanisms, among others, have been proposed. The following sections outline briefly the most popular alternative models.

2.2. Resonating Valence Bond (RVB) Theory

This theory is based upon a model by Anderson,[6] who observed that magnetic ordering occurs within a specific lattice in the presence of a strong electron–electron interaction. In this model, individual electron spins become localized on lattice sites and interact with one another to produce an antiferromagnetic configuration (i.e., alternative up and down spins on adjacent lattice sites). Such an insulating, or valent, configuration is then used as the basis of a further coupling mechanism. For certain optimum lattice conditions (e.g., intersite displacement, charge screening, etc.) a net attraction may result between neighboring spin up and spin down electrons. Hence a dimerized local state could result with each pair of localized carriers having net zero spin (i.e., a bosonlike configuration). If this lattice is then doped by an acceptor ion, for example, holes will appear in the network of localized electron pairs. Providing there is sufficient overlap between the wave functions of electrons associated with each lattice site, a partially filled hole band will result and hence the system will become a conducting liquid of bosons. The equilibrium state of such a configuration, therefore, may be described as a superposition of all possible hole sites within the framework of the lattice (i.e., all of which are equivalent energetically). Consequently the system may be considered as a resonant state between a large number of possible electron–hole configurations associated with particular pairs of lattice sites. As in the BCS case, this boson distribution may condense to a ground state below the Fermi energy and superconductivity will result.

The strength of this model is that at any instant the Cooper pairs are localized to a pair of lattice sites and hence are relatively stable thermodynamically. In practice, however, it has proved difficult to obtain experimentally verifiable predictions from this model and hence its applicability to HiTc systems remains to be proved.

2.3. Bipolaron Model

A second alternative model of the HiTc superconducting mechanism involves a localized coupling of charge carriers to lattice ions (in contrast to the RVB model, which requires the ordering of electrons on specific lattice sites as a precursor to superconductivity).[7,8] The bipolaron model asserts that local polarons (i.e., electron and associated lattice distortion) interact attractively due to a local screening of the Coulomb repulsion between electrons by the lattice to form a single entity (i.e., a Cooper pair). Condensation to the superconducting ground state then occurs in the usual way. This is similar in concept to that of the BCS approach. These theories differ, however, in that the bipolaron model considers a local interaction between two distinct polarons rather than a long-range electron–phonon coupling.

Although the bipolaron model is consistent with a stable high-temperature superconducting state (due to its dependence on the local distortive ability of the lattice), it makes predictions that, as yet, are not observed experimentally. In particular, the predicted presence of a charge density wave due to the presence of such lattice distortions has not been detected in HiTc systems.[8]

While being critical of novel mechanisms, it should be pointed out that superconductivity has proved notoriously difficult to model in the past (the BCS model was developed 46 years after Kamerlingh Onnes discovered the phenomenon). Hence it may be many years before the full potential of such alternative theories of superconductivity is established. This should be assisted by an improvement in materials technology and experimentation, which, at present, remain relatively undeveloped.

3. RECENT DEVELOPMENTS

For 13 years after the discovery of Nb_3Ge,[9] research activity in superconductivity diminished rapidly. In April 1986, however, the German physicists Bednorz and Muller[10] observed the ceramic compound $La_{2-x}M_xCuO_4$ (where M is Ba, Sr, or Ca) to superconduct at temperatures between 30 and 40 K (see Fig. 5). This discovery was to lead to the award of the Nobel prize for physics to its originators in 1987. Subsequently the search for other, higher-temperature superconducting materials became intensified globally and culminated with Chu's group at the University of Houston[11] announcing the discovery of $YBa_2Cu_3O_{7-\delta}$ (YBCO) and a world record transition temperature of 92 K in February 1987. Although this compound is susceptible to the effects of atmospheric degradation, its emergence was particularly significant in that it was the first to superconduct above the boiling point of liquid nitrogen (77 K). A year later other ceramic compounds composed of multiple type II superconducting phases, notably $Bi_2CaSr_2Cu_2O_8$ (BCSCO, $T_c \sim$ 105 K, 80 K, and 60 K)[12] and $Tl_2Ca_2Ba_2Cu_3O_8$ (TCBCO, $T_c \sim$ 125 K, 106 K),[13] were discovered with similar crystallographic structures to the earlier HiTc materials.

Despite their refractory nature, the advent of these exciting type II compounds has revitalized the field of superconductivity and has been instrumental in the renewed search for a room temperature superconductor.

3.1. Crystal Structure and Charge Carriers

By comparing the crystal structures (determined experimentally by x-ray diffraction) of known HiTc superconductors, a variety of common anisotropic features have been identified. Such features include planes of fivefold coordinated copper [designated Cu(II) sites] and oxygen ions separated by an intercalation layer of a unique cation species (X). Additional planes composed predominantly of group 2 metal cations and oxygen then separate each CuO–X–CuO subcell. These features are illustrated by the crystal structure of YBCO, which is shown in Fig. 6. Here the intercalation layer is defined by a plane of yttrium (3+) ions (compared with Ca^{2+} in BCSCO and TCBCO). One-dimensional "chains" of alternate copper (designated Cu(I) sites) and oxygen ions (i.e., in four-fold coordination) and a layer of BaO then divide adjacent CuO planes. In BCSCO and TCBCO, only the analogous Cu(II) sites are occupied by copper ions. In these materials layers of Bi_2O_3 and Tl_2O_3 replace the corresponding Cu(I)—O chains in YBCO, respectively.

Although it was thought originally that electrons carried the supercurrent in BCSCO and TCBCO superconductors, recent thermopower and Hall effect measurements suggest that positive charge carriers (i.e., holes) may be common to all HiTc systems. These holes probably originate from a broad-based oxygen band associated

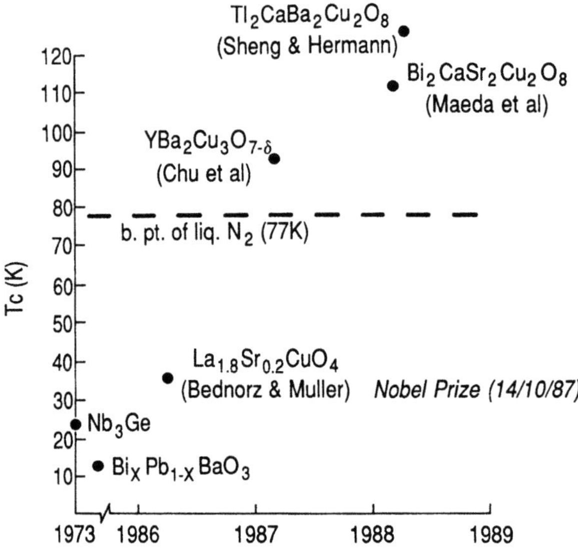

FIGURE 5. Recent developments in superconductivity.

with the Cu(I), in the case of YBCO,[14,15] and the Cu(II) sites in the cases of BCSCO[16] and TCBCO.[17] On-going research is concerned with the degree to which the Cu(II)O planes in YBCO, in particular, may be activated to produce suitable superconducting charge carriers. Since the observed transition temperature of a superconductor is dependent upon the density of charge carriers [proportional to $\mathcal{D}(E_F)^3$ in Eq. (2)],[4] then significantly higher transition temperatures could result if this can be achieved.

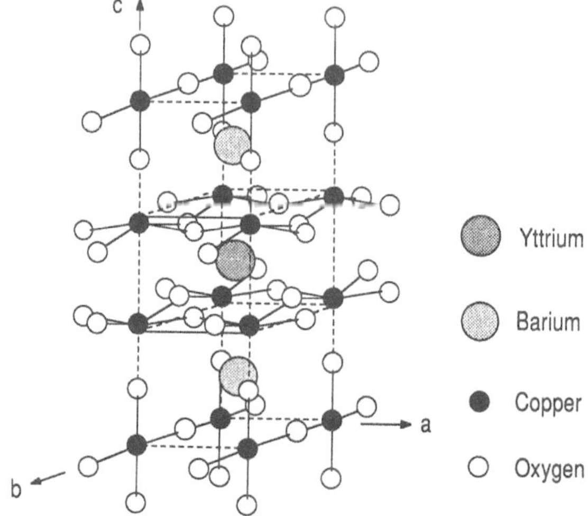

FIGURE 6. The crystal structure of YBa$_2$Cu$_3$O$_{7-\delta}$.

3.2. Fabrication Methods

In general, accepted and well researched methods used for the production of industrial ceramics and thin films (notably piezo- and ferroelectric materials) are applicable to the fabrication of HiTc superconductors.

3.2.1. Ceramics

Similar bulk processing techniques are applicable to all HiTc materials. This is illustrated by the fabrication of YBCO, for example, for which the following procedure may be adopted[18]:

1. Mix Y_2O_3, $BaCO_3$ and CuO powders together in the appropriate mass ratio to yield the desired stoichiometric (1-2-3) composition.
2. Heat mixed powders to 750°C in air to react constituent compounds together chemically (i.e., calcine).
3. Press resultant (black) powder into a pellet.
4. Heat pellet to 950°C in an oxygen environment and hold at temperature for 16 hours (i.e., sinter).
5. Cool pellet slowly to room temperature at a rate of 25°C h^{-1}. Steps 4 and 5 ensure that the oxygen content of the lattice is sufficiently high to produce the required hole charge carriers.

The resulting ceramic is black and typically has a resistance of a few milliohms at room temperature measured between two point contacts placed 1 cm apart on the sample surface (i.e., characteristic of a poor metal). On cooling, a sharp superconducting transition ($\Delta T \sim 1.5$ K) to zero resistivity is observed at about 94 K as shown in Fig. 7. This figure also shows the variation of diamagnetic susceptibility of this material with temperature, which represents an alternative method of detecting a transition to the superconducting state.[18]

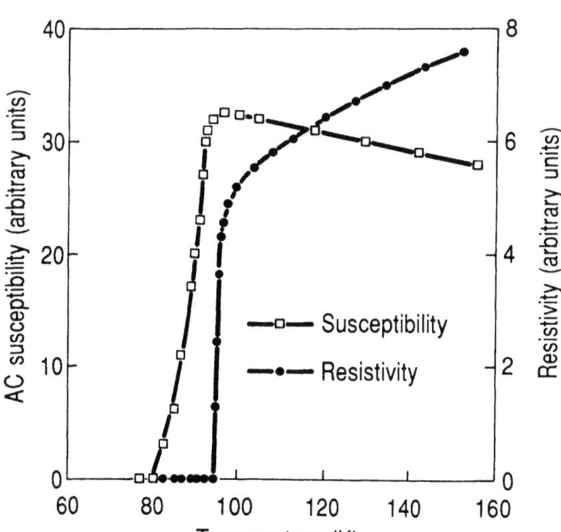

FIGURE 7. The electrical resistivity and diamagnetic susceptibility of $YBa_2Cu_3O_{7-\delta}$[18] as a function of temperature.

Although the methods of fabricating BCSCO and TCBCO are similar to the above process, they differ in that an oxygen-rich sintering environment is not essential in order to achieve superconductivity in the resultant sample. Consequently fabrication of these materials is somewhat simpler than in the case of YBCO.

3.2.2. Thin Films

The ability to fabricate thin films is vital to the realization of practical applications of HiTc superconductors. Thin films of YBCO, BCSCO, and TCBCO have been produced by a variety of standard deposition techniques. In particular, magnetron and ion beam sputtering,[19] laser ablation,[20] and e-beam evaporation[21] have been employed successfully for a variety of single and polycrystalline substrate materials. Of these substrates, MgO and $SrTiO_3$ have proved to be the most appropriate for the deposition of epitaxial thin films, with reduced levels of success being achieved with alumina, sapphire, silicon, and $LaGaO_3$. The latter material does, however, offer a good lattice parameter match with YBCO, in particular. Since this substrate has only been discovered recently, its full potential as a substrate for the deposition of uniform, epitaxial, high critical current HiTc films has yet to be assessed fully. A significant improvement in the quality of HiTc thin films is anticipated with the pending advent of buffer layer technology. In this case a thin dielectric film of a suitable material (from lattice parameter considerations) is deposited in situ on the substrate. This would produce a high-quality surface which should facilitate the subsequent growth of epitaxial films of HiTc material. Again the full potential of this technique has yet to be established.

The postdeposition firing conditions of HiTc thin films are similar in principle to those employed for bulk materials. Annealing times, however, are reduced to the order of minutes rather than hours due to the relatively small film volume. In addition some deposition systems enable in situ firing, which removes the requirement to expose the as-deposited films to the atmosphere. This may be important to the realization of a high-quality superconducting film surface which is a fundamental requirement of many device structures.

The current-carrying capabilities of known HiTc bulk and thin-film materials are compared with common, low-temperature, niobium-based superconductors in Table 1. In general the higher the critical current the better the quality, and hence suitability for potential applications, of the film. It should be noted, however, that although the thallium-based films offer relatively low critical currents, the quoted values were obtained by very little process optimization. This is in contrast to YBCO, which has only recently been deposited with any degree of reproducibility in its superconducting characteristics. From Table 1 it can be seen that significantly higher critical currents may be obtained if the crystallographic c axis can be aligned perpendicular to the substrate surface (see Fig. 6). Epitaxial films, therefore, are likely to offer the best properties for HiTc superconducting applications.

Given the processing requirements outlined in Section 3.2.1, HiTc superconducting ceramics are relatively straightforward to fabricate and have now taken their place on the school laboratory bench alongside more traditional teaching materials such as semiconductors, polymers, and glasses.

TABLE 1. Superconducting Parameters of HiTc and Refractory Materials[a]

Material	T_c (K)	ΔT_c (K)	J_c (film) (A cm^{-2})	J_c (bulk) (A cm^{-2})	Coherence length (Å)	Penetration depth (Å)
$La_{1.8}Sr_{0.2}CuO_4$	36	6	10^4	10^3	<10 c axis ~ 50 basal plane	3000
$YBa_2Cu_3O_7$	94	1	10^6	10^4	10 c axis ~ 30 basal plane	1500
$Bi_2Sr_2CaCu_2O_x$	85	>20	10^4	10^2	<10 c axis ~ 30 basal plane	—
$Bi_2Sr_2Ca_2Cu_3O_x$	110	>20	10^4	10^2	<10 c axis ~ 30 basal plane	—
$Tl_2Ba_2CuO_x$	80	<10	10^5	10^3	<10 c axis ~ 30 basal plane	—
$Tl_2Ba_2CaCu_2O_x$	108	<10	10^5	10^3	<10 c axis ~ 30 basal plane	—
$Tl_2Ba_2Ca_2Cu_3O_x$	125	4	10^5	10^3	<10 c axis ~ 30 basal plane	—
Nb_3Sn	18	$\ll 1$	$>10^6$	—	30	600–900
NbN	17	$\ll 1$	$>10^6$	—	40–70	2000–3000
Nb	9.3	$\ll 1$	$>10^6$	—	100–300	850

[a] Data taken from Refs. 27–34.

3.3. Material Properties and Processing Limitations

Although both high-quality bulk and thin-film HiTc materials can now be produced, a number of material problems remain inherent to these compounds. In the case of YBCO, a loss of oxygen from the lattice degrades its superconducting properties over a period of time. Furthermore this material is susceptible to attack by water vapor and hence suffers from the build up of an impurity phase at its sample surface and grain boundaries on exposure to the atmosphere. This is a particularly significant limitation for the properties of granular YBCO thin films, where individual grain dimensions are typically of an order of magnitude greater than the characteristic penetration depth of this material (i.e., the thickness of the current-carrying surface layer; typically of the order of 10–100 nm).[2] In addition, this impurity layer impairs the usefulness of this material for applications that require a high-quality superconducting surface (e.g., Josephson junction devices). In such devices, the coherence length of the Cooper pairs is necessarily large compared with the width of a thin dielectric layer separating two superconductors. This enables bound pairs to tunnel across the barrier layer without any dissipation of energy. Since the coherence length of Cooper pairs in HiTc systems is probably as short as 5 nm,[22] this places an even greater demand on available thin film and patterning techniques.

The problems associated with the fabrication of thin films and bulk BCSCO are more concerned with the stabilization of the 105 K phase in this multiphase system. It has recently been established, however, that free gaseous permeation of the BCSCO lattice during firing plays an important role in the development of this phase with sintering time.[23] Although this firing constraint is not as critical to the superconducting properties of sintered samples as in the case of YBCO (zero resistance may be observed readily at 80 K for BCSCO samples fired in air), it does identify a common feature between the processing of these materials. To date the criticality of the TCBO sintering environment to its superconducting properties has not been investigated fully. This is due to the acute toxicity of thallium-based compounds, which has effectively limited the extent to which its properties have been researched. Initial results, however, would suggest that this HiTc material is less sensitive to its firing environment than BCSCO.

4. APPLICATIONS

It has been established over the past two decades that the performance of a diverse range of existing and novel solid state devices could be improved by the incorporation of superconducting materials in their design. Practical applications that exploit the properties of traditional superconductors, however, have been limited due to their requirement for a liquid helium coolant which is expensive and difficult to store. The advent of materials that superconduct above liquid-nitrogen temperatures, therefore, has been accompanied by a revival in potential superconducting device applications. These applications fall into four general categories[24]:

1. Zero resistance devices:
   i. Transmission lines;
   ii. Microwave resonators and filters;
   iii. Antennae;
   iv. High-power motors/generators;
2. Josephson devices:
   i. High-speed computer switches;
   ii. SQUIDS (superconducting quantum interference devices);
   iii. Voltage/current amplifiers;
3. Magnetic devices:
   i. Levitating trains;
   ii. Magnetic separators;
   iii. High-field magnets;
4. Other:
   i. Optosensors.

In view of the present limitations of HiTc materials outlined in Section 3.3, however, many of the above applications may not be realized until the mechanical properties, current-carrying capabilities, and surface quality of the HiTc compounds have been improved (ceramics, for example, are brittle, difficult to form into specialized geometries, and suffer from impurity effects at grain boundaries). Specific areas where HiTc superconductivity may take some time to impinge on present-day technology, therefore, include high magnetic field devices and associated high-power applications. Others, such as superdirective, efficient antennae and high-Q microwave cavities and filters, which operate with simple device geometries, have been fabricated and shown to offer a considerable improvement over existing nonsuperconducting techniques.[25,26] This is particularly evident for microwave devices where the size of the energy gap offers potential for wide band, high-frequency applications (i.e., of the order of THz, in principle).[24]

On-going development of thin-film technologies of high-temperature superconductors, which overcome many of the problems associated with bulk materials, should facilitate the fabrication of novel devices. Such devices utilize the ac Josephson effect[4] and include wide band, accurate analog to digital converters, SQUIDS, and frequency generators and counters. Currently there is no competing technology in these device areas.

The most exciting prospect of HiTc research lies in the possible realization of room temperature superconductivity over the next few years. In this case the desired operating temperature of $\sim 2T_c/3$ could be maintained via thermoelectric cooling (i.e., eliminating the need for a liquid coolant). The number of potential applications of a room temperature superconductor would be enormous.

In the meantime the emphasis remains very much on the optimization of the properties of existing HiTc materials, by which progress in this field will be ultimately measured. Given the unprecedented level of interdisciplinary collaboration between both academic and industrial factions that has fueled HiTc research over the past few years, however, expectations in this field remain high. In consequence there is considerable optimism that, given time, HiTc superconductivity will achieve its full potential.

REFERENCES

1. H. Kamerlingh Onnes, *Akad. Wetenschappen (Amsterdam)* **14**, 113, 818 (1911).
2. N. W. Ashcraft and N. D. Mermin, *Solid State Physics* Chap. 34, Holt-Saunders International, New York (1976).
3. C. Kittel, *Introduction to Solid State Physics* Chap. 12, Wiley, New York (1976).
4. A. C. Rose-Innes and E. H. Rhoderick, *Introduction to Superconductivity* Pergamon, New York (1986).
5. J. Bardeen, L. N. Cooper, and J. R. Schrieffer *Phys. Rev.* **106**, 162, (1957); **108**, 1175 (1957).
6. P. W. Anderson, *Science* **235**, 1196 (1987).
7. A. S. Alexandrov, *Phys. Rev. B* **38**, 925 (1988).
8. P. Prelovsek, T. M. Rice, and F. C. Zhang, *J. Phys. C.* **20**, L229 (1988).
9. J. E. Gavaler, *Appl. Phys. Lett.* **23**, 480 (1973).
10. J. G. Bednorz and K. A. Muller, *Z. Phys. B* **64**, 189 (1986).
11. M. K. Wu, J. R. Ashburn, C. J. Torng, P. H. Hor, R. L. Meng, L. Gao, Z. J. Huang, Y. Q. Wang, and C. W. Chu, *Phys. Rev. Lett.* **58**, 908 (1987).
12. H. Maeda, Y. Tanaka, M. Fukutomi, and T. Asano, *J. Jpn. Appl. Phys.* **27**, L209 (1988).
13. Z. Z. Sheng and A. M. Hermann, *Nature* **332**, 139, 623 (1988).
14. T. Penney, S. Von Molnar, D. Kaiser, F. Holtzberg, and A. W. Kleinsasser, *Phys. Rev. B* **38**, 2918 (1988).
15. M. F. Crommie, A. Zettl, T. W. Barbee, III, and M. L. Cohen, *Phys. Rev. B* **37**, 9734 (1988).
16. Y. Lu, Y. F. Yan, H. M. Duan, L. Lu, and L. Li, *Phys. Rev. B* **39**, 729 (1989).
17. J. Clayhold and N. P. Ong, *Phys. Rev. B* **38**, 7016 (1988).
18. R. W. Whatmore, D. A. Cardwell, J. W. Cockburn, A. Patel, P. C. Osbond, L. Dorey, C. J. H. Wort, and F. W. Ainger, *Physica C* **153–155**, 790 (1988).
19. H. Tsuge, S. Matsui, N. Matsukura, Y. Kojima, and Y. Wada, *Jpn. Appl. Phys.* **27**, L2237 (1988).
20. M. Kenai, T. Kawai, M. Kawai, and S. Kawai, *Jpn. J. Appl. Phys.* **27**, L1293 (1988).
21. C.-A, Chang, C. C. Tsuei, C. C. Chi, and T. R. McGuire, *Appl. Phys. Lett.* **52**, 72 (1987).
22. E. Forgan, *Nature* **329**, 483 (1987).
23. D. A. Cardwell, J. W. Cockburn, and R. W. Whatmore, *Supercond. Sci. Tech.* **2**, 132 (1989).
24. D. A. Cardwell, *IEE Rev.* **Feb**, 57 (1989).

25. C. Zahopoulos, W. L. Kennedy, and S. Sridhar, *Appl. Phys. Lett.* **52**, 2168 (1988).
26. S. K. Khamas, M. J. Mehler, T. S. M. Maclean, C. E. Gough, N. McN. Alford, and M. A. Harmer, *Elec. Lett.* **24**, 460 (1988).
27. D. Caplin, *Nature* **335**, 204 (1988).
28. Y. J. Uemura, V. J. Emery, A. R. Moodenbaugh, M. Suenaga, D. C. Johnston, A. J. Jacobson, J. T. Lewandowski, J. H. Brewer, R. F. Kiefl, S. R. Kreitzman, G. M. Luke, T. Riseman, C. E. Stronach, J. R. Kossler, J. R. Kempton, X. H. Yu, D. Opie, and H. E. Schone, *Phys. Rev. B* **38**, 909 (1988).
29. M. B. Salamon, S. E. Inderhees, J. P. Rice, B. G. Pazol, D. M. Ginsberg, and N. Goldenfield, *Phys. Rev. B* **38**, 885 (1988).
30. M. Suzuki, Y. Enomoto, K. Moriwaki, and T. Murakami, *Jpn. Appl. Phys.* **26**, L1921 (1987).
31. P. M. Mankiewich, J. H. Schofield, W. J. Skocpol, R. E. Howard, A. H. Dayema, and E. Good, *Appl. Phys. Lett.* **51**, 1753 (1987).
32. Y. Hidaka, M. Oda, M. Suzuki, Y. Maeda, Y. Enomoto, and T. Murakami, *Jpn. J. Appl. Phys.* **27**, L538 (1988).
33. I. K. Gopalakrishnan, A. M. Umarji, J. V. Yakhmi, L. C. Gupta, R. M. Iyer, and R. Vijayaraghavan, *Mater. Lett.* **5**, 165 (1987).
34. R. J. Cava, R. B. van Dover, B. Batlogg, and E. A. Rietman, *Phys. Rev. Lett.* **58**, 408 (1987).

29

Langmuir–Blodgett Films

D. E. Hookes

1. INTRODUCTION

It was discovered by Katherine Blodgett and Irving Langmuir, in the 1930s, that it was possible to transfer monolayers of fatty acid salts at the air–water interface onto a solid substrate.[1] A monolayer of the fatty acid salt is obtained by depositing an accurately known volume of a solution of the fatty acid, such as stearic acid, in a volatile solvent such as chloroform or *n*-hexane, onto an aqueous subphase containing appropriate divalent cations such as cadmium, calcium, or barium. The molar concentration of the fatty acid solution is also accurately known and the deposition at the air–water interface achieved by touching drops of the solution onto the surface from a syringe. The solution spreads rapidly across the surface as the solvent evaporates leaving a known number of molecules of the fatty acid within an area of the surface defined by a set of movable barriers. The polar carboxyl headgroups of the fatty acids are immersed in the subphase and their alkyl chains remain in the gaseous phase (see Fig. 1). If the subphase has a neutral or mildly alkaline pH the fatty acids will ionize and adsorb the divalent cations to form a salt.

The next step is to compress the monolayer to a compact state by moving the barriers together (see Fig. 1). The compact state is detected by monitoring the surface tension of the monolayer-spread interface. The presence of the monolayer reduces the surface tension by exerting a surface pressure, which is defined as the difference between the surface tension of the free subphase surface, effectively that of water, and that of the monolayer-covered surface. This surface pressure increases rapidly as the molecules start touching together when a compact state is reached. A surface pressure-area/molecule isotherm of eicosanoic(arachidic) acid is also shown in Fig. 1.

The monolayers are deposited from the compact state by slowly dipping a plate through the surface while keeping the surface pressure constant by adjusting the area between the barriers. It was found by Blodgett and Langmuir that this process could be repeated to produce an ordered multilayer structure on the solid surface of the plate. This is illustrated in Fig. 2, which also shows the different multilayer

---

D. E. Hookes • Electrical, Electronic, and Systems Engineering Department, Coventry Polytechnic, Priory Street, Coventry CV1 5FB, U.K.

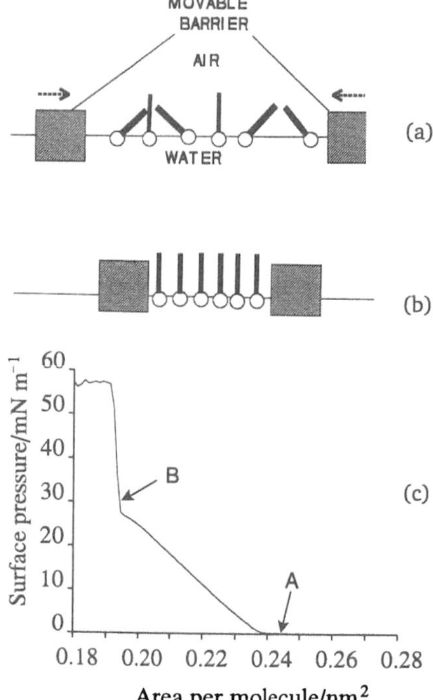

FIGURE 1. The compression of a monolayer of fatty acid at an air–water interface.

types described by Langmuir and Blodgett. X-type are given when deposition only occurs on the downward stroke, Y-type for deposition on both the upward and downward strokes, and Z-type for deposition only on the upward stroke. Today it is realized that, in many cases, the X- and Z-type structures rearrange to give Y-type structures after deposition.[2]

Following this early work there was an hiatus until the revival of interest in the late 1960s with the work of Kuhn and others.[3,4] They used radiation transfer between dyes incorporated in the multilayer structure to both elucidate structural aspects of the LB films as well as to investigate the physics of radiation transfers between dye molecules themselves. In the last decade there has been a veritable explosion of interest in these films, with several international conferences having been held.[5,6,7] (The most recent LB conference was held in Japan in April 1989, but the proceedings had not yet been published at the time of going to press.)

The initial interest in LB films was aroused primarily by their possible role in the microelectronic and optoelectronic device industry. The ability to manipulate and structurally control organic films of known composition at nanometer thicknesses excited a host of speculative proposals. There was no dearth of suggested applications, from mundane passive ones such as their acting as ultrathin resists for photo- and electron beam lithography to exotic proposals for constructing biochips and even biocomputers.

Many of these proposals have yet to appear in commercially available devices. The only commercial application to date has been their use as step gauges for optical measurements. After the initial rush of blood it was realized that a lot more basic

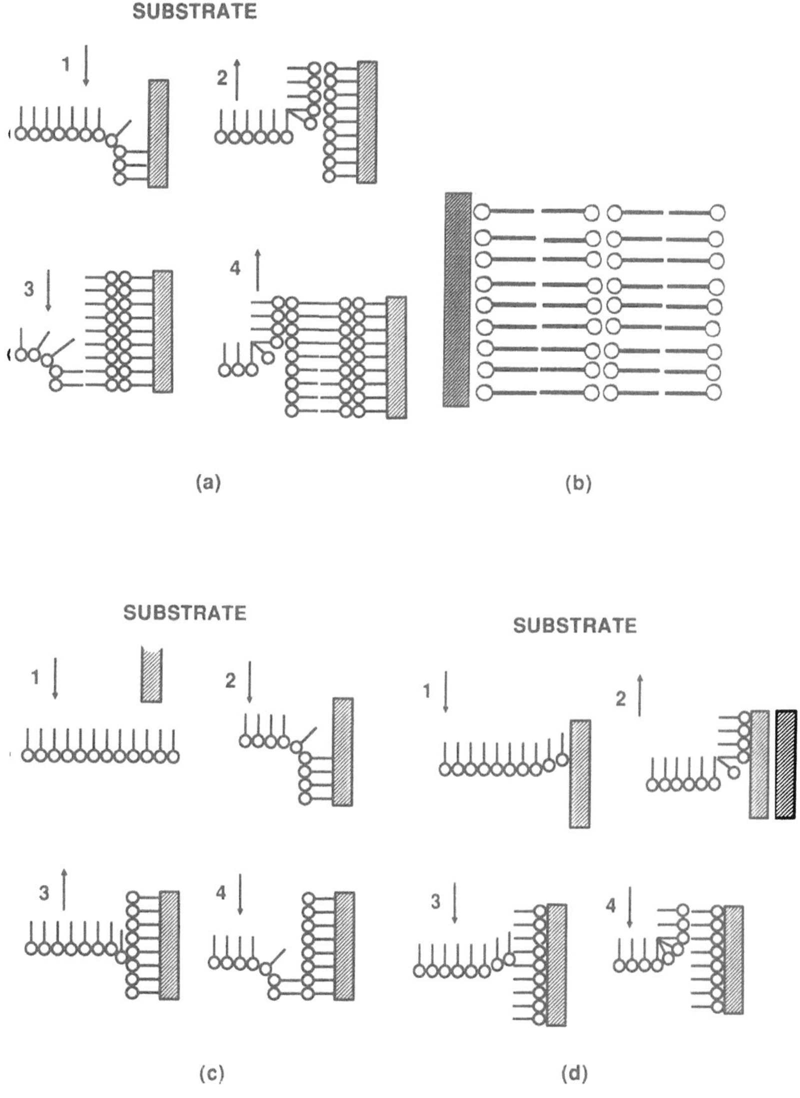

FIGURE 2. The deposition of different types of multilayers: Y-type on a (a) hydrophobic substrate and (b) hydrophilic substrate; (c) X-type; and (d) Z-type.

research on the physical and chemical properties of these multilayer systems was necessary. Even the structure and behavior of monolayers at the air–water interface themselves was not understood in detail despite several decades of investigation.

Chemical sensors and nonlinear optical devices are the currently favored areas of application of LB films, and this is reflected in the papers given at a recent International Conference.[7] Although few of the papers at this conference were directly device-oriented and none reported a practical, commercially available device, the thrust of research clearly has such applications in mind. Almost half of the papers were devoted to fundamental structural and physical investigations of the monolayers at the air–water interface (sometimes misleadingly called "floating LB films") rather than the LB multilayer films themselves. The remaining papers concentrated on the chemical, electrical, and optical properties, the last comprising by far the largest number.

It is proposed, therefore, to review the possible applications in these two areas after a consideration of the general properties of LB films, types of materials that form films, and techniques for the investigation of their structure and properties.

2. GENERAL FEATURES AND PROPERTIES OF LB FILMS

It is worth stating clearly what are the attractive and unattractive features of LB films.

2.1. Attractive Features

- Partial molecular ordering within the layers, complete ordering between layers.
- Known and controllable chemical composition.
- Known and controllable thickness in terms of number of molecular layers.
- Adjustable packing and orientation of molecules within a layer in some cases.
- Preparation does not require high vacuum deposition methods.

2.2. Unattractive Features

- Ordering within each layer is much less easy to determine and control than between layers.
- Film must be composed, predominantly, of surface-active molecules. This can mean that interesting molecules may have to be modified during or after synthesis to give them this property. Usually this means attaching a long-chain alkyl group to some part of the molecule. This may not always be possible or else may alter the properties of the molecule in an undesirable way.
- Imperfections are often present within films. It is now understood that LB films may have a domain structure, surface roughness, and even contain pinholes. The latter limits applications in microelectronics. These imperfections may be due to difficulties in controlling the deposition process precisely, the nature of the substrate surface, and structural instability after deposition.
- Chemical, thermal, and mechanical instability. LB films of small organic molecules are inherently less stable in all of these three properties than films of competing inorganic materials.
- Very high levels of cleanliness are necessary at all stages of the preparation of the films. This means working with the highest-purity solvents, rigorous cleaning procedures for all containers, and working in special clean-rooms.

3. LB FILM MATERIALS

3.1. Monomeric Materials

A wide variety of compounds have been successfully deposited as LB films or else within LB films. It is not possible to give a comprehensive review but to mention only those that show most promise from a device point of view.

● Biological molecules such as enzyme and antibody proteins, chlorophyll, phospholipids, cyclodextrins, aminoacids, and even whole biomembrane molecular assemblies. Proteins are usually adsorbed on traditional monolayers such as fatty acids or phospholipids before deposition on to a solid substrate. Some of these molecules are shown in Fig. 3.

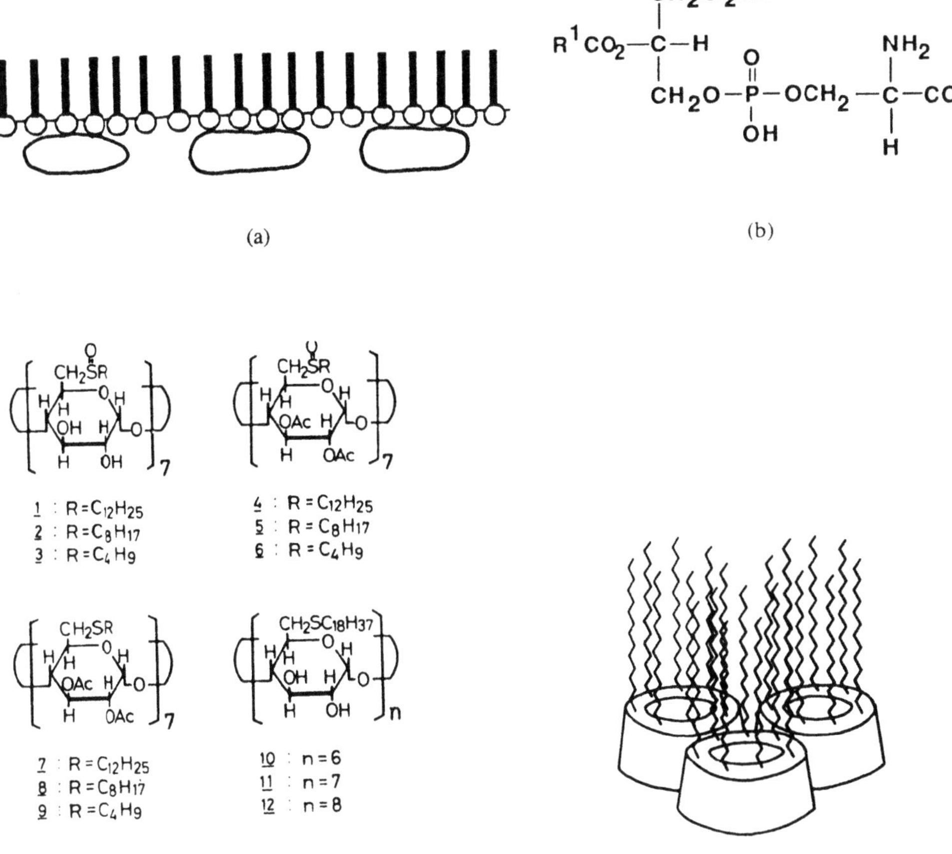

FIGURE 3. Some molecules of biological origin: (a) proteins adsorbed on a lipid monolayer; (b) a typical phospholipid, phosphatidylserine; (c) cyclodextrins, showing chemical structure (left) and packing of molecules (right).

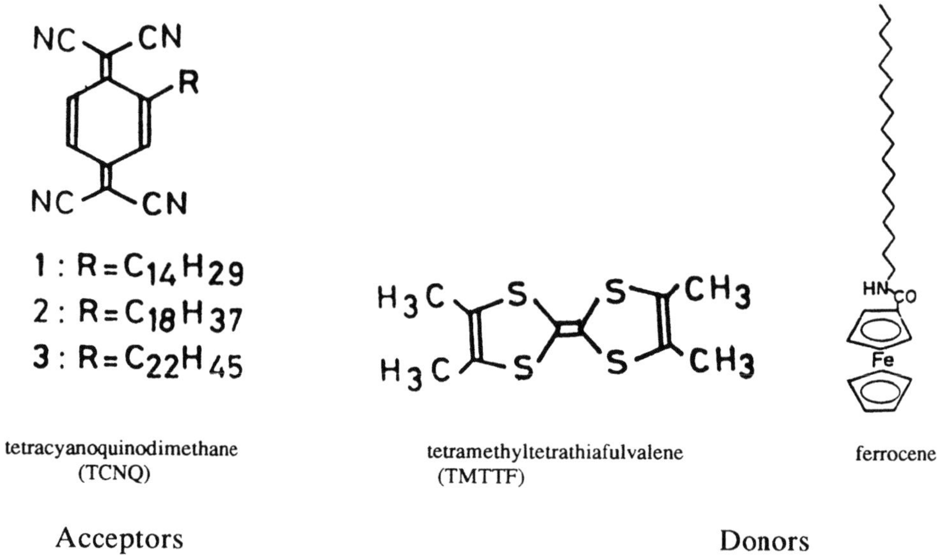

1 : R = C$_{14}$H$_{29}$
2 : R = C$_{18}$H$_{37}$
3 : R = C$_{22}$H$_{45}$

tetracyanoquinodimethane
(TCNQ)

tetramethyltetrathiafulvalene
(TMTTF)

ferrocene

Acceptors   Donors

FIGURE 4. A donor molecule, tetracyanoquinodimethane (TCNQ) (left); acceptor molecules tetramethyltetrathiafulvane (TMTTF) (center) and ferrocene (right).

- Molecules that may form conducting or semiconducting films such as diacetylenes, pyrroles, donor-acceptor charge transfer complexing molecules such as tetramethyltetrathiafulvalene (TMTTF) and ferrocene (donors), and the tetracyanoquinodimethanes (TCNQs) as acceptors (see Fig. 4.).
- Photosensitive molecules such as the cyanine, hemicyanine, merocyanine, rhodamine, and diazo dyes, pthalocyanines, porphyrins, and others (see Figs. 5, 6, and 7). These may be used to obtain nonlinear optical effects or as chemical sensors. Analog photosynthetic systems have also been studied using, for instance, pyrene

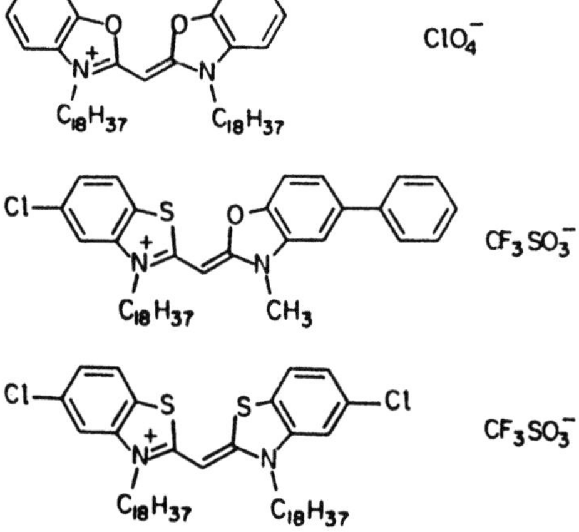

FIGURE 5.   Some cyanine dyes.

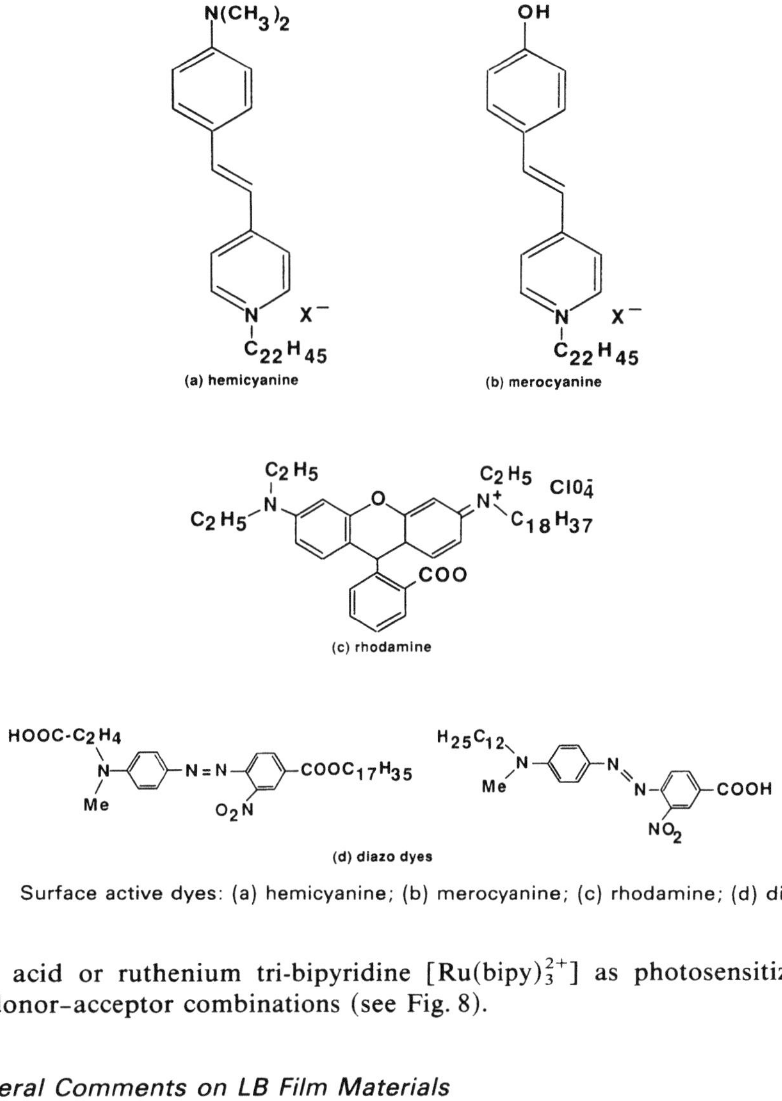

N(CH3)2
C22H45
(a) hemicyanine
OH
C22H45
(b) merocyanine
C2H5
C2H5
CIO4
C18H37
COO
(c) rhodamine
HOOC-C2H4
N
Me
N=N
O2N
COOC17H35
H25C12
N
Me
N
N
NO2
COOH
(d) diazo dyes

FIGURE 6. Surface active dyes: (a) hemicyanine; (b) merocyanine; (c) rhodamine; (d) diazo dyes.

decanoic acid or ruthenium tri-bipyridine $[\mathrm{Ru(bipy)}_3^{2+}]$ as photosensitizers for various donor–acceptor combinations (see Fig. 8).

3.2. General Comments on LB Film Materials

The vast majority of materials used for LB film deposition are relatively small molecular weight "monomers," classically the long-chain fatty acids and their salts and more recently dyes with long alkyl chains attached. It is by no means clear that such materials alone could ever form the basis of practical commercial devices because of their inherent lack of long-term thermal, mechanical, and chemical stability. They would have to have shelf lives of at least one year even if they were used in disposable devices.

3.3. Polymeric Materials

One possible solution to these problems of stability is to use polymeric materials. Some work has therefore been done on polymer monolayers. There are two

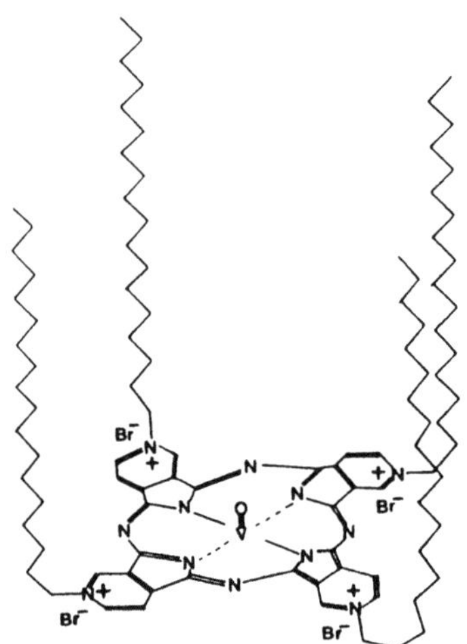

(a) Various pthalocyanines

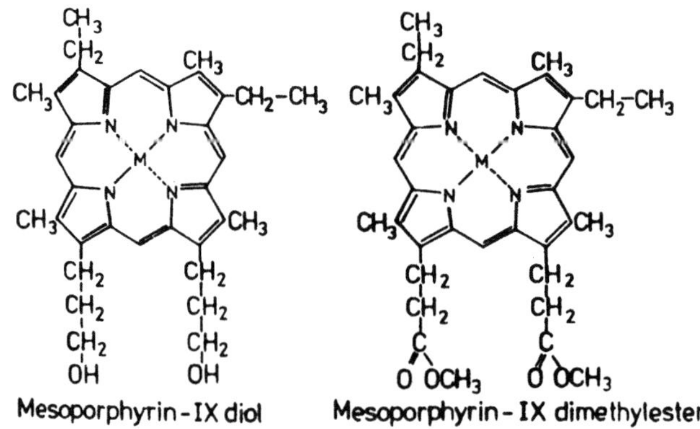

(b) A pthalocyanine derivative

(c) porphyrins

FIGURE 7.   Some pthalocyanines and porphyrins.

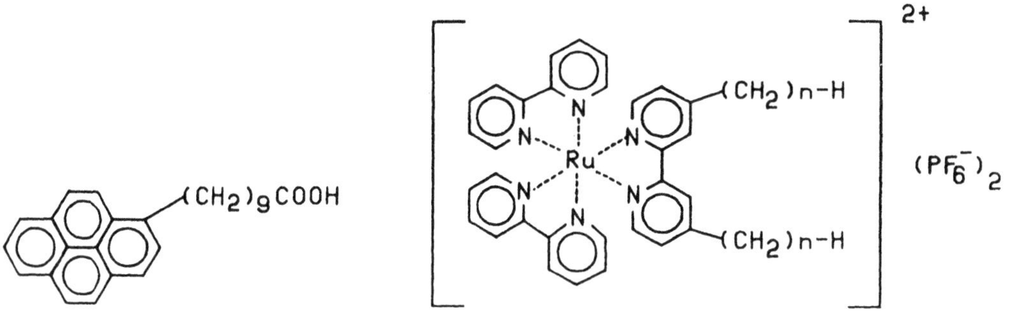

(a) (b)

FIGURE 8. Sensitizing dyes for analog photosynthetic systems: (a) pyrene decanoic acid (PDA); (b) ruthenium tribipyridine[$Ru(bipy)_3^{2+}$].

possibilities. The first is to use polymerizable monomers such as vinyl stearate, octadecyl acrylic acid, ω-tricosenoic acid, pyrroles, and diacetylenes (see Fig. 9) to form LB films and then polymerize them within the multilayer films using, for instance, uv or electrochemical methods. Such polymerized monolayers can be thermally stable up to 200°C. The other approach is to make LB films from preformed polymers, for example, poly(vinyl alcohol), poly(acrylic acid), poly(vinyl stearate), and polypeptides.

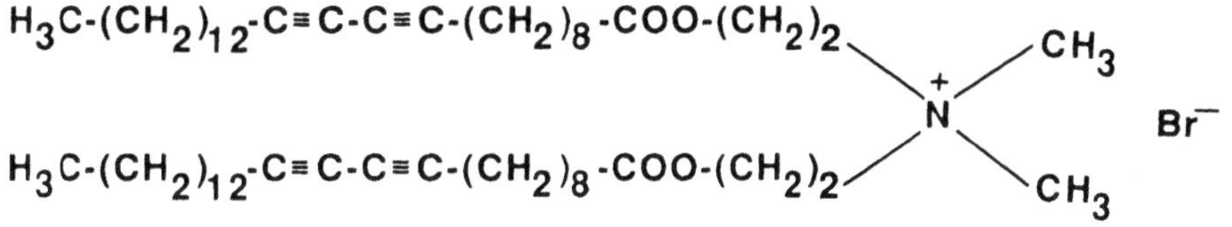

diacetylene

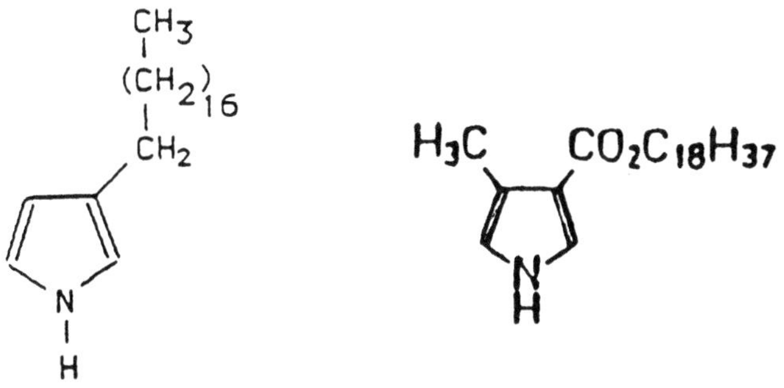

pyrroles

FIGURE 9. Monomer precursors of conducting polymers: diacetylene (upper) and pyrroles (lower).

Further improvements in thermal stability and mechanical strength can be obtained by using polymers containing polar groups, aromatics, and heterocyclic rings. However, polymers containing such groups, for example, polyimides and polyimidazopyrrolene, do not readily form LB films. Techniques have been developed to form films of these materials by heat treatment of LB films formed from polymer precursors such as the octadecyl esters of polyamic acids (see Fig. 10). Such films are found to be stable up to 400°C.

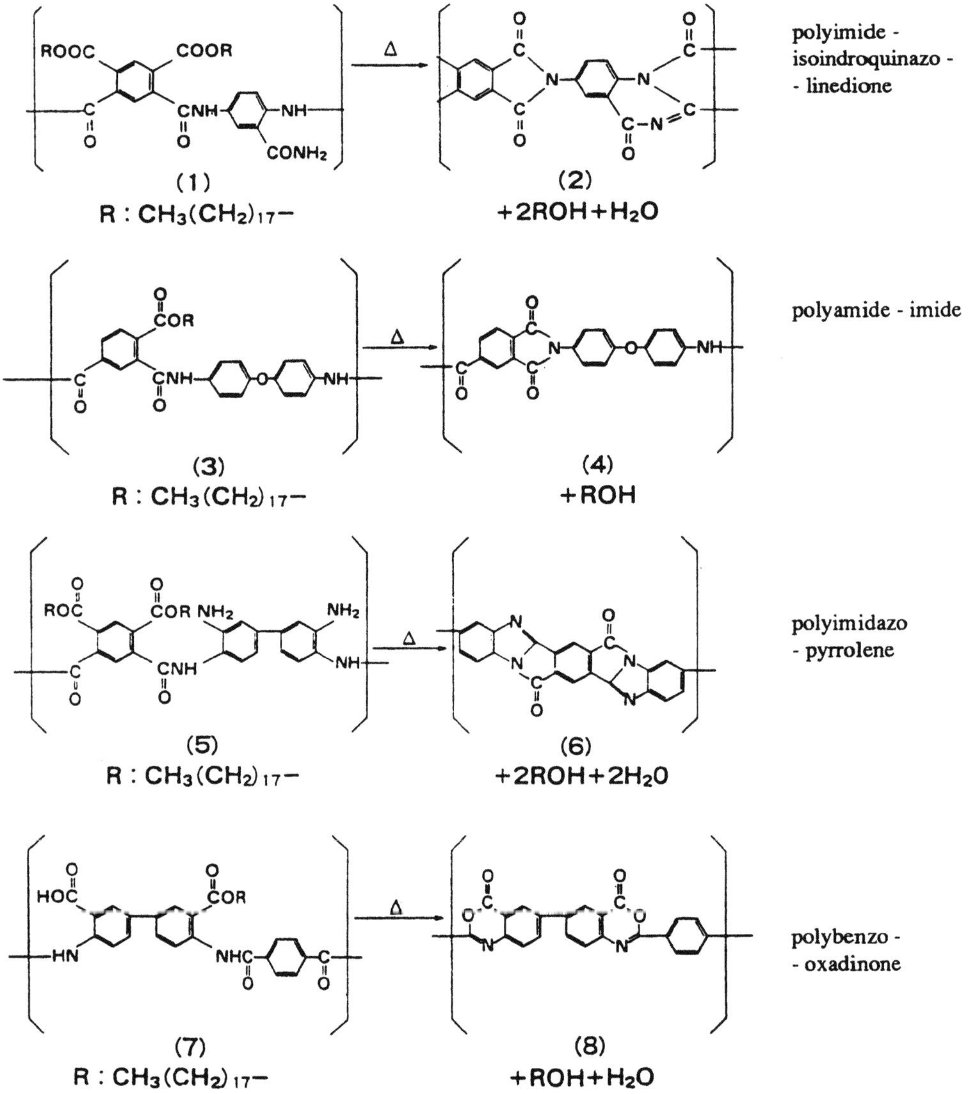

polyamic acids

FIGURE 10. Polymer precursors for heat-stable polymer films.

4. TECHNIQUES FOR INVESTIGATING LB FILMS

In addition to the classical techniques of surface science such as surface pressure and surface viscosity measurements on monolayers at the air–water interface a wide variety of techniques have been applied to the study of the structure and properties of these monolayers and the LB films themselves. The most useful include the following:

4.1. Microscopy

• Light microscopy in its various forms, fluorescence, polarization, and Nomarski interference microscopy.

• Electron microscopy, both transmission (TEM) and scanning (SEM) forms. Of particular interest is the recent use of scanning tunneling microscopy (STM), which allows direct visualization of the surface texture at a molecular level. Only very low-energy electrons are used, usually less than 3 eV, and so electron beam damage is not present as can be the case for TEM.

4.2. Spectroscopic Techniques

• Spectroscopy in its many forms: absorption, fluorescence in the uv, visible, and ir, polarized resonance Raman, Fourier transform ir, total internal reflection, x-ray photoelectron, near edge x-ray fine-structure spectroscopy (NEXAFS), electron spin resonance (ESR), Penning ionization electron spectroscopy (PIES).

4.3. Diffraction Techniques

• X-ray diffraction and reflectivity from single monolayers at the air–water as well as multilayers on solid substrates. Conventional x-ray sources and synchrotron sources have both been used.

• Neutron diffraction and reflectivity also for mono- and multilayers.

• Electron diffraction of LB films transferred to graphite films in conventional electron microscopes for transmission electron diffraction (TED). Reflection high-energy electron diffraction (RHEED) has also been used.

4.4. Miscellaneous Techniques

• Differential scanning calorimetry.
• Ellipsometry.
• Special techniques have also been developed for LB film studies, such as heavy ion decoration methods used in conjunction with TEM. Similarly methods of embedding LB films in epoxy resins for ultramicrotoming into thin ribbons. These are then observed in the TEM.

These techniques have shown that monolayers and LB films are much more complex and varied in their structure than was initially thought. It is now known that they can have a multiphase structure and contain imperfections such as pinholes, vacancies, and dislocations. The inclination of the alkyl chains of fatty acids and their salts with respect to the layer plane can vary considerably. The degree of long-range translational and orientational order can vary also even within a given multilayer system. These factors are affected not just by the chemical nature of the monolayer and the temperature but also by the nature of the substrate.

5. APPLICATIONS OF LB FILMS

5.1. Nonlinear Optics

The increasing use of optoelectronic and purely optical systems for data storage, transmission, and even processing has led to intensive work on finding materials with suitable nonlinear properties. These properties are necessary to enable light beams to be controlled by external electric fields or else by other light beams, and are discussed in several other chapters in this book.

The three types of nonlinearities that are most important are the following:

• Second order nonlinearities (large $\chi^{(2)}$). These can produce second harmonic generation (SHG), frequency up and down conversion, and electro-optic modulation. For example, compact optical disk data storage uses a GaAs laser at about 830 nm. If the frequency were doubled to a wavelength of 415 nm then the spot size would be reduced for higher-density storage and the photon energy available to effect physical and chemical changes in the storage media, e.g., photochromic changes, would be doubled.

• Photorefractive effects. These are the changes in refractive index when light is incident on a material. This is due to induced charge migration. These effects can be used for beam deflection, holography, and wavefront conjugation.

• Third-order effects (large $\chi^{(3)}$). These can be used for third-harmonic generation (THG), and to create optical bistability which can be used for optical switching, logic, and power-limiting.

The reasons for the considerable interest in the use of LB films to fabricate nonlinear optical devices is firstly that organic materials can show nonlinear effects at least two orders of magnitude greater than those of inorganic crystals. Usually they contain molecules that have a strong charge transfer band.[8] The second reason is that the LB technique provides a means of creating the noncentrosymmetric structures necessary to produce a macroscopic $\chi^{(2)}$ from the second-order nonlinear molecular polarizability, β, which is a third-rank tensor.

In fact LB films have been shown to give both second- and third-order effects.[9,10]

LB film SHG investigations have largely concentrated on the merocyanine and hemicyanine dyes, and various azo compounds (see Fig. 6). LB films of layers of these compounds, diluted with eicosanoic(arachidic) acid and usually alternating with layers of ω-tricosenoic acid, have been shown to give strong SHG effects. These effects are maximum if the dye occupies about 50% of the area within a layer, and a saturation of the SHG effect is shown with the number of layers. Alternating layers of two different azo dyes have also been studied and have shown even more enhanced effects (see Ref. 7, Vol. 160, p. 217).

Third-order effects in LB films have been observed in cross-linked polyacetylenes and also in films of hemicyanine and protonated merocyanine dyes. THG has also been reported in films of polysilane.[11,12]

5.1.1. General Comments on Nonlinear Applications

The problems of chemical, thermal, and mechanical stability still present themselves. It is desirable that even higher values for the molecular hyperpolarizabilities be found as well as higher densities for the active nonlinear component. LB films usually have a large fraction of their volume taken up with nonactive aliphatic chains. Film thickness of the order of 1 μm are desirable if they are to influence propagation within waveguides over reasonable pathlengths.

5.2. Chemical Sensors

A convenient scheme for describing a general arrangement for a chemical sensor has been given by Moriizumi (see Ref. 7, Vol. 160, p. 414). This is shown in Fig. 11. Langmuir–Blodgett films containing functions B and C can be deposited on transducers.

Chemical sensors can be classified both by their analyte (gas, ions, organic molecules in solution) and by their transducer mechanism (potentiometric, amperometric, optical, acoustic). A table showing both classifications is given in Table 1 (also from Moriizumi, Ref. 7) and the use or attempted use of LB films in their design is indicated.

5.2.1. Gas Sensors

LB films of various pthalocyanines (Pc) (see Fig. 7) have been widely studies as gas sensors.[13] The conductivity of these films is influenced by some gases, e.g.,

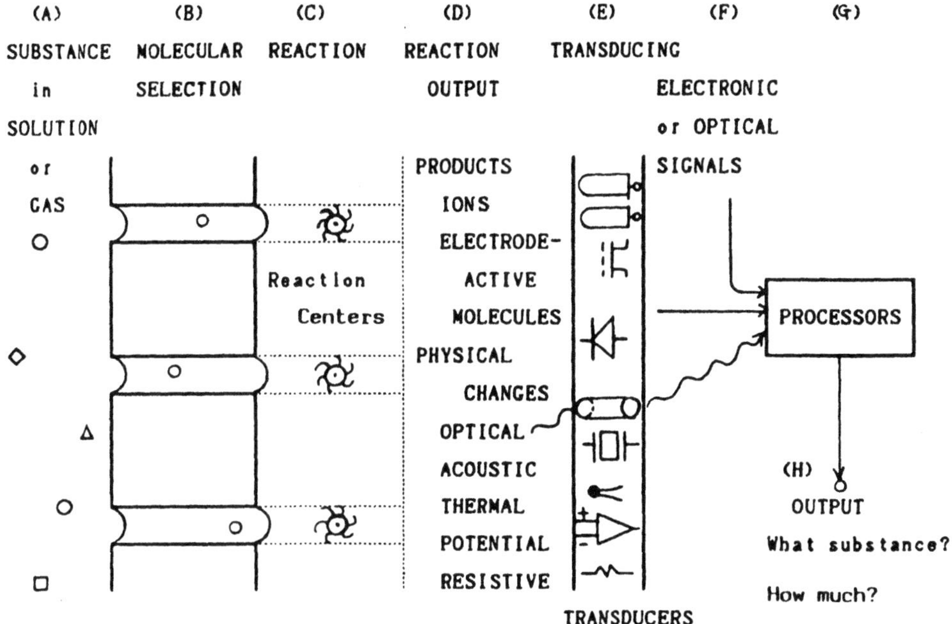

FIGURE 11.   A general scheme for chemical sensors (from Moriizumi[7]).

TABLE 1. A Classification of Chemical Sensors[a]

| | Sensors based on the following transducers | | | | | |
| | Electrochemical devices (electrochemical reaction, interface potential) | | | Resistors (resistance modulation) | Optical devices (absorption refractive index, scattering) | Acoustic devices (mass loading, elastically, viscosity) | |
Species detected	Potentiometric electrodes	Field effect transistors (FETs)	Amperometric electrodes		Fiber, optoelectronic, surface plasmon resonance and other devices	Quartz oscillators	Surface acoustic wave (SAW) devices
Gas molecules	Oxygen electrodes	Pd FETs Suspended gate FETs (SGFETs)	Oxygen electrodes	Chemiresistors[b] (NO_x, Cl_2, NH_3, SO_2 DMMP)	Optical fiber oxygen sensors NH_3 sensors[b]	Humidity, Hg, NH_3 and H_2S sensors[b] Bilayer lipid membranes (BLMs)	NO_2, H_2S and H_2 sensors

Ions	ISEs	pH ion-sensitive FETs (ISFETs[b])			Optical fiber pH sensors		
Organic molecules in a liquid	*Enzyme sensors*	ENFETs[b]	Glucose sensors	Enzyme thermistors	Biophotodiodes (enzyme-luminescence reaction)		Enzyme ultrasonic sensors
	Immunosensors Immuno electrodes[b] BLMs	Immuno-FETs (IMFETs)			SPR immunodetectors luminescent immunosensors	Immuno-oscillators	Immuno-SAW devices
	Others BLM/agar devices[b]				Multichemical sensors[b]	Taste or odor sensors	

[a] From Ref. 7.
[b] LB film sensors.

NO, O, NH, Cl. Other workers have investigated conductivity changes in metal complexes of various porphyrins,[14] and others, charge transfer salts such as TCNQ.[15]

LB films of ω-tricosenoic acid and a silicon Pc deposited on a quartz oscillator have been shown to change the oscillator frequency linearly with concentration in the presence of NH_3 and H_2S gases.[16]

The effect on surface plasmon resonance of Pc LB films deposited on a silver surface in the presence of various gases has also been investigated (Ref. 7, Vol. 160, p. 431). However, at present the selectivity of gas sensors, even using LB techniques, is a problem.

5.2.2. Ion Sensors

Valinomycin, a cyclic polypeptide that specifically binds K ions, has been deposited with cadmium arachidate to detect K ions through conductivity changes.[17] The performance was inferior to conventional ion-selective electrodes (ISEs).

One possibility is to design optical ion sensors using dyes that can selectively bind ions and thus produce a change in their absorption or fluorescence spectrum. These dyes can be deposited in a known orientation in LB films in order to maximize their sensitivity.

5.2.3. Enzyme Biosensors

An enzyme biosensor that detects glucose has been reported with the enzyme glucose oxidase (GOD) adsorbed to monolayers of arachidic acid and up to 10 layers deposited between gold electrodes (Fig. 12 and Ref. 7, Vol. 160, p. 420). It was found to age during the first 5 days and 50 measurements, after which its performance stabilized (for how long is not revealed).

5.2.4. Immunosensors

An anti-bovine serum albumin (BSA) antibody LB film was deposited on a titanium wire and used to detect BSA in concentration range 20–1000 g/ml.[18]

Immunoglobulin G(IgC) antigen was adsorbed onto stearic acid monolayers and layers deposited on iridium oxide (IrOx) electrodes on a glass substrate. IgG antibody was then detected, using an amperometric measurement, in the concentration range from 5×10^{-8} to 1×10^{-6} M.[19]

Immunosensors using acoustic elements such as quartz oscillators and SAW devices have been designed using LB films.

5.2.5. Future Trends in Sensors

5.2.5a. Biomimetic Membrane Approach. Two LB layers can mimic a biological membrane which can show high selectivity to various analytes. A number of different methods of doing this have been reported, usually involving deposition of an LB analog membrane on a microporous surface.

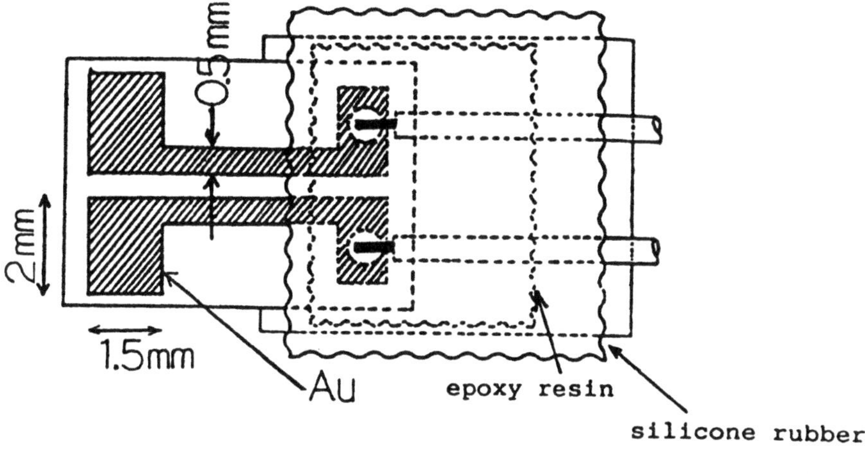

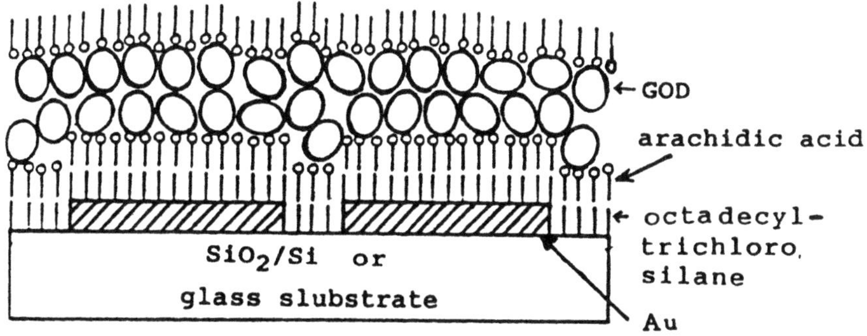

$$C_6H_{12}O_6 + O_2 \xrightarrow{\text{GOD}} C_6H_{10}O_6 + H_2O_2$$

$$\text{GOD} \equiv \text{GLOCOSE OXIDASE}$$

FIGURE 12.   An enzyme-based glucose sensor.

5.2.5b. Transducer Improvements. SAW devices with higher frequencies and smaller losses are being developed. Various improved optical detection methods are being investigated, including surface-enhanced Raman spectroscopy.

5.2.5c. Pattern Recognition. It is thought that natural olefactory cells are not specific but that odor recognition is achieved by pattern recognition from a large number of cells each with a slightly different response to a specific odor. Langmuir–Blodgett films could be a method of producing a range of sensors with small differences of response for given molecules. The responses could be optical, electrical, or acoustic.

6. CONCLUDING REMARKS

Recent advances in preparative and investigative techniques have led to a much better understanding of the structure and properties of LB films. At present device applications are limited to laboratory demonstrations. The possibilities of resolving problems of thermal, mechanical, and chemical stability are, however, very real in light of the improved understanding of film structure and the synthesis of tailor-made materials, especially polymers.

REFERENCES

1. K. B. Blodgett, *J. Am. Chem. Soc.* **56**, 495 (1985).
2. F. Grunfeld, Ph.D., thesis, Univ. London (1983).
3. H. Buecher, K. H. Drexhage, M. Fleck, H. Kuhn, D. Moebius, F. P. Schafer, J. Sondermann, W. Sperling, P. Tillman, and J. Wiegand, *Mol. Cryst.* **2**, 199 (1967).
4. H. Kuhn, D. Moebius, and H. Buecher, in *Physical Methods of Chemistry* Part IIIB, (A. Weisberger and B. Rossiter, eds.), Wiley, New York (1972).
5. Proc. First Int. Conf. on LB films, in *Thin Solid Films* **99** (1983).
6. Proc. Second Int. Conf. on LB films, in *Thin Solid Films* **132, 133** (1985).
7. Proc. Third Int. Conf. on LB films, in *Thin Solid Films* **159, 160** (1987).
8. D. J. Willias, ACS Symp. Ser. 233 (1933).
9. I. R. Girling, N. A. Cade, P. V. Kolinsky, and C. M. Montgomery, *Electron. Lett.* **21**, 169 (1985).
10. F. Kajzar, J. Messier, J. Zyss, and I. Ledoux, *Opt. Commun.* **45**, 133 (1983).
11. G. M. Carter, Y. J. Chen, and S. K. Tripathy, *Opt. Eng.* **24**, 609 (1985).
12. F. Kahzar, J. Messier, and C. Rosilio, *J. Appl. Phys.* **60**, 3040 (1986).
13. S. Baker, G. G. Roberts, and M. C. Petty, *Proc. IEE, Part 1* **130**, 260 (1983).
14. R. A. Tredgold, M. C. J. Young, P. Hodge, C. Chem, and A. Hoorfar, *Proc. IEE Part 1* **132**, 151 (1985).
15. L. Hendron, G. Derost, A. Raudel-Teixier, and A. Barraud, *Proc. Second Int. Conf. Chem. Sens.* Bordeaux, p. 677 (1986).
16. J. F. Ross and G. G. Roberts, *Proc. Second Inf. Conf. Chem. Sens.* Bordeau (1986).
17. A. D. Brown, *Sens. Actuators* **6**, 151 (1984).
18. H. Yamagishi, M. Sriyudthsak, and T. Moriizumi, *Inst. Electr. Inf. Comm. Eng. (Jpn) Tech. Rep.* **24**, p. 19 (1987).
19. T. Katsube and M. Hara, *Int. Conf. Sensors and Actuators* p. 816 (1987).

30

Electrically Conducting Polymers

D. J. Walton

1. INTRODUCTION

Plastics have long been used as insulators throughout the electronics industry, but recently new classes of polymers have been discovered that have appreciable electronic conductivities. The best of these to date has a conductivity only a single order of magnitude below that of copper, while a surprising range of other polymers have conductivities just a few orders further below. This number of conducting polymers suggests that electronic conductivity in such organic systems may be more widespread a phenomenon than has previously been recognized.

These materials are attracting considerable research interest, both industrial and academic; not just because of their electronic conductivity and its theoretical implications for charge transport within organic systems, but also because their structural features give rise to additional properties of ion transport, redox behavior, electrochemical effects, photoactivity, catalytic properties, and electronic junction effects, which are added to light weight, flexibility, nonmetallic surface properties, and other polymeric physical properties. As a result there is considerable effort toward a range of practical applications including nonmetallic conductors, EMI shielding, battery electrodes, electrocatalysis, sensors, switchable membranes, ion-control devices, electronic junction devices, electronic displays, optoelectronic systems, capacitors, information storage and bioactive systems, among others.

This is an exceptional list of attributes and applications for what might at first sight appear to be a simple class of materials, and this chapter comprises an overview of present conducting polymer technology, which must of necessity be restricted in content. Thus synthesis and properties will be briefly described and two major types of polymer, namely, those based, respectively, upon polyacetylene and polypyrrole, will be contrasted. Conduction mechanisms will be discussed briefly but emphasis will be placed on practical aspects.

A feature of these materials is the wide range of parameters that are available for manipulation. Not only are there many suitable polymer structures, but the properties of each may vary with method of preparation and other treatments. In

D. J. Walton • Department of Applied Physical Sciences, Coventry Polytechnic, Priory Street, Coventry CV1 5FB, U.K.

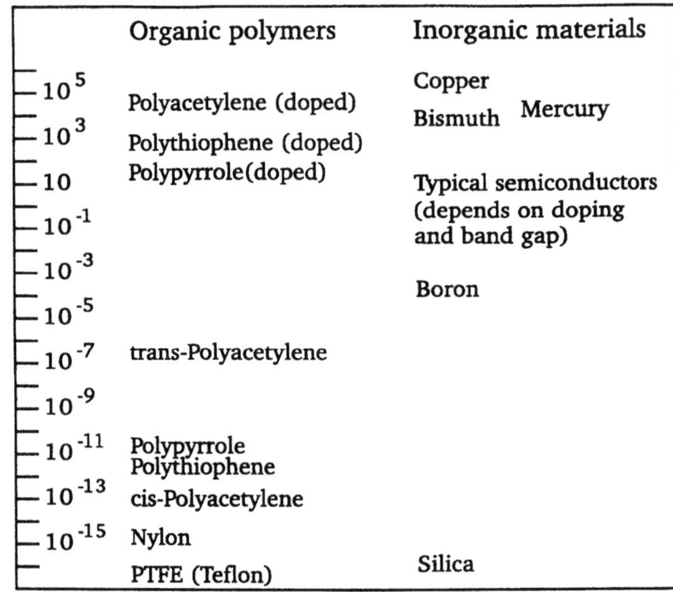

FIGURE 1. Typical conductivities in S cm^{-1} of various organic and inorganic materials (in ambient conditions). Unless indicated, polymers are undoped.

addition these polymers can be obtained in, or converted to, nonconducting forms, which means that a single sample of material may show a conductivity within some 15 orders of magnitude; and the conductivity ranges of several conducting polymers are shown in Fig. 1, contrasted with conventional metals, semiconductors, and insulators. Structural, morphological, and other physical changes may accompany transformations between states of varying conductivivity.

Another important feature of these materials is that they are open to manipulation by normal techniques of organic and polymer chemistry. Thus functionalization of the monomer or of the polymer once formed, or of other species present in the system, may impart additional properties to the system; while typical polymer manipulation includes copolymerization and the formation of composites, blends, and laminates. Several ingenious studies into these aspects will be reviewed.

This great flexibility of technology accounts for the volume of research and the breadth of applications under consideration.

The basic properties of conducting polymers will be addressed by consideration of polyacetylene, which presents in many respects the best-defined system. However, for many practical applications polyacetylene has been superseded by polymers of the polypyrrole class ("heteroaromatic" polymers) on grounds of stability and ease of preparation. Polypyrrole polymers also offer greater opportunities for functionalization, and this is further described below. To emphasize the great effort put into these materials the chapter will close with a statistical analysis of endeavors in the field over a specified time period.

For further details the reader is directed to recent in-depth reviews[1-6] and conferences.[7-8]

2. POLYACETYLENE—THE ARCHETYPAL CONDUCTING POLYMER

2.1. Structure

Electronic conductivity in metals requires delocalized electrons in the conduction band. The equivalent situation in an organic material requires delocalized π-electrons with overlap throughout the structure. Conventional insulative polymers do not have electrons in suitable orbitals; for example, the aromatic π-systems in polystyrene are unable to overlap satisfactorily and the backbone of the polymer comprises Sp^3 hybridized carbon with localized σ-bonds. Conducting polymers found to date have in common a backbone structure that has delocalized π-electrons along the chain. The simplest such alternating double/single bond structure is polyacetylene, but a surprising number of other structures are suitable, and a selection is given in Fig. 2. Of these poly(paraphenylene) and polyphenylene vinylene are similar to polyacetylene in many respects, while polythiophene, polyaniline, and polyisothianaphthene are similar to polypyrrole. The definition of a "polypyrrole class" of polymers is reasonable since a range of polymers display similar principles.

In fact the structures given in Fig. 2 are of the polymers in their insulative forms. Redox perturbation of the energy levels in the delocalized system is necessary to induce conductivity. This process is often termed "doping" and is described further below. A consequence of this is that in their conductive forms the polymers are charged or "doped," while in their insulative forms they are uncharged, "neutral," or "undoped."

2.2. Isomerism

Polyacetylene is not as straightforward as its formula might suggest because of *cis/trans* isomerism, and this is the reason for two conductivity values for the

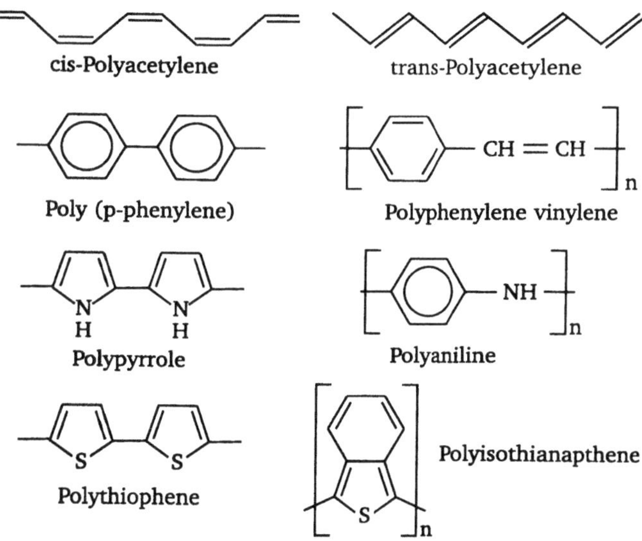

FIGURE 2.   Conducting polymer structures (Uncharged forms).

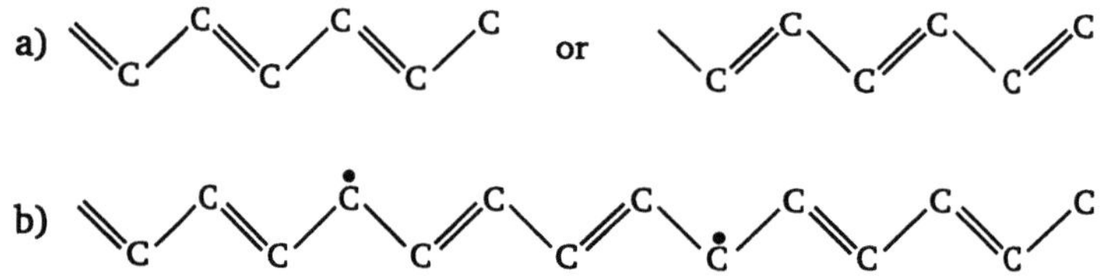

FIGURE 3. Soliton formation in *trans*-polyacetylene; (a) shows the twofold degeneracy of the chain, and (b) the soliton formation (note the inversion of double bonds). (Note that the hydrogen atoms are omitted, as is customary, for clarity.)

insulative uncharged polymer in Fig. 1. *Cis*-polyacetylene is obtained by polymerization at low temperatures ($-77°C$) using conventional stereospecific catalyst systems, and is a typical insulative polymer in its uncharged state. Whereas *trans*-polyacetylene, more thermodynamically stable, is obtained by heating toward 170°C during or after polymerization. It is also obtained during the charging or "doping" process from the *cis* isomer and hence there is only one conductive form of the polymer (though with a range of conductivities depending on preparation procedure). The insulative form of the *trans* isomer is significantly more conductive than the equivalent *cis* form because of twofold degeneracy in the ground state.

Only a symmetrically substituted polymer chain with *trans* configuration has this facility and only polyacetylene to date of possible structures has been prepared. This behavior is shown pictorially in Fig. 3a, where the double-bond "herring-bones" may face in either direction. A consequence of this is that from energetic considerations a soliton may readily be formed with an unpaired electron at each end of an inverted sequence of double bonds. This is shown in Fig. 3b. The solitons have been detected spectroscopically and are considered responsible for the higher conductivity of the *trans* isomer.

2.3. "Doping" and Conduction Mechanisms

Polyacetylene shows three redox regimes. The neutral insulative polymer may be oxidized to give a polycationic chain doped with anions from the oxidizing medium as compensating counterions. Alternatively the polymer may be reduced to give a polyanionic chain doped with cations. This behavior is shown in Fig. 4.

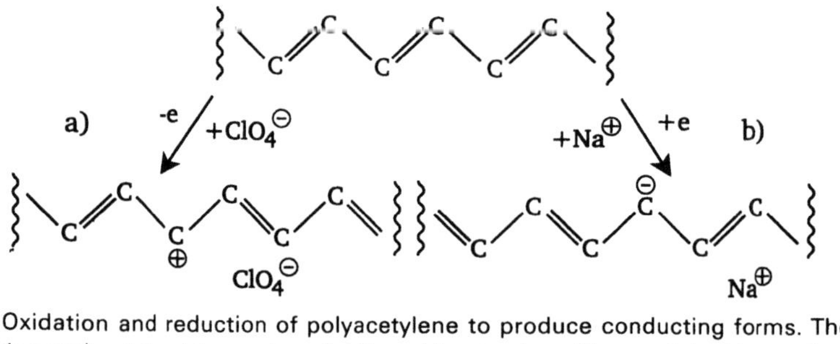

FIGURE 4. Oxidation and reduction of polyacetylene to produce conducting forms. The insulating form is (in this case) oxidized to produce (a) "perchlorate-doped" material with a cationic chain, or reduced to produce (b) "sodium-doped" material with an anionic chain. The conducting forms have a conductivity of about $100 \, S \, cm^{-1}$.

Redox may be carried out either chemically or electrochemically. Chemical oxidants include iodine, nitronium species, or transition metal salts, while suitable chemical reducing agents include sodium naphthenide. Electrochemical methods allow the presence of dopants such as the anions ClO_4^-, or BF_4^-, or more complex species that are not suitable for chemical redox systems but which are stable as electrolytes in an electrochemical cell.

The nature of the dopant can affect conductivity, stability, morphology, and other polymer properties; and the identity of the dopant may not always be straightforward since I_3^-, I_5^-, and I_7^- have all been observed from iodine doping of polyacetylene, while $FeCl_3$ oxidant leads to Cl^- and $FeCl_4^-$. Unlike almost all conducting polymers, even of its own class, polypyrrole shows a surprising tolerance of dopant size, charge, and nature, allowing the use of sophisticated functionalized dopants such as tetrasulfonatophthalocyanine.

This use of the term "dopant" should be distinguished from that in semiconductor physics since considerably higher quantities of dopant ions are employed in conducting polymer systems. An important distinction between polyacetylene and the polypyrrole class of polymers is that to date the latter have only been oxidatively charged, with anionic dopants.

Conductivity, morphology, and a range of properties vary with the doping level of the polymer. A typical curve for change in conductivity of polyacetylene with doping level is given in Fig. 5, which may represent, for example, doping with iodine vapor. Significant quantities of dopant per (CH) unit are required for the highly conductive "metallic" regime, and the doping level may be varied throughout the range.

The charging and discharging of polyacetylene during a doping–dedoping cycle provided an early proposed application for the material as a battery electrode. There are several possible half-cell reactions depending on the other components of the system, but since polyacetylene may be charged by both oxidation and reduction the possibility of an "all-plastic" battery has been investigated. A possible reaction sequence is

$$(CH)_x^{n+} \cdot nClO_4^- \quad | \quad (CH)_x^{n-} \cdot nLi^+ \; \rightleftharpoons \; (CH)_x^0 \quad nLiClO_4 \quad (CH)_x^0$$

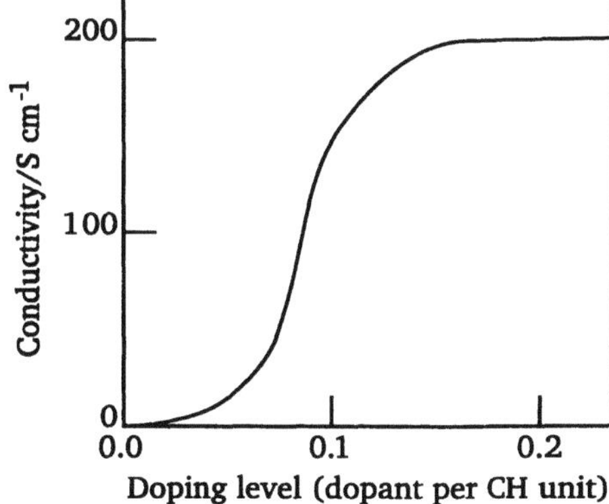

FIGURE 5. Variation of conductivity with doping level for *trans*-polyacetylene.

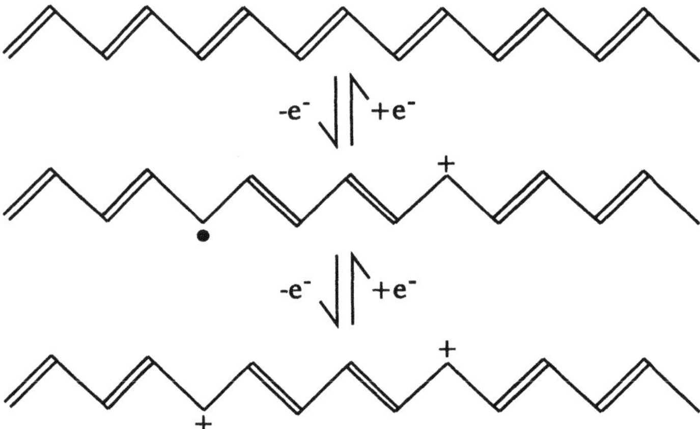

FIGURE 6.   Evolution of polaron and bipolaron states on a polyacetylene chain.

The battery offers a relatively high voltage and energy density, but reproducibility, stability, and cycle life are parameters that require optimization.

Comparison of conductivity with changes in spin or optical and other spectroscopic properties suggests that different conduction processes may operate in different doping regimes. A variety of species have been implicated in conductivity, and for anionically charged (cation-doped) polyacetylene "staging" has been observed similar to that in graphite systems. For cationically charged (anion-doped) polymers, which represents the more general situation, polarons at low doping levels coalescing to bipolarons at high doping levels are considered to be the important species.

A polaron has a positive charge at one end and a free electron at the other, while a bipolaron has two positive charges. This theory explains the observation that highly conductive polypyrrole and similar cationic polymer chains have only low concentrations of free electron spins. This behavior is shown diagramatically in Fig. 6.

The bipolaron can move along a chain and can also change its length (distance between charges).

The parameters of importance in an explanation of conductivity are therefore bipolaron length; conjugation length (which is the distance along a polymer chain before a defect is reached); overall chain length (which affects polymer physical properties); the ease of charge transport from one chain to a nearby one; and the ease of transport between particles or fibrils in a dispersed system. This complexity accounts for some difficulty in the quantification of conduction mechanisms, particularly given the inherently ill-defined nature of polymer systems, with spread of molecular weights, chain lengths, branching, morphologies, and other factors, especially when compared, for example, to inorganic semiconductor lattices. Nonetheless there has been some success in describing band structures for conducting polymers, using data obtained from optical properties and electrochemical data. Thus the effect of increasing doping level upon the absorption spectrum of polypyrrole, for example, or perchlorate, is depicted in Fig. 7 and a pictorial representation of a possible band structure for the polymer is given in Fig. 8.[46]

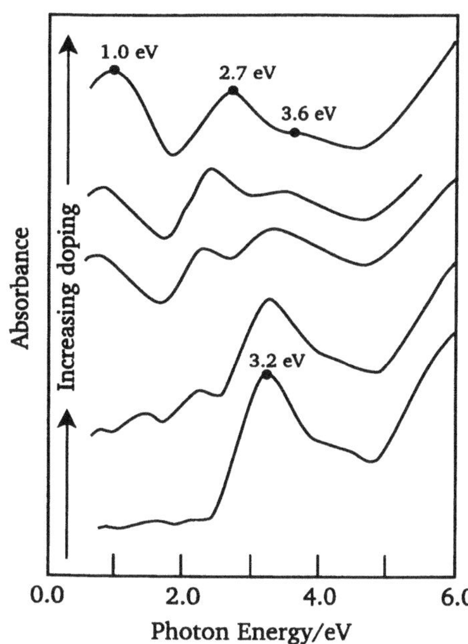

FIGURE 7. Depiction of the optical absorption spectra of polypyrrole films, showing the variation with doping level.

Approximate band gaps have been estimated for several neutral polymers: polyphenylene 3.5 eV, polypyrrole 3.2 eV, polythiophene 2.2 eV, polyacetylene 1.5 eV, and polyisothianaphthene 1.0 eV.

2.4. Preparation and Properties

The recent upsurge of interest in conducting polymers originated from the development of a reproducible route to thin polymer films by Shirakawa,[9] and the discovery that these could be doped to become highly conductive.[10]

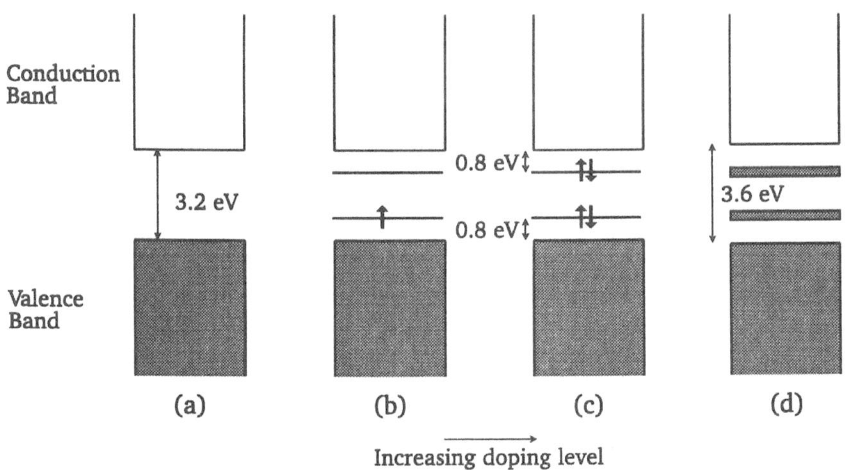

FIGURE 8. Proposed band structure of polypyrrole, and its variation with doping level.

A solution containing a typical Ziegler–Natta type of polymerization catalyst such as $Et_3Al/TiCl_4$ is exposed to acetylene gas, and the polymer forms as a lustrous film across the surface. Altered catalyst concentration and characteristics can lead to formation of the polymer in fine powder form. The films have an open fibrillar morphology which assists the ion-transport processes that accompany redox doping and dedoping. The temperature of production influences the *cis/trans* ratio in the neutral polymer, while isomerization to *trans* accompanies doping.

Polyacetylene is insoluble and infusible, and possesses flexibility and other typical polymeric physical properties. The conductive doped forms of the polymer are quite intractable and possess poorer mechanical properties than the undoped ones, a feature common to most conducting polymers.

Polyacetylene is unstable to oxygen, forming carbonyl and other oxygenated functionalities upon exposure. This limits the applications of this polymer and accounts for the preferment of more stable "heteroaromatic" polymers such as polypyrrole.

The highest conductivity so far obtained from a conducting polymer is 10^5 S cm^{-1}, for iodine-doped polyacetylene prepared by a variation of the Shirakawa route using silicone oil solvent and a pretreated catalyst system.[11] Diminution of Sp^3 carbon atoms, cross-links, and other defects in the polymer chain are thought to account for the enhanced conductivity, which is accompanied by better air stability.

2.5. Processing Improvements—Polyacetylene and Related Polymers

In view of the intractability of conducting polymers there has been considerable effort into the enhancement of processability. For polyacetylene this has involved either the use of copolymers, or the development of precursor route syntheses in which an unconjugated and therefore normally processable polymer is an intermediate in the synthetic route.

For polyacetylene, copolymerization methods have recently been reviewed.[12] Results are interesting although dilution of the conducting polymer with a conventional insulative polymer produces a loss of conductivity. However, both block copolymers of acetylene and isoprene, and graft copolymers of acetylene and butadiene are soluble in toluene.

2.6. Precursor Route Syntheses

At present these are restricted to polymers of the polyacetylene class, and the best known of these is the "Durham" route to polyacetylene developed by Edwards and Feast,[13] which is shown in Fig. 9a. A metathesis polymerization leads to a soluble and processable precursor polymer, which when heated undergoes a retro-cycloaddition reaction to evolve hexafluoroxylene and leave polyacetylene. Other sophisticated chemical schemes have been devised by the Durham group.

The *cis*-polyacetylene will undergo isomerization as usual, and also requires doping to produce conductive forms. The morphology of the Durham polymer is different from the fibrillar Shirakawa material, having a higher apparent density

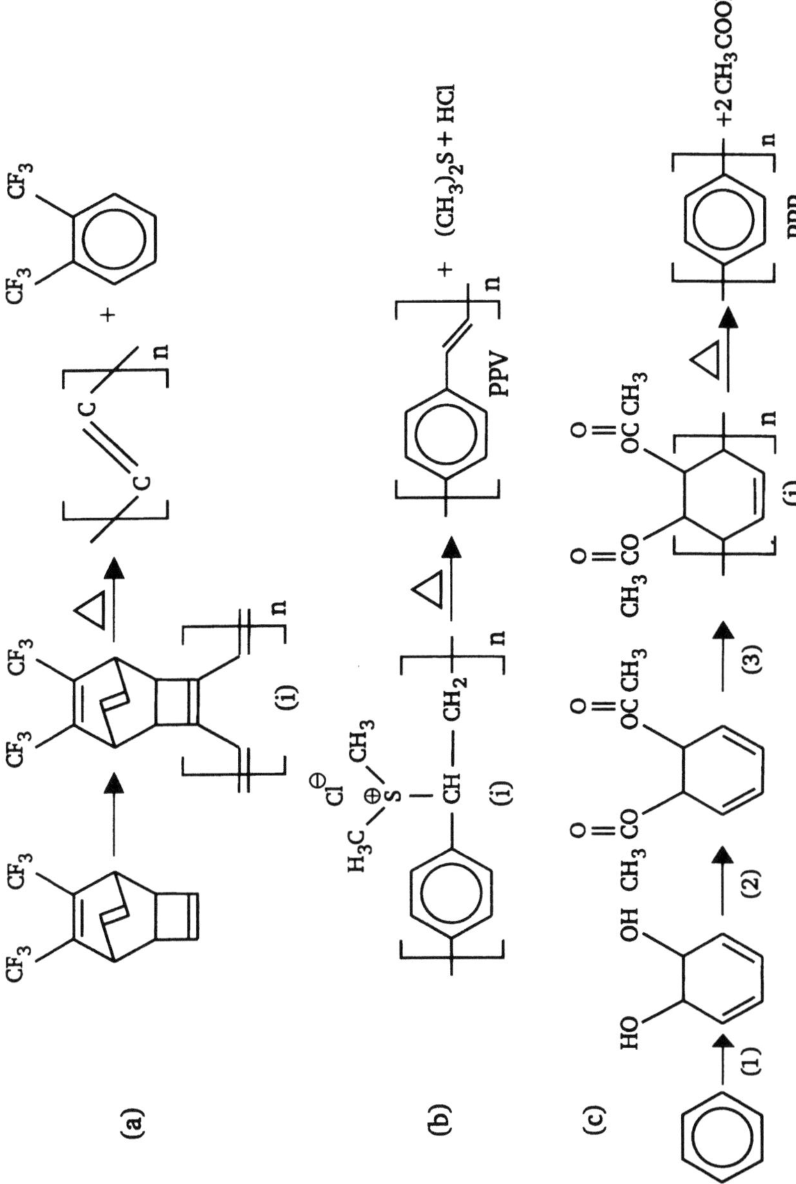

FIGURE 9. Precursor routes to conducting polymers. (a) Polyacetylene (PA): (i) is a soluble, processable precursor. (b) Polyphenylene vinylene (PPV), (i) is a water soluble precursor polymer. (c) Polyphenylene (PPP): (1) bacterial oxidation; (2) exterification; (3) polymerization to give the precursor polymer (i).

and an amorphous structure. Parameters dependent on morphology differ between the two types of polymers, giving subtle distinctions in behavior.

An important result of the Durham synthesis is that by application of a suitable stretching process during the procedure oriented crystalline films of polyacetylene can be obtained with marked anisotropy in conductivity and optical properties.[14] The availability of oriented conducting polymers has produced a significant advance in the technology of these materials.

Other polymers that have been prepared by precursor routes include polyphenylene vinylene, via a water-soluble sulfonium salt presursor polymer[15] (Fig. 9b, which may also be obtained in an oriented anisotropic form; and polyphenylene (Fig. 9c). This latter synthesis, developed by ICI,[16] has an ingenious bio-organic stage in which benzene is converted to 5,6-dihydroxy cyclohexa-1,3-diene by a genetically manipulated organism. Esterification and conventional polymerization give the precursor polymer, which leads to polyphenylene upon heating. Polyphenylene requires powerful oxidants such as AsF_5 to become conductive, but is sufficiently insulative in its neutral uncharged state (cf. Fig. 1) to be used as an insulating coating as it is quite air stable in this form.

The usefulness of the precursor route synthesis is shown by the application of polyacetylene in field-effect transistors and optoelectronic systems.[17,18] Significant results have been obtained; and device fabrication is facilitated by use of the Durham route synthesis.

3. POLYPYRROLE AND THE "HETEROAROMATIC" POLYMERS

3.1. Structure and Preparation

Fine powder pyrrole blacks variously obtained from pyrrole have been known for over a hundred years, but little studied because of their intractability. Interest in the oxidatively formed blacks was rekindled by the discovery that electrochemical oxidation of pyrrole produces a black coating on the electrode that can be peeled off as a free-standing film.[19] Conductivities up to 10^3 S cm^{-1} are obtained.

This coating is polypyrrole. It is insoluble, infusible, and intractable, but importantly is air stable in ambient conditions in its conductive form. The films have mechanical properties similar to a fragile sheet of paper when dry, but are notably flexible and tough when solvent-swollen. The use of chemical oxidants produces fine powders, of importance when preparing blends of polypyrrole with other polymer systems.

The polymer is produced directly in a conductive polycationic form doped with compensating counteranions derived from the oxidizing medium. Electrochemical oxidation allows the use of a wide range of anions as dopants and this is a feature of polypyrrole technology, since the nature of the dopant can significantly influence overall polymer properties, and even induce additional ones.

The polymerization mechanism is shown in Fig. 10 and is not a typical addition (or condensation) type. Initial oxidative electron-transfer produces radical cations that are thought to couple directly, with loss of the protons at the α-ring positions, to give a chain with greater conjugation length that becomes progressively easier to oxidize than the monomer, hence leading to the polymer directly in the oxidized form.

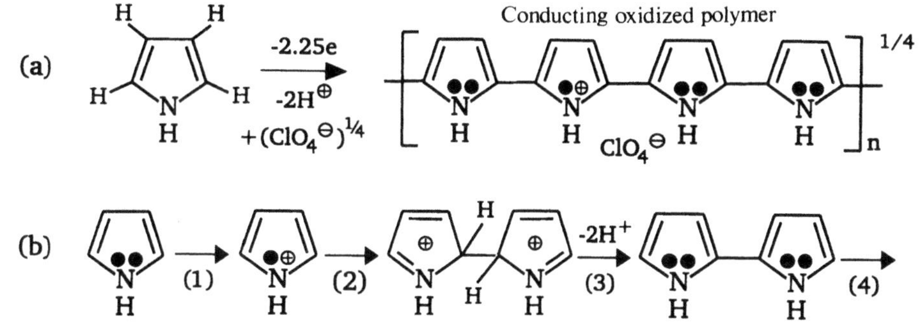

FIGURE 10.   Polypyrrole. Here (a) shows the formation of the conductive oxidized polymer, and (b) the mechanism, which is (1) oxidation to produce a radical cation; (2) dimerization; (3) proton loss; (4) repeat oxidation, coupling, etc.

A striking feature of polypyrrole above other polymers in its class is its tolerance towards oxidation conditions. Not only may monomers with complex substituent groups be made to polymerize, and complex dopant anions be employed, further discussed below, but simple pyrrole polymerization is extremely tolerant of electrolysis conditions, and a list of recently used electrodes, solvents, electrolytes, and other parameters that still lead to successful polymerization is given in Table 1.

Polypyrrole can be prepared on a stainless steel electrode in a beaker open to the air and powered by a battery, and in this respect is the most "user-friendly" conducting polymer.

Careful stretching of solvent-swollen films induces a measure of anisotropy in the sample.

3.2. Other Monomer Systems

The formation of conductive coatings by electro-oxidation of unsaturated organic monomers appears to be a more widely applicable phenomenon than might

TABLE 1.   Polypyrrole—Tolerance of Electrolysis Factors

Electrode materials	• Pb, Au, Ni, W, stainless steel, Hg • ITO-coated glass, n-Si, p-Si, GaP, chalcogenides • Carbon (vitreous, fiber, cloth, paste) • Carbon- or metal-loaded polymers • Other conducting polymers
Solvents	• Water, acetonitrile, methanol, tetrahydrofuran, propylene carbonate, nitrobenzene, butyrolactone, dioxan • Low-temperature molten salts • Solid polymer electrolytes
Electrolyte salts	• Inorganic salts, alkylammonium salts, organic anions, redox complexes, organometallics, macrocyclics, enzymes
Current, density, potential	• Affect polymer properties, but films obtained over wide ranges: Constant potential, constant current, cycled, pulsed techniques

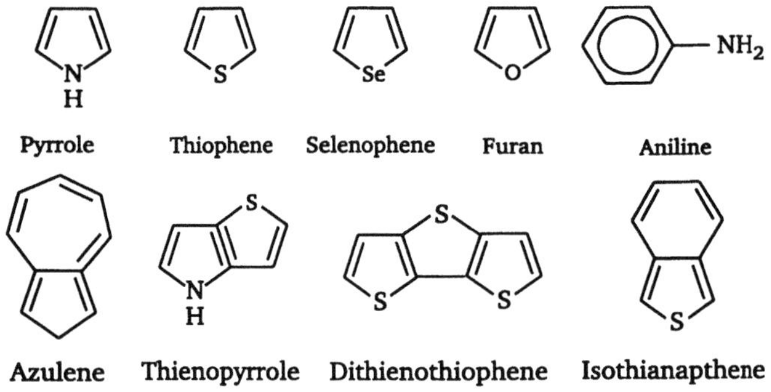

FIGURE 11.   Monomer ring systems.

at first be expected, although the production of free-standing films is the province of polypyrrole and its simple derivatives. The range of monomer ring systems that fall into the "heteroaromatic" category is shown in Fig. 11. Of these polythiophene and polyaniline are of rapidly increasing importance in applications due to their stabilities and other properties, but surprisingly complex systems still give satisfactory polymers for study. Fused-ring heterocycles may be employed, and a heteroatom is not necessary, sihce the carbocyclic azulene will also polymerize. In addition the fused isothianaphthene system produces an almost transparent conductive polymer, of great potential use in optical switches and electro-optic systems.[20] The effect of monomer structure upon polymer band-gap is an important area of conducting polymer research for such applications.

3.3. Redox Properties

Polypyrrole may be reduced to the neutral uncharged form, but unlike polyacetylene cannot be further reduced to a negatively charged chain doped with cations. This behavior is characteristic of the "heteroaromatic" class of polymers.

Reduction causes changes in conductivity, morphology, and also stability; and neutral polypyrrole is unstable in air, which mitigates against this polymer for applications involving redox cycling. Polythiophene and polyaniline, however, are more stable in the reduced forms, and this factor is responsible for increasing interest in these materials.

Color changes also occur upon redox. For polypyrrole a dark green-yellow change is not of obvious application, but polythiophene has a more useful blue-red transformation, and is under consideration for electrochromic displays. The redox of polythiophene is shown in Fig. 12. Polyanilines are also attracting interest as electrochromics.

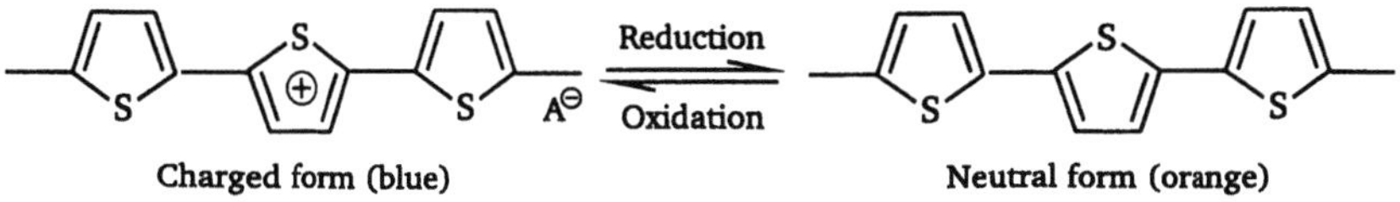

FIGURE 12.   Redox electrochromism of polythiophene.

Reduction of the charged polymer is accompanied by ion transport, either of the dopant anion out of the polymer, or of a cation from the reducing medium into the polymer. This has been used in controlled ion-release systems, in which the bioactive anion glutamate has been expelled from poly-3-methoxythiophene.[21]

Because of differing morphologies in the oxidized and reduced forms, ion mobilities are different, and this has been used in "ion gates" or switchable membranes, to control the flow of ions from one electrolyte solution to another.[22] Redox switching has been exploited in microelectrochemical transistors and similar devices. Signal amplification has been demonstrated for electrode arrays connected by a conducting polymer that can be driven between conductive and insulative forms as part of an electrochemical cell. Poly-3-methyl thiophene and polyaniline have been used with success.[23] In a further sophistication of this system the electrochemical platinizing of the polymer produces an electrocatalytically active Chem FET that can sense for pH, oxygen, or hydrogen.[24]

Response speeds for these devices are relatively slow at present, but efforts to improve this appear to be meeting with success and the full effectiveness of redox applications for conducting polymers may yet become apparent.

3.4. pH–Sensitive Switching

Although not a five-membered heterocycle, polyaniline behaves in many respects as a heteroaromatic polymer. However, because of the pH-sensitivity of the linking nitrogen functionality the polymer can be driven between different forms by acid–base treatment. This is shown in Fig. 13 for the conversion of the significant emeraldine form of the polymer into emeraldine base. Conventional redox switching is also available to this polymer and several representative structural units so formed are also shown in Fig. 13. Polyaniline is of interest for its stability as well as for the opportunities offered by correct manipulations of its structural complexities; but it requires care in synthesis, being markedly less tolerant of preparation parameters and procedure than polypyrrole. To date only simple dopant anions have been employed with this polymer.

4. STRUCTURAL MANIPULATIONS—HETEROAROMATIC POLYMERS

A significant advantage of this class of polymers is the wide range of structural manipulations that can be performed while still producing acceptable materials. These manipulations are shown diagrammatically in Fig. 14.

Thus the monomer may be substituted at ring positions that are not involved in the polymerization reaction [vary R^1 (or R^2 also for pyrrole)]; the dopant ($A^\ominus$) may be varied and also the doping level ($1/N$); while change in the linking group (Y) represents copolymerization.

Such manipulations will alter polymer properties (conductivity, morphology, stability, redox behavior, physical and mechanical properties), and may also be employed to induce specific functionality. Examples are given below.

In addition the polymer may be compounded with another material such as a conventional polymer or an inorganic filler to produce composites, blends, or

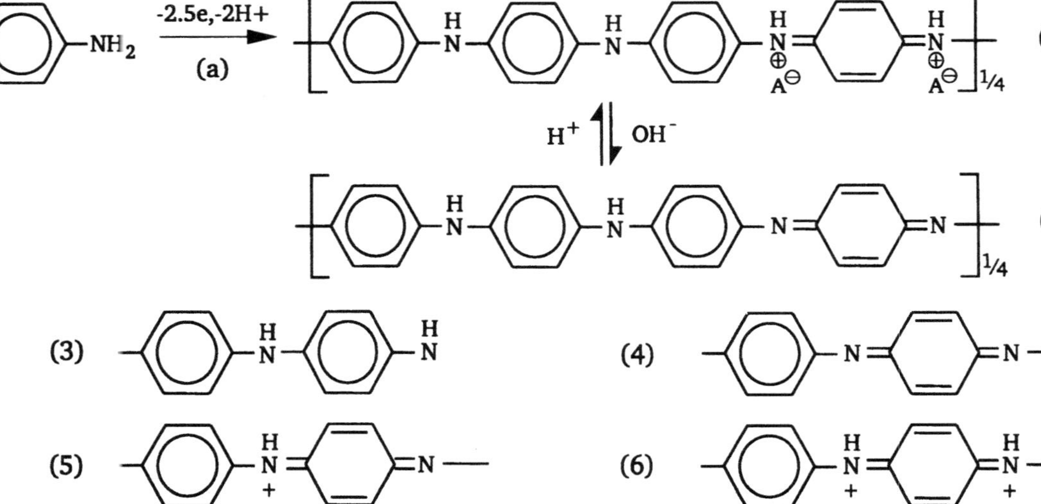

FIGURE 13.   Folyaniline. Here (a) shows the formation of (1) the emeraldine (green, conductive) form in aqueous acid, and the (2) alternative emeraldine base (blue) form. Various repeat units (via proton transfer or redox reactions) are (3) reduced; (4) oxidized unprotonated; (5) oxidized monoprotonated; (6) oxidized diprotonated.

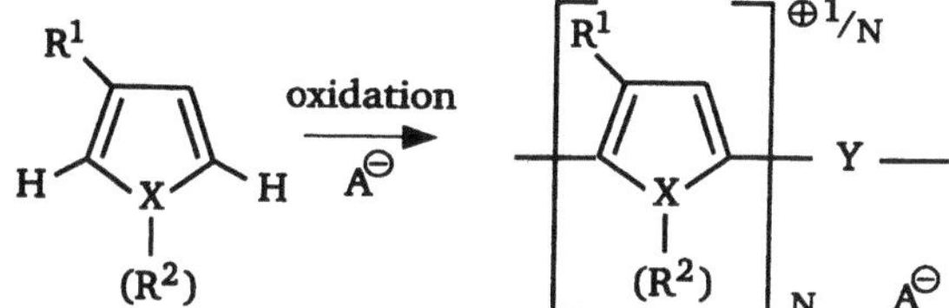

FIGURE 14. Structural manipulations of "heteroaromatic polymers."

laminates. This is a route to enhanced processability, an important aim of workers in the field and relevant to many applications.

4.1. Functionalized Monomers

This demonstrates clearly the great breadth of opportunity offered by systems based on organic compounds. A wide range of functionalized monomers have been examined and a representative sample of pyrrole and thiophene monomers are given in Fig. 15. Surprisingly complex groups may be present, and if the monomer does not homopolymerize directly, acceptable polymers with significant presence of functionality may still be obtained by copolymerization, for example with unsubstituted pyrrole or thiophene.

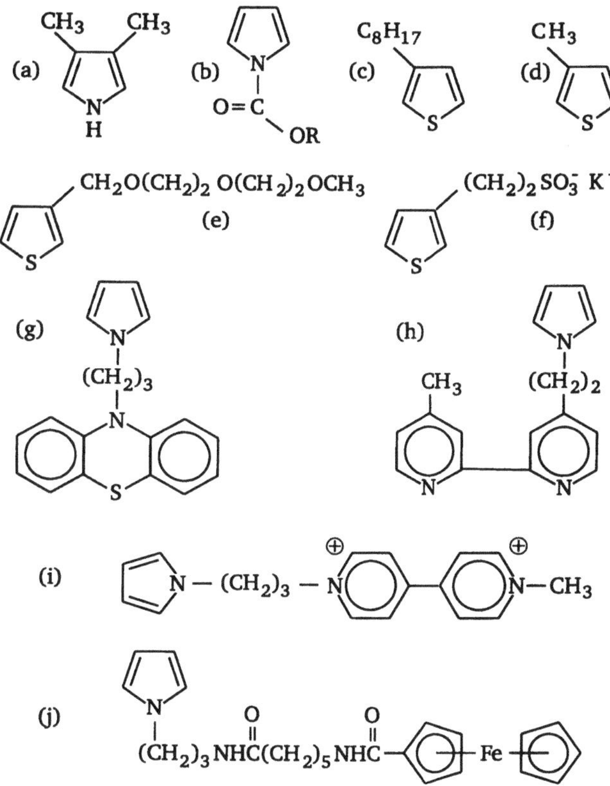

FIGURE 15. Representative functionalized heteroaromatic polymers.

Simple alkyl-substitution at the 3-position enhances solubility of the neutral undoped forms of the resultant polymers, particularly for thiophenes; while for longer alkyl chains (15c), and for alkoxy-substitution (15e) some solubility in the conductive form is obtained.[25] However, some loss of conductivity is observed upon solution and redeposition.

If the substituent side-chain contains a sulfonate group (15f) then the polymers show water solubility in the conducting state,[26] and considerable advancements in the solubility and processability of conducting polymers have recently been made. However, substituted monomers in general lead to polymers with lower conductivity than those from unsubstituted monomers.

Pyrrole is astonishingly tolerant of the size and complexity of groups that may be attached to the ring. Thus a photoactive phenothiazine dye (15g),[27] a bipyridyl coordinating ligand (15h),[28] and viologen (15i)[29] or ferrocene (15j)[30] redox-mediators are examples of specific functionalities that have been employed. Suggested applications for such materials include electronic devices, photoactive systems, sensors, and electrocatalysts.

4.2. Functionalized Dopants

The nature of the dopant anion can markedly affect the properties of a conducting polymer, and even a switch of simple anion from tetrafluoroborate to *p*-toluene sulfonate can influence conductivity, stability, and morphology. The role of the dopant in polymer behavior is not fully clarified and a range of sophisticated dopants may be employed, particularly for polypyrrole. The dopant may impart its own functionality to the system and a representative sample of an extensive range of dopants is given in Fig. 16.

Simple organic sulfonate anions impart good flexibility and strength to polypyrrole without compromising conductivity, but more complex species produce lower

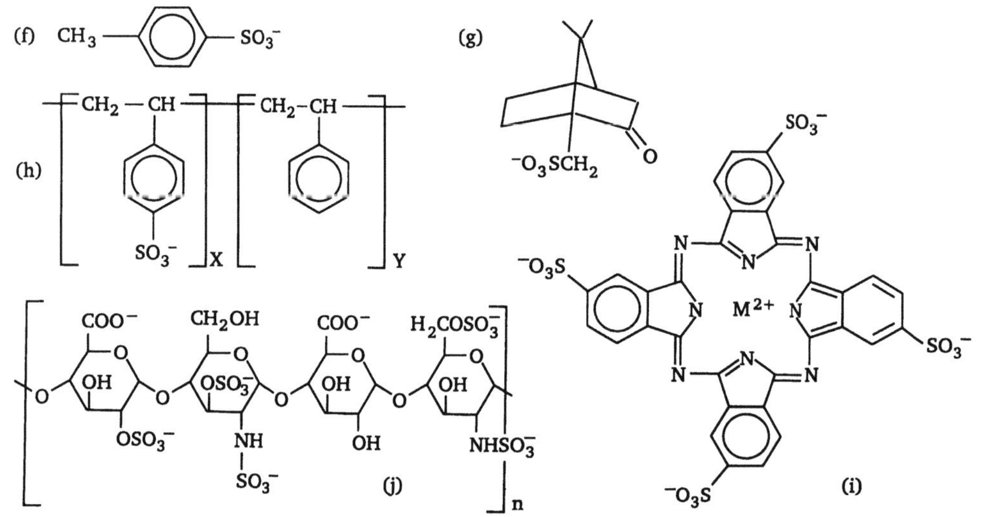

FIGURE 16.   Representative dopants for polypyrrole.

conductivities. Another relatively simple anion is camphor sulfonate (16g), which may also impart chirality to the polymer.[31]

Another system in which the properties of the dopant influence the polymer employs the polystyrene sulfonate anion (16h). In this case the mechanical and physical properties of the conventional polymer chain in the dopant induce a degree of fusibility in the resultant polypyrrole.[32]

Metal complex anions such as tetrachloroferrate (16d), and ferricyanide (16e) have been used, [33] and polypyrrole is so tolerant of extreme dopant characteristics that the very large tetrasulphonated phthalocyanine anion (16i) also produces polymer films of good quality and cohesiveness.[34] This anion is itself a photoactive and electrocatalytic species, and the free-standing films so obtained are purple in color, suggesting that the anion influences the overall optical properties of the system. This anion also enhances polymer stability towards redox.[35]

An ingenious use of dopant in polypyrrole employs heparin (16j), the natural anti-blood-coagulant.[31] Potential applications include artificial nerves and the masking of metal-implant surfaces against blood clotting and rejection. Other bioactive systems are under investigation to exploit the intrinsic "organic" nature of conducting polymers.

4.3. Composites, Blends, and Laminates

The oxidative polymerization mechanism of heteroaromatic polymers is not suited to conventional methods of copolymerization and so attempts to combine the properties of conducting polymers with those of conventional performance polymers have employed technologies of mixing.

Thus if polyvinylchloride, polyvinylacetate, or a similar polymer is first spin-coated as a thin film onto an electrode, then subsequent pyrrole electro-oxidation yields an electroactive film that is a composite of both polymers.[36]

Composites may conveniently be formed by chemical oxidation. Thus the monomer might be imbibed into a carrier polymer (for example, a polyester film), which is then exposed to an oxidant in gaseous or solution form.[37] Alternatively the oxidant may be imbibed first and the heteroaromatic monomer subsequently exposed as vapor or in solution.[38] Inorganic fillers may be coated. In addition polymerization may take place within latex particles.[39]

Considerable effort has been expended in the search for enhanced processability by the production of composites, blends, and laminates; and numerous examples have been reported, particularly in the patent literature. Pyrrole, thiophene, and aniline are the most widely studied monomers. Appreciable conductivities ($>0.1\,\mathrm{S\,cm^{-1}}$) are claimed for some composites, and also semitransparency. Thus a polypyrrole–polyester composite has sufficient qualities of conductivity and transparency for consideration as an electrostatic recording material in reprographics.[40]

Composites show significant improvements in toughness and strength, and may be melt processable. This area of research accounts for a considerable proportion of these publications in the field of conducting polymers directed towards practical applications, and there are a range of different procedures that produce composite materials.

5. CONCLUSIONS

Electrically conducting polymers are novel and versatile materials that exhibit a unique range of properties that impinge upon a host of potential applications within the electronics industry. These materials are attracting considerable interest not only because of their theoretical aspects, but also because of the range of chemical manipulations that are available to alter polymer properties. However, fundamental studies demand well-defined systems, whereas the production of esoteric materials is so rapid that arguably a proper understanding of each lags behind, so that the full usefulness of these materials in practical systems remains to be established.

Several problems associated with the early development of these materials, namely, reproducibility, stability, and processability are being addressed; while the production of oriented samples and of functionalized systems represents recent advances of practical significance.

The diversity of possible applications is exemplified by the observation that conducting polymer coatings protect conventional semiconductor surfaces against photocorrosion,[41] and may protect metallic surfaces against normal corrosion processes [42]; the subtleties of these materials are shown in aspects of their application as sensors, both for gases and in solution. Thus simple polypyrroles and polythiophenes show changes in conductivity upon exposure to both oxidizing and reducing gases (e.g., nitrogen dioxide and ammonia, respectively[43]); and there is a reproducible response in solution to the anion used as dopant.[44] More sophisticated is the chem FET described previously,[24] but an even more ingenious system is the biosensor obtained by the polymerization of pyrrole in the presence of the enzyme glucose oxidase and a suitable redox mediator. The resultant polymer acts as a glucose sensor in which the enzyme still operates satisfactorily despite immobilization in the conducting polymer matrix.[45] The redox mediator may be a ferrocene species, which may be attached directly to the pyrrole ring (structure 15j).[30] Such systems represent a considerable degree of sophistication.

The field is still expanding, and this is demonstrated by the statistics given in the Appendix. These materials offer novel opportunities within electronics technology, and further developments are awaited.

New workers in the field should note the plethora of polymer systems and the opportunities for manipulation. In addition the range of influential parameters provides a source of irreproducibility if mishandled but offers further scope if manipulated correctly. With these considerations in mind it may well prove that the first system studied for a new application may not prove the most appropriate and persistence will be required, and hopefully rewarded.

6. APPENDIX: HETEROAROMATIC POLYMERS— PUBLICATION STATISTICS

The extent of interest in these materials may be shown by the data obtained from a literature search covering the three years 1985–1987 (inclusive), and concerning heteroaromatic polymers alone. The search was not exhaustive—the patent data were obtained using a simplistic keyword system describing a few polymers by name, while the scientific literature search concerned only electrochemical methods of synthesis and was conducted by hand. Nonetheless, the results are instructive, although recent developments in the field are not covered.

Patent data are given in Tables 2-4.

Table 2 concerns patent disclosure by polymer type and main method of preparation and show that although electrochemical synthesis is favored, chemical synthesis is not insignificant for the production of heteroaromatic polymers for practical applications.

Table 3 gives a round breakdown into purpose and application. (Some data appear more than once here—and many claims cover the class generally without specification of monomer systems.)

The geographical distribution of the patent effort is clearly evident from Table 4.

Tables 5 and 6 concern the scientific literature for electrochemical preparations. Table 5 gives a corresponding geographical distribution of effort. Table 6 concerns polymer type. Papers here tend to specify one monomer system more often than in

TABLE 2. Patent Disclosures 1985–87 by Polymer Type

	Electrochemical	Chemical/other
Polypyrrole	77	45
Polythiophene	35	9
Polyaniline	10	17
Polyisothianaphthene	2	1
Polyazulene	2	0
Various	31	24
	157	96

TABLE 3. Patent Disclosures 1985–1987 by Purpose and Application

	Polypyrrole	Polythiophene	Polyaniline	Others mixed systems[a]
Composites, blends	47	7	3	13
Conductivity (main claim)	27	18	9	22
Batteries	16	5	4	15
Antistatic/EMI shielding	16	4	1	3
Photoelectrochemical systems	8	2	2	5
Displays	6	4	0	8
Sensors	9	1	4	4
Capacitors	3	5	5	0
Electronic junction devices	4	5	0	3
Electrocatalysis	4	0	0	0
Information storage	7	0	0	0

[a] Where the claim does not discriminate between monomer systems.

TABLE 4. Patent Disclosures 1985–1987 by Country of Origin, Key-worded as Polymers of Pyrrole, Thiophene, Aniline, Azulene, Isothianaphthene

Japan	177
West Germany	35 (26 to BASF)
USA	28
Sweden	3
France	2
Great Britain	2
Others	3
	253

TABLE 5. Scientific Literature 1985–1987, Electrochemical Polymerization

Japan	92
United States	92
France	52
West Germany	22
Great Britain	22
Italy	17
Sweden	11
China	8
Others	50
	366

TABLE 6. Scientific Literature 1985–1987 by Polymer Type, Electrochemical Polymerization

Polypyrrole	148
Polythiophene	74
Polyaniline	47
Related polymers	15
Several types	38
Unrelated polymers	24
General reviews	20
	366

the patent literature. The preponderance of pyrrole is evident, but is anachronistic. A similar statistical exercise undertaken at the time of writing would show greater interest in thiophene and aniline polymers, although polypyrrole remains the archetype of this class.

The simplistic search procedure and data presentation disguises the wide range of other monomer systems that have been found to electropolymerize to "heteroaromatic" polymers. The properties of many of these have yet to be established, and many offer new attributes of practical benefit. A significant feature of conducting polymer technology is the breadth of opportunity awaiting exploitation.

REFERENCES

1. *Handbook of Conducting Polymers*, Vol. 1 and 2 (T. A. Skotheim, ed.), Dekker, New York (1986).
2. J. W. Chien, *Polyacetylene*, Academic, New York (1984).
3. W. J. Feast, *RAPRA Rev.* **1**, 1 (1987).
4. J. R. Reynolds, *J. Molec. Electron.* **2**, 1 (1986).
5. M. R. Bryce, *Chem. Brit.* **24**, 781 (1988).
6. B. Scrosati, *Prog. Solid State Chem.* **18**, 1 (1988).
7. Proceedings of ICSM Los Alamos Conference, *Synth. Metals* **27**(1-4) (1988); **28**(1-3) (1989); **29**(1-3) (1989).
8. Proceedings of ICSM Kyoto Conference, *Synth. Metals* **17**(1-3) (1987); **18**(1-3) (1987).
9. T. Ito, H. Shirakawa, and S. Ikeda, *J. Polym. Sci. Polym. Chem. Ed.* **12** 11 (1974).
10. C. K. Chiang, Y. W. Park, A. J. Heeger, H. Shirakawa, E. J. Louis, and A. G. MacDiarmid, *J. Chem. Phys.* **69**, 5098 (1978).
11. H. Naarman and N. Theophilou, *Synth. Metals* **22**, 1 (1987).
12. G. L. Baker, *ACS Adv. Chem.* **218**, 271 (1988).
13. J. H. Edwards and W. J. Feast, *Polym. Commun.* **21**, 595 (1980).
14. D. E. White and D. C. Bott, *Polym. Commun.* **25**, 98 (1984).
15. I. Murase, T. Ohnishi, T. Noguchi, and M. Hirooka, *Polym. Commun.* **25**, 327 (1984).
16. D. G. H. Ballard, A. Courtis, I. M. Shirley, and S. C. Taylor, *J. Chem. Soc. Chem. Commun.* 954 (1983).
17. J. Burroughes and R. Friend, *Phys. World* Nov, 24 (1988).
18. Patent Applications assigned to BP, Nos. EP 298628A2 (1987); EP 294061 A1 (1988).
19. A. F. Diaz, K. K. Kanazawa, and G. P. Gardini, *J. Chem. Soc. Chem. Commun.* 635 (1979).
20. H. Yashimo, M. Kobayashi, K. B. Lee, D. Chung, A. J. Heeger, and F. Wudl, *J. Electrochem. Soc.* **134**, 46 (1987).
21. B. Zinger and L. L. Miller, *J. Am. Chem. Soc.* **106**, 6861 (1984).
22. P. Burgmayer and R. W. Murray, *J. Phys. Chem.* **88**, 2515 (1984).
23. E. P. Lofton, J. W. Thackeray, and M. S. Wrighton, *J. Phys. Chem.* **90**, 6080 (1986).
24. J. W. Thackeray and M. S. Wrighton, *J. Phys. Chem.* **90**, 6674 (1986).
25. For example: M. R. Bryce, A. Chissel, P. Kathirgamanathan, D. Parker, and N. R. M. Smith, *J. Chem. Soc. Chem. Commun.* 466 (1987).
26. A. O. Patil, Y. Ikenoue, F. Wudl, and A. J. Heeger, *J. Am. Chem. Soc.* **109**, 1858 (1987).
27. A. Deronzier, M. Essakalli, and J. C. Moutet, *J. Electroanal. Chem.* **244**, 163 (1988).
28. S. Cosnier, A. Deronzier, and J. C. Moutet, *J. Phys. Chem.* **89**, 4895 (1985).
29. C. F. Shu and M. S. Wrighton, *J. Phys. Chem.* **92**, 5221 (1988).
30. N. C. Foulds and C. R. Lowe, *Anal. Chem.* **60**, 2473 (1988).
31. H. Naarmann, *Synth. Metals* **17**, 223 (1987).
32. D. J. Walton, N. Bates, M. G. Cross, and R. Lines, *J. Chem. Soc. Chem. Commun.* 871 (1985).
33. A. Pron, J. Suwalski, and S. Lefrant, *Synth. Metals* **18**, 25 (1987).
34. R. A. Bull, F. R. Fan, and A. J. Bard, *J. Electrochem, Soc.* **131**, 687 (1983).
35. D. J. Walton, Proceedings of ICSM, Tübingen, *Synth. Metals* (to be published).
36. O. Niwa and T. Tamamura, *J. Chem. Soc. Chem. Commun.* 817 (1984).
37. A great many examples are to be found, particularly in the patent literature, e.g., Japanese Patent Application JP63/40211 A2 assigned to Kuraray Co. Ltd. (1986) (*Chem. Abs.* **108**, 21493b).
38. For example, Japanese Patent Application JP63/256617 A2 assigned to Toppan Printing Co. (1988) (*Chem. Abs.* **110**, 106378j).
39. N. Cawdery, T. M. Obey, and B. Vincent, *J. Chem. Soc. Chem. Commun.* 1189 (1988).
40. Japanese Patent Application JP63/15253 A2 assigned to Ricoh Co. Ltd. (1988) (*Chem. Abs.* **108**, 229 627).

41. Reviewed in A. F. Diaz, J. F. Robinson, and H. B. Mark Jr. *Adv. Polym. Sci.* **84**, 113 (1988).
42. U.S. Patent No. 4642331 assigned to B. F. Goodrich.
43. J. Miasik, A. Hooper, and B. C. Tofield, *J. Chem. Soc. Faraday Trans. I* **82**, 1117 (1986).
44. S. Dorg, Z. Sun, and Z. Lu, *J. Chem. Soc. Chem. Commun.* 993 (1988).
45. P. N. Bartlett and R. G. Whittaker, *J. Electroanal. Chem.* **224**, 27, 37 (1987).
46. J. L. Bredas, J. C. Scott, K. Yakushi, and G. B. Street, *Phys. Rev. B* **30**, 1023 (1984).

31

Photochromics for the Future

Harry G. Heller

1. INTRODUCTION

A photochromic organic compound A is a compound that undergoes a major reversible color change on irradiation at an appropriate wavelength (λ_a) to form a more highly colored species B, which undergoes the reverse reaction either thermally or photochemically on irradiation at a second wavelength (λ_b):

$$A \underset{\lambda_b \text{ or heat}}{\overset{\lambda_a}{\rightleftharpoons}} B$$

Usually, compound A is colorless or near colorless with the activating radiation (λ_a) in the near ultraviolet region (e.g., 366 nm) and is close to the absorption maximum of compound A, while the bleaching radiation (λ_b) is preferably near the absorption maximum of species B.

The first organic photochromic compound was reported by ter Meer in 1876,[1] who noted that the explosive potassium salt of dinitroethane changed from colorless to red in sunlight with the reverse change in the dark. In 1956, Hirshberg[2] drew attention to the commercial potential of organic photochromic compounds as an optical memory model, which led to major research programs on photochromic materials particularly in the laboratories of large American and European Companies, to develop organic photochromic compounds suitable for optical information storage in computers. By 1969, many different types of organic photochromic compounds had been investigated, including over 800 photochromic spiropyrans belonging to 46 subclasses.[3] One of the main reasons why no organic photochromic compounds have been developed for commercial applications is fatigue, the insiduous and irreversible photochemical and other degradation reactions that partly destroy the photochromic properties with every color and bleach cycle.

Our research work in recent years has been devoted to the molecular design of organic photochromic compounds with a high intrinsic resistance to fatigue and then tailoring their properties to make them suitable for specific commercial applica-

---

Harry G. Heller • School of Chemistry and Applied Chemistry, University of Wales College of Cardiff, Cardiff CF1 3TP, Wales, U.K.

tions such as optical information storage media for write/read/erase optical disks, and prescription plastic glasses. For optical recording, both colorless and colored forms A and B, respectively, must be thermally stable so that recorded information is not erased by thermal coloring or bleaching. For the photochromic plastic prescription glasses, the requirements are different—namely, that the colorless form must be extremely sensitive to coloring by unfiltered sunlight (i.e., have a very high quantum efficiency for coloring with ultraviolet light), and the colored form must be insensitive to white light (i.e., have a very low quantum efficiency for bleaching). But since it is necessary for the color of the glasses to fade under diffuse daylight conditions, the colored form B must be thermally unstable and fade rapidly at ambient temperatures back to the colorless form A, but not so rapidly that the glasses will not color in sunlight on a warm day. We term compounds possessing this special combination of photochromic properties "heliochromic."[4]

In both the above applications, the photochromic properties must be retained when the organic compound is dissolved in a rigid plastic matrix.

The principles of molecular design and the interesting chemistry of the classes of compounds will be illustrated, but not with examples of the compounds currently under commercial development.

2. THERMALLY STABLE PHOTOCHROMIC COMPOUNDS

To avoid the problems of fatique associated with radical, diradical, or ionic intermediates, we elected to study concerted reactions, namely, electrocyclic reactions involving ring closure of a hexatriene system to a cyclohexadiene system and the reverse process.[5]

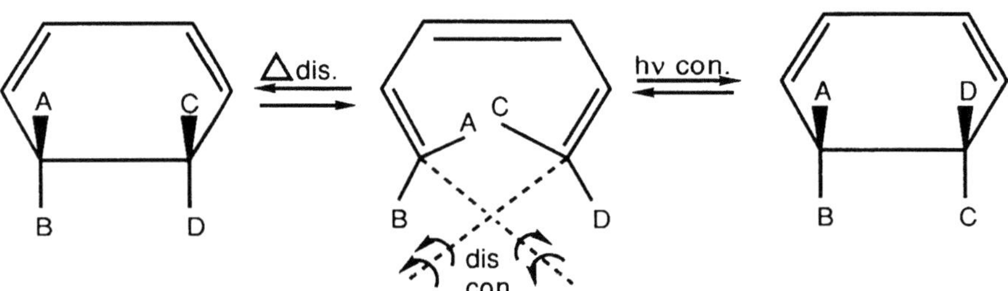

For a $(4n + 2)$ electron system, the thermal process is *disrotatory* and the photochemical process is *conrotatory*.

In this reaction, photochemical ring closure or ring opening occur exclusively in a conrotatory mode while the corresponding thermal processes take place only in a disrotatory mode.

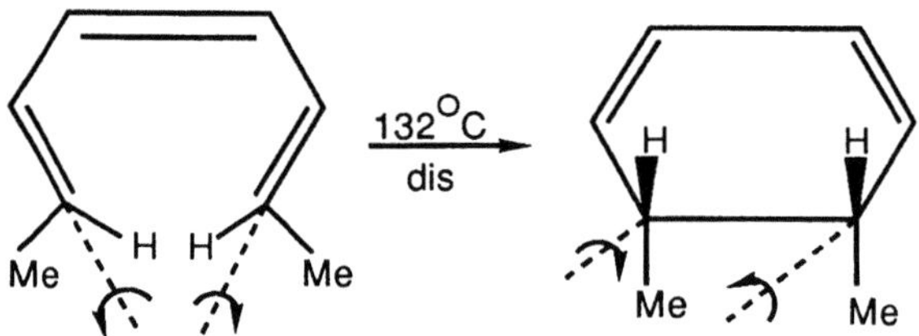

The above illustrative example of an electrocyclic reaction does not involve a color change and is not reversible.[6]

We were able to show that fulgide (1, R = H), initially studied by H. Stobbe,[7] owes its photochromic properties to its photochemical electrocyclic ring-closure to the red 1,8a-dihydronaphthalene derivative (1,8a-DHN) (2, R = H), which undergoes an irreversible 1,5-hydrogen shift, (i.e., fatigues) to the colorless 1,2-dihydronaphthalene derivative (1,2-DHN) (3, R = H).[8] The colored form (2, R = H) reverts back to the fulgide (1, R = H) both thermally at room temperature and on exposure to white light.

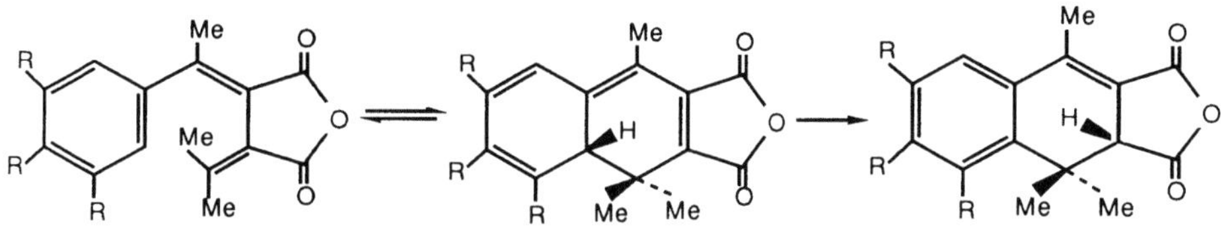

(1)   (2)   (3)

Introduction of methoxy groups into the 3- and 5-positions of the phenyl group (and in this example also into the 4-position for reasons of economy) gave a pale yellow fulgide (1, R = OMe) which photocyclized to give the deep blue 1,8a-DHN (2, R = OMe), which showed a slow thermal 1,5-hydrogen shift to the colorless 1,2-DHN (3, R = OMe). The blue 1,8a-DHN (2, R = OMe) reverts back to the fulgide (1, R = OMe) both thermally or photochemically on exposure to white light.[9] The spectrum of fulgide (1, R = OMe) in toluene before and after irradiation at 366 nm is shown in Fig. 1.

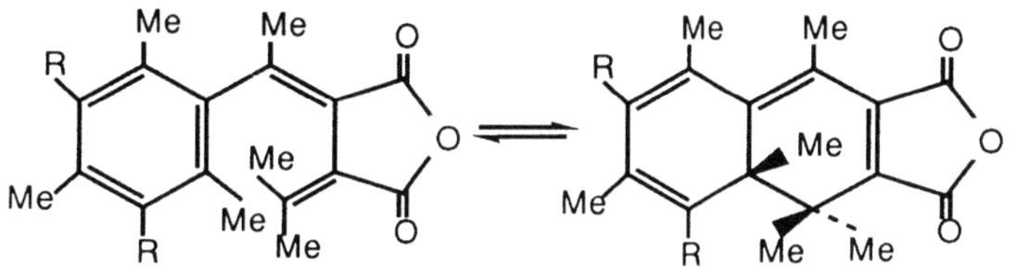

(4)   (5)

To eliminate the 1,5-H shift reaction and at the same time to introduce thermal stability into the colored form, all that was necessary was to introduce methyl substituents into the 2- and 6-positions of the phenyl group (for convenience of synthesis, a methyl group was also introduced into the position 4). Fulgide (4, R = H or OMe) underwent photocyclization in a conrotatory mode to the red 1,8a-DHN (5, R = H) or the purple 1,8a-DHN (5, R = OMe), which underwent the reverse reactions on exposure to white light. The thermal disrotatory ring opening of the 1,8a-DHNs (5, R = H or OMe) was inhibited by the steric interactions of the *cis* methyl substituents in the 1 and 8a positions.[10]

Replacement of the 2,4,6-trimethylphenyl substituent by a 2,5-dimethyl-3-furyl group gave a pale yellow fulgide (6) which cyclized to the thermally stable red 7,7a-dihydrobenzofuran derivative (7,7a-DHBF) (7) on irradiation in toluene at 366 nm.[11]

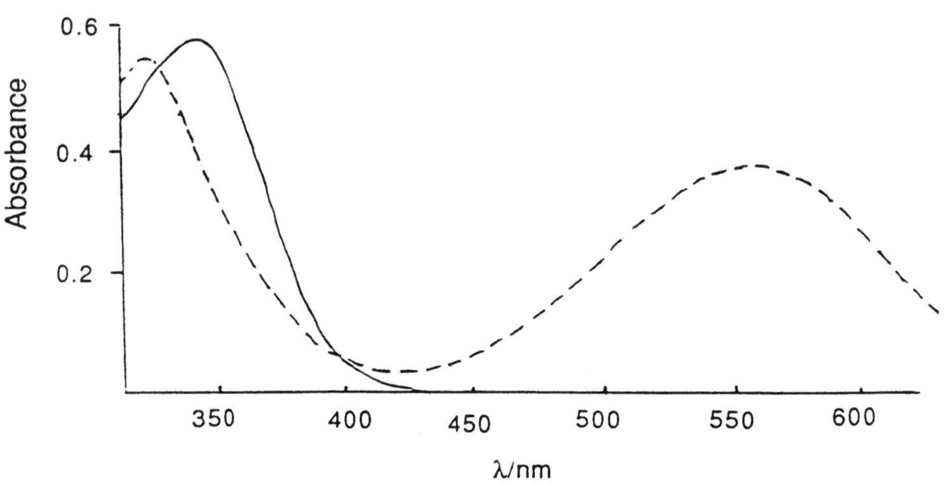

FIGURE 1. Fulgide (1, R = OMe) (7.56×10^{-5} M in toluene) before and after irradiation at 366 nm.

$$\phi_{366}\,0.20 \quad \rightleftharpoons \quad \phi_{514}\,0.055$$

(6)   (7)

The weak absorption by the colored form (7) at 366 nm means that it does not act as an internal filter of the activating radiation, allowing near quantitative conversion of fulgide (6) into the 7,7a-DHBF (7) even at very high concentrations (see Fig. 2). This is an important requirement of many applications, including optical recording. When the colored form absorbs the activating radiation and acts as an internal filter and when the activating radiation also causes the colored form to bleach, conversion into the colored form is not quantitative.

(8)   (9)

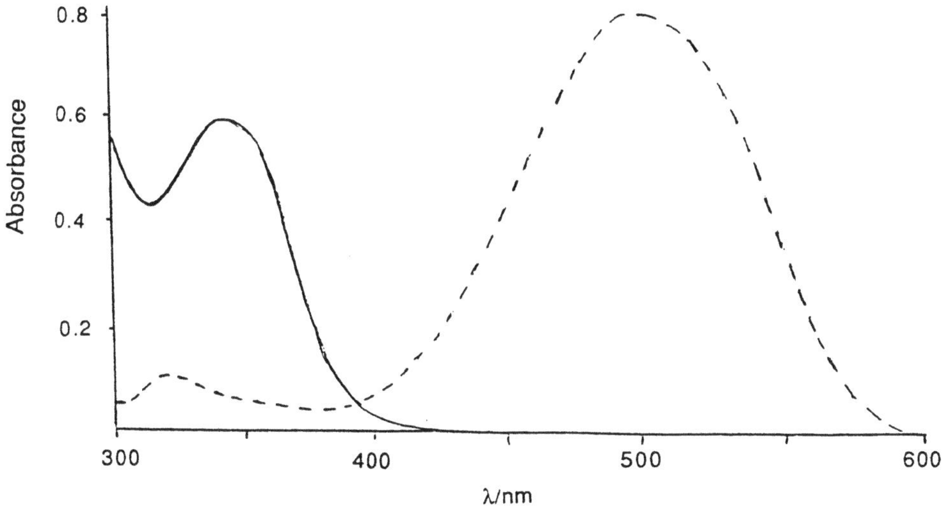

FIGURE 2. Fulgide (6) $(1 \times 10^{-4}\ M$ in toluene) before and after irradiation at 366 nm.

The photochromic compound (8) shows only a 60% conversion into the colored form (9) on irradiation at 366 nm. This example was recently reported by Irie.[12] The spectrum of the photochromic compound (8) before and after irradiation at 366 nm is shown in Fig. 3 and illustrates the effect of absorption of the activating radiation by the colored form (9).

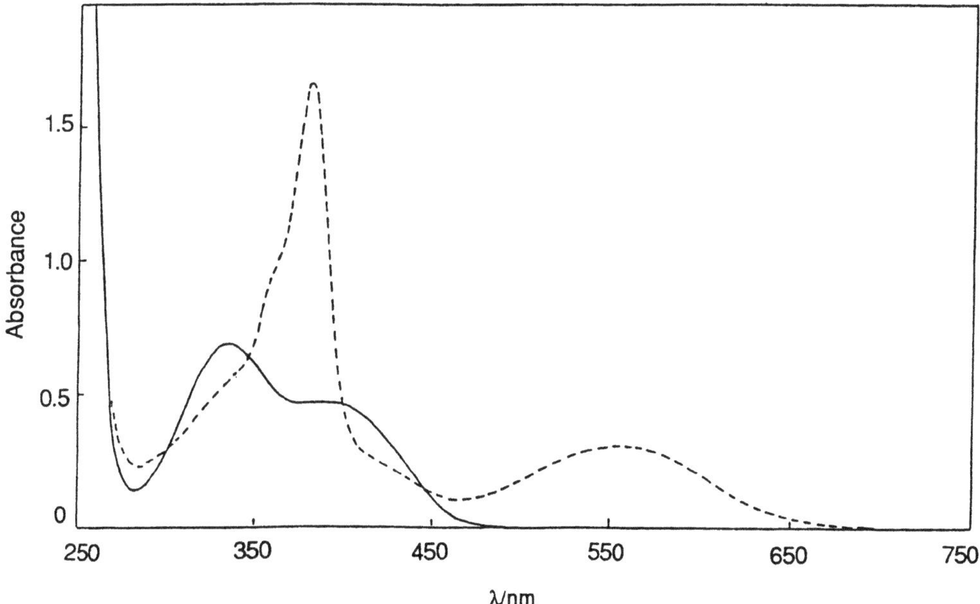

FIGURE 3. Irie's compound (9) $(1 \times 10^{-4}\ M$ in benzene) before and after irradiation at 405 nm.

The photochromic properties of the thermally stable fatigue resistant fulgides can be modified by changing the substituent groups.

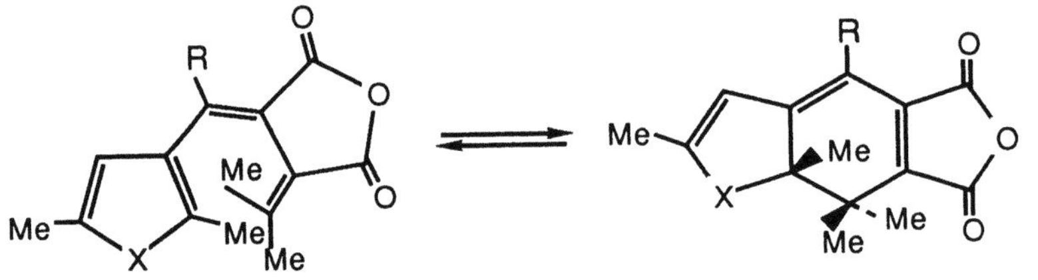

(10) ⇌ (11)

Changing the heterocyclic ring in the fulgide (10, R = Me) from furan to thiophene to pyrrole causes a major color change of the ring-closed form (11) from red to purple to blue (see Fig. 4).[13]

When R is hydrogen, the fulgides (10, X = O, S or NR) are nonphotochromic or only very weakly photochromic.[14] When R is methyl, the fulgides (10, X = O, S or NR) have good photochromic properties with a quantum efficiency for coloring at 366 nm on the order of 20%.[11] When R is an isopropyl group, the quantum efficiency for coloring in the fulgide (10, X = O) is increased to 62%[15] (Fig. 5). When the isopropylidene group in fulgide (10) is replaced by the adamantylidene group as in fulgide (12, R = Me), the bulky inflexible spiroadamantane group in the 7,7a-DHBF (13, R = Me) causes weakening of the 7,7a sigma bond with an increase in the quantum efficiency for bleaching from 5.5% to 30%.[16]

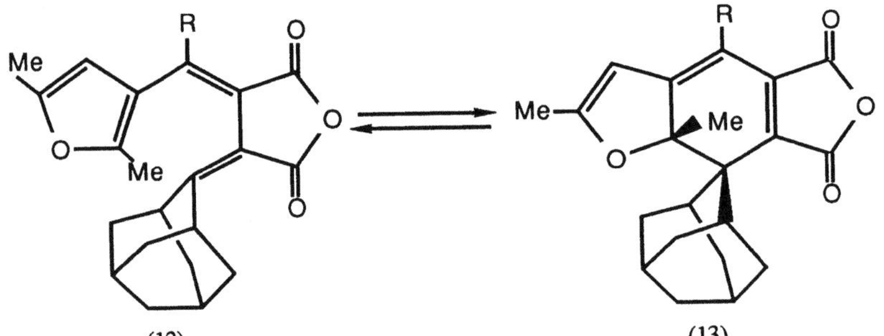

(12) ⇌ (13)

Fulgide (12), having an isopropy group at R and an adamantylidene group, shows a quantum efficiency for coloring of 50% at 366 nm and its colored form (13, R = isopropyl) shows a quantum efficiency for bleaching of 26% at 546 nm[17] (see Fig. 6). This work has also been reported independently by Kurita.

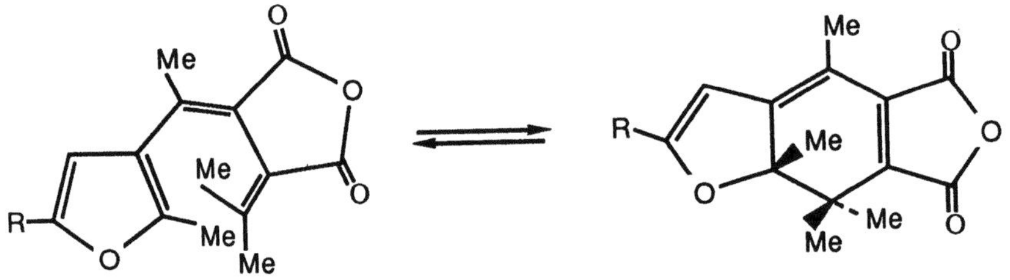

(14) ⇌ (15)

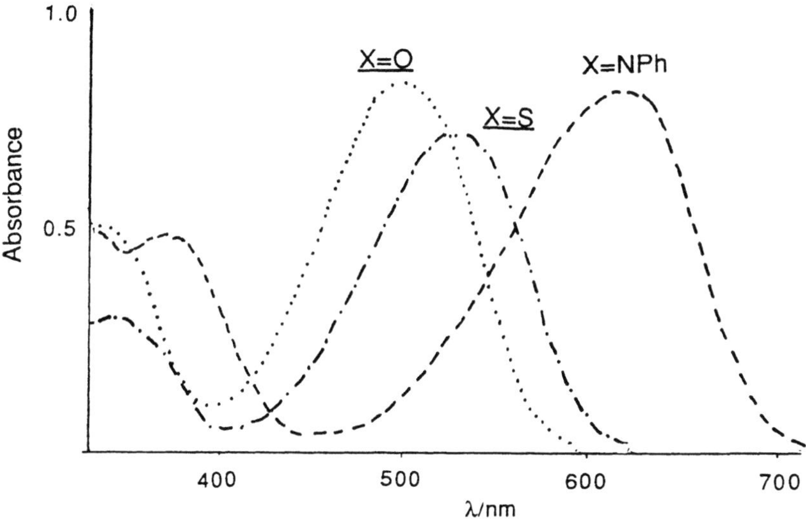

FIGURE 4. Fulgides (10, R = Me, X = O, S or NPh) (1×10^{-4} *M* in toluene) after irradiation at 366 nm.

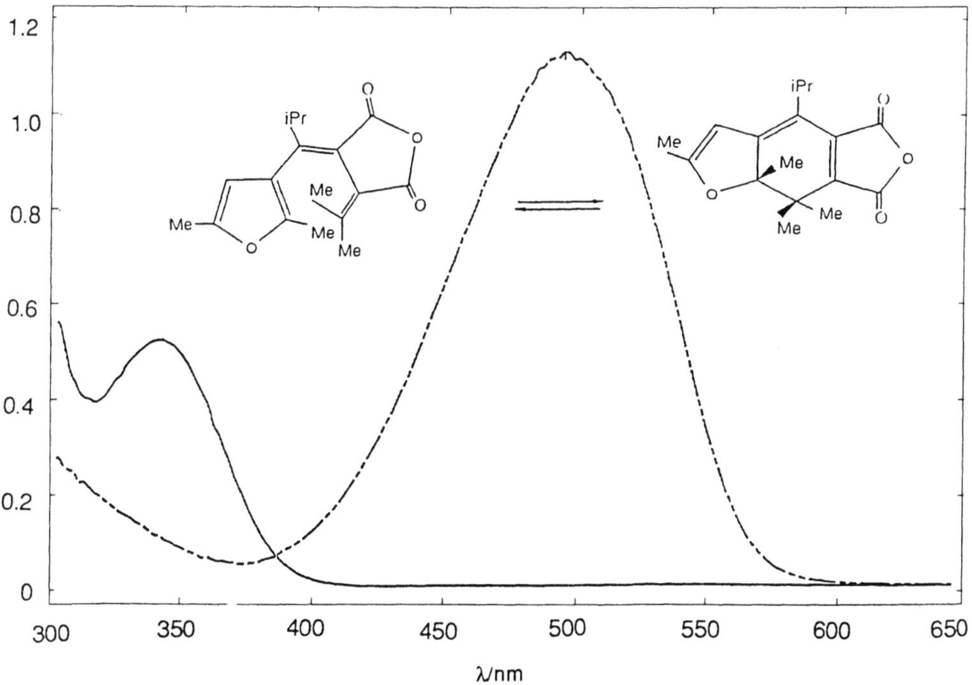

FIGURE 5. Fulgide (10, R = isopropyl) (1×10^{-4} — in toluene) before and after irradiation at 366 nm.

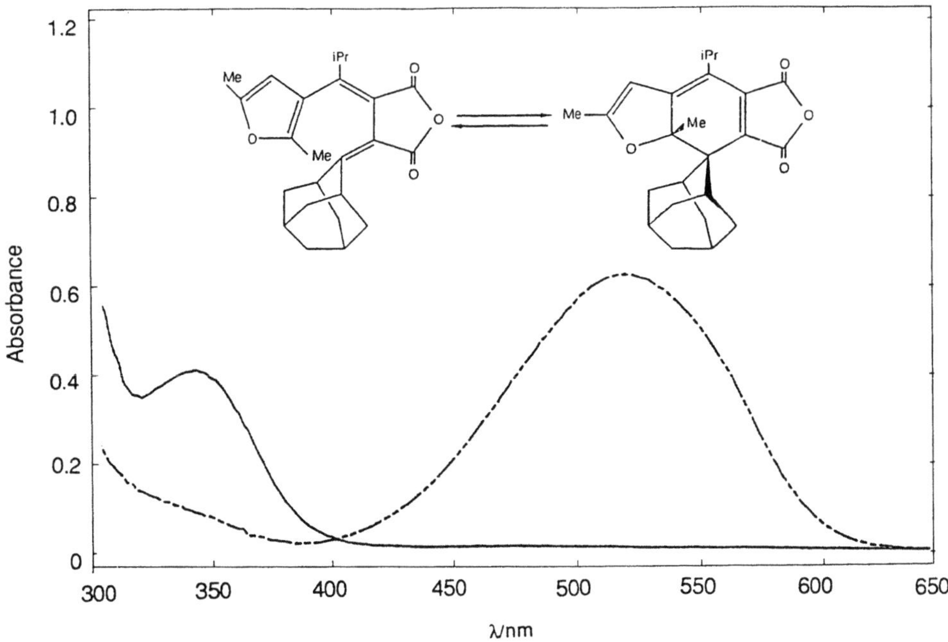

FIGURE 6.   Fulgide (12, R = isopropyl) $(1 \times 10^{-4}\ M$ in toluene) before and after irradiation at 366 nm.

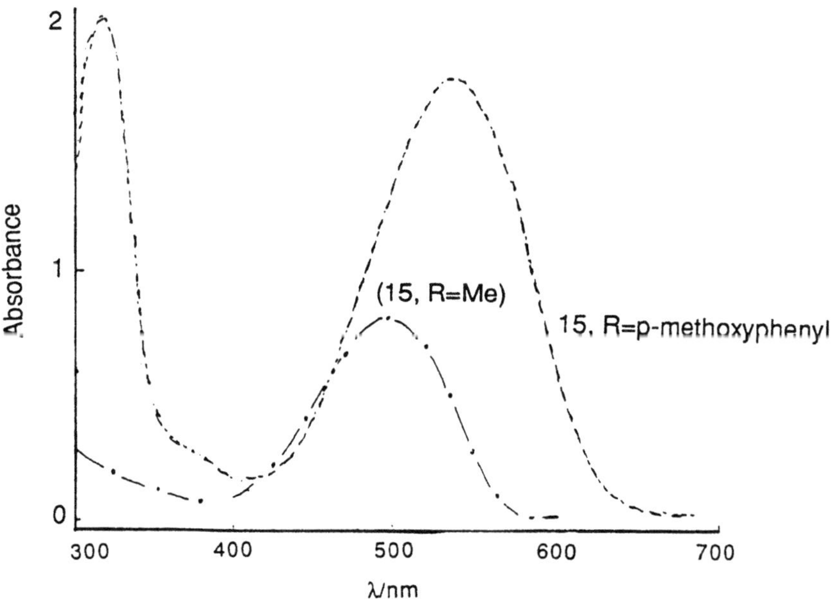

FIGURE 7.   7,7aDHBF (15a) and 7,7aDHBF (15b) $(1 \times 10^{-4}\ M$ solutions in toluene) after irradiation at 366 nm.

Finally, the colored 7,7a-DHBF (15, R = p-methoxyphenyl) shows a major bathochromic shift and a hyperchromic effect compared to the coloured 7,7a-DHBF (15, R = Me). In Fig. 7 it can be seen that 7,7a-DHBF (15, R = p-methoxyphenyl) shows a molar extinction coefficient of 17,800 mole^1 cm^{-1} at 540 nm compared to 8,200 mole^1 cm^{-1} at 496 nm for 7,7a-DHBF (15, R = Me).[17]

3. HELIOCHROMIC COMPOUNDS

It can be seen that the spiroadamantane group has a marked bond-weakening effect in the colored compounds from fulgides, with a resulting six-fold increase increase in the quantum efficiency for bleaching.

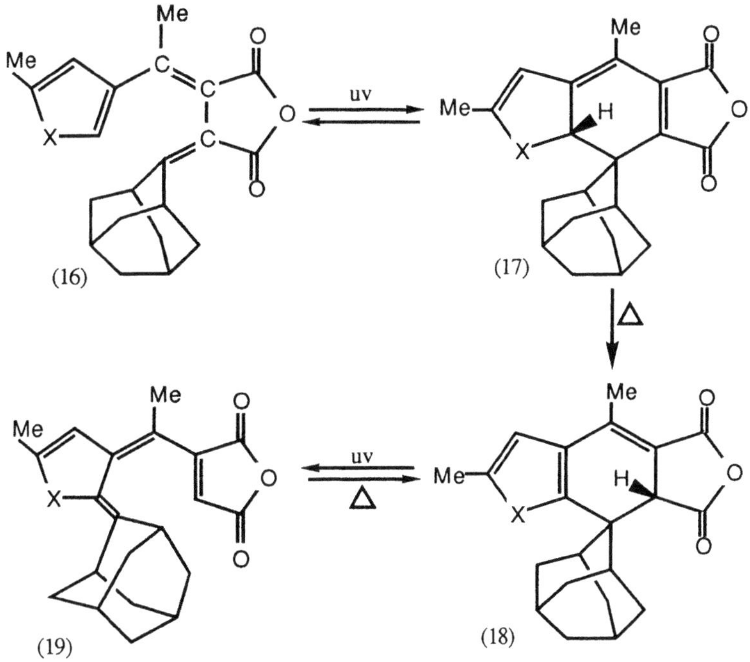

Fulgide (16, X = S) undergoes photochemical ring-closure to the purple 7,7a-dihydrobenzothiophene derivative (7,7a-DHBT) (17, X = S) which undergoes a thermal 1,5-H-shift to the 6,7-DHBT derivative (18, X = S). The 6,7-DHBT derivative (18, X = S), containing the bond-weakening spiroadamantane group, photochemically ring-opens with a high quantum efficiency on exposure to ultraviolet light to give the blue colored form (19, X = S) which has such a low quantum efficiency for bleaching with white light that the 6,7-DHBT (18, X = S) colors rapidly in unfiltered sunlight even when incorporated in CR39 glass. While the photochemical stability of the colored form (19, X = S) is high, it thermally fades rapidly at ambient temperatures in the absence of the activating radiation.

The color and fade characteristics can be modified by molecular tailoring (for example, by benzannellation), by replacement of the thiophene moiety by a furan moiety, or by the introduction of appropriate substituents in the 2- and/or 4-positions.[18]

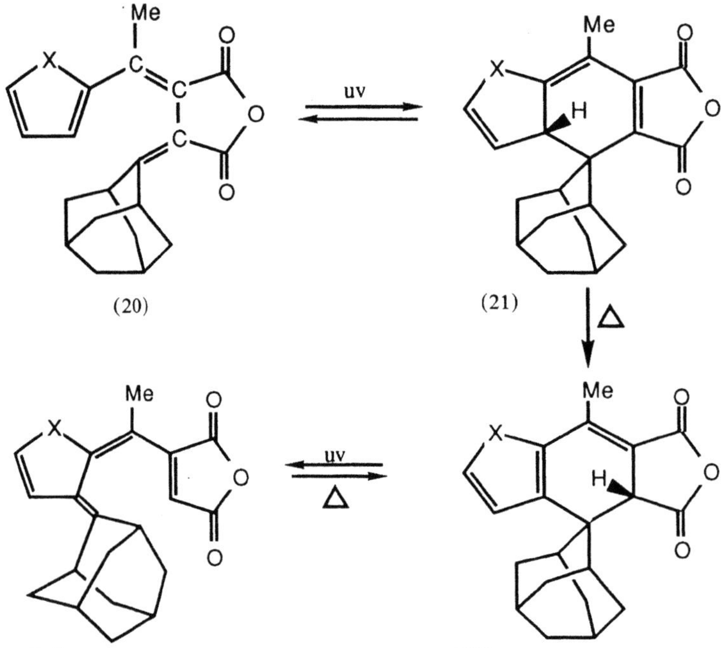

(20)   (21)

(23)   (22)

A further range of different photochromic effects can be attained in the 4,5-DHBT and DHBF derivatives (22, X = S or O), resulting from the thermal 1,5-H shift of the colored forms (21, X = S or O) which, in turn, arose from the photocyclization of the 2-thienyl or 2-furyl fulgides (20, X = S or O). The colors of the colored forms (19, 23) range from red to blue, but it is difficult to synthesize yellow or orange-yellow compounds in these series.[18,19]

4. NAPHTHOPYRAN DERIVATIVES

Photochromic compounds which underwent a colorless to yellow or orange color change were obtained by modifying photochromic naphthopyrans, previously extensively studied by R. S. Becker.[20]

As was to be expected, introduction of a 2-spiroadamantane group into the naphthopyrans (24) and (27) caused weakening of a sigma bond and an increased

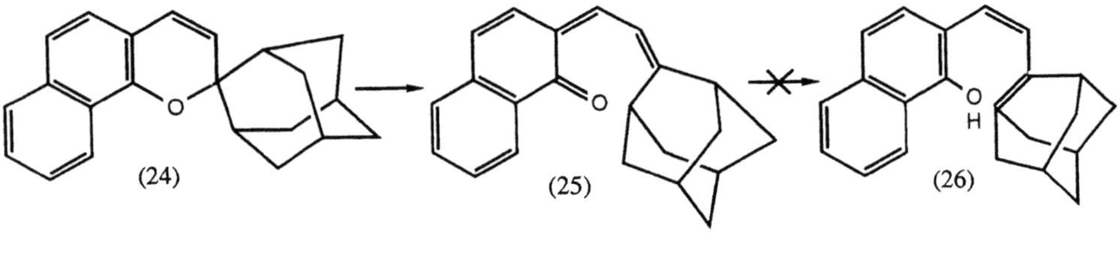

(24)   (25)   (26)

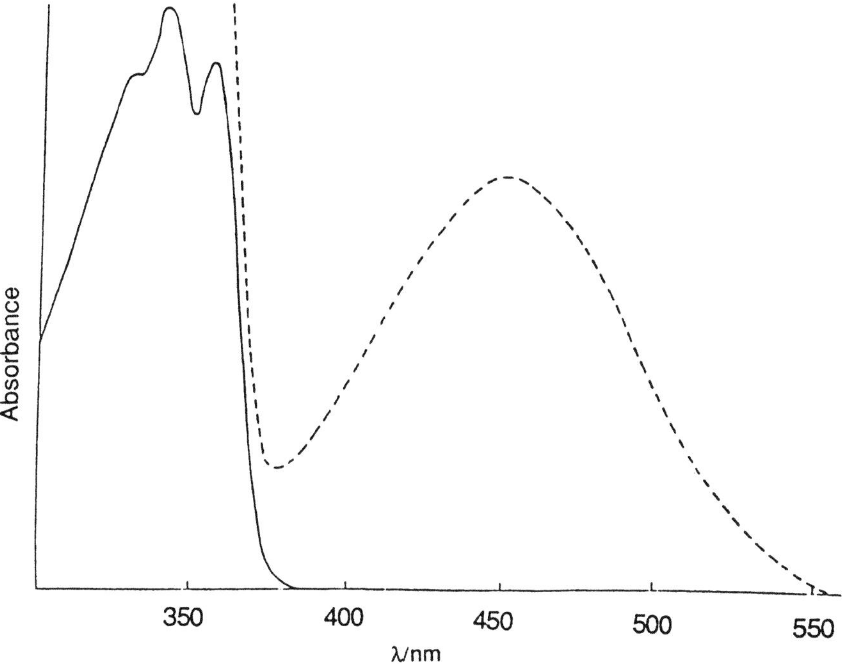

quantum efficiency for coloring. But, equally important, the colored forms (25) and (28) cannot undergo a 1,7-H shift to yield naphthol derivatives (26) and (30), respectively, because the adamantane ring cannot accommodate an endocyclic double bond. The naphthopyrans (24) and (27) are much more resistant to fatigue than the corresponding naphthopyrans with one or two methyl substituents in the 2-position. The colorless naphthopyrans (24) and (27) color to yellow in unfiltered sunlight, indicating a high quantum efficiency for coloring with ultraviolet light and the low quantum efficiency for bleaching of the colored forms with white light. With appropriate modifications, the slow thermal fade can be increased to a rate appropriate for the sunglass application. The spectrum of naphthopyran (24) before and after exposure to unfiltered light from a flash gun (which contains an ultraviolet light component) is shown in Fig. 8.

FIGURE 8. Qualitative spectrum of napthopyran (24) before and after exposure to unfiltered light from a Sunpak 301 flashgun.

If a naphthopyran is required which undergoes a colorless to blue or purple reversible change, this can also be achieved by molecular tailoring. The naphthopyran (30, R = H) with a *p*-diethylaminophenyl group and a methyl group in the 2-positions colors to purple on exposure to ultraviolet light in the presence of white light. The resonance-stabilized colored form (31) does not readily undergo hydrogen abstraction to give a naphthol and

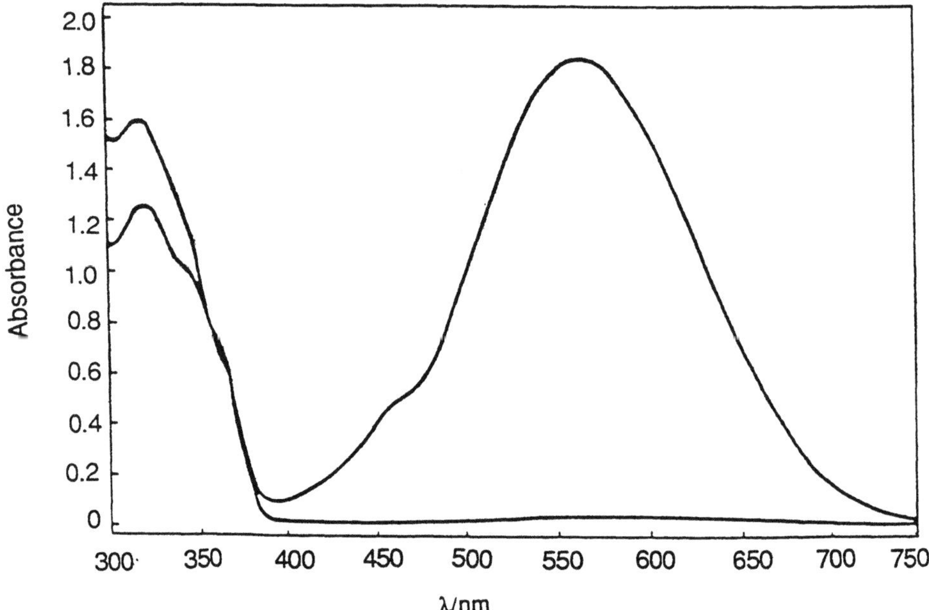

shows a high resistance to fatigue.[21] The spectrum of naphthopyran (30, R = Ph) before and after exposure to unfiltered light from a flash gun is shown in Fig. 9.

I believe photochromic materials suitable for commercial applications are a reality. Academic scientists have the privilege to pursue research, and their reward

FIGURE 9. Qualitative spectrum of napthopyran (30, R = Ph) in chloroform before and after irradiation with unfiltered light from a Sunpak 301 flashgun.

is usually the excitement and pleasure of discovery and invention. Individuals invent but rarely have the resources to innovate. The whole process of productive change that is called innovation carries high financial risk and requires careful strategic planning, team effort, and an unreserved commitment to success, and is usually the company role. The race by major companies to exploit the enormous potential of organic photochromic materials in optical recording, security printing, and variable transmission filters is on, and I predict that the compounds described as photochromics for the future will soon become part of the technology of today.

REFERENCES

1. E. ter Meer, Über Dinitroverbindungen der Fettreihe, *Annalen* **181**, 1–22 (1876).
2. Y. Hirshberg, Reversible formation and eradication of colors by irradiation at low temperatures. A photochemical memory model, *J. Am. Chem. Soc.* **78**, 2304–2312 (1956).
3. G. H. Brown, *Photochromism, Techniques of Chemistry*, Vol. 3, Wiley Interscience, New York (1971).
3. Y. Hirshberg, Reversible formation and eradication of colors by irradiation at low temperatures. A photochemical memory model, *J. Am. Chem. Soc.* **78**, 2304–2312 (1956).
4. H. G. Heller, S. N. Oliver, I. Tomlinson, and J. Whittall, Patent 0246 114 (1987).
5. R. B. Woodward and R. Hoffmann, Conservation of orbital symmetry. *Ac. Chem. Res.* **1**, 17–22 (1968).
6. E. N. Marvell, G. Caple, and B. Schatz, Thermal valence isomerizations. Stereochemistry of the 2,4,6-octatriene to 5,6-dimethyl-1,3-cyclohexadiene ring closure, *Tetrahedron Lett.* 385–389 (1965).
7. H. Stobbe, Die Fulgide, *Annalen* **380**, 1–129 (1911).
8. R. J. Hart and H. G. Heller, Overcrowded molecules—Part 7: Thermal and photochemical reactions of photochromic (E)- and (Z)-benzylidene(diphenylmethylene)succinic anhydrides and imides. *J. Chem. Soc. Perkin Trans.* 1, 1487–1492 (1972).
9. H. G. Heller, P. J. Darcy, S. Patharakorn, R. D. Piggott, and J. Whittall, Photochromic systems—Part 1: Photochemical studies on (E)-2-isopropylidene-3-[1-(3,4,5-trimethoxyphenyl) ethylidene]succinic trihydride and related compounds, *J. Chem. Soc. Perkin Trans.* 1, 315–319 (1986).
10. H. G. Heller and R. M. Megit, Overcrowded molecules—Part 9: Fatigue-free photochromic systems involving (E)-2-isopropylidene-3-(mesitylmethylene)succinic anhydride and N-phenylimide, *J. Chem. Soc. Perkin Trans.* 1, 923–927 (1974). H. G. Heller and H. Gonzenbach, unpublished results.
11. H. G. Heller and J. R. Langan, Photochromic heterocyclic fulgides— Part 3: The use of (E)-α-(2,5-dimethyl-3-furylethylidene (isopropylidene) succinic anhydride as a simple convenient chemical actinometer, *J. Chem. Soc. Perkin Trans. II*, 341 (1981).
12. M. Irie and M. Mohri, Thermally irreversible photochromic systems. Reversible photocyclization of diarylethene derivatives. *J. Org. Chem.* **53**, 803–808 (1988).
13. H. G. Heller, Organic fatigue-resistant photochromic imaging materials, *IEE Proc.* **150**, 209–211 (1983).
14. H. G. Heller, S. A. Harris, W. Johncock, S. N. Oliver, P. J. Strydom, and J. Whittall, Photochromic heterocyclic fulgides. Part 4. The thermal and photochemical reactions of (E)-isopropylidene-[α-2- and 3-(3-thienyl)ethylidene]succinic anhydrides and related compounds, *J. Chem. Soc. Perkin Trans.* 1, 957–961 (1985).
15. Y. Kurita, Y. Yokoyama, T. Goto, T. Inoue, and M. Yokoyama, Fulgides as efficient photochromic compounds. Role of the substituent on furylalkylidene moiety of furyl fulgides in the photoreaction. *Chem. Lett.* 1049–1052 (1988).
16. H. G. Heller, New fatigue-resistant organic photochromic materials, The Royal Society of Chemistry Special Publication No. 60. *Fine Chemicals for the Electronics Industry.* pp. 120–135 (1986).
17. H. G. Heller, B. Helliwell, and D. Wood, unpublished results.
18. H. G. Heller, S. N. Oliver, J. Whittall, W. Johncock, P. J. Darcy, and C. Trundle, Photochromic compounds and their uses in photoreactive lenses, E.P.A. 0 140540 (1984).
19. T. Tanaka, S. Imura, and Y. Kida, Photochromic compounds, E.P.A. 0 316179 (1988).
20. R. S. Becker and J. Kolc, Proof of structure of the coloured photoproducts of chromenes and spiropyrans, *J. Phys. Chem.* **71**, 4045–4047 (1967).

Materials for Sensing Flammable and Toxic Gases

P. T. Moseley

1. INTRODUCTION

A widely accepted definition of a chemical sensor is "a device that allows the transduction of a chemical composition into an electrical signal that can be used for alarm, monitoring, or control purposes." The development of chemical sensors has largely been concerned with the fluid states of matter: Sensors in both the liquid and gas phases are used to monitor industrial and biological quantities and to guard against the development of various hazards arising from sudden accidents or persistent social negligence.

There is currently a large and growing demand for chemical sensors, and a key element in the technology is the exploitation of materials with specific physical and chemical properties. The contribution made by materials science to the development of chemical sensors can be well illustrated by reference to the particular case of gas detection and it is this contribution that is the focus of both this chapter and of the one that follows.

2. GENERAL REQUIREMENTS OF GAS SENSORS

Gas sensors are used in a range of circumstances and for a wide variety of purposes. However, the gases that are most commonly targets of monitoring systems may be grouped in three major categories:

- Oxygen, which is measured in the course of combustion control (internal combustion engines and boilers) and for monitoring breathable atmospheres.
- Flammable gases, which are monitored in an atmosphere that is largely air. The need here is to avoid fire and explosion, typically in the oil, coal, and gas industries.
- Toxic gases, which are also monitored in an atmosphere that is largely air.

P. T. Moseley • Electrical Ceramics and Gas Sensors, AEA Industrial Technology, Harwell Laboratory, Oxfordshire OX11 0RT, U.K.

Toxic gases are a major concern in many heavy industries but also in homes if combustion appliances are incorrectly adjusted.

The present chapter deals with materials aspects of the monitoring of the latter two categories: flammable gases and toxic gases. These two categories are distinct from the first in that the atmosphere in which both flammable and toxic gases are monitored is normally air and an important mode of detecting such gases will be by making use of their reactions with oxygen.

The chapter by Yates deals with the selection of materials for the monitoring of oxygen.

Although most, if not all of the gases of interest can be measured accurately by infrared spectroscopy or mass spectrometry, devices based on these techniques are usually too expensive or too sophisticated for general use. Most gas sensor development effort has, therefore, focused on solid state devices or electrochemical cells because they produce an electrical signal that can be simply related to the concentration of the species of interest. The solid state devices developed to date depend on surface chemical reations of the gas to be measured so that there are many similarities to the behavior of catalysts.

Sensors are required to be reliable, robust, low cost, selective, and sensitive according to application. This latter factor is important to bear in mind because different gases present different hazards. Table 1 shows, for a number of gases, the concentrations at which they become a hazard: a time-weighted average value for toxic gases and a lower explosive limit for flammable gases. Also shown are the *natural* background levels of the various gases and the normal urban values.[1] Table 1 thus defines the range, from base line to hazard level, over which the various gases need to be monitored.

The background levels can rise above the values given for local reasons, for example, methane can rise to 30 parts per million or more near marshes and rivers and carbon monoxide to similar values close to operating internal combustion engines.

TABLE 1. Background and Hazard[a] Concentrations of Various Gases in Air Mixtures

Gas	Natural background (ppm)	Urban background (ppm)	Lower explosive limit (ppm)	8-h time waited average (ppm)
Methane	1.6	2.0	50 000	
Other hydrocarbons	0.1	1.7	20 000	
Carbon monoxide	0.15	2.0		50
Nitrogen dioxide	0.15	0.25		3
Sulfur dioxide	0	0.15		2
Ammonia	0	0.05		25
Hydrogen sulfide	0	0.01		10

[a] Only the most likely hazard is given here, although it is possible for some gases to pose both explosive and toxic problems.

3. SENSOR TYPES

3.1. Catalytic Calorimeter

The device most commonly used for the monitoring of flammable gases in industry is based on the catalytic sensor or pellistor. It consists of a platinum wire supporting a bead of refractory oxide with a surface layer of noble metal catalyst.[2] The sensor is operated at a sufficiently high temperature to ensure rapid combustion of any flammable gas molecules present in an air ambient. The sensor is effectively a microcalorimeter, measuring gas concentration by monitoring changes in the resistance of the wire resulting from temperature increases caused by combustion. The resistance of the wire is constantly evaluated in a bridge circuit with a deactivated pellistor acting as one of the other arms of the bridge.

The sensor is not selective, but in fact this is very useful. The response depends on the product of the concentration of a flammable gas and its heat of combustion. The sensor thus offers a different response to different gases at a single concentration, expressed on a volume percent basis, but gives a universal measure of all flammable gases expressed as a percentage of the lower explosive limit (LEL). This measure of "explosiveness" of an atmosphere is often more important than the identification of the individual components of the atmosphere. Figure 1 shows the variation in the size of the signal versus the concentration of a flammable gas in air.

In some applications, the sensitivity of pellistor sensors can be seriously impaired by contaminants in the atmosphere, which can poison the catalyst. When methane is being detected the sensor can be reversibly poisoned (the term "inhibited" is sometimes used) by halogenated hydrocarbons. More generally, the presence of lead- and phosphorus-containing vapors and silicones leads to an irreversible poisoning which shortens the sensor's life.

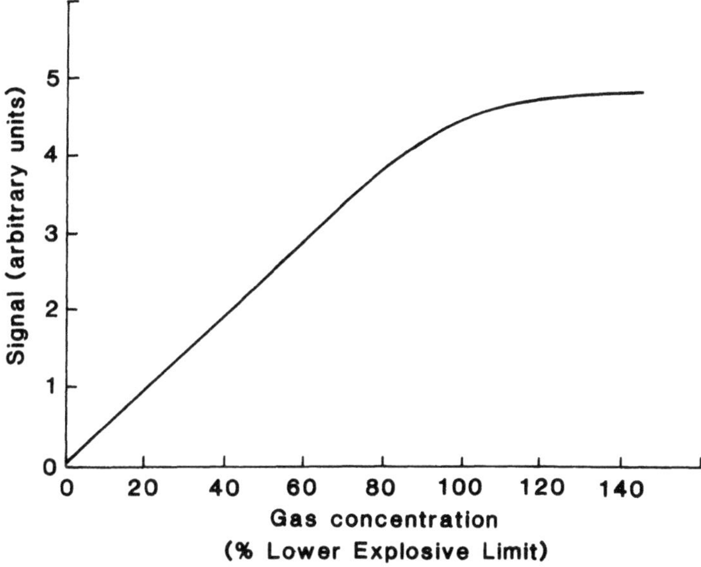

FIGURE 1.   Signal versus concentration for the catalytic sensor.

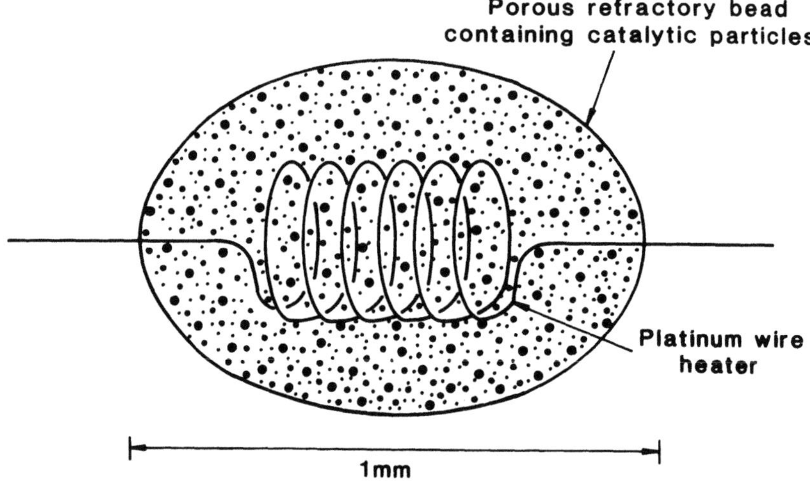

FIGURE 2. Porous catalytic detector for enhanced poison resistance in the detection of flammable gases.

The poison resistance of a catalytic sensor depends generally on four factors[3]: the nature of the poisoning agent, the type of gas being sensed, the choice of precious metal catalyst, and the porosity of the catalytic bead.

This last factor has been exploited successfully in the development of a variety of catalytic detector with a greatly enhanced poison resistance in comparison with the original pellistor.[4] The improved sensors have the platinum wire embedded in a bead comprising an active catalyst dispersed throughout a porous oxide (often γ-alumina) substrate (Fig. 2). The origins of the improved poison resistance of this sensor are threefold:

- The precious metal catalyst is selected to have a high intrinsic resistance to poisoning. When this selection is made consideration is given to the specific susceptibility of catalytic sites to adsorption of the poison in question, to poison-induced surface reconstruction, and to compound formation.
- When the total area of catalytic surface available is sufficiently high that the rate of reaction is limited by diffusion of reactants to the surface there is effectively "spare" surface available and the observed rate of poisoning effects is reduced.
- The third effect of using a porous bead is that the pore size can be selected so as to limit access of the poison molecules to the catalyst while allowing free access of the reactants. This discrimination is particularly effective against high molecular weight poisons.

Catalytic sensors are used in many industrial applications such as in coal mines and in oil and gas production. However, the need for frequent recalibration and their relatively high operating temperature are significant shortcomings.

Future materials development for catalytic sensors is likely to focus on improving long-term stability and reducing the temperature of operation by the design of more effective catalytic systems.

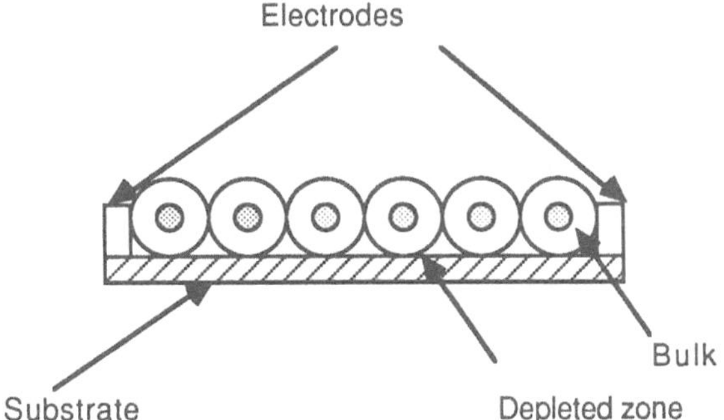

FIGURE 3.   Schematic of a porous thick film of semiconducting oxide.

3.2. Semiconductor Gas Sensors

An alternative means of detecting relatively small (a few parts per million up to a few percent) concentrations of gases in air is by monitoring the effect of their reactions at the surface of a semiconducting material.[6] Semiconducting gas sensors are simply resistors that change their value with changing composition of the atmosphere. They are normally evaluated in simple dc circuits and thus offer the prospect of low cost when manufactured in very large numbers.

3.2.1. Inorganic Devices

The first type of semiconductor gas sensor to reach the market was based on tin dioxide and operates at a temperature (between 300° and 500°C) where no bulk change in stoichiometry occurs but reactions that alter the measured resistance of the material take place readily on the material's surface.

There are two extreme configurations of microstructure that have been used. The first case, shown in Fig. 3, is of a relatively thick porous ceramic sensing element, which may take the form of a bead or a thick layer. The second is shown in Fig. 4 and is a thin film between a set of screen-printed interdigitated electrodes on a resistive substrate.

In both cases a depletion zone is generated by the formation of negatively charged oxygen ions on the surface, which abstract electrons from the near surface

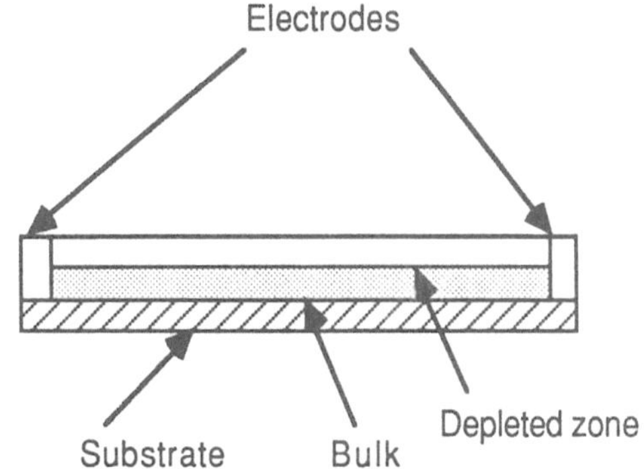

FIGURE 4.   Schematic of a thin film of semiconducting oxide.

region of the oxide. Since these oxides are n-type semiconductors this process raises the apparent resistivity of both forms of device to a level much higher than it would be in an oxygen-free atmosphere. The process of trapping charge at the oxide surface generates a large potential barrier between the gas space and the bulk of the material. This is important because the conductivity, σ, of the material between the electrodes depends on the number of charge carriers, n, as well as their mobility, μ, and their charge, Ze.

$$\sigma = nZe\mu \tag{1}$$

In order for gas reactions at the oxide surface to manifest themselves as changes in the conductance between the electrodes there will have to be a transfer of charge between the surface states and the conduction band of the oxide and the surface potential barrier will exercise control over this transfer. The situation is shown schematically in Fig. 5,[7] in which the upper figure shows the electron energy as a function of distance.

The size of the potential barrier is related to the amount of charge trapped on the surface, N_t, and to the density of electron donors in the bulk, N_d. The potential energy difference between an electron in the space charge region and an electron in the deep bulk is defined in terms of the barrier voltage $V(x)$ as $eV(x)E_c(x) - E_{cb}$, where $V(x) = \phi_b - \phi(x)$ (ϕ being an electric potential). The (positive) charge density in the space-charge region is eN_d. Thus Poisson's equation may be written (in one dimension) as

$$\frac{d^2 V(x)}{dx^2} = -\frac{eN_d}{\varepsilon\varepsilon_0} \tag{2}$$

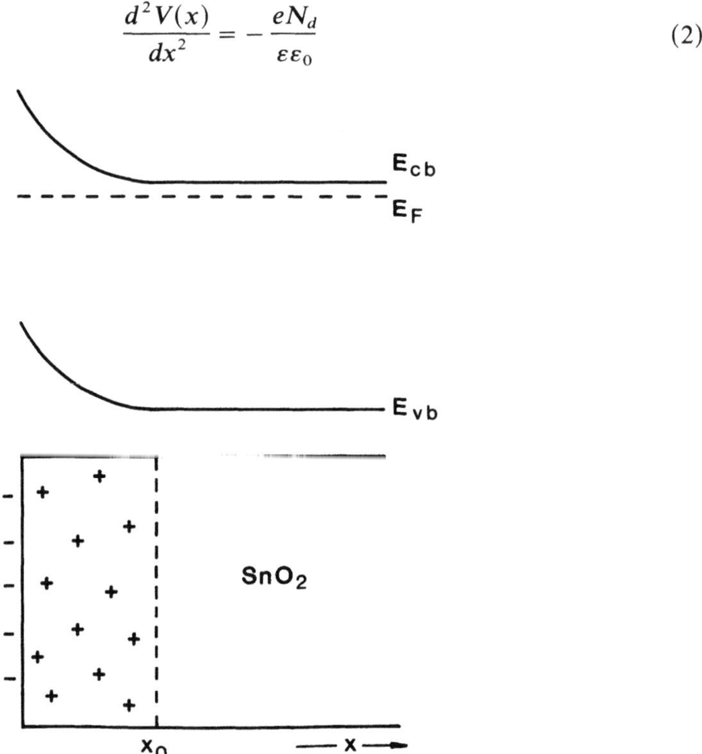

FIGURE 5.   Schematic of the tin dioxide surface in air.

Integrating (2) with the boundary condition $dV/dx = 0$ at $x = x_0$ (where x_0 is the limit of the space-charge region), we obtain

$$\frac{dV(x)}{dx} = -\frac{eN_d}{\varepsilon\varepsilon_0}(x - x_0) \qquad (0 < x < x_0) \tag{3}$$

Integrating again, with the boundary condition $V = 0$ at $x = 0$, yields

$$V(x) = -\frac{eN_d}{\varepsilon\varepsilon_0}\left[\left(\frac{x^2}{2} - xx_0\right) + \frac{x_0^2}{2}\right] \tag{4}$$

Using the further boundary condition $V = V_s$ at $x = 0$ and the charge neutrality condition $N_t = N_d x_0$ (where N_t is the total barrier charge per unit area) yields the well-known Schottky relation:

$$V_s = \frac{-eN_t^2}{2\varepsilon\varepsilon_0 N_d} \tag{5}$$

The height of the surface potential energy barrier (for electrons) can now be expressed as

$$-eV_s = +\frac{e^2 N_t^2}{2\varepsilon\varepsilon_0 N_d}$$

The energy barrier to changes in the charge carrier density of the bulk of the material following chemisorption processes at the material's surface is proportional to the square of the surface charge.

The maximum response of tin dioxide to most reducing gases lies in the range 300–500°C (Fig. 6) and the magnitude of the response is a logarithmic function of gas concentration. As in the case of the pellistor a simple circuit is sufficient for evaluating the response of this type of sensor.

Tin dioxide sensors characteristically respond to a wide range of reducing gases (Fig. 7a). For many applications, however, it is important that the sensor be selective between gases, and a great deal of development effort is aimed at improving the specificity of semiconducting gas sensors without loss of sensitivity or simplicity.

There are several ways in which the selectivity can be enhanced:

1. Operating Temperature. The magnitude of the response reaches a maximum value (e.g., Fig 6) at a temperature that is characteristic of the rate of reaction of a particular gas on the oxide surface. It is thus possible to obtain a degree of selectivity between reactive gases, such as hydrogen sulfide, and relatively inert gases, such as methane, by judicious choice of temperature.

2. Material Porosity. By constructing the sensing element of a thick layer of porous material it is possible to arrange for reactive gases to be oxidized in the outer layers so that more stable gases are selectively sensed by electrodes sited in the innermost region.

3. Alternative Materials. Among semiconducting oxides a very wide range of electronic and catalytic characteristics is found. Since these are properties on which

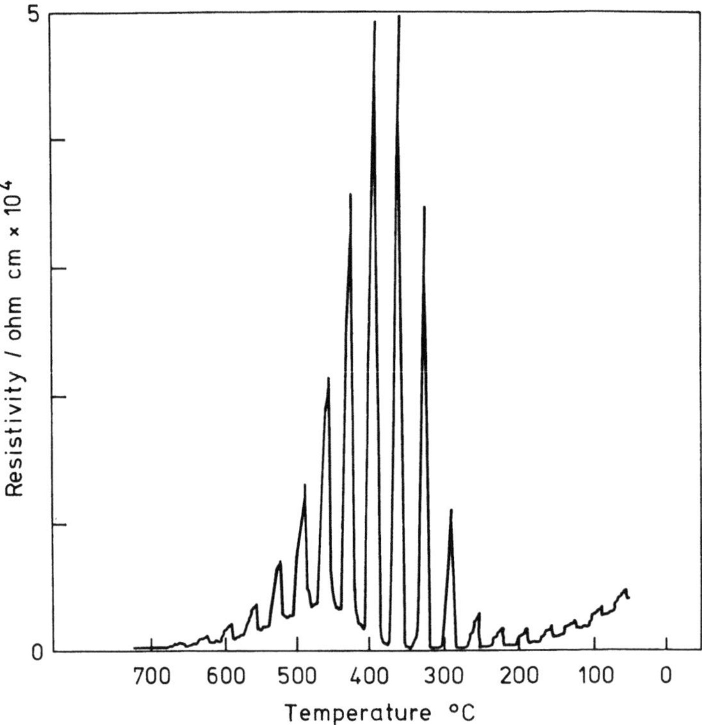

FIGURE 6.   Temperature dependence of the gas response of tin dioxide. Reproduced by permission of the Royal Society of Chemistry.

gas reactions depend it is not surprising that some of these materials (Fig. 7b) exhibit a far more selective pattern of gas responses than tin dioxide (Fig. 7a). In general the sign of the resistivity response depends on whether the semiconductor is an n- or p-type and on the nature of the gas reaction taking place on the surface.

   4. Dopants, Catalysts and Coatings. Because the response of resistance-modulating gas sensors originates in the near-surface region, the addition of second phase materials produces profound effects on both the sensitivity and the temperature of peak responses. Dopants, catalysts, and coatings have all been investigated and utilized in devices. Surface catalyst additions to tin dioxide[9] can induce responses to hydrogen at 100°C or below (Fig. 8). There is a strong parallel between the gas sensor function and the heterogeneous catalysis of gas reactions and there are several precious metals that modify the gas sensor function of tin dioxide.

   5. Use of Single Crystals. In consideration of the mode of operation of a polycrystalline ceramic gas sensor account must be taken of reactions taking place both at grain boundaries and at the surface of individual crystallites. One degree of simplification can be achieved by using a single crystal of semiconducting oxide instead of a ceramic, and this choice can affect the specificity of the resulting sensor. In a series of tests with gas mixtures in air,[10] a gas sensor based on a single crystal of zinc oxide showed resistance responses to carbon monoxide and hydrogen but not to methane. The single crystal sensor is thus more selective than are sensors based on polycrystalline zinc oxide which respond to all three gases.

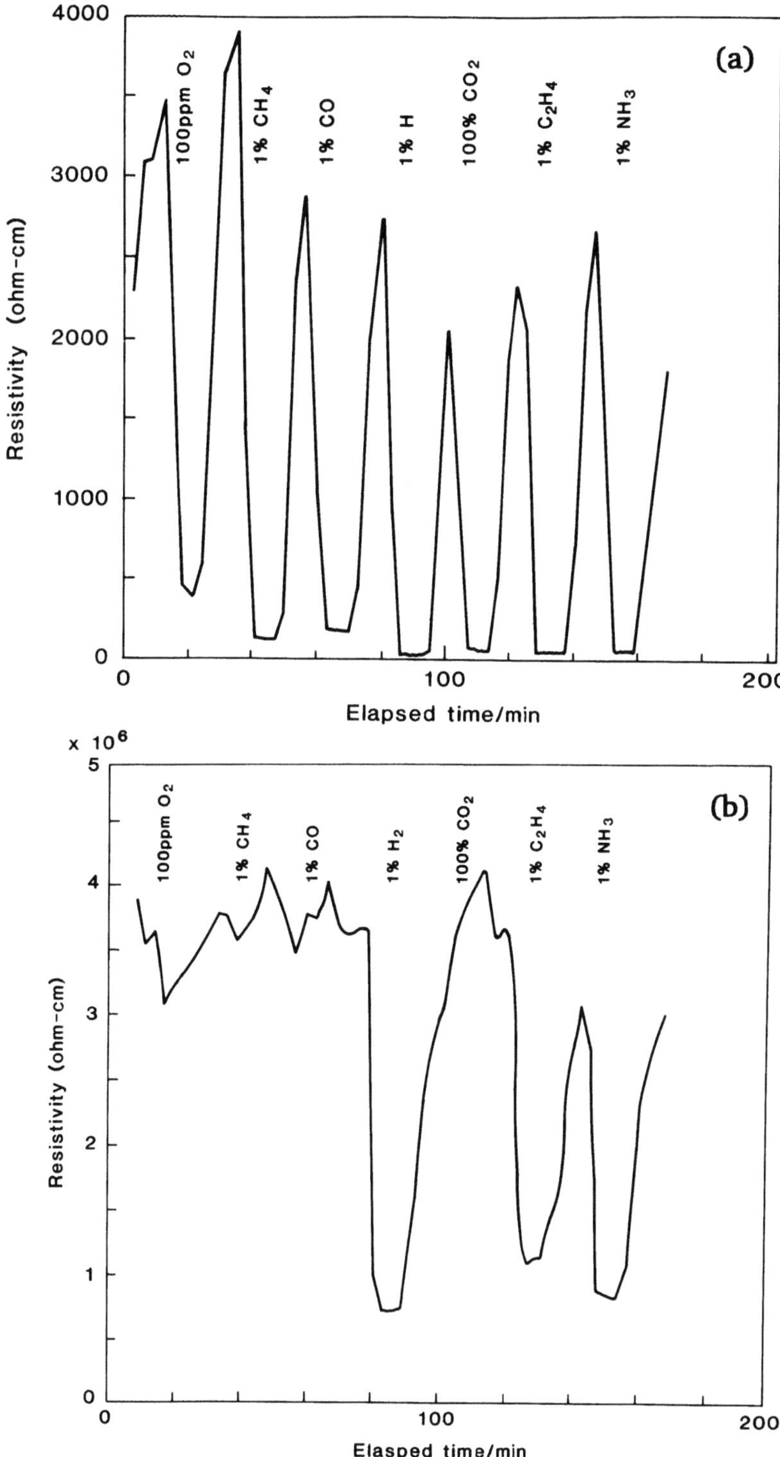

FIGURE 7.   Semiconductor exposed to a variety of gases in sequence. (a) The nonspecific response typical of tin dioxide to a range of reducing gases. (b) The more specific response of titanium niobate.

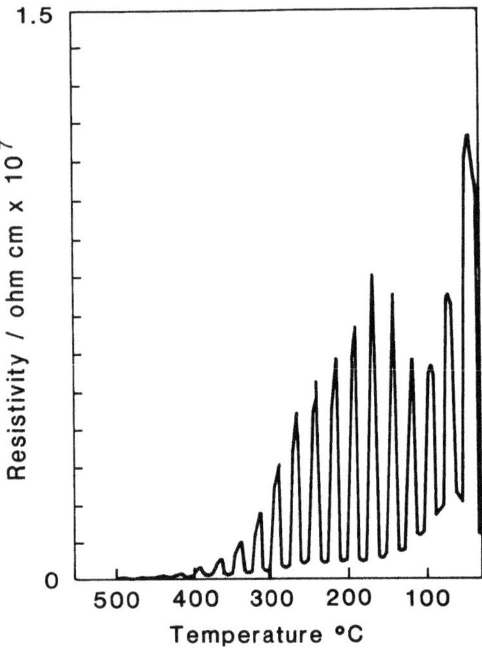

FIGURE 8. The response of a SnO_2/Pt sensor to hydrogen in air as a function of temperature. The resistance modulation shown is a consequence of the periodic (fifteen minutes out of every thirty) introduction of 1% hydrogen into an air ambient.

3.2.2. Organic Devices

A growing number of organic materials has been proposed for use in semiconducting gas sensors. Films of metal phthalocyanines can be readily set down across electrodes on alumina substrates by vacuum evaporation to form devices that are extremely sensitive to nitrogen dioxide.[10] At operating temperatures in the range 100–170°C the sensors can detect nitrogen dioxide at concentrations from 1 part per billion to 10 parts per million in air (Fig. 9). The resistance responses of these sensors are consistent with a p-type conduction mechanism in the phthalocyanines. The chemisorption of strongly electron-accepting gases on the surface leads to an increase in the hole concentration in the solid and hence to an increase in conductivity. The sensors are rather specific for strongly oxidizing gases. Thus chlorine produces a significant response, but there is little interference from hydrogen or carbon monoxide.

Other heterocyclic metal complexes are likely to prove useful for specific tasks, as are polymers such as polypyrrole. Pyrrole can be electropolymerized from a solution containing a suitable counterion to produce a p-type semiconducting polymer that is sensitive to a number of toxic gases at room temperature. Figure 10 shows the response of tetrafluoroborate-doped polypyrrole to three concentrations of ammonia in air.[12] Polypyrrole ammonia sensors can be made rather insensitive to interference from carbon monoxide and hydrocarbons. Operation at room temperature represents an advantage in terms of reduced power drain compared with oxide semiconductor sensors, but, since gas reaction within the sensor is a thermally activated process, response and recovery times are longer than for heated sensors.

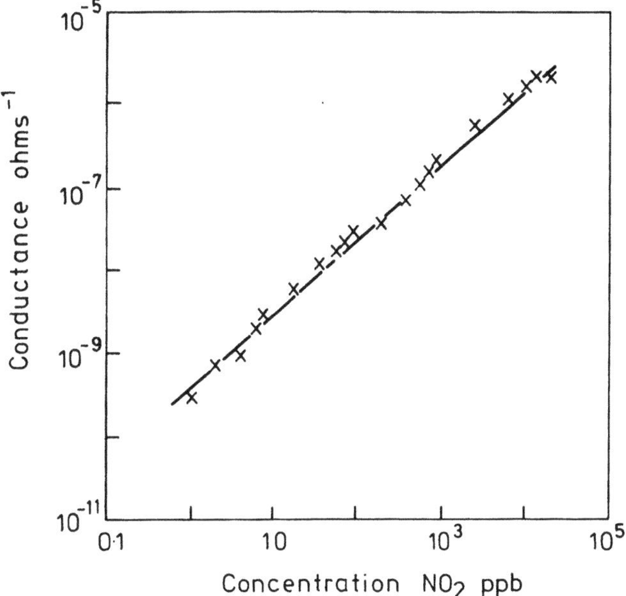

FIGURE 9.   Concentration dependence of the response of lead phthalocyanine to nitrogen dioxide.[11] Reproduced by permission of Elsevier Sequoia SA.

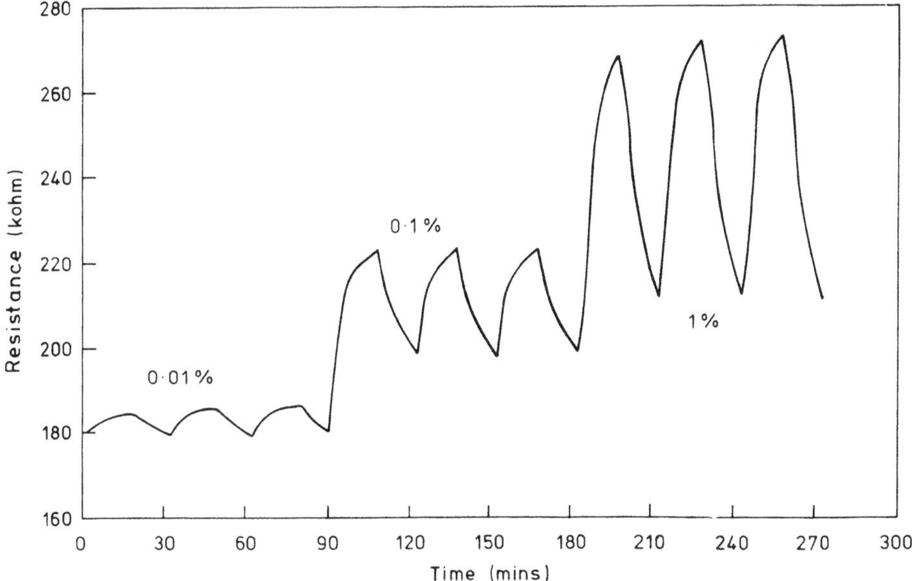

FIGURE 10.   Response of a polypyrrole sensor to three concentrations of ammonia in air.[12] Reproduced by permission of the Royal Society of Chemistry.

3.2.3. Model Response Mechanisms

There are clearly numerous materials, microstructure, and temperature options for consideration in the course of sensor design, and a review of the reactions that could possibly contribute to the measured signal is an important part of the prediciton of a sensor's likely characteristics. A number of different types of gas reaction could, in principle, give rise to changes in the carrier density in the semoconductor.

Case 1: Reaction with Oxygen Species Adsorbed on the Surface. The surface conductivity of semiconducting oxides such as tin dioxide in air is dominated by the presence of surface oxygen ions such as O_2^-.[6] Reactions of such species with reducing gases such as carbon monoxide or hydrocarbons generally form neutral products which are desorbed. The charge from the trapped ions is then obliged to return to the conduction band of the oxide, resulting in a marked increase in conductivity. An example of such a reaction would be

$$2CO + O_2^- \rightarrow 2CO_2 + e^-$$

This is the mechanism most usually invoked to account for the behavior of tin dioxide sensors with reducing gases.

Case 2: Direct Reaction. Oxidising gases such as chlorine and nitrogen dioxide clearly cannot take part in reactions similar to that shown in Case 1 but nevertheless do give responses on tin dioxide. Their reaction is likely to be similar to the one involved in the formation of surface oxygen ions. Such gases can be ionosorbed on the surface of semiconductors to form negative ions, by abstracting electrons from the material. This process causes an increase in the resistance of *n*-type conductors and a decrease in the resistance of *p*-type conductors. An example of such a reaction would be

$$Cl_2 + 2e^- \rightarrow 2Cl_s^-$$

where Cl_s^- is a surface chlorine ion.

The responses of the organic semiconductor gas sensors are likely to be of this type.

Case 3: Detection via the Products of Combustion. A number of gases can combust on a hot oxide surface in air to form product gases which have a different electronic reaction on the surface from that of the original target gas. An example would be ammonia, which, under normal circumstances, would produce a resistance decrease according to a reaction scheme similar to that shown in Case 1, but following oxidation on the surface, will produce nitric oxide, with subsequent formation of nitrogen dioxide, which produces a resistance increase according to a reaction scheme such as that outlined in Case 2. As a result there are a number of confusing reports in the literature of a variable response of semiconductors to ammonia in air.

Case 4: Responses Transmitted Through Metals. Many semiconductor gas sensors are provided with a second phase of precious metal particles which was originally employed in an attempt to encourage gas sensitive reactions to take place at lower operating temperatures. If the metal has a high work function then it can become involved in a significant electronic interaction with the semiconducting oxide. The result is that the oxide becomes depleted of electrons once more but that the effect is no longer dependent upon the presence of oxygen. Instead the process can be encouraged to reverse by the direct adsorption of gases on the *metal.* This sort of process gives rise to responses at room temperature such as that shown in Fig. 8.

4. SUMMARY

It is clear that any one of several potential reactions can be exploited as a means of sensing gases. Both the catalytic detector and the original tin dioxide sensor operate at elevated temperatures in air-based atmospheres where there is a need to detect and monitor hazard levels of explosive and toxic gases. In these circumstances the ready availability of oxygen provides the key to reaction mechanisms that are important in the detection process.

There are other circumstances in which the direct interaction of gas molecules with the detector surface can result in a usable signal.

The sensors described above are simple, generally of low cost, and are sold in millions. Their use is expected to grow in both industrial and domestic applications. Enhanced capabilities of future generations of sensors will depend on the development of optimized materials and on key fabrication techniques. As with integrated circuit technology, microfabrication of gas sensors offers a significant unit cost reduction, a reduced size, and improved reproducibility in comparison with bulk devices, although individual sensors, no matter how they are fabricated, may still be used discretely. Individual testing and calibration may also still be required.

Once silicon-based microfabrication techniques become standard, there will be an advantage in placing multiple sensors on the same chip to permit monitoring of multicomponent gases. Nevertheless, although reference is sometimes made to the concept of the "artificial nose" in the technical literature, it will be clear from the foregoing that such a degree of sophistication is not yet a commercial proposition.

REFERENCES

1. P. L. Hanst, in *Fourier Transform Infrared Spectroscopy: Applications to Chemical Systems* Vol. II (J. R. Ferraro and L. J. Basile, eds.), Academic, New York (1979).
2. A. R. Baker, Electrically heatable filaments, U.K. Patent 892,530 (1962).
3. S. J. Gentry and P. T. Walsh, Poison-resistant catalytic flammable-gas sensing elements, *Sensors Actuators* **5**, 239–251 (1984).
4. D. W. Dabill, S. J. Gentry, N. W. Hurst, A. Jones, and P. T. Walsh, Catalytic combustible Gas Sensors, U.K. Patent 2,083, 630 (1982).
5. L. L. Hegedus and R. W. McCabe, Catalyst poisoning, *Cat. Rev. Sci. Eng.* **23**, 377 (1981).
6. D. E. Williams, in *Solid State Gas Sensors* (P. T. Moseley and B. C. Tofield, eds.), pp. 71–123, Adam Hilger, Bristol and Philadelphia (1987).
7. P. Romppainen and V. Lantto, The effect of microstructure on the height of potential barriers in porous tin dioxide gas sensors, *J. Appl. Phys.* **63**, 5159–5165 (1988).

8. L. V. Azaroff and J. J. Brophy, *Electronic Processes in Materials*, McGraw-Hill, New York (1963).

9. J. F. McAleer, P. T. Moseley, J. O. W. Norris, D. E. Williams, and B. C. Tofield, Tin dioxide gas sensors. Part 2. The role of surface additives, *J. Chem. Soc. Faraday Trans. 1* **84**, 441–457 (1988).

10. B. Bott, T. A. Jones, and B. Mann, The detection and measurement of CO using ZnO single crystals, *Sensors Actuators* **5**, 65–73 (1984).

11. B. Bott and T. A. Jones, A highly sensitive NO_2 sensor based on electrical conductivity changes in phthalocyanine films, *Sensors Actuators* **5**, 43–53 (1984).

12. J. J. Miasik, A. Hooper, and B. C. Tofield, Conducting polymer gas sensors, *J. Chem. Soc. Faraday Trans. 1* **82**, 1117–1126 (1986).

33

Exploiting Semiconducting Oxides for Automotive Exhaust Gas Oxygen Sensors

A. Yates

1. INTRODUCTION

The use of gas sensors for monitoring and controlling combustion processes to improve efficiency and reduce emissions has long been established.[1] Generally to achieve control of combustion processes the air-to-fuel ratio of the process needs to be determined. As the air-to-fuel ratio is related to the oxygen concentration in the exhaust gases, oxygen has become a critical gas for sensors to measure.

Traditionally oxygen sensors were based on electrochemical cells made of stabilized zirconia. However, with developments in semiconducting materials, semiconductor sensors offer alternatives that have significant advantages in specific applications. Devices based on these materials are becoming increasingly common. This chapter aims to briefly describe the principles of operation of both electrochemical and semiconductor oxygen sensors with reference to stabilized zirconia, titania and recently developed barium tantalum ferrate materials. The limitations and advantages of each type of sensor and material will be identified with reference to the use of the materials for internal combustion engine fuel control.

2. PRINCIPLES OF OPERATION

2.1. Electrochemical Oxygen Sensors

An electrochemical sensor operates by detecting a difference in chemical potential of a given compound in two regions that are physically separated by the sensor. The chemical potential of a gaseous species is related to its partial pressure. Furthermore, the device can only fulfil this sensing role if the species of interest is

---

A. Yates • Lucas Automotive Limited, Advanced Engineering Centre, Dog Kennel Lane, Shirley, Solihull, West Midlands B90 4JJ, U.K.

readily transformed by chemical reaction into some other species, which will migrate through solid regions of the sensor.

As zirconia (ZrO_2) is a material commonly exploited for these types of sensors, it will be used to exemplify the principle of operation. Several requirements need to be met by the material. It is essential that the zirconia is not permeable to gaseous oxygen so that any difference in oxygen chemical potential can only be equalized by migration of oxygen ions through the zirconia lattice. It is also necessary to have both surfaces of the zirconia coated in a porous coating of a material which can readily donate or accept electrons. Normally this is a metal such as platinum. Finally it is necessary for the zirconia to have been doped with a metal oxide in which the metal is of a lower valency than zirconium. Typically oxides of calcium, magnesium, or yttrium are used. The effect of this doping is twofold: Firstly it stabilizes the cubic phase of zirconia, which is normally only stable at temperatures above 2200°C. As a consequence doped zirconia is often referred to as stabilized zirconia. Secondly the dopant produces a large number of oxygen vacancies in the lattice. This greatly improves the diffusion of oxygen ions in the solid.

When the oxygen partial pressures on the two sides of the zirconia ceramic are not equal, electrochemical reactions take place. Gaseous oxygen is absorbed on the surface of the zirconia exposed to the higher oxygen concentration. The metallic electrode layer on this surface donates electrons to produce oxygen ions, which combine with oxygen vacancies in the lattice of the zirconia.

These reactions lead to a rapid redistribution of the oxygen vacancies throughout the ceramic lattice, resulting in the reverse processes taking place at the other ceramic/electrode surface (i.e., oxygen is expelled from the lattice).

If the two electrodes on either side of the zirconia remain isolated, charge will build up on them: positive charge on the surface in contact with the higher oxygen partial pressure, negative charge on the other. The equilibrium electromotive force (EMF) is determined by the oxygen chemical potential gradient. The voltage measured is therefore related to the difference in oxygen partial pressure across the zirconia sensor. (An alternative view is that the equilibrium results from the balance of the ion diffusion current by drift in the resultant field.)

It is important to note that for the device to be useful the zirconia must have a high level of oxygen ion conductivity. This is only reached at about 300°C, and as such represents the first constraint upon the usefulness of this device. In addition, for a quantitative evaluation of the oxygen partial pressure to be made, it is necessary for the ceramic to be an ionic conductor only. For stabilized zirconia, this criterion is known to be fulfilled between 300 and 900°C. Under these conditions, and in the presence of equal total gas pressures on both sides of the ceramic, the thermodynamics of the process can be simplified to yield the expression shown in Eq. (1), which results from application of the Nernst equation:

$$\text{EMF} = \frac{RT}{nF} \ln\left(\frac{p_1}{p_2}\right) \tag{1}$$

where EMF is the voltage generated by the cell, R is the universal gas constant, F is the Faraday constant, n represents the number of electrons involved in the transfer across the cell of one molecule of gas (4 in this case), T is the absolute temperature,

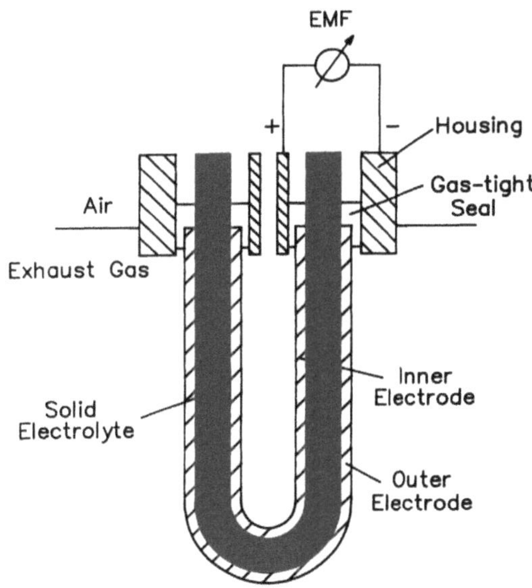

FIGURE 1. Schematic diagram of a solid electrolyte gas sensor.

p_1 is the oxygen partial pressure of the test environment, and p_2 is the oxygen partial pressure of the reference chamber.

Consequently, if the oxygen concentration on one side of the cell is constant and known (i.e., air) then the oxygen concentration of the test environment can be calculated from the measurement of the voltage produced at the temperature of operation.

The basic requirement of physically separating the two gaseous media has resulted in numerous designs, of which the use of a closed end tube of stabilized zirconia with the inside of the tube open to air (Fig. 1) is most commonly employed.[2,3] Generally platinum is coated onto both surfaces acting as electrode, and catalyst. Devices of this type are successfully being utilized in the analysis of flue gases from various combustion processes,[4,5] detection of oxygen in molten metals,[6,7] and in the control of fueling in automotive applications by monitoring the oxygen content of the exhaust gases (which will be discussed in more detail later).

2.2. Semiconducting Oxide Oxygen Sensors

It is important to distinguish between the two types of semiconducting oxide sensors. One type, which has been known for many years for the detection of reducing and toxic gases,[8] operates at comparatively low temperatures between 200 and 500°C. This class of sensor depends on the chemisorption of gaseous species onto the sensor surface. This process results in a change in the charge carrier density in the oxide lattice, which can be measured. The other type of sensor operates at much higher temperatures (up to 900°C) where chemisorption is negligible and changes in charge carriers result from changes within the bulk of the material. It is this latter type of semiconductor sensor that is more commonly employed to sense oxygen that will be discussed here.

A metal oxide must exhibit a range of nonstoichiometry for it to behave as an oxygen sensor. This nonstoichiometry can result from one of the following defects[9]: oxygen vacancies, metal interstitials, metal vacancies, or interstitial oxygen atoms. The first two defects result in oxygen deficient (excess metal) oxides with respect to the stoichiometric composition, the latter two, metal deficient (excess oxygen) oxides. A general equation as shown in Eq. (2) can be written. This shows that at equilibrium, by the law of mass action, the composition of the metal oxide (MO_x) depends on the partial pressure of oxygen in the gas phase:

$$MO_x \rightleftharpoons MO_{x-y} + (y/2)O_2 \qquad (2)$$

A change in composition of the oxide will cause a change in the average oxidation state of the metal ion to maintain electroneutrality in the system. Negative electrons or positive holes are produced, which, if not strongly trapped at particular atomic sites will be mobile causing a change in electrical conductivity. The relationship of this change in conductivity with oxygen partial pressure is also dependent on temperature and activation energy of the semiconductor material:

$$\sigma = A \exp(-E_a/kT)[p(O_2)]^{\pm 1/n} \qquad (3)$$

where σ is the electrical conductivity, A is a constant, E_a is the activation energy for conduction in the solid, k is the Boltzmann constant, T is the absolute temperature, n is determined by the nature of the chemical equilibrium between oxygen and the sample, and $p(O_2)$ is the oxygen partial pressure.

Titania (TiO_2) is a material that can be used as an oxygen sensor of this type. It is an n-type semiconductor (electrical conductivity decreases with increasing oxygen partial pressure) and is oxygen deficient when nonstoichiometric. The defect structure model is unclear, as although doubly charged oxygen vacancies are proposed to predominate, aliovalent impurities in the structure have significant impact.[10] Impurities in the structure even at very low levels (70 ppm) can cause titania to go through an n- to p-type transition, with increasing oxygen concentration. Also they influence markedly the sensitivity of the conductivity to changes in oxygen partial pressure [n in Eq. (3)], which have been measured between -4 and -6. Most of these effects are caused by impurities with lower valency than titanium (i.e., less than 4) such as aluminum, and doping with impurities with higher valencies such as niobium have been shown to improve the devices.[11]

The activation energy for conduction (E_a) for titania is around 1.6 eV, which is considered high. This has a significant impact on the changes in conductivity caused by temperature fluctuations. The conductivity in air at 850°C for sintered polycrystalline titania is between 10^{-3} and 10^{-4} (Ω cm)$^{-1}$, depending on purity.

Barium tantalum ferrates ($BaFe_{1-y}Ta_yO_{3-x}$ where $0 < y < 1/2$) represent another group of semiconducting oxide materials that have been developed for use as oxygen sensors (by the Materials Development Division A.E.R.E. Harwell and Lucas Advanced Engineering Centre). A range of materials exist that have properties that are particularly interesting in specific applications.

These materials are p-type semiconductors (conductivity increases with increased oxygen partial pressure) and are oxygen deficient when nonstoichiometric.[12] The defect structure model appears to be one of interacting

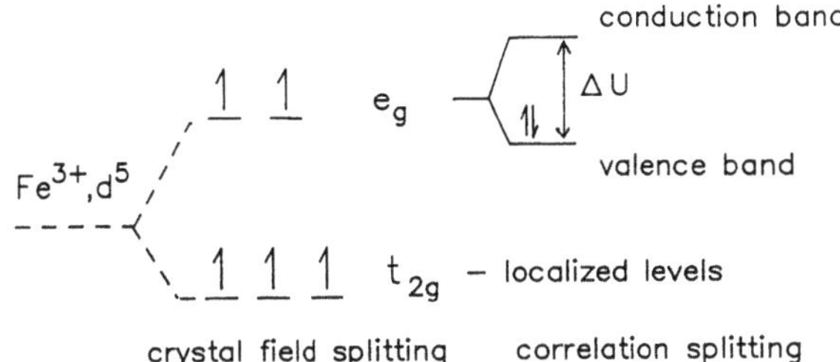

FIGURE 2. Bonding scheme showing the splitting of the energy levels for iron d orbitals and electron population.[20]

vacancies, and conductivity changes are related to variation in stoichiometry resulting from changes in the Fe^{3+}/Fe^{4+} ratio.

The conductivity is due to holes in the valence band and the change with temperature and oxygen pressure is due to a variation in the mobility of the carriers and their numbers. This can be rationalized by drawing a bonding scheme for the iron d electrons (Fig. 2). Conduction is supposed to occur via "narrow" d bands. The splitting of the e_g level into two bands gives a band gap. This is presumed to be the band gap of the fully reduced forms ($\Delta U = 2E_a$). Oxidation to Fe^{4+} gives a hole in the valance band with mobility restricted by the narrowness of the band and dependant on temperature. The energy ΔU decreases as the counter cation (Ba, Sr, Ca, La) becomes smaller and more highly charged. If ΔU is too small then thermal activation of electrons to the conduction band swamps out any noticeable effect on the number of carriers. Therefore barium or strontium is the preferred cation.

The interesting property of this group of materials is that above 650°C these materials have extremely low activation energies of conduction (E_a in some instances zero), which makes the change in conductivity almost insensitive to temperature fluctuations. The conductivities of the materials change with oxygen partial pressure, with values of n [in Eq. (3)] being measured as $+5$ and being unaffected by small levels of impurities in the lattices. The conductivity of barium tantalum ferrate in air at 850°C is typically 0.1–1 $(\Omega\,cm)^{-1}$ depending on the value for Y in the composition $BaFe_{1-y}Ta_yO_{3-x}$.

The construction of devices using semiconducting metal oxide sensors, unlike the electrochemical type, does not need to allow for the use of a reference atmosphere. The basic requirements for this type of sensor are high porosity, to aid gas diffusion, and inert electrodes for measuring changes in conductivity. Additionally, a heater and temperature measurement method (such as a platinum resistance thermometer (PRT) or thermocouple) is usually employed to allow the temperature of the device to be controlled. Thick films or pellets of the selected semiconducting oxides with screen printed or embedded electrodes are commonly used as the sensor element.

3. APPLICATION OF OXYGEN SENSORS IN AUTOMOTIVE EXHAUST GAS SENSING

There is a requirement in the automotive sector to control the combustion process of internal combustion engines to reduce emissions and improve fuel economy. One method by which this can be achieved is to measure and control the air/fuel ratio (AFR) being supplied to the carburetor.

If the exhaust gas emission from an internal combustion engine are plotted against air/fuel ratio the concentrations will typically look like those shown in Fig. 3. It can be seen from this plot that oxygen concentrations in the exhaust gases give the most unambiguous measure of the AFR. Therefore the sensing of oxygen concentrations in the exhaust is an effective way to provide control.

Conventionally, electrochemical cells based on zirconia have been used for controlling the fueling near the stoichiometric (chemically correct) mixture (AFR approximately 14:1). A catalyst made of noble metals located further down the exhaust stream further reduces emissions by reducing oxides of nitrogen (NO_x) to nitrogen and oxidizing carbon monoxide and hydrocarbons to water and carbon dioxide (so-called three-way catalysts). However, this approach has some serious limitations: Firstly the sensor can only effectively be used to control at the stoichiometric mixture, where equilibrium oxygen concentrations change dramatically due to reactions taking place on the active surface of the sensor.[13,14] This necessitates the use of a three-way catalyst to reduce emissions. This in turn makes the system costly. Secondly, contamination seriously impairs the sensor output and catalyst efficiency.

Exhaust gas oxygen sensors which could be used in the excess air (lean) region of combustion by providing a measurement over a wide range of air/fuel ratios are required. This would eliminate the need for expensive three-way catalysts, as emissions are significantly reduced if the engine runs sufficiently lean. At the same time they would provide flexibility in the control philosophies.

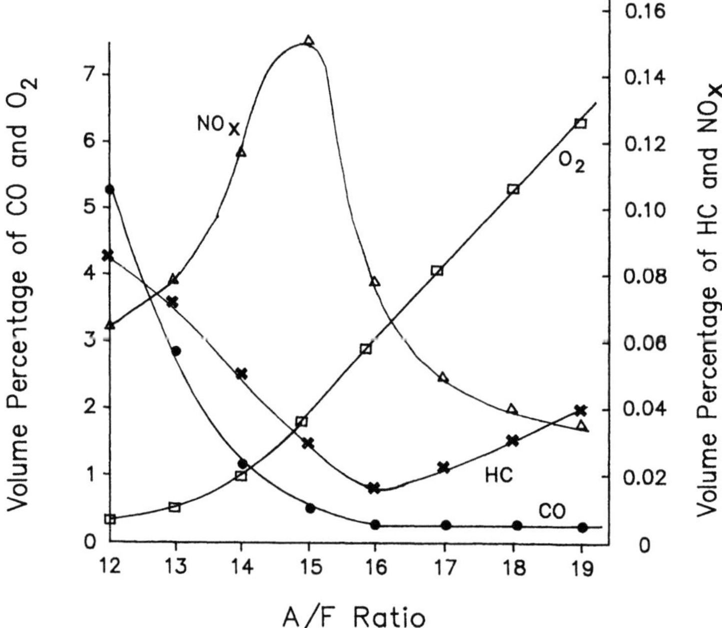

FIGURE 3. Automotive exhaust gas composition versus air/fuel ratio.

Apart from being able to detect the level of oxygen in the exhaust, the sensor also needs to respond rapidly to changes in the exhaust oxygen concentration to allow changes in the fueling to be made at the correct time. This has implications on the sensor material, the mechanical design, and the location. The sensor needs to be sited as close as possible to the exhaust manifold to reduce delays caused by gas transport. This means the sensor will be operating in gases fluctuating in temperatures between 350 and 850°C. The sensor material therefore must be capable of functioning at the maximum exhaust temperature. The sensor will also be exposed to high-velocity particulates, and so mechanical protection is required to prevent erosion. This must be achieved without significantly impairing the gas transport to or through the sensor material.

Semiconducting metal oxides offer potential sensor materials which can be used in this environment and which will allow measurement of oxygen concentrations. Titania (TiO_2) has been investigated as an automotive sensor for use at the stoichiometric point,[15] as a cost effective alternative to the electrochemical sensors based on zirconia. It was recognized from these investigations that it should be possible to develop titania as a lean (excess air) region sensor for automotive exhaust gases. During development, however, a number of serious limitations were encountered when attempting to use titania. As with other semiconducting oxides its electrical conductivity is realted to temperature and oxygen partial pressure according to Eq. (3).The sensitivity of the changes in conductivity due to temperature is dependant also on the activation energy for conduction (E_a). In the case of titania, this is high, which has the effect of making the conductivity very sensitive to small changes in temperature. It can be calculated that for sufficiently accurate control of the oxygen partial pressure, at the evaluated temperatures in the exhaust, the temperature of the titania must be controlled to within ± 3°C at 850°C. With automotive exhaust gas temperature varying so dramatically, this proves extremely difficult to achieve. Potential solutions to the temperature sensitivity problem for titania have been attempted with limited success.[16] Another limitation when using titania is the transition from *n*- to *p*-type behavior that occurs in the lean region. This confuses the oxygen partial pressure measurement still further.

A device construction used successfully to measure the output from titania is shown in Fig. 4. It consists of an alumina substrate with the titania sensing element, platinum electrodes, and a platinum resistance thermometer on one face, and a thick film heater on the other. A typical output from the titania sensor at various temperatures at two different oxygen concentrations as measured in a controlled laboratory environment is shown in Fig. 5.

The limitations of titania as an oxygen sensor for the measurement of automotive exhaust gas oxygen required the development of other semiconducting oxides for this application. Development concentrated on identifying materials that were responsive to oxygen, but that were not sensitive to temperature (i.e., have a low activation energy for conduction). A number of compounds based on the perovskite crystal structure were reported in the literature as having low activation energies of conduction. It was also known that alkaline earth ferrates and ferrites have highly defective lattices with ordered or partially ordered arrays of oxygen vacancies. Barium tantalum ferrates were identified in preliminary studies to have properties that would be desirable for sensors in the required automotive application. The

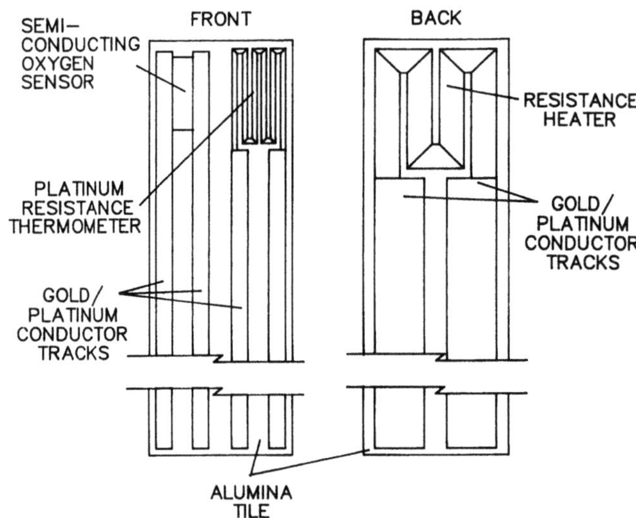

FIGURE 4. An example of a semiconducting exhaust gas sensor design.

materials produced a significant sensitivity to changes in oxygen partial pressure and more importantly had very low activation energies of conduction. This meant that although temperature control would still be required, the degree of control could be significantly relaxed and reliable and accurate measurement of the oxygen partial pressure could still be made.

The device design used to evaluate the sensor was similar to that shown in Fig. 4. The electrodes were usually gold and the sensor material was a sintered pellet attached to the substrate via gold leads. A typical output from the sensor material at various temperatures at two oxygen concentrations as measured in a controlled

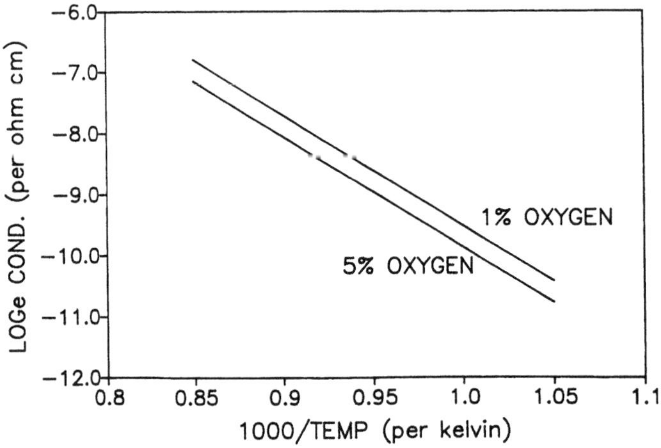

FIGURE 5. Typical output from titania.

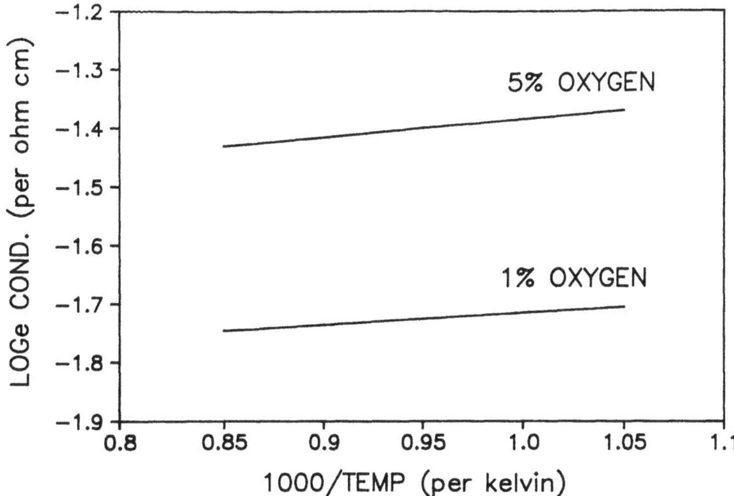

FIGURE 6.   Typical output from a barium tantalum ferrate material.

laboratory environment is shown in Fig. 6. If the output is compared to that of titania in Fig. 5 then the advantages of barium tantalum ferrate as a sensor material in this application can be seen.

Devices using barium tantalum ferrate pellets have been run in vehicle trials and gave useful signals. However, some potential drawbacks have been highlighted. Firstly the materials do not respond to changes in oxygen as quickly as required, and there appears to be a poisoning effect caused by sulfur dioxide in the exhaust. Further development of these materials is continuing.

4. CONCLUSIONS

Semiconducting oxide sensors offer a cost effective and accurate means of detecting and measuring oxygen concentrations. The automotive application described for exhaust gas oxygen sensors is extremely demanding, with sensors having to perform at high and suddenly fluctuating temperatures in a corrosive gas stream. The materials have shown that they can be used even in this harsh environment. However, it must be remembered that a great number of other applications exist where conditions are a lot less severe and where semiconducting oxides have great potential. The advantages described for barium tantalum ferrates will still be applicable for these other applications and make it an extremely interesting material for exploitation in this field. It should also be mentioned, with sensors having to perform at high and suddenly fluctuating temperatures in a corrosive gas stream, that developments in the solid electrolyte type of sensor have also been taking place and a number of sensors for automotive exhaust gas oxygen sensing in particular have been reported.[17-19]

The choice of sensing principle will largely be determined by the application, but semiconducting metal oxide sensors offer a cost effective, accurate, and reliable method for measuring oxygen and many other gases. For these reasons they will be widely exploited in an increasing number of applications in the future.

REFERENCES

1. D. E. Williams and P. McGeehin, R. Chem. Soc. Specialist Periodical Reports on Electrochemistry, Vol. 9, pp. 246–250 (1984).
2. D. S. Eddy, Physical principles of the zirconia exhaust gas sensor, *IEEE Trans. Vehicular Technol.* **VT-23**(4), 125–128 (1974).
3. H. U. Gruber and H. M. Wiedenmann, Three years field experience with the lambda-sensor in automotive control systems, Robert Bosch GmbH, Society of Automotive Engineers Report No. 800017 (February 1980).
4. H. C. Lord, In stack monitoring of gaseous pollutants, *Australian Proc. Eng.* **7**(2), 27, 29–31 (February 1979).
5. D. L. May, Cutting boiler fuel costs with combustion controls, *Chem. Eng.* **82**(27), 53–57 (1975).
6. P. Roy, Metal slag gas reactions and processes, *Electrochem, Soc.* 919 (1975).
7. P. Smith, Int. Conf. Liquid. Met. Technology in Energy Prodn. Proceedings p. 631 (1976).
8. J. Watson and R. A. Yates, A solid state gas sensor, *Electron. Eng.* **May,** 47–57 (1985).
9. P. Kofstad, *Nonstoichiometry Diffusion and Electrical Conductivity in Binary Metal Oxides*, Wiley, New York (1972), pp. 1–14.
10. P. Kofstad, *Nonstoichiometry, Diffusion, and Electrical Conductivity in Binary Metal Oxides*, Wiley, New York (1972), pp. 139–152.
11. J. D. Baumard, *J. Chem. Phys.* **67**(3), 857–860 (1977).
12. D. E. Williams, *Solid State Gas Sensors*, Adam Hilger, London (1987), pp. 75–80.
13. C. T. Engh, Development of the Volvo Lambda-Sond system, Society of Automotive Engineers Report No. 770295 (February 1977).
14. W. S. Fleming, D. S. Howarth, and D. S. Eddy, General Motors, sensor for on vehicle detection of engine exhaust gas composition, Society of Automotive Engineers Report No. 730575 (May 1973).
15. E. F. Gibbons, A. H. Meitzler, L. R. Foote, P. J. Zacmandis, and G. L. Beuudain, Ford Motor Co., Society of Automotive Engineers Report No. 750224 (February 1975).
16. M. J. Esper, E. M. Logothetis, and J. C. Chu, Ford Motor Co., Titania Exhaust Gas Sensor for Automotive Applications, Society of Automotive Engineers Report No. 790140 (January 1979).
17. T. Kamo, Y. Chujo, T. Akatsuka, and J. Nakano (Toyota Motor Corp.), and M. Suzuki (Nippondenso Co. Ltd.), Lean mixture sensor, Society of Automotive Engineers Report No. 850380 (March 1985).
18. R. Hetrick, W. A. Fate, and W. C. Vassell, Ford Motor Co., Oxygen Sensing by Electrochemical Pumping, Society of Automotive Engineers Report No. 810433 (April 1981).
19. S. Soejima and S. Mase, NGK Insulators Ltd., Multilayered Zirconia Oxygen Sensor for Lean Burn Application, Society of Automotive Engineers Report No. 850378 (March 1985).
20. B. C. Tofield, The study of electron distributions in organic solids, *Progress in Inorganic Chemistry* (*S. J. Lippard, ed.*), *Wiley, New York* (1976).

34

Field-Effect Chemical Sensors

J. R. Dodgson

1. INTRODUCTION

The surface field effect is the modulation of the band energies, and hence the carrier density, in a semiconductor surface by an electric field applied to the surface. The first MOSFETs based on this principle were made in the 1950s, but sensitivity of the insulator to chemical contamination (mainly by sodium ions) meant that it was another ten years before they could be fabricated reliably. Great effort has been made in the semiconductor industry to exclude chemical effects, but when taking place in a sensing layer overlying the active area, usually replacing the standard gate electrode, such effects can be used to advantage in chemical sensors for species in both gas and solution phases. Field-effect sensors have advantages over their conventional counterparts (where these exist) of small size, solid-state construction, and, with appropriate sensing material deposition techniques, ease of mass production; also, in cases where signals are small and conventional transducer impedances are high, of in situ impedance conversion and amplification.

The first field-effect sensor, the ion-selective field-effect transistor (ISFET), was described by Bergveld in 1970.[1] The field is now large; general reviews of field-effect-based chemical sensors have been written by Janata and Huber[2] and more recently Sibbald[3,4]; a full account of ISFETs has been given by Bergveld and Sibbald.[5] Field-effect-based chemical sensors have been made using MOS capacitors[5] and transistors, and also Schottky-[6] and gate-controlled diodes.[7] In this chapter the emphasis is on the properties of the materials used to confer chemical sensitivity, and only FET-based devices will be considered. The FET structure and sensor types based on interface and bulk properties of sensing materials are outlined in Section 2, and then individual devices, materials, and mechanisms of response are described briefly, concentrating on the basic ion- and gas-sensitive FETs. The aim is to give an idea of the principles involved and more details, especially of transient response, can be found in the references. Section 3 will describe FET-related sensors in which ISFETs are used to detect nonionic analytes; the ISFET structure is modified to incorporate gas-permeable or enzyme reaction membranes and in some cases titration devices, the target species being sensed indirectly by means of auxiliary

---

J. R. Dodgson • THORN EMI Central Research Laboratories, Dawley Road, Hayes, Middlesex UB3 1HH, U.K.

chemical reactions. The application of field-effect-based chemical sensors in analytical instrumentation is discussed briefly in Section 4. General references for each section are given after the heading.

2. FET SENSOR TYPES: SURFACE/BULK EFFECTS

2.1. FET Structure[8]

The structure of an *n*-channel insulated gate FET (IGFET) is shown in cross section in Fig. 1. The threshold voltage V_t beyond which the silicon surface becomes strongly inverted and source-drain conduction takes place is given by

$$V_t = \phi_{ms} - \frac{(Q_f + Q_D)}{C} + 2\phi_f - \frac{Q_i}{C} - \frac{Np}{\varepsilon_0} \tag{1}$$

$$Q_i = \frac{1}{\alpha} \int_0^d x q_i(x)\, dx \tag{2}$$

where ϕ_{ms} is the gate metal-silicon work function difference, d and C are the thickness and capacitance per unit area of the insulator, Q_f is the fixed (positive) charge associated with the Si/SiO_2 interface, Q_D is the maximum charge in the Si depletion layer, and ϕ_f is the potential difference between the Fermi level and intrinsic level in the silicon. Q_i is the effective charge at the Si surface due to an ionic charge distribution $q_i(x)$ in the insulator. The last term is the effective addition to V_g arising from a two-dimensional dipole layer at the gate–insulator interface:

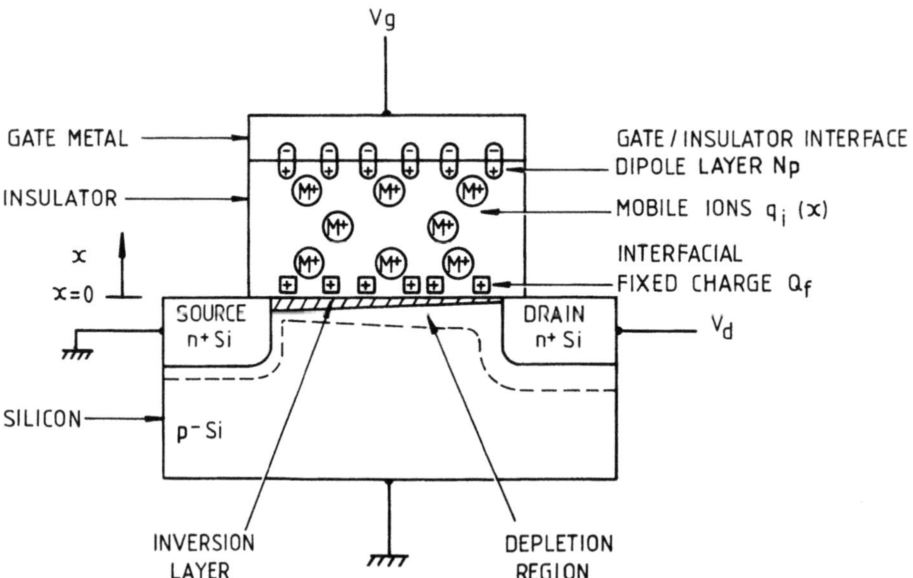

FIGURE 1.   Cross section through the IGFET structure showing charges in the insulator and gate-insulator dipole layer.

N is the density per unit area of dipoles of moment p. The drain current I_d is given for unsaturated and saturated conditions by

$$I_d = \mu C \frac{W}{L} [(V_g - V_t)V_d - \tfrac{1}{2}V_d^2] \qquad \text{(unsaturated)} \qquad (3)$$

$$I_d = \mu C \frac{W}{2L}(V_g - V_t)^2 \qquad \text{(saturated)} \qquad (4)$$

where W and L are the width and length of the channel and μ is the mobility of electrons in the channel. These are commonly used simplified expressions and ignore, for instance, the large serial resistance of the long source and drain diffusions usually found in ISFETs to allow easy encapsulation of the contacts. FET characteristics are nonlinear and temperature dependent; compensating interface circuitry or differential measurement techniques are used to give linear, thermally stable outputs from devices used in FET sensors. Examples are given in Refs. 2–5.

2.2 Interface and Bulk Effects: Sensor Types

From the above it can be seen that a FET incorporating a chemically sensitive membrane material in contact with the insulator will respond to changes in any of C, ϕ_{ms}, Q_i, N, or p of the combined layer, and in the case of devices where the sensing material is in contact with a solution in place of a metal gate contact, to any change in the potential between the material and the solution. Of these, C and Q_i are bulk properties; ϕ_{ms}, N and p are properties that can be changed by surface effects; potential differences between membrances and liquids are primarily surface effects. The main types of field-effect sensor to be described in the following sections rely on one or more of these effects; in most cases changes in one parameter only are sought, and changes in others are secondary effects often described as drift or instability.

ISFETs ideally rely on surface effects: the primary response is to differences in electrochemical potential of an ion between the solution and sensing material, with interfacial and diffusion potentials being set up in the surface layers of the material; for fast response these surface layers should be thin. Changes in bulk charge distribution $q_i(x)$ where present in the sensing layer cause slow response and drift.

Gas-sensitive FETs (GasFETs) rely on surface and bulk effects: in hydrogen-sensitive thick palladium gate devices the response arises from dissociative adsorption of hydrogen or hydrogen-containing gases on the catalytic gate surface, followed by diffusion of hydrogen atoms through the bulk, which acts as a very selective filter. The hydrogen atoms are then detected at the metal–insulator interface, either from a change in work function of the metal (formation of a dipole layer) or an interfacial potential. Suspended-gate GasFETs (SGFETs) are surface effect devices that sense a dipole layer resulting from gas adsorption onto the gate surface. Both mechanisms have been applied to describe porous-gate GasFETs sensitive to ammonia and other gases.

Humidity-sensitive FETs (HumFETs) rely on changes in the capacitance of a slightly hygroscopic gate layer on water absorption, a bulk effect.

Of these, the first three types respond with a change in V_t; HumFETs show a change in transconductance dI_d/dV_g also from the appearance of C in Eq. (3) and (4).

2.3. Ion-Selective Field-Effect Transistors

2.3.1. Ion-Selective Electrodes and ISFET Structure

ISFETs are a development from conventional ion-selective electrodes (ISEs) and much of the work on the latter is applicable to both. A comprehensive review of ISEs and their applications is given in the books by Covington[9] or Koryta.[10] ISEs are an established technology capable of very stable, reproducible response, but they are often bulky, needing a solution inner contact, fragile and expensive. Much of the impetus for ISFET development has come from their desirability in biomedical analysis: their size and solid-state construction allows in vivo use in catheters, small sample volume allows frequent ex vivo sampling in intensive care or under anaesthetic, and their low cost allows disposability after blood contact.

The principle of ISE measurements is illustrated in Fig. 2a. An ISE consists of an ion-selective membrane with an inner contact, either a solution and reference electrode or in special cases, a solid contact. An interfacial potential E_i is set up by equilibration of ions between the test solution and the membrane; this is measured relative to the potential E_{ref} of a reference electrode in the test solution. The cell potential $E = E_i + E_{in} - E_{ref}$; E_{ref} and the potential of the inner contact E_{in} are presumed constant. For equilibrium of a single ion I of charge $+z_i$ across the interface, E_i is given by the Nernst equation:

$$E_i = E_0 + 2.303 \frac{RT}{z_i F} \log \frac{a_i(s)}{a_i(m)} \tag{5}$$

where E_0 is a constant term arising from the differences in the standard chemical potentials of I in the two phases, $a_i(m)$ and $a_i(s)$ the activities of I in the membrane and solution, R the gas constant, and F the Faraday.[11] For a properly functioning ("Nernstian") ISE membrane $a_i(m)$ is constant and $E_i = E'_0 + k \log a_i(s)$, with $k = 2.303 RT/z_i F = 59.2/z_i$ mV/decade change in activity at 25°C. The activity a_i in solution is related to the concentration m_i by the activity coefficient γ_i, $a_i = \gamma_i m_i$; γ_i is a function of the concentration of the solution. In many cases, for example in medicine, activity values are required; if concentrations are needed γ_i must be calculated (Ref. 12; Ref. 9, Vol. I, Chap. 4).

ISEs often have a high impedance (>10 MΩ) and MOSFET input buffer amplifiers are used as in Fig. 2a. ISFETs are a direct derivation from this combination; the ISFET cell is shown in Fig. 2b, and the ISFET structure in Fig. 3: the MOSFET gate is replaced by the combination of ion-selective membrane, solution, and reference electrode. A dual dielectric structure including an Si_3N_4 layer, which acts as a barrier to ion motion, is often used to prevent ionic contamination of the SiO_2. Si_3N_4 and SiO_2 are themselves (inferior) pH sensor materials; a separate ion-selective membrane is most often used. A bias potential V_r is applied to the reference electrode; taking V_r as fixed, any changes in membrane potential E_i

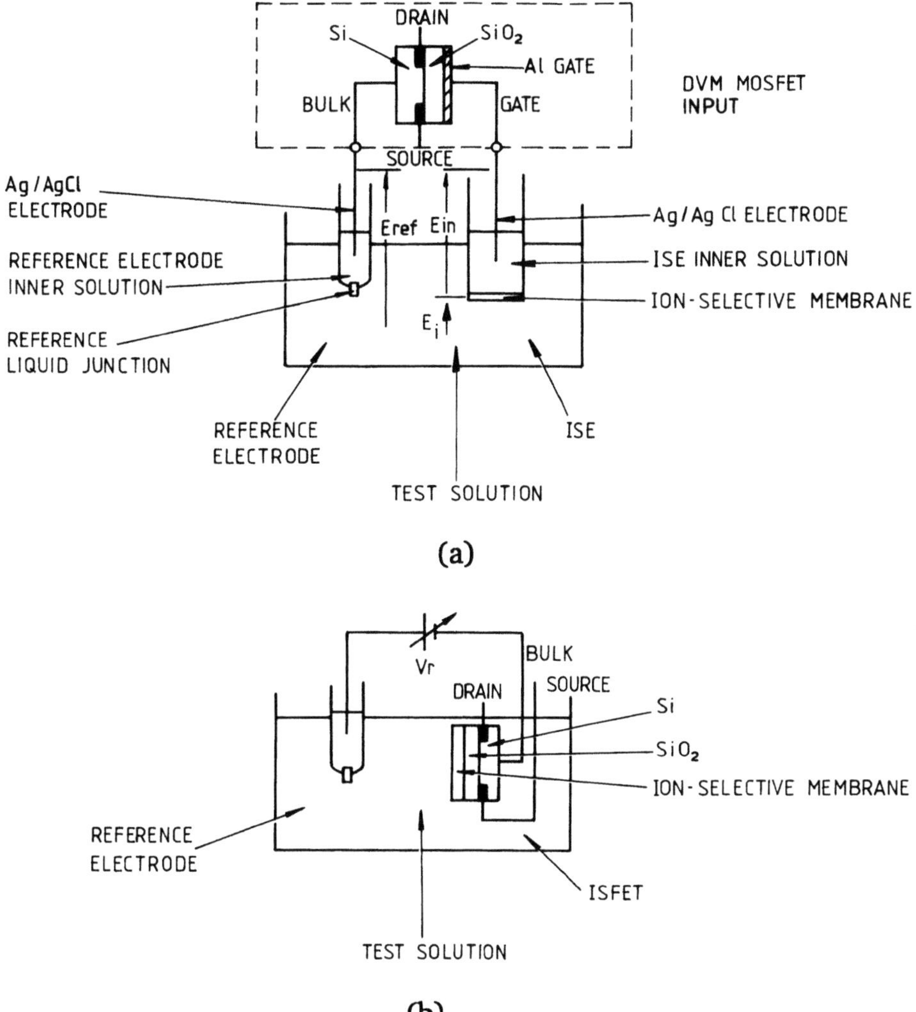

FIGURE 2.   (a) A conventional ISE cell and associated DVM MOSFET input stage. (b) The corresponding ISFET cell.

modulate the effective gate voltage and hence I_d. In terms of the transfer characteristics $I_d(V_r)$ of the ISFET, E_i is a component of the threshold voltage:

$$V_t = E_{\text{ref}} - E_0' - k \log a_i(s) - \frac{\Phi_{si}}{e} - \frac{(Q_f + Q_D)}{C} + 2\phi_f - \frac{Q_i}{C} \qquad (6)$$

where Φ_{si} is the silicon work function. It can be seen from the above that reliable functioning of both ISFETs and ISEs relies on a stable, constant reference electrode potential.

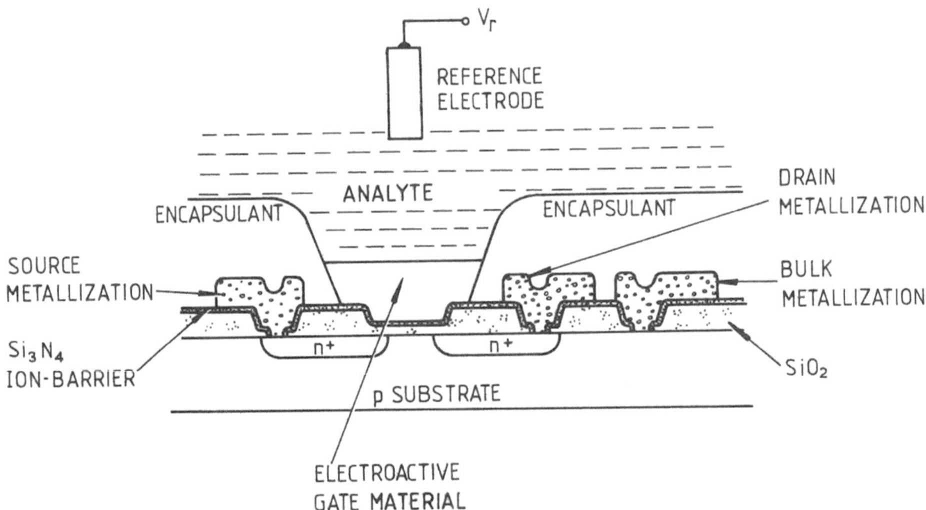

FIGURE 3. Cross section of a dual-dielectric SiO_2/Si_3N_4 ISFET structure.

2.3.2. Ion-Selective Membrane Materials

The requirements of an ISE membrane are that it rapidly establishes an ionic equilibrium across the solution–membrane interface and a concentration-dependent interfacial potential as in Eq. (5), and that it is an electrical conductor. Conventional ISE membranes may be divided into three classes (Ref. 9, Vol. I, Chap. 1): glasses, sparingly soluble inorganic salts, and organic membranes containing ion-complexing agents (ligands); all three types have been used with ISFETs. The conductivity requirement is relaxed in the case of ISFET membranes; the close proximity of the membrane and channel means that the coupling between the two is essentially capacitive, and some glasses and pH-sensitive metal oxides have been used that have resistivities far too great for ISE application. However, V_t depends through Q_i in (1) on the distribution $q_i(x)$ of any mobile charge in the membrane, so if this is present the response rate will be limited by charge motion in the bulk membrane, a situation not found in ISEs; for fast response, ISFET membranes must be either insulators or good conductors.

Glasses.[13,14] Glass electrodes are widely used for pH measurements (Ref. 12, Chap. II; Ref. 14). They are also used for sensing metal and other monovalent cations,[13] e.g., Ag^+, NH_4^+, and to a lesser extent alkaline earth cations, transition metal cations (Cu^{2+}, $Fe^{3+}Pb^{2+}$), and nitrate anions. The vast majority of electrode glasses are silicates; these consist of a random network of Si atoms linked approximately tetrahedrally by Si-O-Si oxygen bridges, broken up by lower-valency alkali metal atoms in $\equiv SiO(-)M(+)$ groups (Fig. 4a). The alkali metal ions are loosely bound to the nonbridging oxygens, leading to ionic conductivity. They can exchange with other ions from solution (s), a typical case being for H^+ entering a sodium silicate glass (g):

$$\equiv SiONa(g) + H^+(s) \;\rightleftharpoons\; \equiv SiOH(g) + Na^+(s) \qquad (7)$$

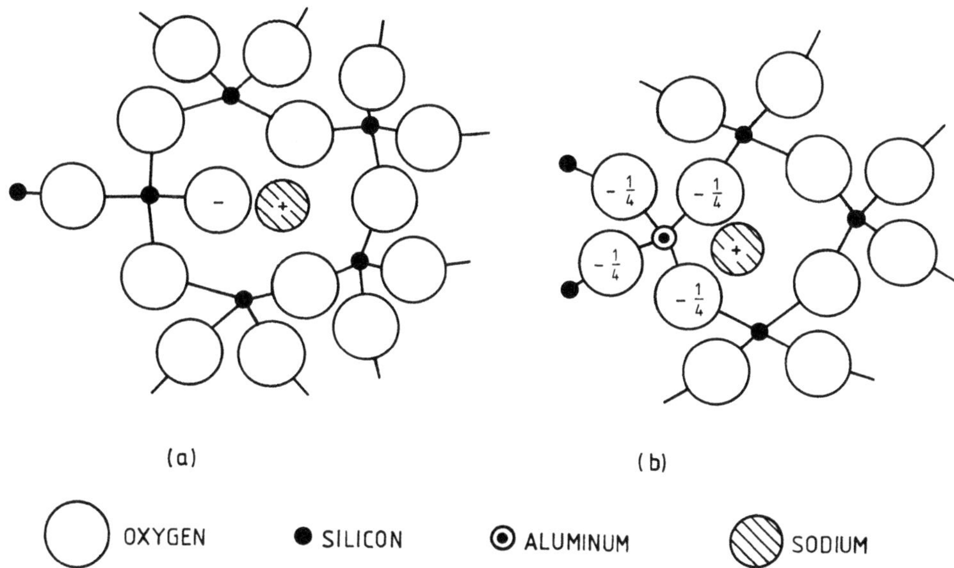

FIGURE 4. Two-dimensional schematic representation of the structure of silicate glass (after J. O. Isard, Ref. 13, Chap. 3). (a) The $\equiv$SiO$^-$Na$^+$ site in sodium silicate glass; (b) the (AlOSi)$^-$Na$^+$ site in sodium aluminosilicate glass. The negative charge is shown delocalized over the four oxygen atoms neighboring the aluminum ion.

and in addition the network can be hydrolyzed:

$$\equiv\text{Si-O-Si}\equiv + \text{H}_2\text{O} \rightarrow \equiv\text{SiOHHOSi}\equiv \tag{8}$$

Both these processes contribute to the formation of a leached and hydrated surface layer containing$\equiv$SiO$^-$ sites for ion association:

$$\equiv\text{SiO}^-(\text{g}) + \text{H}^+(\text{s}) \rightleftharpoons \equiv\text{SiOH}(\text{g}) \tag{9}$$

Alkali silicate glasses are highly pH-selective as the $\equiv$SiO$^-$ site in reaction (9) is physically small, with a high field strength, and binds the small (when unhydrated) H$^+$ ion with much greater energy than larger alkali metal ions.[13] The hydrated layer can act as a diffusion barrier, leading to slow response; practical pH glasses contain extra components to limit hydration, for example, alkaline earth metals such as calcium or barium, which form bridging groups $\equiv$Si-O-Ca-O-Si$\equiv$, which are resistant to reactions (7) and (8); the glass Na$_2$O(22%)CaO(6%) SiO$_2$(72%)-(Corning "015") was for a long time the standard for pH electrodes. Lithium silicates show lower sodium interference at high pH and are the basis of modern pH glasses.[14] If trivalent ions such as boron or aluminum are added to an alkali metal silicate glass a negative site AlSiO$^-$ is formed on leaching (Fig. 4b); this has a lower field strength and confers sensitivity to alkali cations at neutral or basic pH. Selectivity varies with composition through the density with which the Si, Al and O atoms are packed around the site: sodium aluminosilicates such as NAS11-18 (Na$_2$O 11 mole%,

Al_2O_3 18 mole%, SiO_2 71 mole%) for Na^+ and NAS27-4 for K^+ were established by Eisenman[13]; lithium aluminosilicates are now used for Na^+ (Ref. 9, Vol. I, Chap. 5).

Glass membranes have been applied to ISFETs by various techniques, e.g., chemical vapor deposition (CVD) and dip-coating/polymerization from mixed alkoxides,[15] both of which involve high-temperature processing, which must be done before contact metallization; rf sputtering, though with degradation of device characteristics by plasma exposure,[16] and fast-atom beam sputtering. The last method has the advantage that irradiation of the device gate is low and it requires no annealing steps at temperatures that will damage contact metallization. It can therefore be used for sequential deposition of different glasses on multichannel devices.[17]

Insulating Inorganic Membranes (Ref. 5, Chap. 3; Ref 15). The use of insulating inorganic membranes such as SiO_2, Si_3N_4, Al_2O_3, and Ta_2O_5 is one of the more important developments associated with ISFETs in that it allows miniature pH sensors with much faster responses than are achievable with comparable ISEs. The oxide surface forms proton association sites as in Eq. (9) and is hydrolyzed as in Eq. (8); the depth of hydration will depend on the oxide but is considered to be negligible on Ta_2O_5 and Al_2O_3, and appreciable on SiO_2 and the surface SiO_2 layer on Si_3N_4, leading to more rapid pH responses from the former two. The deeper hydration layer on the latter is probably of more open structure, allowing cation ingress, association, and interfering response. A comparison of the four materials above has been made by Matsuo and Esashi.[15] Deposition is usually by CVD or sputtering. These oxides respond (ideally) only to pH; attempts to modify SiO_2 surfaces by ion implantation to give cation response have met with some success,[18] the advantage being greater compatibility with standard processing.

Inorganic Salts (Ref. 9, Chaps. 8 and 9). An inorganic salt MX will respond to both M^+ or X^- ions by direct exchange of the ion between the solution and sites in the solid; examples are fluoride electrodes of single-crystal LaF_3 or electrodes of $Ag/AgCl$ for Ag^+ or Cl^-. The silver halide electrodes form good metal contacts and so there has not been great pressure to use them as ISFET membranes; sputtered LaF_3 films have been made.[19] The ability of ISFETs to use insulating membranes means that virtually any insoluble salt could be used that has sufficient surface ion exchange with solution and which can be deposited by a suitable thin-film technique. This is an area for future research.

Organic ion-exchanger and neutral carrier membranes (Ref. 9, Chaps. 6 and 7; Ref. 10, Chap. 6. These membranes offer the widest range of target ions; they consist of an organic species capable of binding ions selectively, immobilized in a water-insoluble membrane, most often plasticised PVC. Organic ion exchangers are charged species which bind ions of opposite sign by means of selective electrostatic interaction as described for glasses; examples are calcium dialkylphosphate salts for Ca^{2+} or tetradodecyl ammonium nitrate for NO_3^-.[20] Neutral carriers are long chain, or often ring organic molecules which contain ether groups at intervals along their structure capable of coordinating with cations in their vicinity. The flexibility

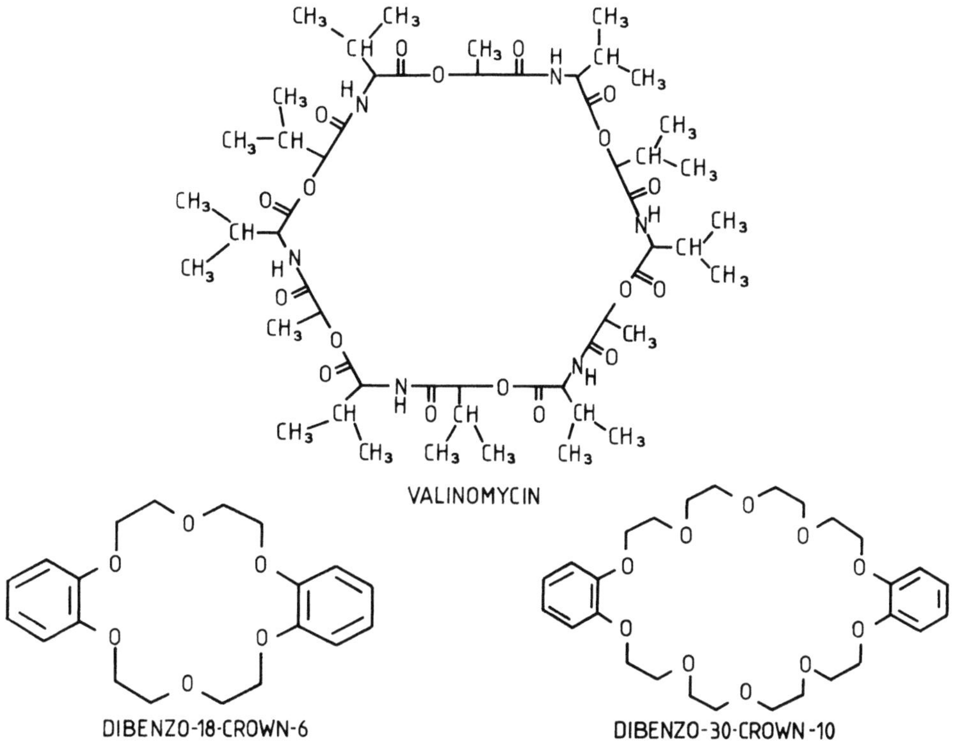

FIGURE 5. Cyclic ligands for K^+ (after Ref. 9, Vol. 1, Chap. 6). Dibenzo-18-crown-6 has a ring size close to that of the K^+ ion; dibenzo-30-crown-10 and valinomycin are larger and bind by wrapping around the ion. Valinomycin is more selective for K^+ than dibenzo-30-crown-10, which is more selective than dibenzo-18-crown-6. Valinomycin is said by R. P. Buck to fit K^+ "like the seams around a tennis ball."

of the chain or ring allows a change in conformation of the molecule to accommodate the ion, giving very selective, reversible binding. The most frequently used example is valinomycin, a naturally occurring antibiotic (see Fig. 5a), which binds potassium with very high selectivity, most significantly over sodium, and is the basis of most K^+ selective ISFETs and ISEs. Synthesized ligands such as crown ethers (Fig. 5b) show selectivity for monovalent cations that fit the ring size; noncyclic ligands are used in membranes for divalent cations.[21]

PVC membranes were first used on ISFETs in 1975, with valinomycin.[22] Deposition is usually by casting from solution. The main problems are to ensure good adhesion to the device surface and to devise a rapid method of depositing membranes on the wafer scale, possibly patterned for multichannel devices. Single and multifunction ISFETs sensitive to Na^+, K^+, and Ca^{2+} have been reported for clinical application[23]; the membranes were deposited by solvent casting in individual encapsulation pools over the gate regions. Valinomycin has been incorporated with some success into photoresist films which offer the possibility of wafer-level patterning.

2.3.3. Ion-Exchange Theory of Equilibrium Response

Two theories are used to describe the equilibrium response of ISFETs. The ion-exchange theory developed by Nicolski and Eisenman[13,24] is an extension of the Nernst equation (5) to more than one ion, describing membrane response in terms of an ion-exchange equilibrium as in Eq. (7). It relies on an assumption of electroneutrality in the bulk membrane far from the interface and so applies to membranes that are thick relative to the interfacial region in which the electrical potential is changing. This includes conventional ISE membranes and most glass or polymeric membranes on ISFETs. The second, the site-binding theory (Ref. 4, Chap. 3), does not assume electroneutrality: it considers explicitly the association reactions between ions and sites on a charged surface, such as on minimally hydrated oxides. The same ideas have recently been extended to three-dimensional hydrated layers.[25] The ion-exchange theory gives a clearer picture of the mixed response to more than one ion and will be outlined below.

The equilibrium response of a membrane to the presence of one ion I is given by Eq. (5). If two monovalent cations I^+ and J^+ are present in the solution and/or membrane, electroneutrality (with a fixed concentration C_0 of the negative sites) imposes an ion exchange equilibrium (whether or not I^+ or J^+ are original components of the membrane):

$$I^+(s) + J^+(m) \rightleftharpoons I^+(m) + J^+(s) \tag{10}$$

$$K_{ij} = a_i(m)a_j(s)/a_i(s)a_j(m) \tag{11}$$

where (s) denotes the solution, (m) the membrane, a_i and a_j the activities, and K_{ij} is the equilibrium constant. In ion exchangers where two ions are competing for one type of site, the activities are found empirically to be related to mole fractions X_i, X_j of the ions by a power law[24]:

$$a_i(m) = [X_i(m)]^{n_{ij}}, \qquad a_j(m) = [X_j(m)]^{n_{ij}} \tag{12}$$

where C_i and C_j are the concentrations in the membrane; $C_i + C_j = C_0$; $X_i = C_i/C_0$; $X_j = C_j/C_0$ and n_{ij} is a parameter specific to the particular ions I^+, J^+. Equations (5), (11), and (12) give the Eisenman equation:

$$E_{ij} = E_0 + kn_{ij} \log[a_i^{1/n_{ij}} + (K_{ij}a_j)^{1/n_{ij}}] \tag{13}$$

Equation (13) describes the phase boundary potential between the membrane and solution. There will be a difference between the activities of I^+ and J^+ at the surface and those in the bulk of the glass, which leads to a diffusion potential. It has been shown[13] that this changes only the value of K_{ij} in (13). In most cases, K_{ij} can be regarded as an empirical parameter describing ISFET or ISE performance and precise assignment of the origin of potential is unimportant. Behavior of many membranes on ISFETs is described well by Eq. (13); Fig. 6 shows an atom-beam sputtered Na-glass (NAS11-18) device responding to Na^+ and K^+ in a tris buffer at pH 7.9. It should be noted that the membrane parameters K and n are empirical functions of the activities of all the ions present; for the Na/K selectivity calibration in Fig. 6 they will vary with pH.

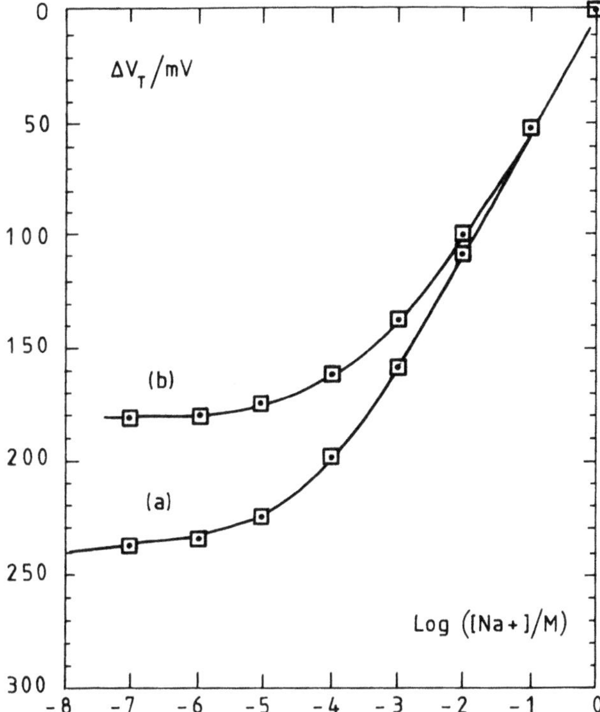

FIGURE 6. Na^+ and K^+ response of an ISFET with a NAS11-18 glass membrane deposited by atom beam sputtering: change in threshold voltage ΔV_t against log(Na^+ concentration). (a) Na^+; (b) Na^+ and 0.1 M K^+, both in 150-mM tris buffer pH 7.9. ———, experiment; $\square$, theory from Eq. (13) with slope $s = 54$ mV/log[Na^+]; $K_{NaH} = 3 \times 10^3$; $n_{NaH} = 1.7$; $K_{NaK} = 4.5 \times 10^{-3}$; $n_{NaK} = 2.0$. The slope s differs from k in (13) as the plot is on a concentration rather than activity scale; the membrane response is also slightly sub-Nernstian.

2.4. Gas-Sensitive FETs

Gas-sensitive FETs (GasFETs) are IGFET devices in which the standard gate metallization is replaced with a catalytic metal (or other material) capable of adsorbing the gas of interest or its dissociation products (normally hydrogen). The resulting change of gate metal–insulator interface properties is sensed by the FET structure. Unlike ISFETs they have no analogous preexisting sensor type. The first GasFETs, reported by Lundstrom *et al.* in 1975,[26] had thick palladium gate metallization and were responsive to hydrogen. Variations on the structure and metallization have since given response to other hydrogen-containing gases such as H_2S, arsine, ammonia, unsaturated hydrocarbons, and also polar gases such as carbon monoxide and alcohol vapors. Recent reviews by Lundstrom *et al.* deal with the physical aspects[27] and applications[28] of GasFETs; earlier publications give details of the development of hydrogen sensitive devices.[29] Applications have been suggested in areas including leak detection using H_2, detection of fires in which H_2 or NH_3 are combustion products, monitoring of anaerobic bioreactor processes and in biosensors operating by detection of gaseous products from enzyme reactions. The three basic gate structures, thick, nonporous palladium, thin and porous (mainly Pd, Pt, or Ir), and suspended gates will be described in the following sections.

2.4.1. Palladium-Gate GasFETs Sensitive to Hydrogen[27]

Nonporous palladium gate GasFETs (Fig. 7a) respond selectively to hydrogen in the 1 ppm–several percent range; they are usually operated at elevated temperatures (150–200°C) to reduce response time and to desorb water from hydrogen adsorption sites on the Pd surface. The response is a reversible negative threshold voltage shift of several hundred millivolts with increasing hydrogen concentration; a typical response is shown in Fig. 8. The mechanism of operation proposed by Lundstrom and co-workers is that hydrogen molecules are adsorbed and dissociated by the Pd surface; a proportion diffuse rapidly through the film to the Pd–insulator interface where they form a dipole layer, which acts to reduce the threshold voltage in accordance with the last term in Eq. (1). Back-reactions take place which desorb hydrogen atoms as H_2, and in the presence of oxygen, as H_2O. The low mobility of other gases in Pd confers hydrogen selectivity; the gate, which is typically 100–500 nm thick, a presents a negligible diffusion barrier to hydrogen atoms at the temperatures of operation.

A change in the fermi level in the bulk metal with hydrogen absorption could also cause a response. The bulk resistivity of Pd films increases on absorption of appreciable quantities of hydrogen, but it is found that the resistivity increases after the initial V_t response occurs, and continues to rise after V_t has saturated, suggesting that the hydrogen responsible for the V_t response resides at the surface of the film, and a surface effect model is more appropriate. However, very large V_t shifts have been seen to be associated with bulk phase changes at high hydrogen concentrations.[27]

The response of GasFETs is determined by the chemical reactions occurring on the catalytic surface; possible reactions for hydrogen on a Pd-gate device will be described as illustration. The H atoms H_i at the Pd/insulator interface are assumed to be in rapid equilibrium with those at the Pd surface, H_a:

$$H_a \underset{k_{-i}}{\overset{k_i}{\rightleftharpoons}} H_i \tag{14}$$

Assuming Langmuir adsorption, i.e., all sites are equivalent, gives at steady state

$$\frac{d\theta_i}{dt} = 0 = k_i\theta_H(1 - \theta_i) - k_{-i}\theta_i(1 - \theta_H) \quad \Rightarrow \quad \frac{\theta_i}{1 - \theta_i} = K_i\frac{\theta_H}{1 - \theta_H} \tag{15}$$

where θ_i, θ_H are fractional hydrogen coverages at the interface and surface respectively, the (fast) bulk diffusion rates are incorporated into the rate constants k_i and k_{-i}, and $K_i = k_i/k_{-i}$ is the equilibrium constant. In an inert atmosphere the surface hydrogen coverage is determined only by

$$H_2(g) \underset{k_{-1}}{\overset{k_1}{\rightleftharpoons}} 2H_a \tag{16}$$

similarly to Eq. (15):

$$\frac{d\theta_H}{dt} = 0 = k_1\alpha P(H_2)(1 - \theta_H)^2 - k_1\theta_H^2 \quad \Rightarrow \quad \frac{\theta_i}{1 - \theta_i} = K_i\left[\frac{k_1}{k_{-1}}\alpha P(H_2)\right]^{1/2} \tag{17}$$

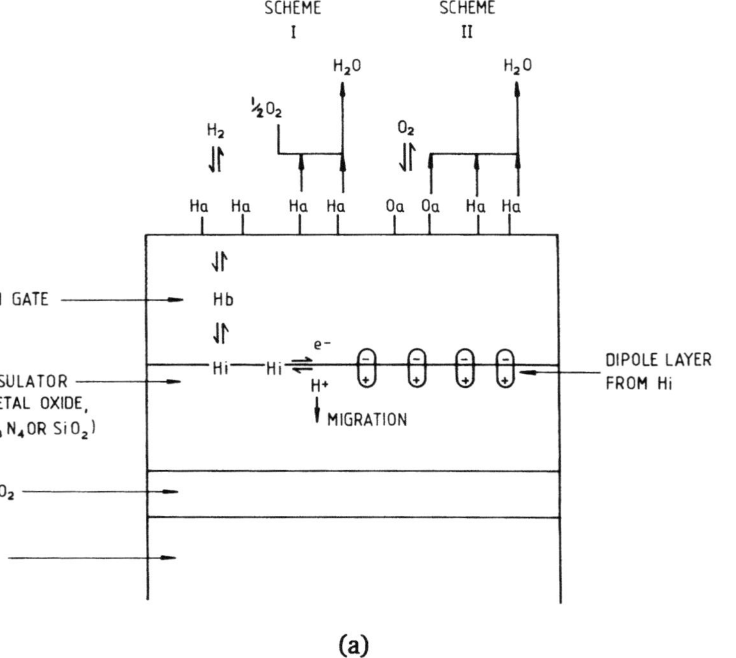

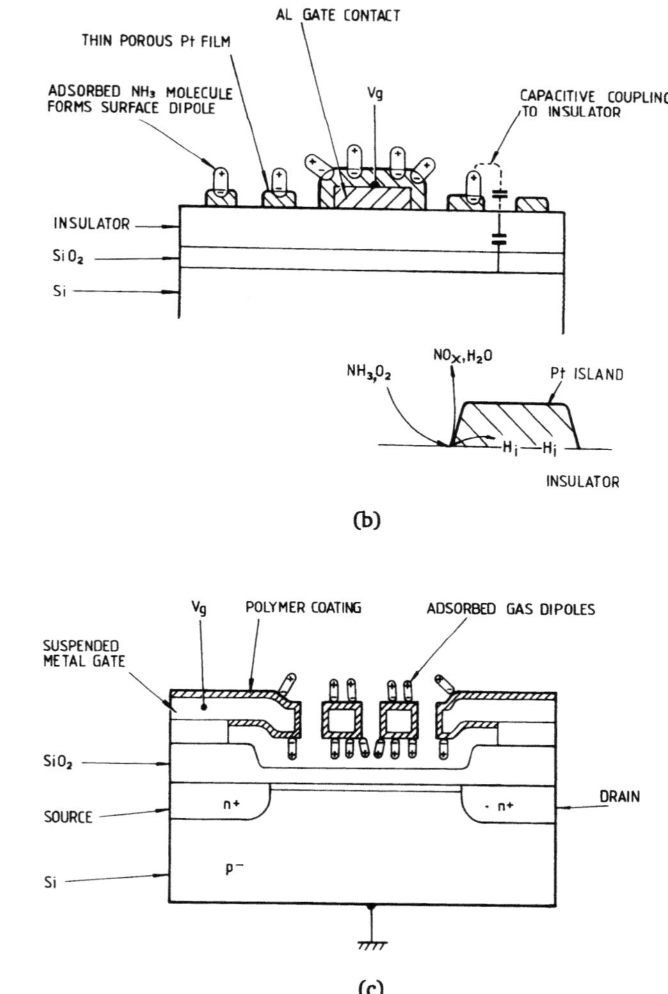

FIGURE 7. GasFET gate structures and response mechanisms. (a) Nonporous Pd-gate structure and two possible schemes for the surface oxidation reaction. (b) Upper: porous gate structure: dipole adsorption layer model of NH_3 response; lower: NH_3 dissociation model: NH_3 is adsorbed dissociatively at the Pt/insulator interface and oxidized to water and nitrogen oxides NO_x; a proportion of the hydrogen atoms escape oxidation and diffuse into the metal. (c) Suspended gate structure.

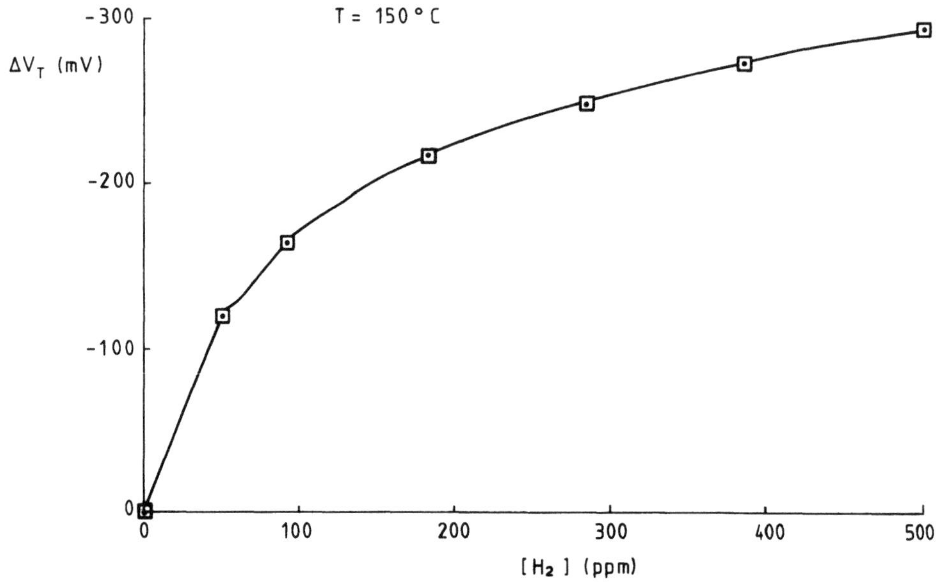

FIGURE 8.   Change in threshold voltage ΔV_t of a nonporous palladium gate, Si_3N_4/SiO_2 insulator GasFET in response to hydrogen in air at 150°C (I. Robins, THORN EMI).

where α is a constant and $P(H_2)$ is the partial pressure of hydrogen. In the Lundstrom model the hydrogen atoms at the Pd–insulator interface are polarized by adsorption on the metal–insulator interface in the same way as those at the gas–metal interface. The shift in V_t is given by

$$dV_t = \frac{N_p}{\varepsilon_0} = dV_{t(\max)}\theta_i \qquad (18)$$

where N is the interface hydrogen concentration, p the dipole moment of the adsorbed atom, and $dV_{t(\max)}$ the maximum V_t shift seen when all the interface sites are filled, found experimentally by extrapolation. Equations (17) and (18) describe the response well at moderate and high $P(H_2)$.

In an atmosphere containing oxygen, reactions between hydrogen and oxygen may take place as shown in Fig. 7a. There are two extreme possibilities for the reaction scheme: (I) that adsorbed hydrogen atoms react directly with gas phase oxygen molecules (the Eley–Rideal mechanism[30]), with the possibility of hydrogen adsorption sites being blocked by adsorbed dissociated oxygen atoms; or (II) that oxygen first undergoes dissociative adsorption and the reaction occurs between adsorbed hydrogen, oxygen, and OH species (the Langmuir–Hinshelwood mechanism[30]). The real situation is probably an intermediate. The following reactions take place on the Pd surface:

$$H_2(g) \underset{k_{-1}}{\overset{k_1}{\rightleftharpoons}} 2H_a \qquad (16)$$

$$O_2(g) + 2H_a \xrightarrow{k_2} 2OH_a \tag{19}$$

$$H_a + OH_a \xrightarrow{k_3} H_2O \tag{20}$$

$$O_2(g) \underset{k_{-4}}{\overset{k_4}{\rightleftharpoons}} 2O_a \tag{21}$$

$$O_a + 2H_a \xrightarrow{k_s} H_2O(g) \tag{22}$$

with (19) and (20) dominating the water production in scheme (I) and (22) (assumed to be a third order reaction in Ref. 27 and 29) in scheme (II). The back reactions in (19), (20), and (22) are neglected as being energetically unfavorable. Steady state rate equations similar to (15) and (17) may be written for the surface coverages of H, OH, and O: the solution gives the form of the dependence on oxygen partial pressure[27]; that most commonly seen in high-temperature operation corresponds to scheme (I):

$$\frac{\theta_i}{1 - \theta_i} = K_i \left[\frac{k_1 \alpha P(H_2)}{k_{-1} + 2k_2 \beta P(O_2)} \right]^{1/2} \tag{23}$$

where β is a constant and $P(O_2)$ the partial pressure of oxygen. k_{-1} is very much less than $2k_2\beta P(O_2)$ in air. Equations (18) and (23) again give a good description of behavior. At lower temperatures ($<75°C$) an increased contribution from reaction scheme (II) occurs and the $P(O_2)$ dependence deviates from (23); the $P(H_2)^{1/2}$ dependence is the same. This, and transient behavior as described by mechanisms (I) and (II), is discussed in Ref. 27 and 29.

An alternative mechanism proposed by Robins et al.[31] relies on the same surface reactions but a different origin of the interfacial potential. They find dV_t to be better related logarithmically to $P(H_2)$ and suggest that the Pd–insulator interface is operating as an electrochemical half-cell analogous to the hydrogen gas electrode, with a redox equilibrium between atomic hydrogen in the metal and protons or hydroxyl groups in the insulator. The overall reaction is $H_2(g) \rightleftharpoons 2H^+(i)$, where (g) is the gas phase and (i) the insulator. The interfacial potential will then be given by a Nernst equation similar to Eq. (5):

$$E = E_0 - 2.303 \frac{RT}{zF} \log \frac{a_{H2}(g)}{a_{H^+}(i)} \tag{24}$$

where E_0 is a notional standard potential for the redox reaction, $z = 2$ is the number of electrons transferred in the raction, $a_{H^+}(i)$ is the activity of H^+ in the insulator, $a_{H2}(g)$ is that of H_2 in the test gas, and R, T, and F have their usual meanings. The slopes of $E/\log(a_{H2})$ were found to be less than theoretical for devices with Si_3N_4 insulators and greater for those using SiO_2. This was explained as arising from H^+ migration into the insulator in the case of SiO_2, this being blocked by Si_3N_4.

The precise nature and energy distribution of the interfacial adsorption sites in the dipole layer model have not been elucidated. The Langmuir isotherm (which assumes a constant adsorption energy) used in the above fits behavior well at moderate $P(H_2)$ in inert or oxygen atmospheres; at low hydrogen partial pressures in inert atmospheres a logarithmic $dV_t/P(H_2)$ dependence is seen, suggesting sites with a distribution of adsorption energies.[27,31] The nature of the insulator affects the response strongly, as might be expected by considering that the dipole layer lies outside the metal. Pd/SiO_2 devices show slow response to $P(H_2)$ steps, hysteresis, and drift; the response has been found to consist of two parts, a fast dipole layer formation followed by a slow negative V_t change attributed to H^+ diffusion into the oxide,[32] with the unstated implication of the existence of an interfacial potential as described above. Devices using Al_2O_3, Ta_2O5, or Si_3N_4 layers in contact with the palladium show smaller, faster responses and greater stability than those with SiO_2 only[33]; diffusion of protons into these materials is much lower (cf. ISFET membranes) and the response is considered to arise from the dipole layer (or interfacial potential) alone.

2.4.2. Porous Gate Devices Sensitive to Ammonia and Other Gases

If the catalytic metal gate is made thin and porous, GasFETs show a response also to ammonia, unsaturated hydrocarbons, alcohols, and carbon monoxide. The response is a large negative shift in threshold voltage as for hydrogen; larger responses are seen from platinum and iridium gates than for palladium. The response increases with temperature over the range 40–200°C for ammonia and unsaturated hydrocarbons; the hydrogen response is independent of temperature from 100 to 200°C. The sensitivity depends on the discontinuous nature of the film, having areas of exposed insulator: 25 nm-thick evaporated Pt gates are discontinuous and respond to both hydrogen and ammonia while atom-beam sputtered gates of the same thickness are continuous and respond only to hydrogen.[34] CO sensitivity has been achieved with devices having 2–5-μm holes patterned photolithographically in Pd gates.[35,36]

Two operating mechanisms have been proposed. The first involves catalytic dissociation of the gas, e.g., ammonia, a proportion of the hydrogen atoms diffusing through the gate metal to be detected as before, the remainder being oxidized at the surface.[37] The interface between the metal and the exposed gate insulator is presumed to be the active site in promoting dissociative adsorption of the gas molecules without rapid complete oxidation of the products (Fig. 7b). Ross *et al.*[34] were able to describe the response of ammonia GasFETs in these terms with 1.5 hydrogen atoms per NH_3 molecule being detected. In the second mechanism[35,30] changes in surface potential (work function) of the metal gate resulting from gas adsorption are coupled through the air capacitance in the pores to the insulator surface (Fig. 7b). Oxidation reactions play a part in determining the coverage of adsorbed species.

The role of each is not clear cut. Dissociative chemisorption of ammonia is known to take place on silica-supported Pt catalysts at temperatures below 200°C[34] and would be expected to increase with temperature in agreement with the observed response; a response relying solely on coverage of adsorbed NH_3 would not be expected to rise with temperature. The response to ammonia is much larger than that to hydrogen; this may suggest a different mechanism to be in operation,[28] or that the oxidation process for hydrogen is more efficient relative to dissociation than that for ammonia on supported Pt. The slower recovery of the gate film in air after exposure to ammonia than after exposure to hydrogen [38] may arise from blocking of Pt surface sites for hydrogen oxidation in the latter case by less easily oxidized NH_3 molecules or partial oxidation products. Support for the adsorption model comes from the CO response of perforated gate devices in dry air,[36] ruling out hydrogen-forming reactions between CO and water. It is possible that both mechanisms play a part in the response to hydrogen-containing gases, possibly in different temperature regimes, and further work is needed to determine which is the more appropriate model for these devices.

2.4.3. Suspended-gate GasFETs

Suspended gate GasFETs (SGFETs)[39] rely on a similar adsorption mechanism though the capacitive coupling is more obvious in this case. The structure is as in Fig. 7c; adsorption of molecules onto the perforated suspended metal gate changes the gate work function and hence the threshold voltage. An attribution of the effect to formation of a dipole layer on the gate surface, or change in the bulk fermi level of the film, or both, has not yet been made. The gate may be coated with an adsorption-enhancing layer: for instance, polypyrrole films electropolymerized in situ are used in Ref. 39 to detect lower aliphatic alcohols with a degree of selectivity.

2.5. Humidity-Sensitive FETs[40]

Humidity sensors have applications from control of humidity in air-conditioned or medical environments such as incubators, to prevention of condensation in electronic equipment. A review of the field of humidity sensing is given in Ref. 40. Conventional sensors are based on either conductivity or capacitance changes of hygroscopic materials such as porous ceramics, hygroscopic polymers, or ionic electrolytes absorbed in porous media. Both types of measurement lead to specific interfacing needs: capacitance sensors in particular have high impedances, and careful design is needed to avoid stray capacitance. This problem can be avoided by incorporating the sensing layer in the gate structure of a humidity-sensitive FET (HumFET), as in Fig. 9a. A porous evaporated gold gate electrode (200 Å) allows moisture access to a spin-coated, patterned polymer layer ($\sim 1\ \mu$m thick). Suitable polymers include cross-linked PVA, polyimide, ethyl cellulose, and cellulose acetate butyrate; when dry these polymers have dielectric constants of around 3–6: absorption of water (with dielectric constant ~ 79 at 25°C) gives a large increase in capacitance, mainly between 30% and 90% RH. The capacitance and relative humidity value are temperature dependent and a sensing diode is included on the chip. The bulk properties required of the polymer are that it should have an open structure, should not show a high degree of hydrogen bonding to water to reduce hysteresis, and should not swell excessively on absorption, thus preventing problems with adhesion and breakage of the contact to the gate electrode. The latter problem can be avoided by using a structure shown in Fig. 9b, where a standard metallized FET gate is coupled to the driving gate contact via the sensing layer and a common top electrode.

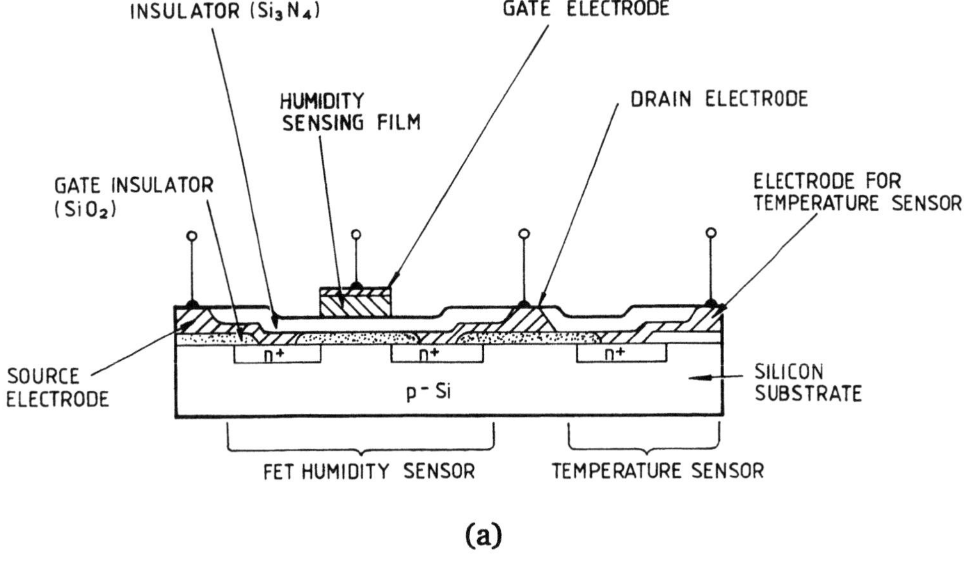

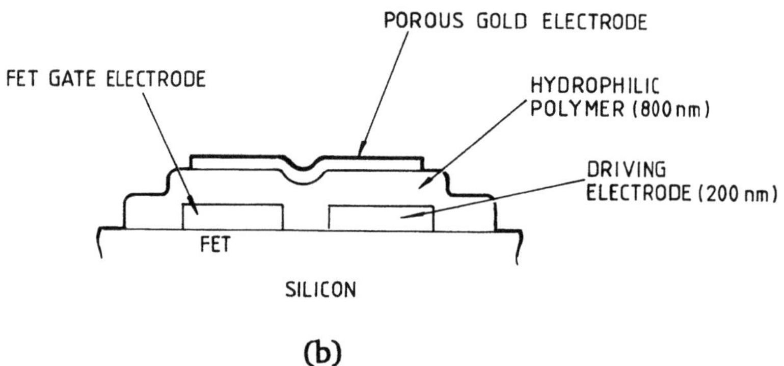

FIGURE 9.   HumFET structures. (a) Porous gate electrode over the humidity sensitive polymer layer. (b) Driving electrode coupled to the FET gate via the humidity sensitive layer and a common floating porous electrode.

3. FIELD-EFFECT RELATED SENSORS

The "related" sensors to be described in this section use ISFETs to detect nonionic analytes indirectly via the products of chemical reactions carried out in the vicinity of the sensor; this may be with reagents immobilized in a chemically sensitive layer incorporated in the device, or protons produced coulometrically by an integral titration electrode, or both. Examples of the first type are Severinghaus-type membrane-modified ISFET sensors, in which acid or basic gases in solution diffuse through a membrane to alter the pH of a standard solution in contact with the sensor,[41] and also enzyme-modified ISFETs (or EnFETs). These devices have conventional forerunners in Severinghaus electrodes[42] and enzyme-modified ISEs,[43] respectively. The second type differ from the other devices described in this chapter as the output is a current, the end point of the titration being determined by the ISFET; examples are microtitration pH sensors.[44] pH-static EnFETs[45,46] are a combination of the two types.

3.1. Enzyme-Modified ISFETs

The structure of an EnFET is shown in Fig. 10a. A conventional pH-ISFET is overlaid by a membrane in which an enzyme is immobilized, either chemically by covalent bonding to groups in the membrane, or physically by entrapment in a polymer matrix, such as polyHEMA or bovine serum albumin (BSA) cross-linked by glutaraldehyde. The ISFET senses the pH change resulting from the reaction between enzyme and substrate. A reference device is often incorporated to measure the pH of the solution.

A much studied example is the determination of glucose from reaction with the enzyme glucose oxidase:

$$\beta\text{-D glucose} + O_2 + H_2O \xrightarrow{\text{glucose oxidase}} H_2O_2 + \text{D-gluconic acid}$$

$$\xrightarrow{\text{dissociation}} H^+ + \text{gluconate} \tag{25}$$

The pH change at the sensing surface depends on the relative rate of the reaction and diffusion processes shown in Fig. 10b. As the substrate (glucose) diffuses into the membrane it reacts to form protons: the higher the reaction rate relative to the substrate diffusion rate, the greater the proportion of substrate which is converted before it reaches the sensing surface and the larger is the signal. In competition with this is loss of protons by diffusion into the bulk solution and by association with buffer species, the rate of which depends on the proton diffusion rate and the concentration and diffusion rate of the buffer respectively. Steady state EnFET response has been described by Eddowes[47] and Caras et al.[48] A tradeoff exists between the size and speed of the response: for a given enzyme loading the response will increase with membrane thickness as this increases the proportional convertion of substrate, but the response time will increase also as establishment of the steady state diffusion profile of substrate will be slower. Devices have been made with membrane thicknesses varying between <10 and >200 μm; thick membranes offer large signals and also longer lifetime, as there is capacity to allow for loss of membrane activity through physical loss of enzyme, or loss of its catalytic activity; thin membranes give fast response at the expense of signal size and lifetime. The membrane thickness and enzyme activity are crucial parameters of the device, and difficulty in controlling these leads to irreproducibility in the response of devices at present. The response also depends strongly on the buffer capacity and bulk pH of the solution. The presence of buffer species will reduce pH changes arising in the enzyme reaction. Bulk pH affects the response in two ways: firstly as the buffer capacity is itself pH dependent, and secondly, as the rate of diffusional loss of protons to solution increases with the difference in pH between the membrane and

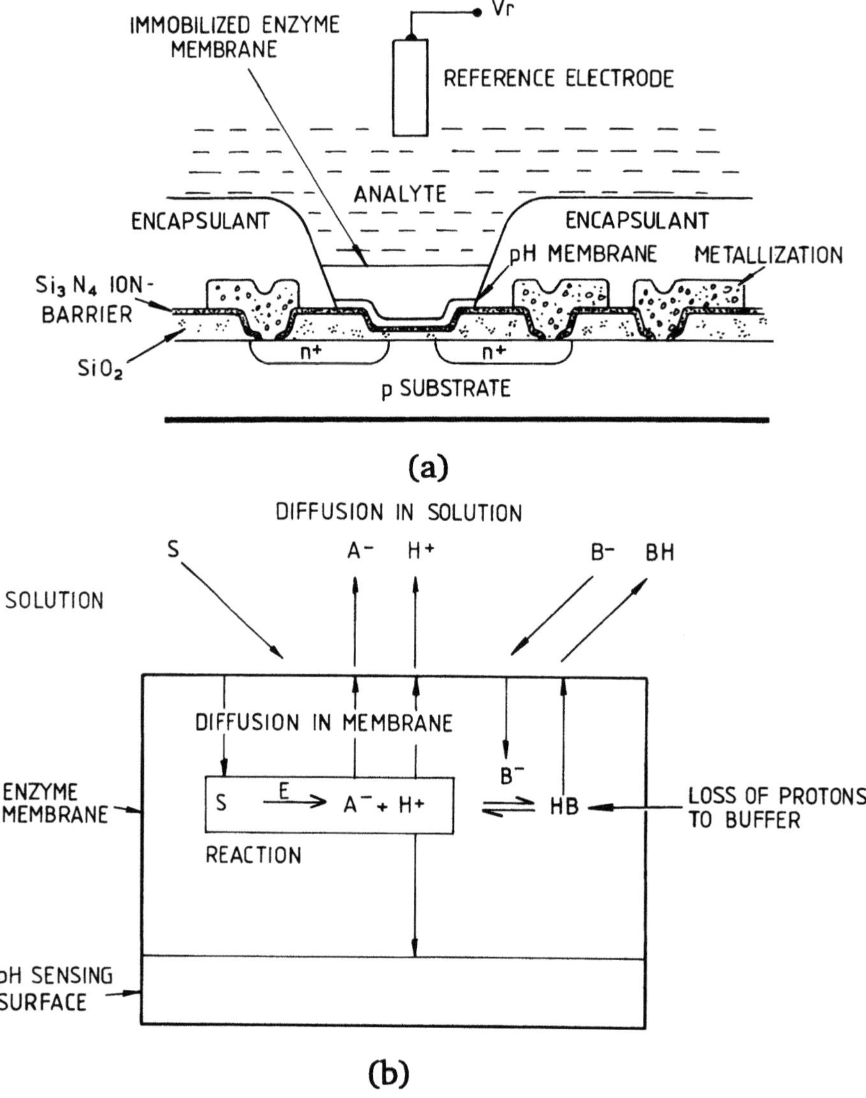

(a)

(b)

FIGURE 10.   (a) EnFET structure (b) Diffusion and reaction processes in EnFET operation. A substrate S is converted by enzyme E to a dissociated product acid HA. Protons associate with a buffer species B^-

the solution.[47] A typical response from a glucose oxidase EnFET operated in various background buffer concentrations is shown in Fig. 11.

Apart from glucose, EnFETs have been reported for, e.g., penicillin, acetylcholine, and urea. As for ISFETs, the utility of EnFETs largely depends on whether active membranes can be deposited reproducibly using wafer-level techniques. Examples are immobilization of urease (for urea) in photopatternable PVA, or by photoresist lift-off patterning of cross-linked BSA-glutaraldehyde membranes containing urease or glucose oxidase. These techniques have the disadvantage of low

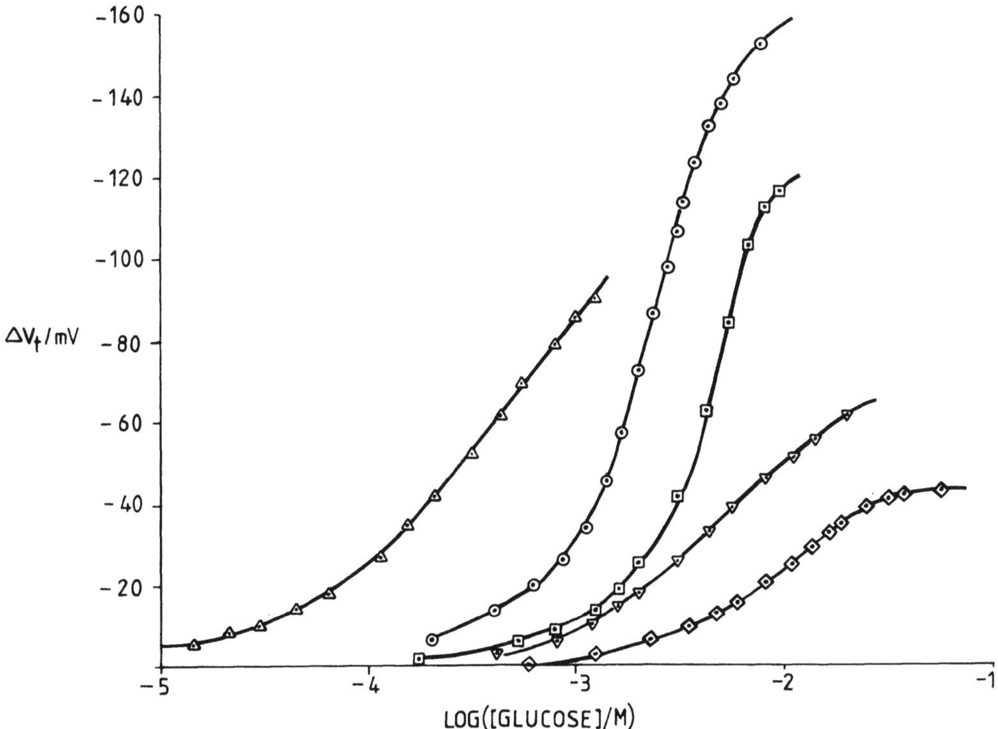

FIGURE 11.   Response of a glucose oxidase membrane EnFET to glucose in phosphate buffer + 0.1 *M* KCl at pH 7.2: change in threshold voltage ΔV_t against log(glucose concentration). Buffer concentrations are △, 0 mM; ⊙, 2.5 mM; ▽, 5 mM; ◇, 10 mM (M. J. Eddowes, THORN EMI).

membrane activity of the very thin (1–10 μm) spun-on films, but response can be fast. A review of the development of glucose oxidase and urease based EnFETs and comparison of membrane deposition techniques is given in Ref. 46.

3.2. Integrated ISFET/Coulometric Titration Sensors

Acid (protons) or base (hydroxyl ions) can be added to aqueous solutions coulometrically by passing current between a pair of noble metal electrodes, protons being generated at the anode and hydroxyl ions at the cathode:

$$H_2O \;\rightarrow\; 2H^+ + 2e^- + \tfrac{1}{2}O_2 \qquad (anode)$$

$$H_2O + \tfrac{1}{2}O_2 + 2e^- \;\rightarrow\; 2OH^- \qquad (cathode)$$

Two devices have been proposed recently that use this principle, incorporating a titration electrode next to a pH ISFET gate, and a second distant counter electrode in the solution. One is a titration-based sensor for dissolved CO_2[49] (Fig. 12a), in which pH changes in a bicarbonate electrolyte arise from CO_2 diffusion and are titrated back to the starting pH under ISFET control, the charge passed being related

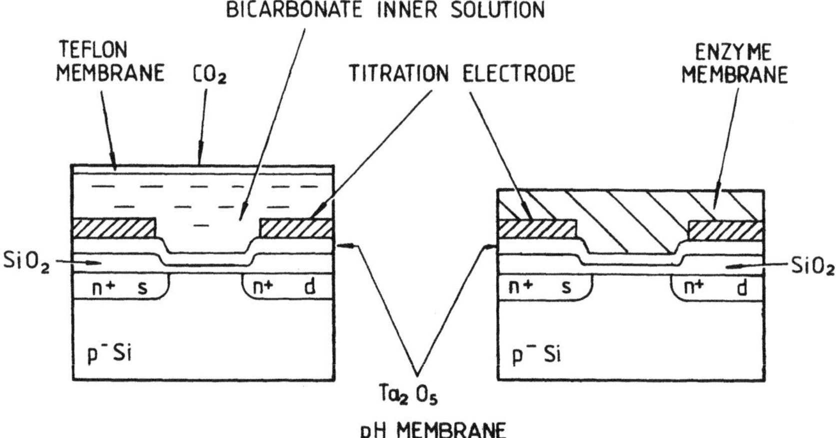

FIGURE 12. ISFET sensors with integrated titration electrodes. Current is passed between the titration electrode and a distant counter electrode (not shown). (a) A potentiometric titration ISFET sensor for CO_2 (after Ref. 48). (b) Coulometric titration EnFET structure. Feedback control sets the titration current to neutralize pH changes generated in the enzyme reaction. The pH at the sensing surface is maintained at that of the bulk solution, measured by a second, unmodified, ISFET (after Ref. 45).

to CO_2 concentration. An advantage of this method is that slope calibration of the ISFET is less important. The second device is a pH-sensitive EnFET (Fig. 12b) in which feedback-controlled titration is used to maintain the pH in the vicinity of the enzyme membrane constant at that of the bulk solution, as measured by a reference pH ISFET.[45,46] The effect of buffer species and bulk pH on the titrating ion is the same as that on the ions generated in the enzyme reactions; the same proportion of both are lost by association with buffer and diffusion into solution. Consequently the titration current for a given substrate concentration is independent of these factors. The current is proportional to the rate of generation of protons (or hydroxyl ions) in the enzyme reaction and so will be linear with substrate concentration,[45] as compared with the pH relation complicated by buffer effects seen for conventional EnFETs (Fig. 11). The problem that the response rate is limited by slow diffusion of substrate into the membrane still remains, however.

4. CONCLUSIONS

This chapter has outlined the various types of FET-based chemical sensor and their advantages over their conventional counterparts, but despite these advantages the former have yet to achieve successful commercialization. ISFETs in particular have received much attention for their application to biomedical monitoring, but several technological developments remain to be made. For example, reproducible wafer-scale deposition is possible for solid-state membranes, but improved techniques are needed for polymer membranes for ion sensitive or enzyme-based devices, and also for production-scale encapsulation; conventional reference electrodes are still necessary: microfabricated reference electrodes have been produced but have not yet shown an adequate combination of stability and lifetime; the causes of drift in both ISFET and GasFET response is still not completely understood. Progress is being made in these areas, however, and none should be regarded as posing a fundamental problem for the application of field effect sensors.

Naturally field-effect sensors will succeed only in so far as they offer significant advantages over existing technology, and it is important to recognize that this applies to the system of which the sensor is a part as well as the sensor itself. Some drawbacks of present sensors have been bypassed by system design; for instance, ISEs in commercial medical analyzers can have response times that are long relative to the desired measurement time, but in a system that ensures reproducible conditions a transient response measured after a fixed time is adequate. In much microsensor research efforts have been made to achieve high levels of integration to reduce costs and size, such as integrated ISFET and interface circuitry, reference electrodes and devices, and integrated EnFETs.[4] While such approaches are useful for the development of simpler analytical instrumentation and possible in vivo use, wider applications of the new devices may be (at least initially) as improved components in more conventional systems. For instance, the use of NH_3 sensitive GasFETs in a flow injection analysis (FIA) system for enzyme-based analysis has been reported to be promising.[50] As a second example, much work has been done on improving the sensitivity and response time of EnFET devices (Section 3.1), but the compromise between the two, and difficulty with reproducible deposition of the enzyme layer, still exist. By using separate pH ISFETs in a FIA system with an immobilized enzyme column, a fast response without compromised signal size can be achieved in urea determination.[51] New sensors of these types will allow such analysis systems to be constructed as single (possibly disposable) components, an extension of the idea of integration from the sensor to the system as a whole.

REFERENCES

1. P. Bergveld, Development of an ion-sensitive solid-state device for neurophysiological measurements, *IEEE Trans. Biomed. Eng.* **BME–17**, 70–71 (1970); P. Bergveld, Development, operation and application of the ion-sensitive field-effect transistor as a tool for electrophysiology, *IEEE Trans. Biomed. Eng.* **BME–19**, 342–351 (1972).
2. J. Janata and R. J. Huber, Chemically sensitive field-effect transistors, in *Ion -Selective Electrodes in Analytical Chemistry* (H. Freiser, ed.), Vol. 2, Chap. 3, Plenum Press, New York (1980).
3. A. Sibbald, Chemical-sensitive field-effect transistors, *IEE* Proc. I **130**, 233–244 (1983).
4. A Sibbald, Recent advances in field-effect chemical microsensors, *J. Mol. Electron.* **2**, 51–83 (1986).
5. P. Bergveld and A. Sibbald, Analytical and biomedical applications of ion-selective field-effect transistors, in *Wilson and Wilson's Comprehensive Analytical Chemistry* (G. Svehla, ed.), Vol. 23, Elsevier, Amsterdam (1988).
6. J. N. Zemel, B. Keramati, and C. W. Spivak, Non-FET chemical sensors, *Sensors Actuators* **1**, 427–475 (1981).
7. I. Lauks, Polarisable electrodes: Part 2, *Sensors Actuators* **1**, 393–403 (1981).
8. S. M. Sze, *Physics of Semiconductor Devices*, 2nd. ed., Wiley, New York (1981).
9. A. K. Covington (ed.), *Ion-Selective Electrode Methodology*, Vols. I and II, CRC Press, Boca Raton, Florida (1979).
10. J. Koryta, *Ion-Selective Electrodes*, Cambridge University Press, Cambridge (1975).

11. P. W. Atkins, *Physical Chemistry*, 3rd ed., Chapter 12, Oxford University Press, Oxford (1986).
12. R. G. Bates, *Determination of pH, Theory and Practice*, Wiley, New York (1973), Chap. 3.
13. G. Eisenman (ed.), *Glass Electrodes for Hydrogen and Other Cations*, Marcel Dekker, New York (1967).
14. G. Johannson, B. Karlberg, and A. Wikby, The hydrogen-ion sensitive glass electrode, *Talanta* **22**, 953–966 (1975).
15. T. Matsuo and M. Esashi, Methods of ISFET fabrication, *Sensors Actuators* **1**, 77–96 (1981).
16. T. Harbinson, Ph. D. thesis, Dept. Physical Chemistry, University of Newcastle-upon-Tyne, U.K. (1986).
17. J. R. Dodgson, T. Harbinson, J. E. A. Shaw, and A. Sibbald, unpublished work.
18. T. Ito, H. Inagaki, and I. Igarashi, ISFETs with ion-sensitive membranes fabricated by ion implantation, *IEEE Trans. Electron. Devices* **ED-35**, 56–63 (1988).
19. W. Maitz, I. Meierhofer, and L. Muller, Fluoride-sensitive membrane for ISFETs, *Sensors Actuators* **15**, 211–219 (1988).
20. A. K. Covington and A. Sibbald, Ion-selective field-effect transistors (ISFETs), *Phil. Trans. R. Soc. London* **B316**, 31–46 (1987).
21. W. E. Morf, *The Principles of Ion-Selective Electrodes and of Membrane Transport*, Chapter 12, Elsevier, Amsterdam (1981).
22. S. D. Moss, J. Janata, and C. C. Johnson, Potassium ion selective FET, *Anal. Chem.* **47**, 2238–2243 (1975).
23. A. Sibbald, P. D. Whalley, and A. K. Covington, A miniature flow-through cell with a four function ChemFET IC for simultaneous measurement of K^+, H^+, Ca^{2+}, and Na^+, *Anal. Chim Acta.* **159**, 47–62 (1984).
24. G. Eisenman, Cation-selective glass electrodes and their mode of operation, *Biophys. J.* **2**, Part 2, Suppl., 259–323 (1962).
25. J. Sandifer, Theory of interfacial potential differences: effects of adsorption onto hydrated (gel) and nonhydrated surfaces, *Anal. Chem.* **60**, 1553–1562 (1988).
26. K. I. Lundstrom, M. S. Shivaraman, and C. M. Svensson, A hydrogen sensitive Pd-gate MOS transistor, *J. Appl. Phys.* **46**, 3876–3881 (1975).
27. I. Lundstrom, M. Armgarth, and L. G. Petersson, Physics with catalytic metal-gate chemical sensors, *CRC Crit. Rev. Solid-State Mat. Sci.* **15**, 201–279 (1989).
28. I. Lundstrom, M. Armgarth, A. Spetz, and F. Windquist, Gas sensors based on catalytic metal gate field-effect devices, *Sensors Actuators* **10**, 399–423 (1986).
29. I. Lundstrom, Hydrogen sensitive MOS structures. Part 1: Principles and applications *Sensors Actuators* **1**, 403–427 (1981); Part 2: Characterisation, *Sensors Actuators* **2**, 105–138 (1981–2).
30. R. P. H. Gasser, *An Introduction to Chemisorption and Catalysis by Metals*, Chap. 8, Clarendon Press, Oxford (1985).
31. I. Robins, J. F. Ross, and J. E. A. Shaw, The logarithmic response of palladium-gate metal insulator semiconductor field effect transistors to hydrogen, *J. Appl. Phys.* **60**, 843–845 (1986).
32. C. Nylander, M. Armgarth, and C. Svensson, Hydrogen induced drift in Pd gate metal oxide silicon structures, *J. Appl. Phys.* **56**, 1177–1188 (1984).
33. K. Dobos, M. Armgarth, G. Zimmer, and I. Lundstrom, The influence of different insulators on palladium-gate metal–insulator–semiconductor hydrogen sensors, *IEEE Trans. Electron Devices* **ED–31**, 508–510 (1984).
34. J. F. Ross, I. Robins, and B. C. Webb, The ammonia sensitivity of platinum-gate MOSFET devices: Dependence on gate electrode morphology, *Sensors Actuators* **11**, 73–91 (1987).
35. D. Krey, K. Dobos, and G. Zimmer, An integrated CO-sensitive MOS transistor, *Sensors Acutators* **3**, 169–177 (1982/3).
36. K. Dobos and G. Zimmer, Performance of CO-sensitive MOSFETS with metal-oxide semiconductor gates, *IEEE Trans. Electron Devices* **ED–32**, 1165–1169 (1985).
37. F. Winquist, A. Spetz, M. Armgarth, C. Nylander, and I. Lundstrom, Modified palladium metal-oxide-semiconductor structures with increased ammonia gas sensitivity, *Appl. Phys. Let.* **43**, 839–841 (1983).
38. A. Spetz, M. Armgarth, and I. Lundstrom, Hydrogen and ammonia response of metal–silicon dioxide–silicon structures with thin platinum gates, *J. Appl. Phys.* **64**, 1274–1283 (1988).
39. M. Josowicz and J. Janata, Suspended-gate FET modified with polypyrrole as alcohol sensor, *Anal Chem.* **58**, 514–517 (1986).

40. N. Yamazoe and Y. Shimizu, Humidity sensors: Principles and applications, *Sensors Actuators* **10**, 379-399 (1986).

41. K. Shimada, M. Yano, K. Shibatani, Y. Komoto, M. Esashi, and T. Matsuo, Application of catheter-tip ISFET for continuous in vivo measurement, *Med. Biol Eng. Comput.* **18**, 741-745 (1980).

42. M. Riley, Gas-sensing probes, Chap. 1 in Ref. 9, Vol. II, pp. 1-21.

43. R. K. Kobos, Potentiometric enzyme methods, in *Ion-selective Electrodes in Analytical Chemistry* (H. Freiser, ed.), Vol. 2, pp. 1-84, Plenum Press, New York (1980).

44. B. H. van der Schoot and P. Bergveld, An ISFET-based microlitre titrator: integration of a chemical sensor-actuator system, *Sensors Actuators* **8**, 11-22 (1985).

45. G. K. Chandler and M. J. Eddowes, Enzyme-mediated pH-sensitive FET devices: Problems of the non-linear response, *Sensors Actuators* **13**, 223-229 (1988).

46. B. H. van der Schoot and P. Bergveld, ISFET-based enzyme sensors, *Biosensors* **3**, 161-186 (1987/88).

47. M. J. Eddowes, Response of an enzyme-modified pH-sensitive ion-selective device; Analytical solution for the response in the presence of pH buffer, *Sensors Actuators* **11**, 265-274 (1987).

48. S. D. Caras, J. Janata, D. Saupe, and K. Schmitt, pH-Based enzyme potentiometric sensors, Part 1: Theory, *Anal. Chem.* **57**, 1917-1920 (1985).

49. B. van der Schoot and P. Bergveld, Coulometric sensors, the application of a sensor-actuator system for long-term stability in chemical sensing, *Sensors Actuators* **13**, 251-263 (1988).

50. F. Winquist, A. Spetz, M. Armgarth, and I. Lundstrom, Biosensors based on ammonia sensitive metal-oxide-semiconductor structures, *Sensors Actuators* **8**, 91-101 (1985).

51. G. K. Chandler, J. R. Dodgson, and M. J. Eddowes, ISFET-based enzyme sensors for urea: Enzyme-modified ISFETs and column-immobilised enzyme flow injection analysis, *Analyt. Proc.* **26**, 154-156 (1989).

Index

Note: Generally, chemical symbols rather than names have been used for materials (e.g., GaAs); not all possible permutations have been indexed [for example, (GaAl)As has been indexed, but not (AlGa)As].